PETROLOGY

The

Study

of

IGNEOUS

SEDIMENTARY

METAMORPHIC

Rocks

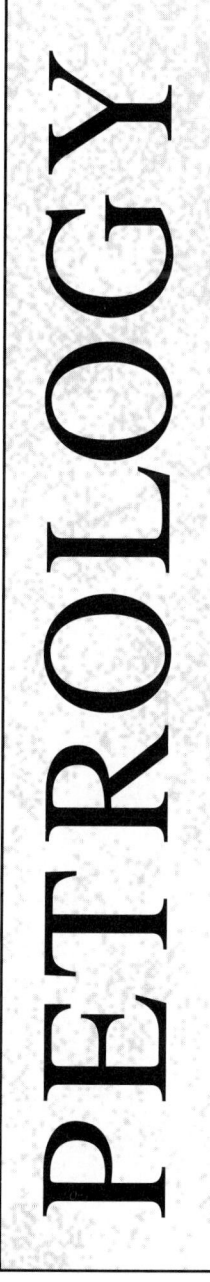

PETROLOGY

The
Study
of
IGNEOUS

SEDIMENTARY

METAMORPHIC

Rocks

Loren A. Raymond
Appalachian State University

WCB

Wm. C. Brown Publishers

Dubuque, IA   Bogota   Boston   Buenos Aires   Caracas   Chicago
Guilford, CT   London   Madrid   Mexico City   Sydney   Toronto

Book Team

Editor *Jeffrey L. Hahn*
Developmental Editor *Mary Hill*
Production Editor *Kay J. Brimeyer*
Designer *Jeff Storm*
Art Editor *Carla Goldhammer*
Photo Editor *John C. Leland*
Permissions Coordinator *Mavis M. Oeth*

Wm. C. Brown Communications, Inc.

Vice President and General Manager *Beverly Kolz*
Vice President, Publisher *Jeffrey L. Hahn*
Vice President, Director of Sales and Marketing *Virginia S. Moffat*
Vice President, Director of Production *Colleen A. Yonda*
National Sales Manager *Douglas J. DiNardo*
Marketing Manager *Jane Ducham*
Advertising Manager *Janelle Keeffer*
Production Editorial Manager *Renée Menne*
Publishing Services Manager *Karen J. Slaght*
Royalty/Permissions Manager *Connie Allendorf*

Wm. C. Brown Communications, Inc.

President and Chief Executive Officer *G. Franklin Lewis*
Senior Vice President, Operations *James H. Higby*
Corporate Senior Vice President, President of WCB Manufacturing *Roger Meyer*
Corporate Senior Vice President and Chief Financial Officer *Robert Chesterman*

Copyedited by *Martha Morss*

Original illustrations by *Tom Terranova*

Cover Photos:

Petrology: The Study of Igneous, Sedimentary, and Metamorphic Rocks:
Background—Letraset/Photone Vol. 1
Inset Top—© R. Dahlquist /Superstock/East Rift Zone Kilauea, Island of Kauai, Hawaii
Inset Middle—© Kerrick James Photography/Sandstone "Hogdoos," Bryce Canyon, Utah
Inset Bottom—© Doug Sherman/Geofile/Migmatite Exposed in the Canadian Shield near White River, Ontario
Volume I: Igneous Petrology: © R. Dahlquist/Superstock/East Rift Zone Kilauea, Island of Kauai, Hawaii
Volume II: Sedimentary Petrology: © Kerrick James Photography/Sandstone "Hogdoos," Bryce Canyon, Utah
Volume III: Metamorphic Petrology: © Doug Sherman/Geolfile/Migmatite Exposed in the Canadian Shield near White River, Ontario

The credits section for this book begins on page 727 and
is considered an extension of the copyright page.

A Times Mirror Company

Library of Congress Catalog Card Number: 93–73526

ISBN **Petrology,** casebound, 0–697–00190–3
ISBN **Igneous Petrology,** paper binding, 0–697–23692–7
ISBN **Sedimentary Petrology,** paper binding, 0–697–23691–9
ISBN **Metamorphic Petrology,** paper binding, 0–697–23690–0

Printed in the United States of America by Wm. C. Brown Communications, Inc.,
2460 Kerper Boulevard, Dubuque, IA 52001

10 9 8 7 6 5 4 3 2 1

Dedicated to
Margaret
whose patience and love during eleven years of
lost vacations,
and whose typing and encouragement
over that time,
helped make this book possible
and to
Matt
who sacrificed significant quality time
while he was growing up.

BRIEF CONTENTS

CONTENTS

Part I   Introduction

Part II   Igneous Rocks

Contents

Contents

Contents

Contents

LIST OF TABLES

PREFACE

Petrology is a required subject of study in a majority of geology programs in the United States and is an integral part of geology curricula in Canada and elsewhere in the world. The subject merits this important position because rocks make up much of the Earth. Furthermore, many subdisciplines in geology incorporate petrologic knowledge. Thus, petrology provides the foundation for the study of the Earth and its history.

Available petrology texts designed for middle-level undergraduate students (sophomores and juniors) take a variety of approaches. Some concentrate on principles, others deal only with igneous and metamorphic rocks, and still others tend to emphasize the descriptive aspects of the field (petrography). Only a few books incorporate recent advances in crystal growth studies, isotopic analyses, chemical petrology, sedimentary environmental analysis, and petrotectonics, as well as basic petrographic information.

This text, designed for the middle-level undergraduate geology major, incorporates both fundamentals and information on recent advances in our understanding of igneous, sedimentary, and metamorphic rocks. It provides an overview of the field of petrology and a solid foundation for more advanced studies. For each class of rocks—igneous, sedimentary, and metamorphic—I describe textures, structures, mineralogy, chemistry, and classification as a background to discussing representative occurrences and petrogenesis (rock origins). I have not tried to summarize all occurrences, but the explanatory notes at the end of each chapter provide additional sources the student can use to expand his or her understanding and knowledge.

Advances in petrologic knowledge come about through the efforts of individuals and groups of individuals. Individuals do not operate in a vacuum; rather, they work within a social and political environment. Whether or not an idea is accepted or has an impact on a specific scientific milieu is influenced by the quality of the idea (e.g., its ability to explain phenomena) and the quality of the data that support it; by the social, scientific, and political climate; and by the reputation of the scientist presenting the idea. Certain scientists, more than others, influence the development of a discipline. For this reason, I have chosen to cite, in the text, scientists who have been particularly influential, individuals who have made unique contributions, and a few whose work has advanced or updated earlier contributions of significance. Others who also have made significant contributions or have

added to the body of knowledge in a specific area are cited in the notes. Not all of those cited have been correct in their advocacy of particular hypotheses, but each has influenced petrology by contributing major ideas or significant data or by spurring others to prove his or her ideas wrong. In citing a limited number of scientists in the text, my aim has been to make the text more readable while still including sufficient references that may also serve as a model of proper referencing and good scholarship for students.

Igneous rocks are introduced first. Here, as with the other classes of rock, the rock types that are most important volumetrically in the crust are given the greatest emphasis. Because great debates arise from time to time about the origins of these most common rock types (e.g., basalts, andesites, granites), I have chosen to discuss each rock type individually rather than as part of a tectonic association, suite, or clan, as some other texts do. Where it is essential, I do discuss associated rocks.

Conflicting points of view on rock origins are included in many chapters. The inclusion of these controversies is important for two reasons. First, they provide the student with historical background that makes contemporary theory more meaningful. One can see ideas that have come before, see where mistakes have been made (science is a self-correcting enterprise), and see why certain ideas have been abandoned. Some scientists have reinvented the wheel (independently created anew an idea that was formerly evaluated and rejected) because of a lack of knowledge of petrologic history. Second, contemporary controversies reveal that there are many unresolved questions and there is much petrologic work to be done.

Sedimentary rocks are divided into siliciclastic types (mudrocks, sandstones, conglomerates), biochemically and chemically precipitated types (e.g., limestones, cherts), and allochemical types (e.g., limestone conglomerates). Because classifications vary and devotees of one classification or another are often strongly committed to the classification and descriptive practices they follow, various classifications are presented and discussed. Mudrocks and wackes (sandstones), especially turbidites, are given more attention than is common in petrology texts, because of the abundance of mudrocks and the tectonic and economic importance of the wackes. Sedimentary facies analysis is introduced to provide the student with an understanding of the interrelationships among various sedimentary rock types. A chapter on weathering, transportation, and diagenesis link the chapters on sediments and their environments of deposition to those dealing with specific sedimentary rock types.

Metamorphic rocks are presented in the same general fashion as the other classes of rocks, with descriptive aspects preceding chapters detailing the petrogenesis of particular suites of rocks. The facies series concept of Miyashiro is used as a basis for subdividing metamorphic rock suites and for describing the distribution of these suites in metamorphic terranes. Eclogites and cataclasites are given special attention because, although the former are considered to be representative of some mantle rocks and the latter are widely distributed, both are underdescribed in most petrology texts.

The epilogue places all of the rock types into appropriate petrotectonic assemblages representing various plate tectonic sites. Plate tectonics is first discussed in the introduction and is reemphasized in many of the rock chapters where petrogenesis is cast in a plate tectonic framework. Thus, the epilogue synthesizes the broad aspects of petrogenesis.

Important terms are set in bold print in the text where they are first defined and used. Definitions are also provided in a glossary. As a teacher, I subscribe to the view that one must know the vocabulary of a subject in order to be able to effectively communicate in that subject, just as one must know the vocabulary of a foreign language if one desires to communicate in that language. Students are commonly frustrated when they make the transition from texts, which define terminology, to the professional literature, where a knowledge of terminology is assumed. The significant number of defined terms used in this text should help students with this transition.

The language (syntax and word choice) used in the text ranges from simple to moderately complex, with the intent of developing in the student an increased vocabulary and an increased facility with the English language. The presentation of ideas follows a similar (simple to complex) pattern in many of the chapters. In addition, later parts of the text assume an understanding of some earlier parts. A list of Latin words and abbreviations used in the text (e.g., *sensu lato*), as well as other abbreviations precedes Part I of the book. This too is presented in the hope of increasing the general literacy of students of geology. Also in To the Student on page xxi is a list of units of measure used in the text.

I am indebted to a large number of individuals and institutions for assistance in bringing this book to completion. I wish to thank the William C. Brown Company, including former editors Robert Stern and Ed Jaffe and present editors Jeff Hahn, Bob Fenchel, Lynne Meyers and Cathy Di Pasquale, for making this book possible. Appalachian State University provided periodic support for my endeavors, including one semester of paid leave. I am especially indebted to all of those individuals who helped increase my understanding and knowledge of petrology, notably Professors R. N. Abbott, D. O. Emerson, C. V. Guidotti, M. E. Maddock, I. D. MacGregor, E. M. Moores, R. L. Rose, S. Skapinsky, C. H. Stevens, S. E. Swanson, and F. Webb. A number of colleagues generously shared thoughts, time for reviews, and reprints of their work.

REVIEWERS

Samuel E. Swanson
University of Alaska–Fairbanks

Stephen A. Nelson
Tulane University

Michael Smith
University of North Carolina–Wilmington

Daniel A. Textoris
University of North Carolina

Gail Gibson, Director
Math and Science Education Center
University of North Carolina–Charlotte

Stephan Custer
Montana State University

David Lumsden
Memphis State University

Robert Furlong
Wayne State University

Jad D'Allura
Southern Oregon State College

Gunter K. Muecke
Dalhousie University

Dexter Perkins
University of North Dakota

Calvin Miller
Vanderbilt University

Steven P. Yurkovich
Western Carolina University

Barbara Lott
University of North Carolina

Richard Heimlich
Kent State University

Edward Stoddard
North Carolina State University

PUBLISHER'S NOTE TO THE INSTRUCTOR

Petrology: The Study of Igneous, Sedimentary and Metamorphic Rocks, has been developed to fit the special needs of your petrology course. There are several binding options that have been developed to ensure that you, the instructor, have your students purchase the material you choose to teach. Since many professors do not teach the petrology of all rock types, you can purchase the material relevant only to your particular petrology class.

Petrology: The Study of Igneous, Sedimentary and Metamorphic Rocks, by Loren Raymond is available as a complete, casebound textbook or as paperback custom separate of each individual rock type. You can purchase these textbooks individually, or have them packaged together to make a custom teaching package. To purchase the custom separates, please order the appropriate custom separate from the following table, or contact your local Wm. C. Brown Representative.

Binding Option	Description
Petrology: The Study of Igneous, Sedimentary and Metamorphic Rocks ISBN 0–697–00190–3	The complete petrology textbook covering all rock types.
Igneous Petrology ISBN 0–697–23692–7	Parts one, two, and five from *Petrology: The Study of Igneous Sedimentary, and Metamorphic Rocks*
Sedimentary Petrology ISBN 0–697–23691–9	Parts one, three, and five from *Petrology: The Study of Igneous Sedimentary, and Metamorphic Rocks*
Metamorphic Petrology ISBN 0–697–23690–0	Parts one, four, and five from *Petrology: The Study of Igneous, Sedimentary, and Metamorphic Rocks*

If you are using one of our customized textbooks from our *Petrology Series* by Loren Raymond, please note that only a portion of the table of contents applies to your textbook.

FROM THE AUTHOR

All errors in understanding or knowledge are my own responsibility. A large number of students challenged me during my teaching, discovered sources during their own library research, or otherwise contributed to the successful completion of this project. Among those who provided particular assistance on specific projects were Paul Dahlen, Vickie Owens, Elizabeth Stevens, and Susan Wilson. Paul Dahlen assisted with photographic work. Tom Terranova, who prepared the line art for the book, provided inspiration for developing high-quality illustrations. Matt Raymond provided scale in some photographs and gave up much "quality time" over the eight-year period that I was writing the text. Margaret Raymond provided, love, support, and many hours of typing time, all of which were of immeasurable help in the completion of this project. To all of the above-named individuals and organizations, I give my heartfelt thanks.

TO THE STUDENT

Units of Measure Used in the Text and Conversion Factors

Pressure (P)

1 Gpa (Gigapascal) = 10^9 Pa (Pascal) = 10 kb (kilobars) = 10^4 bars

Temperature (T)

°K (degrees Kelvin) = °C (degrees Celsius) + 273°

Length (l)

1 km (kilometer) = 10^3 m (meters) = 10^5 cm (centimeters) = 10^6 mm (millimeters)

Area (A)

1 km^2 (square kilometer) = 10^6 m^2 (square meters) = 10^{10} cm^2 (square centimeters)

Volume (V)

1 km^3 (cubic kilometers) = 10^9 m^3 (cubic meters) = 10^{15} cm^3 (cubic centimeters)

Mass (m)

1 kg (kilogram) = 10^3 grams

Density (ρ)(rho)

1 kg/m^3 (kilograms per cubic meter) = 10^{-3} g/cm^3 (grams per cubic centimeter)

Acceleration of Gravity (g)

0.0098 km/sec^2 = 9.8 m/sec^2 = 980 cm/sec^2

Time (t)

1 b.y. (billion years) = 10^3 m.y.(= 10^3 ma)(million years) = 10^9 years ~ 3.16×10^{16} seconds

Phase Rule and Phase Diagrams

P = number of phases

C = least number of components necessary to define a system

F = the number of degrees of freedom = the variance

X = composition

List of Chemical Symbols Used in Text

Al—aluminum
Ar—argon
B—boron
Ba—barium
Be—beryllium
C—carbon
Ca—calcium
Ce—cerium
Cr—chromium
Cs—cesium
Eu—europium
F—fluorine
Fe—iron
Ga—gallium
Gd—gadollium
H—hydrogen

He—helium
K—potassium
La—lanthanum
Li—lithium
Lu—lutetium
Mg—magnesium
Mn—manganese
Na—sodium
Ni—nickel
Nd—neodymium
O—oxygen
Os—osmium
P—phosphorous
Pb—lead
Pr—praseodymium
Rb—rubidium

Re—rhenium
S—sulfur
Si—silicon
Sm—samarium
Sr—strontium
Ta—tantalum
Ti—titanium
Th—thorium
U—uranium
V—vanadium
W—tungsten
Y—yttrium
Yb—ytterbium
Zr—zirconium

List of Common Abbreviations and Prefixes Used in the Text and Their Meanings

blasto—to bud; to sprout; hence to form anew in a metamorphic rock

cf.—compare to

e.g.—for example

et al.—and others

i.e.—that is

in situ—in place

inter—between

intra—within

iso—the same

sensu lato—in the broad sense

sensu stricto—in the strict sense

Aleutian Arc

Muskox Intrusion•

CANADA

Duke Island

Bay-of-Islands Ophiolite

St. Simon, Quebec

Lake Superior (Sediments)

Onawa, ME•

Waterville (SE Maine Metamorphic Belt)

•Salmo, B.C.

Cascade Range

Columbia River Plateau

•Shonkin Sag

•Stillwater

•Yellowstone

•Tin Mountain, SD

Michigan Basin Salina Group

St. Peter Sandstone

Rattlesnake Pluton

No. Appalachian Orogen

Appalachian Basin

Cincinnati Arch

Franciscan Complex, Great Valley Group

Green River Formation

•Notch Peak, UT

UNITED STATES

Southwestern Virginia (Nebo; Ben Clark's Farm)

Carbona Quad. and Del Puerto Ophiolite

Coast Ranges

•Yosemite

Navajo Sandstone

•Buck Creek Dunite

So. Appalachian Orogen

Crestmore, CA

Grand Canyon

Rio Grande Rift

•Laborcita Formation, NM

Brevard Zone

•Magnet Cove, AR

Bahamas

Atlantic Ocean

Pacific Ocean

Texas Shelf

Gulf of Mexico

MEXICO

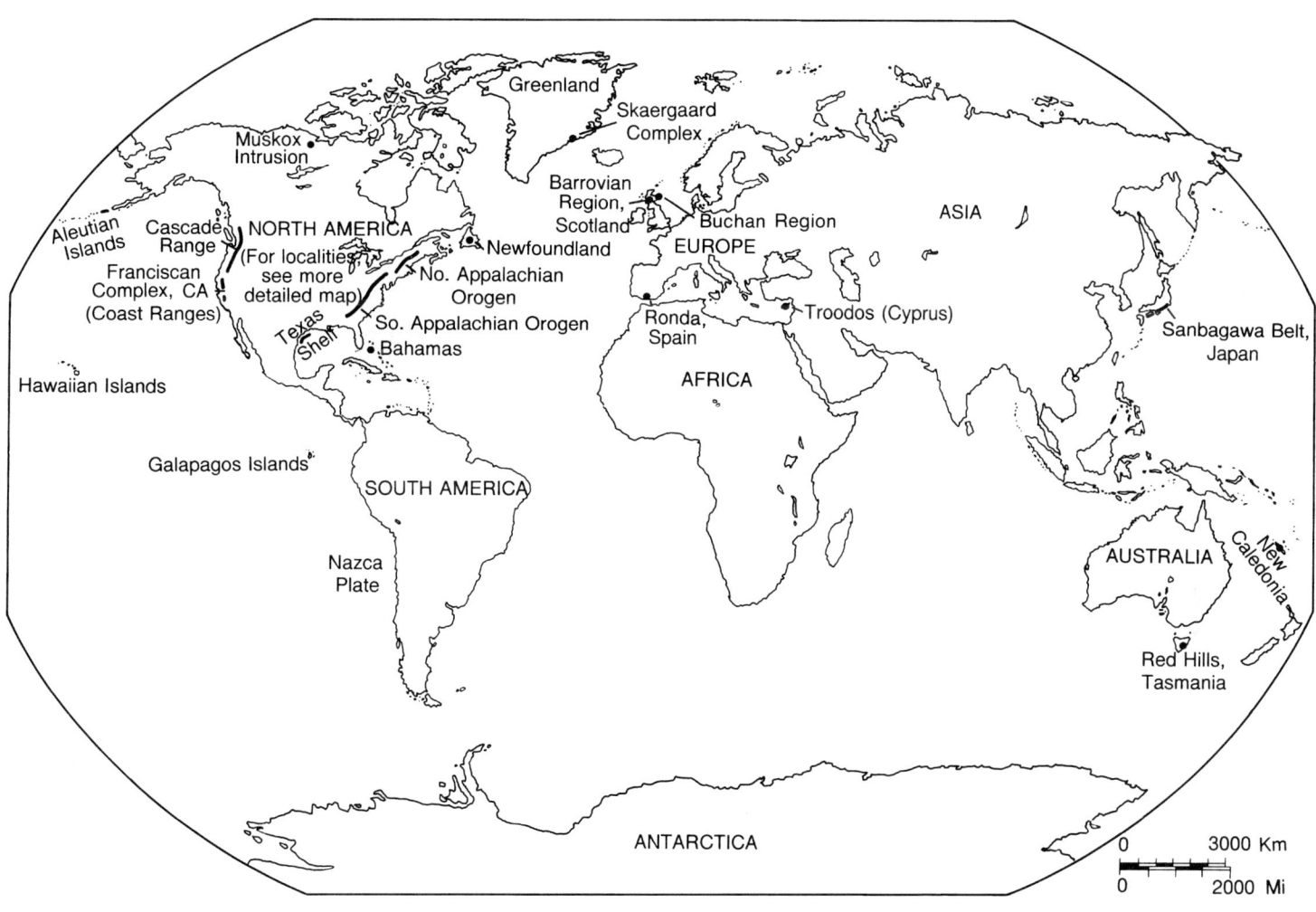

Greenland

Skaergaard
Complex

Muskox
Intrusion

Barrovian
Region,
Scotland

Buchan Region

ASIA

Aleutian
Islands

Cascade
Range

NORTH AMERICA

(For localities
see more
detailed map)

Newfoundland

EUROPE

Sanbagawa Belt,
Japan

Franciscan
Complex, CA
(Coast Ranges)

No. Appalachian
Orogen

Texas
Shelf

So. Appalachian Orogen

Ronda,
Spain

Troodos (Cyprus)

Hawaiian Islands

Bahamas

AFRICA

Galapagos Islands

SOUTH AMERICA

AUSTRALIA

New
Caledonia

Nazca
Plate

Red Hills,
Tasmania

ANTARCTICA

| 0 | | | | 3000 Km |
| 0 | | | | 2000 Mi |

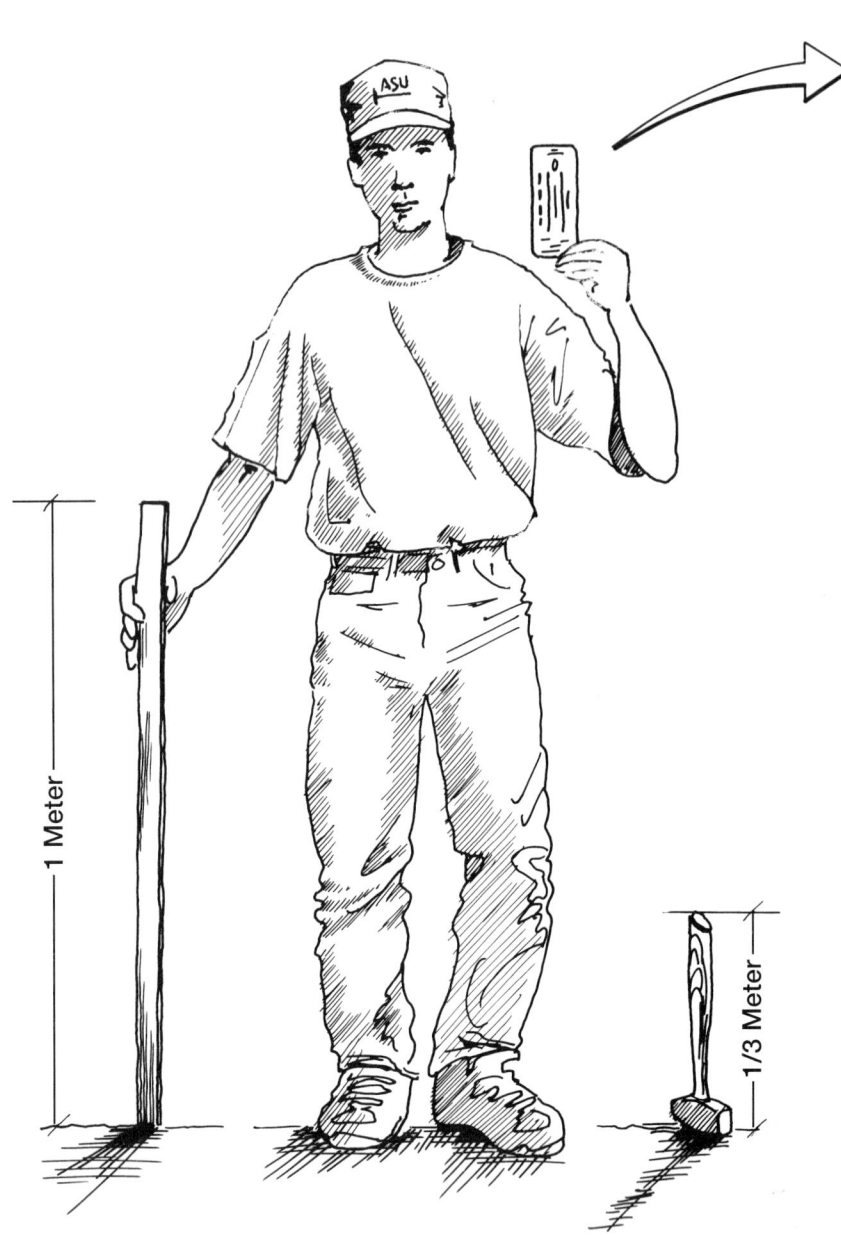

1 Meter

1/3 Meter

GSA
PHOTO SCALE/FOCUS GUIDE

THE GEOLOGICAL SOCIETY OF AMERICA · 1888 ·

Avoid incorrect exposures by taking meter readings before placing scale in picture area. Fine-focus on GSA seal.

CM

IN

Grain Size Scale

1   2   3   4   5

mm

PETROLOGY

The

Study

of

IGNEOUS

SEDIMENTARY

METAMORPHIC

Rocks

Introduction

Geology is the study of the planet Earth, including its composition, origin, and history. Most of the solid part of the Earth is composed of rocks. Because the Earth is composed of rocks, an understanding of their nature, composition, origin, and histories is central to understanding Earth history. This book is an introduction to the study of rocks.

PART I

View of the Three Sisters volcanoes, Oregon, from the top of Mount Bachelor. Mount Hood is in the distance to the north.

1

Rocks and Earth Structure

INTRODUCTION

A **rock** is a solid aggregate of mineral grains, or a solid, naturally occurring mass composed of mineral grains, glass, altered organic matter, and combinations of these components. Rocks may be studied in a variety of ways. **Petrology** is the term used to denote the overall study of rocks, including **petrography** (also called lithology), the study of the description and classification of rocks, and **petrogenesis,** the study of the histories and origins of rocks.[1] Petrography is basically an observational science. Features such as color, mineralogy, and texture are studied and used as a basis for subdividing the naturally occurring, continuous range of rock types into groups to which petrologists assign names. In some cases, petrogenetic information is also used as the basis for subdivision and classification. Petrogenetic studies combine various experimental and theoretical analyses and use inductive and deductive reasoning to arrive at conclusions about rock origins and histories.

THE THREE CLASSES OF ROCKS

All rocks may be assigned to one of three major classes—igneous, sedimentary, or metamorphic. **Igneous rocks** are rocks that form by crystallization (freezing) of melted rock materials at temperatures well above those standard at the Earth's surface. The melt, commonly referred to as **magma,** may or may not be entirely molten, as substantial amounts of solids, both crystals and rock fragments, as well as gases, occur in magmas.

Sedimentary rocks form under surface conditions and consist of accumulations of (1) chemical and biochemical precipitates; (2) fragments or grains of rocks, minerals, and fossils; or (3) combinations of these kinds of materials. Like igneous rocks, some sedimentary rocks result from crystallization, but in sedimentary environments crystallization occurs from an aqueous solution under surface or near-surface conditions of pressure and temperature. Most sedimentary rocks develop through cementation, compaction, and recrystallization of accumulated mineral, rock, and fossil fragments. In these rocks, minerals that constitute cements also crystallize from aqueous solutions.

Metamorphic rocks are rocks that formed originally as igneous or sedimentary rocks but which have been changed mineralogically, texturally, or both—without undergoing melting—in response to heat, pressure, directed stress, or chemically active fluids or gases. The pressures and temperatures that induce the changes in mineralogy and texture are generally rather different from standard surface conditions.

Nevertheless, a continuous range of conditions exists from those that prevail during sedimentation through those that produce melting at the highest grades of metamorphism.

The distinctive characteristics of each class of rocks result from the processes that form the rocks. Sedimentary rocks are typically layered, because they form by settling of materials through water or air. Igneous rocks may be structureless or they may show features that reflect movement and crystallization of hot fluids. Metamorphic rocks commonly show an alignment of grains that gives a flaky, banded, or layered appearance to the rock, and such rocks show folds or other features that reflect plastic flow. These distinctive characteristics are described in subsequent chapters dealing with the respective rock types. Mineral associations typical of each class of rocks are listed in appendix A.

The student should realize that nature provides a continuous array of rock properties. One rock type may grade into another, and the subdivisions imposed by petrologists on the continuous natural order are boundaries arbitrarily placed at seemingly significant and reasonable locations within the continuum. The boundaries between igneous and sedimentary rocks, sedimentary and metamorphic rocks, and metamorphic and igneous rocks are diffuse.

ROCK DISTRIBUTION IN THE EARTH

The Earth is a solid body, except for the outer core (figure 1.1) and some relatively small local spots within the upper mantle and crust, which are liquid. Much of the solid material is metamorphic rock. This is so because much of the rock of the inner core, mantle, and crust has been changed in texture or mineralogy in response to the high pressures, temperatures, and stresses that exist everywhere but at or near the Earth's surface. Most of these metamorphic rocks are inaccessible to study. Many of the liquids (magmas) that form in the upper crust or mantle rise to higher levels in the crust, where they crystallize as igneous rocks. Sedimentary rocks form at or near the surface. These, plus igneous and metamorphic rocks that become exposed at the surface through erosion, are accessible, have a direct impact on human endeavors, and have been studied the most.

On the land, sedimentary rocks total 66% of the exposed rocks (Blatt and Jones, 1975). Crystalline rocks, which make up the remaining 34% of the exposures, are about equally divided into igneous and metamorphic types. Below the oceans, the areal distribution of sedimentary rocks and sediments is not precisely known, but recent oceanographic work reveals that most of the ocean floor is coated with a thin veneer of sediment and sedimentary rock. Beneath the sedimentary cover, igneous and metamorphic rocks dominate the oceanic crust.

Considering the volume of the Earth as a whole ($1.083 \times 10^{21} m^3$), the core comprises about 16.2% by volume, the mantle makes up 83.2%, and the crust amounts to only 0.6% of the volume.[2] Of the crust, about 4.8% is sedimentary materials,[3] which means that those materials comprise only about 0.029% of the total volume of the Earth. These facts should be remembered in reading this book, since the text deals almost exclusively with the petrology of the rocks of the crust and upper mantle.

EARTH STRUCTURE AND PETROTECTONIC ASSEMBLAGES

Traditionally, the Earth has been recognized as having a series of layers (figure 1.1). These are defined on the basis of geophysical phenomena, with each layer distinguished from the next by a zone of marked change in seismic wave velocity called a discontinuity.[4] A solid inner core of iron-rich rocks, with a variety of probable impurities, is separated from

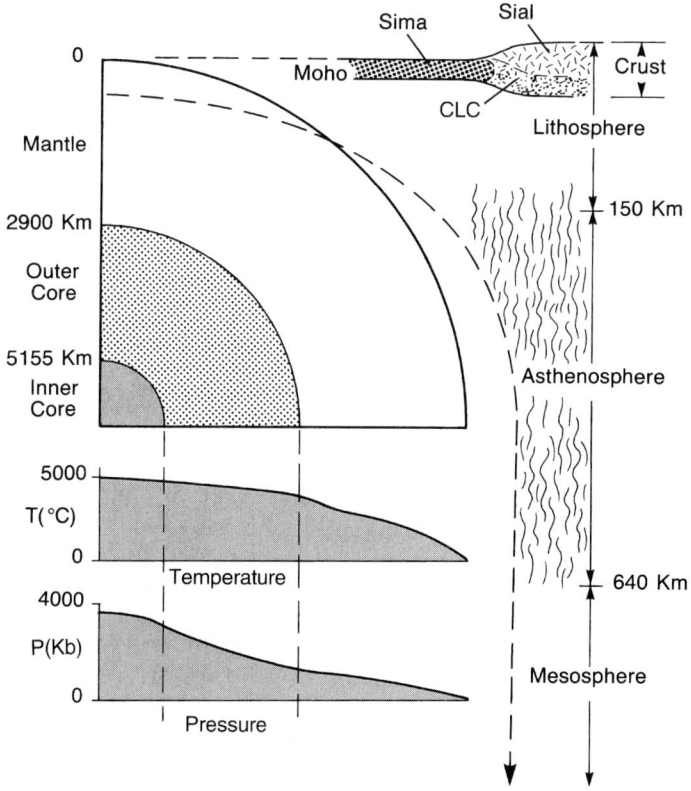

Figure 1.1 Generalized sketch of the interior of the Earth, showing major subdivisions delineated using seismic data. Pressure and temperature curves show the changing conditions towards the center of the earth. CLC = complex lower crust. *(Numerical data from Bolt, 1982)*

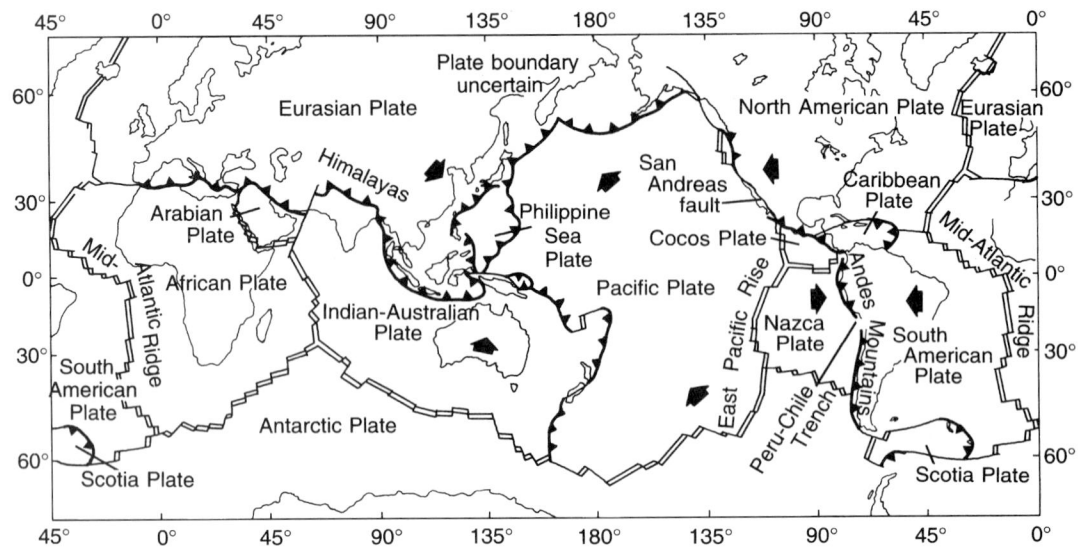

Figure 1.2 World map showing major lithospheric plates. Double lines indicate spreading centers, single lines on plate boundaries are transform faults, and heavy barbed lines are subduction zones and other zones of convergence. *(Source: After W. Hamilton, U. S. Geological Survey.)*

a liquid outer core by the Lehmann Discontinuity. Outward from the outer core, and separated from it by the Gutenberg Discontinuity, is the mantle, a layer composed of magnesium-rich rocks. The Mohorovicic Discontinuity or **Moho** separates the mantle and the crust. The crust is dominated by feldspars and other silicate minerals. Locally, the crust consists of *silicon-*, *iron-*, and *magnesium-rich* rocks, which are referred to as *simatic* rocks. The oceanic crust is simatic. In contrast, the continental crust is dominated by *silicon-* and *aluminum-rich* minerals and is said to be *sialic.* The lower continental crust is quite complex and consists of a variety of igneous and metamorphic rocks of both simatic and sialic character (Berckhemer, 1969; Finlayson, 1982; Rudnick and Taylor, 1987).[5]

Pressure and temperature increase with depth in the Earth, as shown in figure 1.1. Near the surface, the **geothermal gradient** (the increase in temperature with depth) is pronounced. In the mantle, however, the slope of the temperature curve is less steep. Although the average gradient is about 20° C/km, geographic variations in the geothermal gradient exist. Over volcanically active areas, such as in the volcanic island arcs, the temperature increases at a rate of 30–50° C/km or more. In contrast, near ocean trenches, the rate of increase may be as low as 5–10° C/km. Continental areas away from tectonically active zones have intermediate gradients.

The pressure in the Earth is given by the expression

$$P = \rho gh$$

where

P = pressure
ρ = density
g = acceleration of gravity
h = height of the column of overlying rock (depth of burial).

Average crustal values for density yield pressure gradients on the order of 0.1 Gpa/3.3 km (1 kb/3.3 km), a value useful for making quick estimates of pressure within the crust.

Increased emphasis has been placed on a threefold subdivision of the crust and mantle as a result of tectonic studies conducted during the past three decades. An upper rigid zone, the **lithosphere,** consisting of the crust and upper mantle, is separated from a lower rigid zone composed of mantle rocks, the **mesosphere,** by a plastic zone called the **asthenosphere** (figure 1.1).[6] The asthenosphere consists of partially melted mantle materials. The lithosphere is fragmented into several major and minor pieces called **plates,** the boundaries of which are marked by zones of earthquake activity (figure 1.2).

Lithospheric plates move, producing plate interactions along three types of boundaries (figure 1.3). At **spreading centers,** or divergent boundaries, plates pull apart and new crust

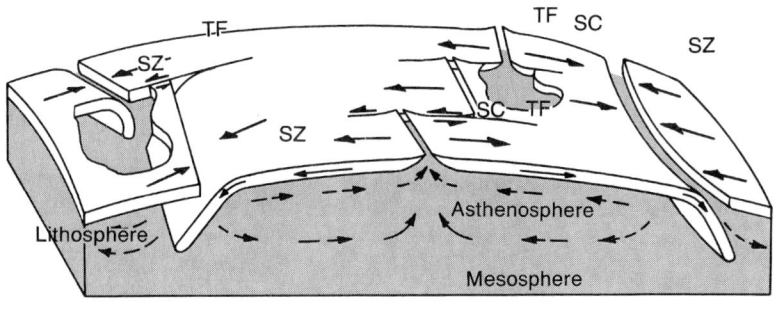

Figure 1.3 Schematic block diagram showing the three types of plate boundaries—spreading centers (SC), subduction zones (SZ), and transform faults (TF).
(Modified from Isacks et al., 1968.)

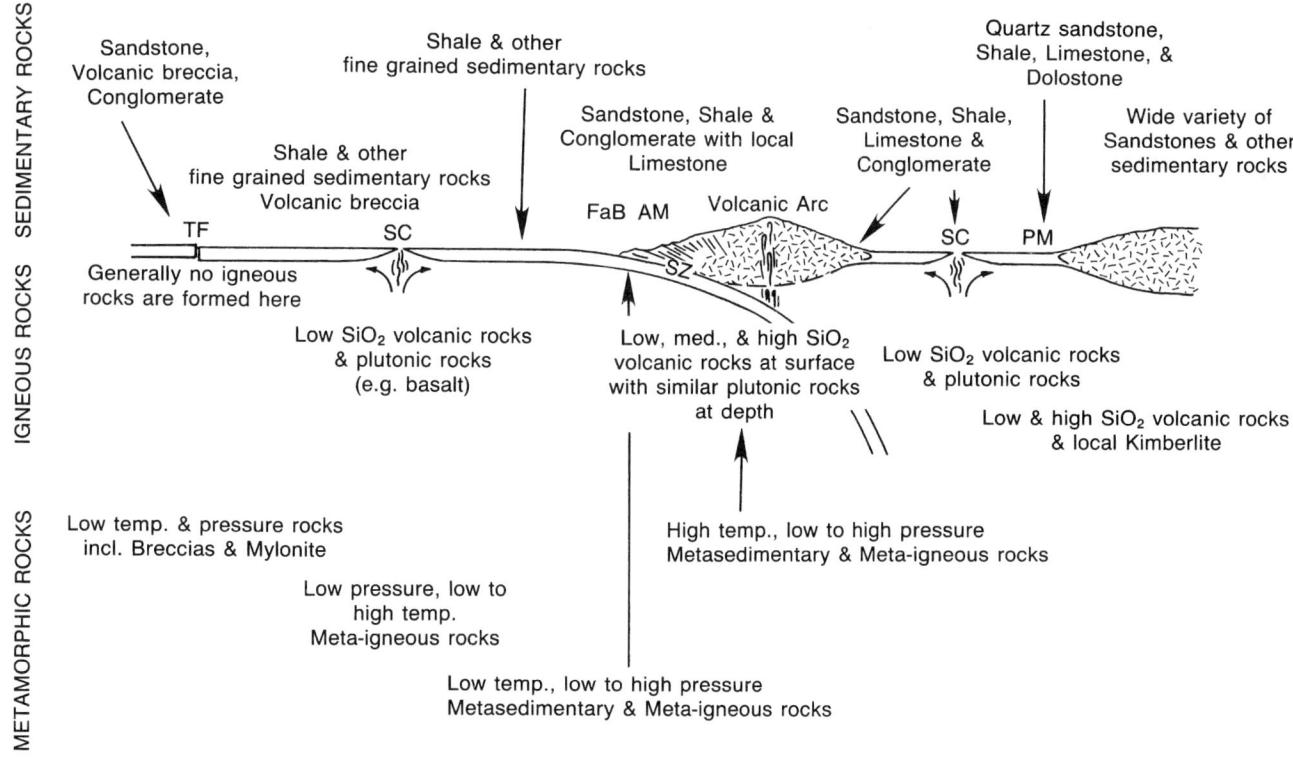

Figure 1.4 Diagram showing petrotectonic assemblages at various sites. FaB = forearc basin, SC = spreading center, SZ = subduction zone, TF = transform fault, AM = active margin, PM = passive margin.

is formed. **Subduction zones** represent convergent boundaries, where plates collide, with one plate descending beneath another. **Transform faults** mark shear boundaries, where plates slide past one another. Sites where three plate boundaries come together are called **triple junctions.** Each type of plate boundary, as well as triple junctions and sites within plates, gives rise to distinctive suites of rocks called **petrotectonic assemblages** (Dickinson, 1971) (figure 1.4). The interaction between petrologic and tectonic processes at a given plate boundary gives the rocks forming at the boundary distinctive characteristics valuable for recognizing the rocks formed at

that type boundary in the geologic past. A detailed review of petrotectonic assemblages indicative of each environment is presented in the epilogue. As a basis for discussion in the following chapters, a brief, simplified summary is presented here.

At spreading centers, one or two types of volcanic rock form and these are accompanied by various sedimentary and metamorphic rocks. Initiation of the spreading process results in the formation of graben or rift zones. On continents, these zones serve as basins for the accumulation of sand and gravel. Intrusion of magma into these sediments results in local metamorphism of low-pressure, high-temperature type. Extrusion of magma produces lava flows and other volcanic features. In ocean-floor spreading centers, volcanism is followed by circulation of seawater through hot volcanic rocks and into surrounding rocks, resulting in low-pressure chemical alteration or metamorphism. Sedimentation in the oceanic areas produces thin layers of very fine-grained sediment. Locally, however, gravels derived from the volcanic rocks are deposited.

Convergent plate boundary zones exhibit a wide range of rock types. Volcanic rocks of diverse compositions develop in volcanic mountain chains, called **arcs,** that form on the overriding plate. At depth, magma intrudes to form masses of siliceous igneous (granitoid) rock. The invading magmas cause high-temperature, low-to-high-pressure metamorphism in the surrounding rocks. Sediments derived from the erosion of arc rocks are shed both away from the plate boundary and towards it. At the plate boundary, a trench or an elongate basin forms at the collision zone and collects sediments not trapped in the arc-trench gap (also called the forearc basin) (figure 1.4). Sediments deposited on oceanic crust in the trench are then either accreted onto the overlying plate or subducted (J. C. Moore, 1989). The subducted sediments and underlying oceanic crust and mantle either become metamorphosed by the high-pressure, low-temperature conditions prevailing immediately arcward of the trench and are accreted to the overlying plate or they are carried to greater depths by the descending plate (Dewey and Bird, 1970; Ernst, 1970; 1973).

Shearing at transform fault boundaries forms metamorphic rocks. Because the rocks are deformed rapidly, they may show extensive breaking or stretching of mineral grains. Seawater percolating into such shear zones aids in the metamorphic process.

Triple junction assemblages are not easily recognized, as they consist of composite assemblages representing the various types of plate boundaries that form the triple junction. Nevertheless, some triple-junction assemblages have been preserved and recognized in the rock record.

Intraplate (within-plate) volcanism and sedimentation, like plate boundary processes, yield a variety of products. Silica-poor rocks dominate the volcanic and plutonic suites within oceanic plates. A variety of fine-grained sedimentary rocks also forms here. Within continents, at intraplate sites, unusual rocks, including diamond-bearing types, form along with a diversity of common volcanic rock types. Continental sites of sediment formation are quite varied, as are the resulting rocks. Distinct sediments form in river, lake, glacial, and other environments. Within the plates, along the tectonically quiet coastal areas, deposits of sediments yield sandstone shale, and limestone.

Because the processes and sites of rock formation vary, the petrotectonic assemblages vary. The characteristics of these assemblages and the included rocks provide the data for understanding petrogenesis, a major goal of petrology.

SUMMARY

The Earth is divided into crust, mantle, and core, with the crust and upper mantle comprising the rigid lithosphere. The lithosphere is divided into plates bounded by transform faults, subduction zones, and spreading centers. Pressure and temperature increase through the lithosphere and towards the center of the core.

The three classes of rocks—the igneous rocks, the sedimentary rocks, and the metamorphic rocks—all occur in the crust. Sedimentary rocks form a thin veneer covering 66% of the continental crust, but the igneous and metamorphic rocks are dominant volumetrically. All three classes of rock occur in petrotectonic assemblages that reflect the plate setting of their petrogenesis.

EXPLANATORY NOTES

1. Definitions of terms in this book generally follow those of the *Glossary of Geology* (Bates and Jackson, 1980, 1987). However, I have made some modifications and additions. Important (boldfaced) terms are defined in the glossary of this text.
2. Units of measure (e.g., meters, grams) used in this book are primarily SI units, although commonly used alternatives (°C for °K) are used in some cases. These units are listed in the front section of the book.
3. Ehlers and Blatt (1982, p. 6).
4. This section on the geophysical properties of the earth is based on review works, especially Wyllie (1971a), Ringwood (1975, 1979), and Bott (1982). For information on the core, for example, see Bott (1982, p. 147) and Ringwood (1979, p. 41ff.). Additional information is available in standard geophysics texts.
5. Also see Touret and Dietvorst (1983), Griffin and O'Reilly (1987), Dodge, Lockwood, and Calk (1988), and Bohlen and Mezger (1989).
6. The asthenosphere is discussed in simple terms by D. L. Anderson (1962), and its limits are shown by the work of Kanamori and Press (1970) and many others. See reviews of the seismic data in Wyllie (1971a, ch. 3), Bott (1982), and the summary discussion in Bolt (1982).

PROBLEMS

1-1. Using the generalized formula (0.1Gpa/3.3 km), calculate the pressures expected at 10 km, 20 km, and 30 km in the crust. Next, using the formula for pressure and assuming an average density of 2.70 gm/cc, calculate the pressure at 20 km. Compare your answers derived by the two calculations for 20-km depths.
2-2. Calculate the temperatures expected in the crust at a depth of 20 km, assuming (a) a geothermal gradient of 5° K/km, (b) a geothermal gradient of 20° K/km, and (c) a geothermal gradient of 50° K/km.

Metamorphic Rocks

Metamorphic rocks characterize mountain belts and pervade the continental shields, marking the roots of long-eroded mountain belts. In the mantle, they are ubiquitous. These rocks form through transformations of pre-existing igneous, sedimentary, and metamorphic rocks brought on by changes in the prevailing intensive variables and fluids. The processes of transformation are not generally visible at the Earth 's surface and must be re-produced in the lab. In nature, the transformation is not always complete. That incompleteness is a fortunate circumstance in that it allows the petrologist to see beyond the most recent events and into the past history of the rock.

PART IV

Swirled patterns in metamorphic rocks, Indian Peaks Wilderness Area, Colorado.

23

Metamorphism and Metamorphic Rock Textures and Structures

INTRODUCTION

Banded and swirled patterns typify the metamorphic rocks (see page 471). These are rocks that have undergone a change (*meta-*) in their form (morph) from preexisting igneous, sedimentary, or metamorphic progenitors. Metamorphic rocks, which are common in the cores of mountain ranges, bear the imprints of heat, pressure, chemically active fluids, and deformation. Like the igneous and sedimentary rocks from which they are derived, metamorphic rocks have histories reflected in their textures and mineral compositions.

The purposes of this chapter are to define metamorphism, to discuss the agents that cause it, and to describe the structures, textures, and mineral compositions of metamorphic rocks. Through these features, the conditions of metamorphism are recognized. Yet, if the history of the metamorphic rock is to be fully understood, one must also understand the chemistry of the rock and recognize any minerals, structure, and texture remaining from the premetamorphic parent rock or **protolith** (literally, first stone). Together, all of these data enable petrologists to decipher metamorphic rock histories, which may be complex and incompletely recorded.

DEFINITIONS: METAMORPHISM AND METAMORPHIC ROCKS

Metamorphism is a process or set of processes that affect rocks in such a way as to produce textural changes, mineralogical changes, or both, under conditions in the Earth between those of diagenesis and weathering (at the lower limit) and melting (at the upper limit).[1] Processes of textural change that may occur *without accompanying mineralogical change* are of two types: recrystallization and cataclasis. **Cataclasis** is the crushing and breaking of grains in rocks. It may affect any type of rock. **Recrystallization,** however, is a process of reorganization of crystal lattices and intergrain relationships through ion migration and lattice deformation, without accompanying breaking of grains. Recrystallization occurs most commonly in monomineralic rocks, such as pure limestone, quartz arenite, or dunite. It may also occur where a directed stress acts on a rock under conditions of pressure (P), temperature (T), and composition (X) for which existing minerals in the rock are stable. **Neocrystallization** is the process that results in the formation of new minerals that did not previously exist in the metamorphic rock. Recall that equivalent processes occur during diagenesis. Thus, metamorphism is similar to diagenesis, but encompasses only those processes that occur *beyond* the near-surface (low-*P*), low-temperature, and low-stress limits of diagenesis.

Metamorphic rocks are rocks with textures, minerals, or both, that reflect cataclasis, recrystallization, or neocrystallization in response to conditions that differ from those under which the rock formed and that lie between those of diagenesis and anatexis. Like igneous rocks and sedimentary precipitates, metamorphic rocks have features that are adjusted to or are adjusting to specific physical and chemical conditions. As a consequence, they are amenable to study using principles of physics and physical chemistry, such as the Phase Rule.

AGENTS AND TYPES OF METAMORPHISM

The *agents* of metamorphism are pressure, temperature, directed stress, and chemically active fluids. In general, when a rock is transferred to a new set of conditions from the set of conditions under which it formed, it will have been transferred from a condition of stability to one of instability.[2] Thus, the minerals, the texture, or both, are out of equilibrium. Provided adequate energy is available, changes will occur in the rock to bring the minerals and texture into equilibrium under the new set of conditions. Metamorphic rocks exist at the surface of the Earth, even though they are not at equilibrium there, because there has been insufficient energy to transform the minerals and textures of the metamorphic rocks back into those stable at the surface.

Pressure

Stress (σ) is defined as force per unit area ($\sigma = F/A$). Pressure is a uniform stress.[3] It confronts the rock equally in all directions. Trapped fluid phases, such as H_2O and CO_2, may create pressure and such pressures are referred to as P_{fluid}, P_{H_2O}, or P_{CO_2}, whichever is appropriate. Alternatively, pressure is created by the load of the overlying rock and is referred to as P_{load} or lithostatic stress.

Recall that, as a "rule of thumb," P_{load} in the crust increases by about 0.1 Gpa for every 3.3 km of burial (0.1 Gpa = 1 kb = 3.3 km). Pressures of metamorphism range from less than 0.1 Gpa up to the enormous pressures of tens of Gigapascals present in the deep mantle and core.[4] Because rock masses exposed at the surface are derived only from the crust and uppermost mantle, however, most metamorphic petrologists are primarily concerned with pressures in the range of 0.1 Gpa (1 kb) to about 1.5 Gpa (15 kb).

Deviatoric Stress

In addition to pressure, directed or **deviatoric stresses,** stresses acting in particular directions and exceeding the (mean hydrostatic) stress, commonly affect rocks during metamorphism. Such stresses may act (1) along a line in opposite directions away from a point, producing tension (figure 23.1a), (2) along a line in opposite directions towards a point, producing compression (figure 23.1c), or (3) in opposite directions along different lines, yielding a force couple and producing attendant compression, tension, and shear (figure 23.1e). In each case, distinctive textures and structures may develop (figures 23.1b, d, and f).

Foliation, the most typical characteristic of metamorphic rocks, is a planar feature.[5] It results primarily from the parallel to subparallel alignment of inequant mineral grains, such as mica or amphibole grains. Though the details of the development of foliation in polymineralic rocks have been investigated experimentally only in recent years[6] and are subject to debate,[7] it is generally agreed that deviatoric stress induces the alignment of minerals that gives rocks their foliation. Thus, deviatoric stress plays a critical, yet commonly overlooked, role in the formation of many metamorphic rocks.

Temperature

The temperature conditions under which metamorphism occurs are relatively well defined, but variable. At the absolute upper limit, temperatures of metamorphism are bounded by those of the solidus of dry ultramafic and ultrabasic rocks (figure 23.2). This lies between about 1200° C and 2000° C, depending on the pressure and composition of the rock.[8]

Because melting occurs at markedly different temperatures in rocks of different composition, the upper limit of metamorphism is different for each different bulk composition. For granitic or quartz-feldspar-mica rocks, the minimum upper limit of metamorphism is the solidus for wet granite, which may be as low as 600° C (figure 23.2) (Huang and Wyllie, 1981; C. R. Stern and Wyllie, 1981). Between the melting curves of wet granite and dry ultramafic rock is a zone in which rocks will undergo either metamorphism, partial melting, or complete melting—depending on the rock composition, the pressure, the temperature, and the composition and quantity of the fluid phase present.

The lower temperature limit of metamorphism is not quite as well defined. Diagenesis and weathering end and metamorphism begins where new minerals form that are not stable at or very near the surface. The reactions that give rise to such minerals are commonly considered to begin to occur at about 100° C, though it is possible that some might occur at slightly lower temperatures. Alt et al. (1986) showed that zeolites and prehnite form in the oceanic crust

473

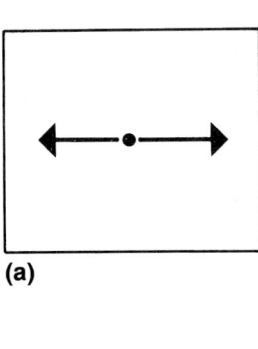

(a)

(b)

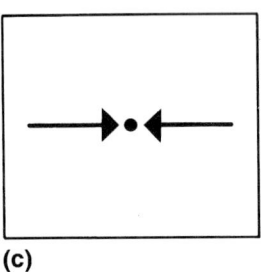

(c)

(d)

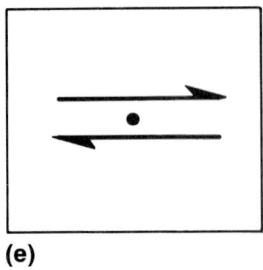

(e)

(f)

Figure 23.1 Stress and metamorphic structures. (a) Force vectors (arrows) showing configuration of stress for conditions of tension. (b) Boudinage produced by tension (extension) acting on a tabular granitoid dike, Highway 321, between Boone and Blowing Rock, North Carolina. (c) Force vectors (arrows) showing configuration of stress for compression. (d) Folds in amphibole gneiss, produced by compression, Ashe Metamorphic Suite, northwestern North Carolina. (e) Force vectors (arrows) showing configuration of stress for a shear couple. (f) Tension fractures (vein filled) produced by a shear couple acting on meta-arkose of the Grandfather Mountain Formation, Boone, North Carolina. Scale in (b), (d), and (f) is DNAG scale.

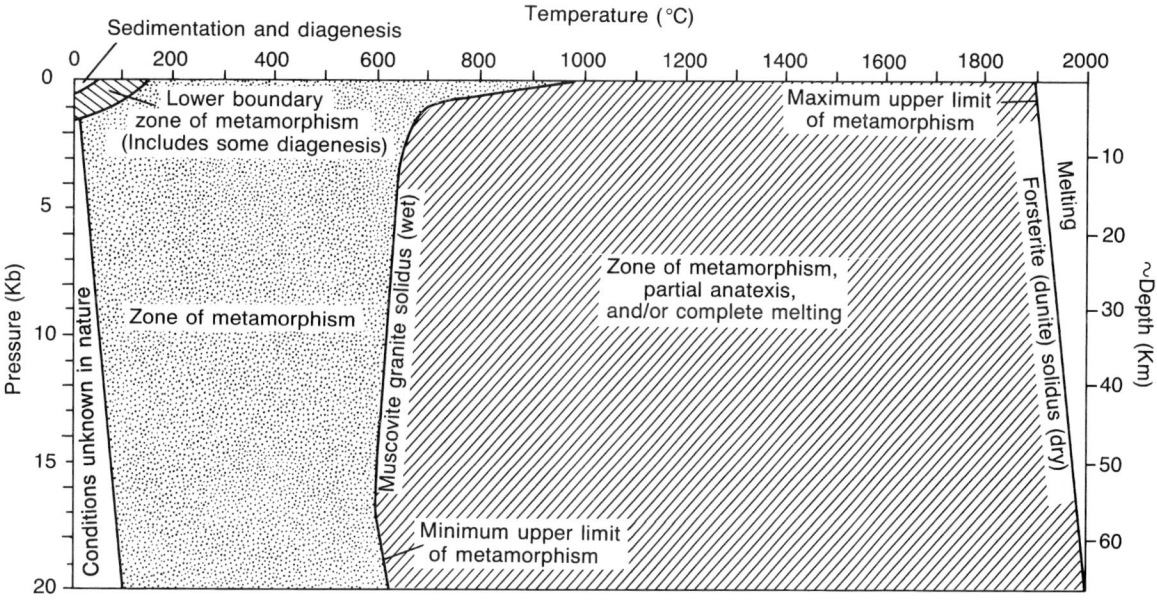

Figure 23.2 Pressure-temperature grid showing the P-T limits of metamorphism. The forsterite solidus is based on Bowen and Anderson (1914), B. T. C. Davis and England (1964), and Ohtani and Kumazawa (1981). The muscovite granite solidus is from Huang and Wyllie (1981).

at temperatures as low as 100° C, whereas weathering occurred in the same rocks at temperatures below about 50° C. Most common metamorphic rocks apparently were formed under temperatures between 100° C and 750° C. Nevertheless, metamorphism is known to have occurred, in unusual cases, at temperatures outside these limits.[9]

The main sources of the heat that results in temperature changes in metamorphic rocks are (1) increases in pressure with depth, (2) radioactive decay, (3) deformation, and (4) migrating magmas. It is well known that temperature increases with depth in the Earth (see chapter 1). In general, this increase results both from the temperature increase that accompanies increased pressure and from radioactive decay.[10] In addition, small to large regions may be heated by the heat escaping from magmas that have migrated into the rocks. Such is obviously the case where rocks have been metamorphosed adjacent to an intrusion. It is less obvious, though equally true, that the numerous intrusions that penetrate into actively forming mountain belts bring with them the heat that pervades the mountain root, causing regional heating and associated metamorphism (Lux et al., 1986). Locally, thermal effects also result from frictional shear heating along fault zones.[11]

Chemically Active Fluids

Except for certain localized areas in which very high temperature conditions or impermeable rocks prevail, rock masses contain a fluid phase. The volatile-rich phase is called fluid because at nearly all metamorphic P-T conditions, the volatile phase is in a supercritical state. In such a state, a distinction between gas and liquid cannot be made.[12]

Several kinds of evidence suggest that, in fact, a fluid phase does exist in rocks during metamorphism.

1. Fluid inclusions occur in metamorphic minerals (Touret, 1971; Touret and Dietvorst, 1983; Craw, 1988; Nwe and Grundmann, 1990).
2. Formation of metamorphic phases that include CO_2 (e.g., calcite), H_2O (e.g., muscovite), or other components (S, N_2, F, Cl, B) requires the presence of a fluid phase (Ferry and Burt, 1982).
3. Whole-rock analyses show that metamorphic rocks developed under conditions of high P and T are depleted in volatile components relative to rocks formed under lower-grade conditions (Ferry and Burt, 1982).
4. Isotopic studies indicate the involvement of fluids during metamorphism (Losh, 1989; Nesbitt and Muehlenbachs, 1989).
5. The presence of veins in metamorphic rocks suggests that fluids were present during metamorphism (Bucher-Nurminen, 1981; Walther and Orville, 1982; Yardley, 1983; Nishiyama, 1989).
6. Active metamorphism is occurring today in active geothermal regions (Muffler and White, 1969; Rona et al., 1983; Charles, Buden, and Goff, 1986).[13]
7. Metamorphic reactions commonly involve dehydration and decarbonation reactions that yield a fluid phase (Bowen, 1940; Walther and Orville, 1982; Mohr, 1985).[14]

Usually the fluid phase is dominated by H_2O, but CO_2, CH_4, N_2, Cl, S, B, Na, K and other components may be present. In some cases, species other than H_2O are dominant.[15]

In a given rock body, the fluid phase may be in equilibrium with the solid phases that comprise the rock. If, however, that fluid phase subsequently changes in composition, disequilibrium between the rock and fluid results. The rock will adjust through mineralogical changes, textural changes, or both, in order to re-equilibrate. Fluid phases that have changed become "active" with respect to the rock and will interact with it.

A number of events may result in the activation of a fluid. Fluids may be induced to migrate from one rock mass to another as a result of temperature, pressure, or stress changes. Fluid migration of this kind activates the fluids *if* the new rock into which the fluid migrates has a composition different from that from which the fluid came. Such activated-fluid migrations result in alteration zones associated with ore-bearing intrusions and in other metamorphic changes (Ferry, 1983a).[16]

Activation may also occur *in situ*. Igneous intrusions in a region may add *new fluid or components* to the fluid phase, changing its composition and thereby activating it (Burnham, 1959). Some alteration zones around plutons are produced by such a process. Finally, metamorphic reactions themselves, induced by regional or local changes in pressure or temperature, may result in changes in the chemistry of the fluid phase. Such a process may activate the fluid by producing chemical potential gradients, i.e. gradational differences in chemistry between different locations in the rocks (Ferry, 1983a; Grambling, 1986). Altered fluids produced in this way may also migrate into other rocks and effect metamorphism in them.

Types of Metamorphism

Metamorphism is usually subdivided into a variety of types on the basis of the chemical nature of the metamorphism, the dominant agent of metamorphism, and/or the area or volume of rock affected.[17] In the last case, metamorphism is subdivided into local and regional types. **Local metamorphism** is metamorphism that affects relatively small volumes of rock (less than 100 km^3). **Regional metamorphism** typically affects thousands of cubic kilometers of rock.

Subdivision of the types of metamorphism on the basis of the dominant metamorphic agent yields several subtypes (table 23.1). Where temperature is dominant, metamorphism is local and is called **contact metamorphism,** because such metamorphism occurs in country rocks at and near their contact with an igneous rock mass. At shallow levels in the crust, at low pressures (LP), contact metamorphism may be referred to as LP-contact metamorphism. At deeper levels, where pressure becomes a factor the metamorphism may be referred to as MP-contact metamorphism at medium pressures or HP-contact metamorphism at high pressures. HP-contact metamorphism is rare,

Table 23.1 Types of Regional and Local Metamorphism

Dominant Agent	Local	Regional
Pressure	—	Static
Deviatoric stress	Local dynamic (includes impact)	Regional dynamic
Temperature	LP-contact	—
Chemically active fluids	Local metasomatic (including alteration)	Regional metasomatic
Temperature + pressure ±deviatoric stress ±chemically active fluids	MP-contact HP-contact	Dynamothermal

because high temperatures usually accompany high pressures, so that at depth, katazonal intrusions cannot supply enough heat to effect noticeable changes in the already hot country rocks. Where rare intrusions occur within accretionary complexes in subduction zones (Carden et al., 1977), HP-contact metamorphism may develop.

Metamorphism induced primarily by deviatoric stress is called **dynamic metamorphism.** In this book, local dynamic and regional dynamic types are distinguished. Local dynamic metamorphism develops along discrete fault zones, in metamorphic core complexes, and in the areas associated with meteorite impacts (e.g., at Meteor Crater, Arizona).[18] Regional dynamic metamorphism occurs in the mantle and in developing mountain belts—particularly in accretionary complexes at convergent plate margins—where deviatoric stress is distributed over large regions. Such regional distributions of deviatoric stress at moderate to high temperatures result in regional mylonite belts, and at low temperatures yield tectonic melanges.[19]

Pressure may also be the principal agent of metamorphism at the regional scale. Such metamorphism is called **static metamorphism.**[20] Static metamorphism is considered to occur at depth, in thick piles of sedimentary rock in continental and forearc basins, in trenches, and in sedimentary prisms along passive continental margins. Structural burial, produced where thick sections of rock are thrust upon other rock masses, also results in static metamorphism.[21] Local metamorphism caused primarily by pressure does not occur under natural conditions.

Chemically active fluids produce a kind of metamorphism, called **metasomatism,** that is dominated by chemical changes. Near pluton contacts, local metasomatism may be important.[22] Metasomatism is usually referred to as *alteration*, where it is associated with ore deposits. On the regional scale, the importance of metasomatic processes in the production of metamorphic rocks is debatable (see chapter 11). Some geologists *do* consider some types of regional metasomatism to be significant, referring to them as regional alteration, granitization, or basification.[23]

Dynamothermal metamorphism, the most widely distributed type of metamorphism, is metamorphism induced primarily by a combination of pressure and temperature. In some texts, the terms *regional* and *dynamothermal* are used interchangeably. This is an outdated practice derived from earlier works in which *only two* major kinds of metamorphism were recognized, thermal (contact) and regional (dynamothermal) (Harker, 1932).

In all forms of regional metamorphism—dynamothermal, regional dynamic, regional metasomatic, and static metamorphism—fluids are important. In some cases of dynamothermal metamorphism, fluids may play a role that is essentially equal to that of pressure and temperature (Grambling, 1986). In contrast, it may be the absence or scarcity of fluid in some static to regional dynamic terranes that controls the distribution of phases (B. A. Morgan, 1970; Raymond, 1973b).

Metamorphism may also be subdivided on the basis of the nature of the chemical processes that occur. **Isochemical metamorphism** is metamorphism in which there is no change in the bulk chemistry of the **domain,** or rock volume, being considered. Changes in water content, however, are commonly ignored. **Allochemical metamorphism,** in contrast, is metamorphism in which there *is* a change in the bulk chemistry of the domain being considered. Metasomatism is an allochemical metamorphic process.

In the above definitions, the phrase "of the domain being considered" is critical. The **domain** refers to the volume of rock under consideration. Definition of the domain, which may range from less than one cubic millimeter to tens of cubic kilometers or more, is an essential first step in evaluating whether or not a particular metamorphic event is isochemical or allochemical. For example, in figure 23.3, metamorphism of domain A, which consists of all of the material within the limits of the outer ellipse (including that in domain B), is isochemical, because the total content of chemical species (represented by the dots) in the domain is unchanged by the metamorphic process. In contrast, if we consider domain B, the volume of the small circle within domain A, metamorphism is allochemical. Chemical species have migrated within the larger domain to concentrate in the smaller one. In the case of domain B, the total content of the chemical species within the domain changes as a result of the metamorphic process; that is, the chemistry of domain B is changed. Clearly, then, definition of the domain is prerequisite to considerations of allochemical metamorphism in general and metasomatism in particular.

Two additional terms used to describe the metamorphic history of terranes are prograde and retrograde metamorphism. Traditionally, *prograde metamorphism* has referred to metamorphism that progressed from lower to higher temperatures. In practice, such a change is recognized where minerals stable at

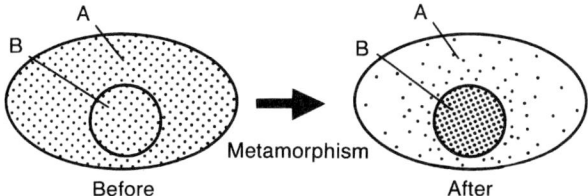

Figure 23.3 Diagram showing two domains (A and B) before and after metamorphism. Chemical species are indicated by dots. The total chemistry of domain A (which includes B) does not change during metamorphism; thus, the metamorphism is isochemical. During metamorphism the total chemistry of domain B changes; thus, the metamorphism of domain B is allochemical. Metasomatism has occurred in domain B.
(Modified from Bayly, 1968, p. 225)

lower temperatures are only partially replaced by those stable at higher temperatures. Alternatively, an increase in temperature across a region may be reflected by a series of mineral zones characterized by minerals that are stable at progressively higher temperatures. *Retrograde metamorphism* has traditionally been considered to be re-metamorphism that progressed from higher to lower temperatures (Harker, 1932, p. 193). Usually, retrograde metamorphism is reflected by the incomplete or pseudomorphic replacement of minerals stable at higher temperatures by those stable at lower temperatures.

The view that prograde versus retrograde metamorphism is temperature-dependent is supported by the fact that many mineral reaction curves have steep slopes (figure 23.4a). Changes in pressure, even of large magnitude, result in relatively few mineralogical changes in some systems. For example, path $a \rightarrow a'$ in figure 23.4a is a classical prograde path and crosses several univariant reaction curves, whereas curve $b \rightarrow b'$ crosses few reaction curves.

Experimental study of many systems over the past half century has led to the delineation of large numbers of mineral reaction curves (see appendix C). The slopes of these curves are variable. Consequently, we now recognize that curves in P-T-X space—representing important metamorphic reactions—slope in such a way that similar appearing mineralogical changes may result from all of the following:

1. An increase in temperature at constant pressure (path $a \rightarrow x$; figure 23.4b),
2. A decrease in pressure at constant temperature (path $b \rightarrow x$),
3. Combined decreases in P and T (path $c \rightarrow x$),
4. An increase in T with a concomitant decrease in P (path $d \rightarrow x$),
5. An increase in T with a concomitant increase in P (path $e \rightarrow x$), and
6. A change in the fluid phase (where fluid is involved in the reaction).

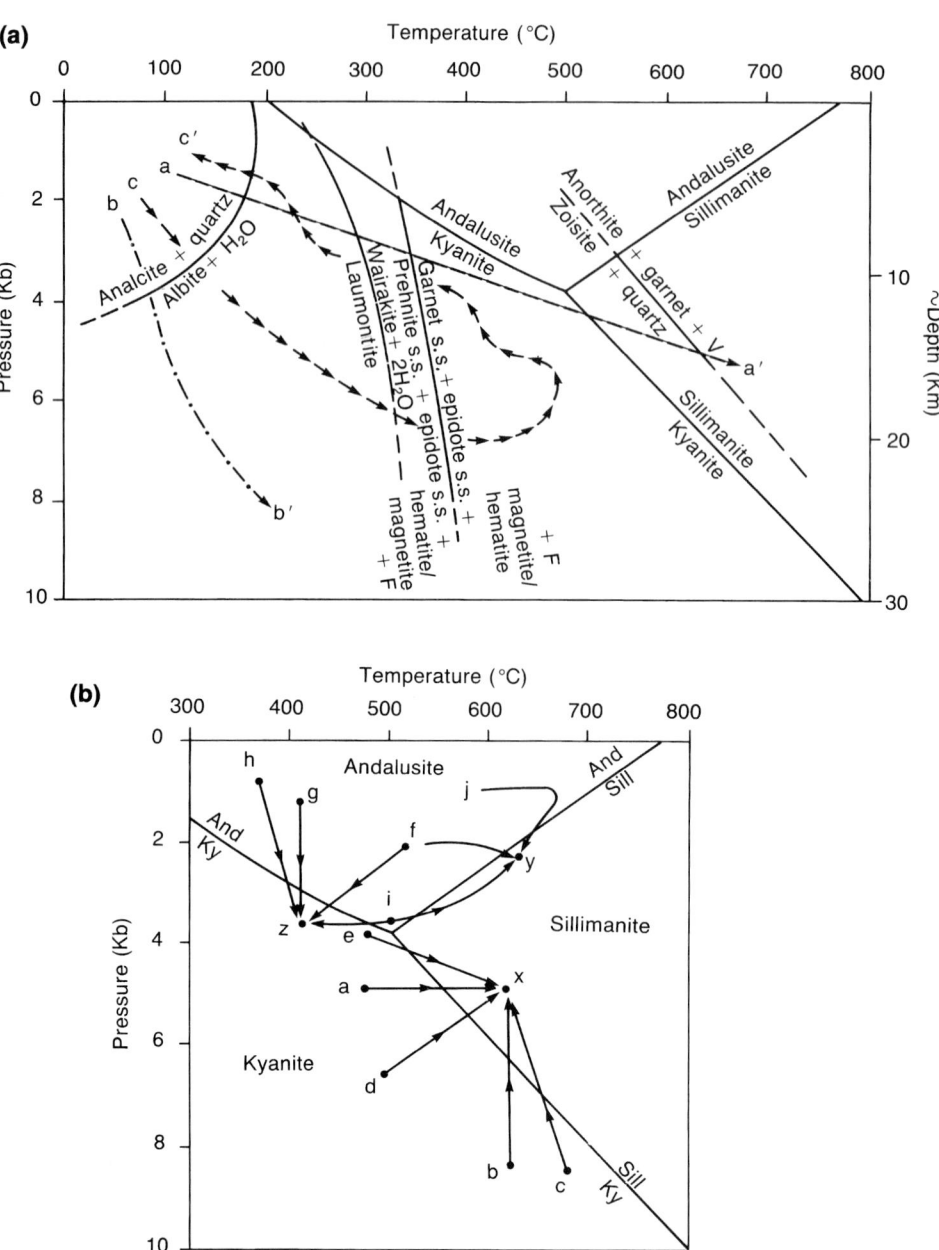

Figure 23.4 Pressure-temperature plots showing the locations of various experimentally determined univariant curves and selected metamorphic paths. Metamorphic paths (P-T-t paths) are curves that represent the set of all successive positions in pressure-temperature (P-T) space occupied by a rock as it undergoes metamorphism in the earth over time (t). (a) P-T-t paths a→a′ and b→b′ are prograde paths that cross many and few mineral reaction curves, respectively. Path c→c′ is a representative path that includes both prograde and retrograde parts. (b) P-T plot of part of A showing three sets of prograde and retrograde paths that yield the same mineralogical change.

[Analcite-albite curve from A. S. Campbell and Fyfe (1965), laumontite-wairakite curve from Liou (1971b), prehnite-garnet curve from Liou, Kim, and Maruyama (1983), aluminum silicate curves from Holdaway (1971), zoisite + quartz = anorthite + garnet curve from Boettcher (1970).]

Thus, a change from one mineral assemblage to another (e.g., from an andalusite-bearing assemblage to a sillimanite-bearing assemblage, in figure 23.4b), thought earlier to represent a simple increase in temperature, may reflect any number of path histories (e.g., f→y, i→y, or j→y). In some cases, decreases in temperature with concomitant increases in pressure (path f→z) may result in the *same* mineralogical change as that produced by either an increase in P at constant T (path g→z) or an increase in both temperature and pressure (path h→z). Consequently, the terms prograde and retrograde must be used with caution.

In this book, **prograde metamorphism** is defined as metamorphism that follows a path generally *away* from surface temperature and pressure conditions (STP). Where fluid phases induce the metamorphic changes, the metamorphism is considered prograde *if* the new assemblages mimic those that are produced by a P-T path that progresses away from STP conditions. In figure 23.4b, paths such as $a{\rightarrow}x$, $e{\rightarrow}x$, and $h{\rightarrow}z$ are prograde. **Retrograde metamorphism** is metamorphism that follows a path generally towards STP conditions. Again, where a fluid phase is the primary agent of metamorphism, the metamorphism is considered retrograde *if* the path history mimics one that approaches STP conditions. Paths such as $c{\rightarrow}x$ and $i{\rightarrow}z$ are retrograde. In nature, rocks probably follow irregular prograde-retrograde paths through P-T space like path $c{\rightarrow}c'$ in figure 23.4a. Dated paths of this kind are called **P-T-t paths** (pressure-temperature-time paths) (England and Thompson, 1984; Spear and Peacock, 1989).

STRUCTURES AND TEXTURES OF METAMORPHIC ROCKS

In many cases metamorphic rocks are easily recognized in the field on the basis of their distinctive structures and textures. Many are foliated. Like most sedimentary rocks and some igneous rocks, metamorphic rocks may have layers. Metamorphic layering is usually distinguished by the fact that the layers are the result of, or are enhanced by, the alignment of phyllosilicates, inosilicates, or other inequant mineral grains.

Metamorphic rocks are also distinctive because of the presence of associated **mesoscopic structures** (handspecimen-to outcrop-scale structures), such as folds, veins, and rock cleavage.[24] Ductile shear zones, boudin, and other structures that are common in outcrops of metamorphic rock, further aid in distinguishing between metamorphic rocks and sedimentary or igneous rocks.

Structures

Structures are those features that characterize handspecimen or larger masses of rock. They differ in scale from *textures*, which are microscopic to small-scale mesoscopic features of the rock produced by grain shapes, grain sizes, grain orientations, grain distributions, and intergrain relationships. Many textures are *penetrative* (i.e., they pervade all parts of a handspecimen), whereas most structures are not penetrative. The distinction between structures and textures is not clear-cut, and one of the most important metamorphic structures, cleavage, bridges the boundary between the two types of feature.

In metamorphic rocks, metamorphic structures result from deformation. Rocks with a fabric that reflects a history of deformation are called **tectonites** (Turner and Weiss, 1963). Here, the word **fabric** refers to *all* of the structural and

textural features of a rock that together define its geometrical character.[25] Inasmuch as most metamorphic rocks exhibit structures and textures that are the product of deformation, most are tectonites. Some contact-metamorphic rocks, as well as other undeformed rocks produced by metasomatism and nontectonic processes, are not tectonites.

Tectonites are of two major types. Those with fabrics produced predominantly by flow in the solid state (i.e., by ductile deformation with recrystallization) and characterized by *oriented* mineral grains are called S-tectonites or L-tectonites (Sander, 1970; Turner and Weiss, 1963).[26] S-tectonites have planar fabrics, whereas L-tectonites have lineated fabrics. Tectonites with fabrics that were produced primarily by brittle fracture or shear along a pervasive set of anastamosing, subparallel, or parallel surfaces have been called SF-tectonites (Raymond, 1975). These tectonites are characterized by commonly slickensided, mesoscopic, parallel to subparallel fractures that are dominantly *independent* of the overall arrangement and orientation of mineral grains or rock fragments within the rock.[27]

Not all structures in metamorphic rocks are the product of either deformation or metamorphic processes. Because metamorphic rocks had sedimentary or igneous protoliths, they may retain structures from those parent rocks. Such remnant primary structures are called **relict structures** or palimpsest features.[28] The relict structures that are commonly retained in metamorphic rocks are those less intricate structures that are visible because of differences in composition or grain size. Bedding, cross-bedding, and grading are the typical relict structures in metasedimentary rocks. Primary layering, amygdaloidal structure, and pillow structure are among the most frequently encountered relict igneous structures.

Cleavage

Rock **cleavage,** the tendency of rocks to break along parallel to subparallel surfaces, is the most widespread metamorphic structure (figure 23.5). Cleavage reflects either the textural alignment of mineral grains, called **preferred orientation,** or the subparallel arrangement of discontinuities (abrupt changes in physical properties) in a rock. Traditionally, cleavage was classified into a variety of types that are assignable to two main categories—fracture cleavage and flow cleavage.[29] The two types correspond to the two main types of tectonites. Thus, SF-tectonites are characterized by fracture cleavage and S-tectonites are characterized by flow cleavage.

In an effort to avoid use of terms that imply origin, C. M. Powell (1979) proposed that cleavage types be classified according to the physical character of the cleavage. Accordingly, cleavage is divided into two main types—continuous cleavage and spaced cleavage—on the basis of the spacing of cleavage domains or surfaces (figure 23.6a). *Cleavage domains,* consisting of zones of minerals with preferred orientation, separate *microlithons,* consisting of less cleavable material.

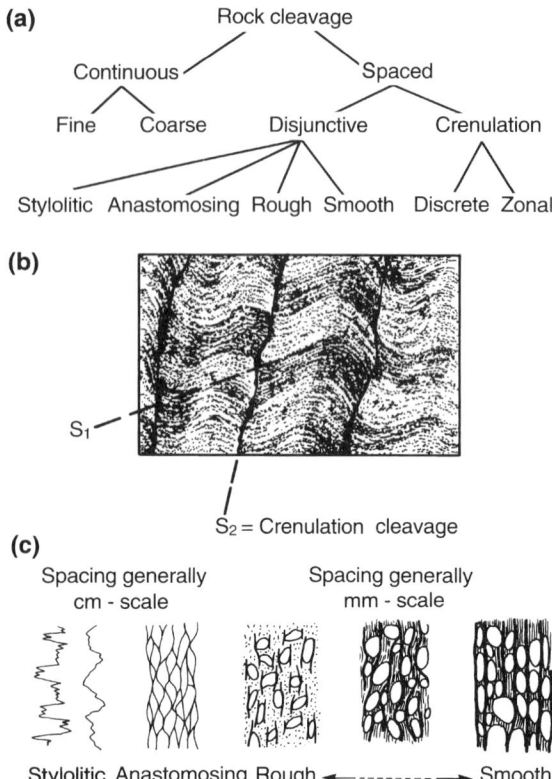

Figure 23.5
Continuous cleavage (approximately vertical) cutting bedding (inclined) in metamorphosed Precambrian metasediments, Wallace, Idaho. DNAG scale is in lower left (see arrow).

Cleavage that has a spacing of <0.01 mm *or* that results from a penetrative preferred orientation of phyllosilicates or other minerals is called **continuous cleavage.** Continuous cleavage is further subdivided into fine and coarse types based on the grain size (fine = <0.1 mm, coarse = >0.1 mm). Cleavage with domain spacings of >0.01 mm and which lacks a microscopic penetrative fabric is **spaced cleavage.** Spaced cleavage is subdivided into disjunctive and crenulation types. Crenulation cleavage is cleavage imposed on rocks that already possess a planar fabric, such as a preexisting cleavage (figure 23.6b). Disjunctive cleavages, of which there are four types (figure 23.6c), are cleavages in rocks lacking a preexisting planar fabric. Anastamosing and some smooth disjunctive cleavages correspond, to some degree, with fracture cleavage, as defined above, whereas rough and some smooth disjunctive cleavages, crenulation cleavage, and continuous cleavage are approximately equivalent to "flow cleavage".

The origin of the various types of cleavage is a subject about which much has been written.[30] Briefly, the origins are as follows. Stylolitic disjunctive- and crenulation-spaced cleavages result from solution under pressure (pressure solution) and movement of materials (diffusion) out of the cleavage domain (Gray, 1979a; Wanless, 1979).[31] Residues of less soluble materials left in the domain mark the cleavage surface.

Relatively little work has been done on the formation of anastamosing, disjunctive-spaced cleavage, in spite of the fact that such cleavages occur widely in scaly clays and other matrix materials of melanges. Some anastamosing cleavage appears to have developed as a result of gravity-induced shearing in submarine and subaerial slide deposits. Alternatively, such cleavage may be induced by tectonic stresses. Detailed studies suggest that mechanical rotation of phyllosilicate grains into the plane of the cleavage and

Figure 23.6 Morphology and morphological classification of rock cleavage. (a) Morphological classification of rock cleavage (after C. M. Powell, 1979). (b) Sketch of microscopic view of crenulation cleavage in phyllite. Crenulation cleavages are marked by both microfolds in preexisting foliation and the concentrations of opaque minerals (black). (c) Shape and spacing of cleavage in the four types of disjunctive spaced cleavage.
(After C. M. Powell, 1979)

cataclasis of grains are important processes responsible for the development of this type of cleavage (J. C. Moore et al., 1986; Lucas and Moore, 1986, Lash, 1989).

The origins of rough and smooth disjunctive-spaced cleavages and continuous cleavages have been the subject of the most debate and intensive study.[32] The origin of these cleavages, in part referred to as "slaty cleavage," has been variously attributed to one or more of the following processes:

1. Mechanical rotation of phyllosilicate grains into the cleavage plane,
2. Pressure solution, accompanied by rotation of residual phyllosilicate grains into the plane of the cleavage, and
3. **Syntectonic recrystallization,** that is, dissolution of detrital phyllosilicates and other minerals and recrystallization or neocrystallization of phyllosilicate and other grains, parallel to the cleavage direction, during deformation (J. C. Maxwell, 1962; Kanagawa, 1991).

480

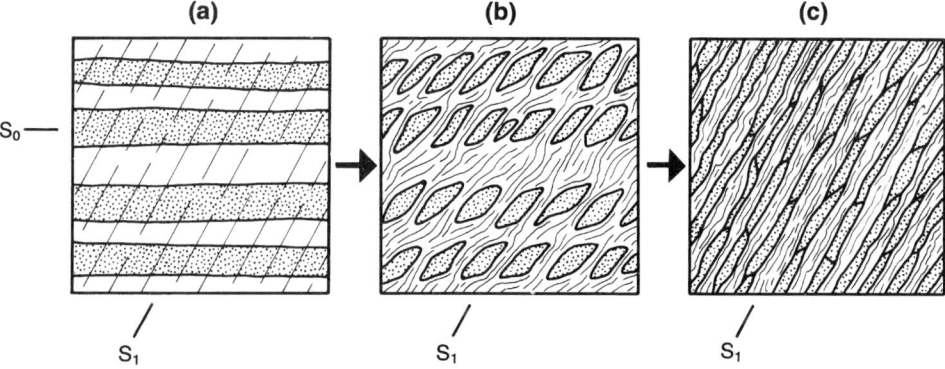

(a) **(b)** **(c)**

S_0 —

S_1 S_1 S_1

Figure 23.7 Schematic sequential diagrams showing transposition of bedding (S_0) along a cleavage (S_1). (a) Bedding transected by incipient cleavage.

(b) Separation, rotation, and deformation of bedding segments. (c) New compositional bands parallel to cleavage.

All may play some part in the development of cleavage in rocks at specific localities. Yet, a growing body of evidence indicates that syntectonic recrystallization may be the dominant process of cleavage formation in rocks with well-developed cleavage, especially those of higher metamorphic grade (W. J. Gregg, 1985; J. H. Lee et al., 1986).

Layers and Transposition of Bedding

Metamorphic rocks commonly have mesoscopic layers. The layering is compositional banding defined by differences in mineral composition and texture (e.g., grain size). Physically, metamorphic layers may be meters to millimeters thick. Layers that are less than 1 mm thick are called microlayers, whereas layers larger than 1 m in thickness are designated megalayers. Layers that have thicknesses between these limits define the structure called **gneissic structure.**[33] Gneisses are banded rocks in which alternate layers are composed of different minerals.[34]

The compositional layers observed in metamorphic rock may have a variety of origins. Some bands represent primary layering (i.e., relict bedding) in the sedimentary rock from which the metamorphic rock was derived. In contrast, faulting may interleave layers of rock of different compositions that are later metamorphosed to produce a banded rock. In other cases, banding may represent dikes or veins that invaded a host rock and were later metamorphosed along with it. In still other cases, banding results from a process of chemical migration called metamorphic differentiation (discussed below).

New compositional banding in metamorphic rocks may also result during cleavage formation through **transposition** of bedding or other planar features. In this process, the layering is transected at a significant angle by a cleavage (labelled S_1 in figure 23.7a).[35] As the cleavage (S_1) develops, the bedding is separated into segments that are physically rotated or deformed by flow, or both, so that elongate lenses that parallel the cleavage direction are created (figure 23.7b). Continued

cleavage development may cause the lenses to merge yielding new compositional bands that strike and dip in a direction different from that of the original layering (figure 23.7c).

Other Structures

A number of other structures are important locally in metamorphic rocks. Among these are folds, kink bands, boudins, mullions, rods, faults (including ductile shear zones), joints, and veins. Folds, kink bands, some boudins, mullions, rods, and ductile shear zones are produced by ductile deformation.[36] Some boudins, brittle faults, and joints result from brittle deformation. Veins and (commonly) compositional banding represent the effects of chemical migration of materials.

Folds are bends in planar structures of the rock (figure 23.8a). Folds may occur in bedding, foliation, veins, or other features, including previously existing folds. In terms of their scale, they range from microscopic structures to macroscopic structures covering tens or hundreds of square kilometers. Folds are formed as a result of compressional or shearing stresses acting parallel or at an angle to rock layers.

Kink bands are small, abrupt folds developed in rocks that already have a fabric (figure 23.9a). They develop from compressive stresses directed at an angle to the preexisting fabric (Dewey, 1965; Cobbold, Cosgrove, and Summers, 1971; P. F. Williams and Price, 1990). Most commonly, kink bands are found in fine-grained rocks such as phyllite, but they also occur in schists and gneisses.

Boudin (French for sausage) are cylindrical masses of rock, originally part of a single bed or layer that has been stretched and pulled apart.[37] The cylinders in true boudins lie side by side. Three-dimensional exposures of boudin are uncommon and typically one sees cross sections that appear as a series of crude ellipses or rectangles with rounded corners.

Mullions and rods are similar to boudins in that they are long, cylindrical structures (G. Wilson, 1953, 1982). **Mullions** are columns, 2 cm to 2 or more meters in diameter,

(a)

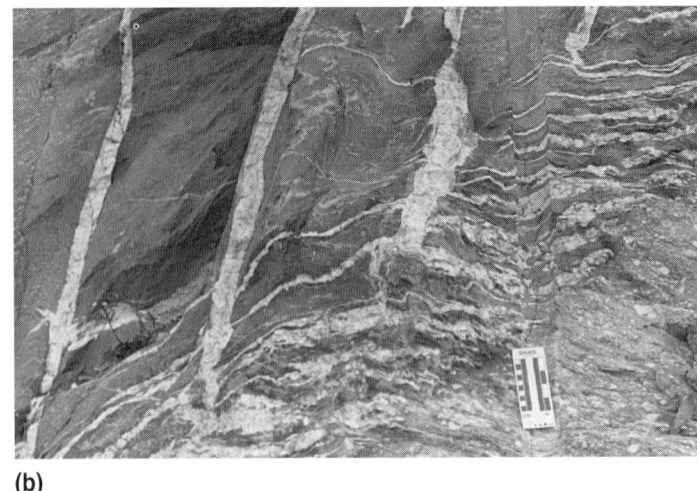

(b)

(c)

(d)

Figure 23.8 Photographs of structures in metamorphic rocks. (a) Mesoscopic folds in gneiss and schist, Franciscan Complex, Jurassic-Cretaceous, Laytonville Quarry, south of Laytonville, California. (b) Quartz veins in meta-quartz arenite of the Grandfather Mountain Formation, Neoproterozoic, Highway 321, south of Boone, North Carolina. (c) Melange Franciscan Complex Jurassic-Cretaceous, exposed at Goat Rock, south of Jenner, California. Notice the football-shaped boudin at the lower left, the folds in the lower center, and the boudins at the right center. Blocks in the melange include serpentine quasimylonite, metachert, metawacke, and metabasalt. The beach is at the lower right. (d) Laminated mylonite of the Linville Falls Fault Zone, Linville Falls, North Carolina. Here, Mesoproterozoic semischistose granitoid rocks of the Cranberry Gneiss are thrust over younger, Neoproterozoic metaquartz arenite and phyllite of the Chilhowee Group.

composed of the country rock of the metamorphic terranes in which they occur. The exterior of the columns are angular to rounded and are commonly polished or striated parallel to their length. In cross section, they may exhibit folded internal layering. The origins of mullions are varied, as there are several types of mullions.[38] **Rods** are similar to mullions, but they are composed of segregated or introduced material, (i.e., dike or vein material), such as quartz.

Joints are fractures along which there has been no significant movement parallel to the plane of the structure. In metamorphic rocks, joints commonly occur in sets that have been filled by veins (figure 23.8b). Joints commonly develop from primary or secondary tensional stresses resulting from stored stress in the rocks (N. J. Price, 1966), but may also develop from active shear or tensional stresses.

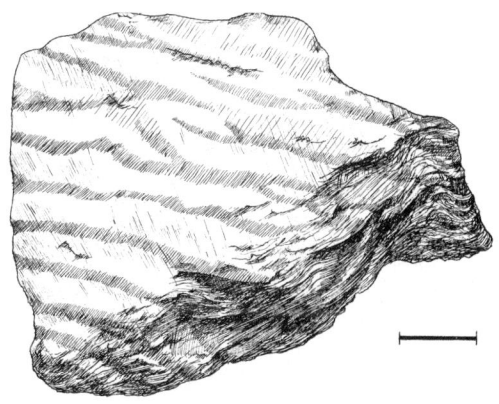

(a)

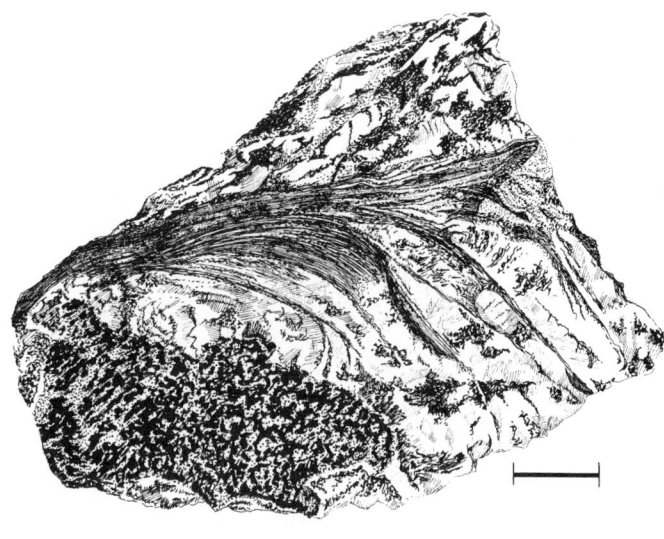

(b)

. . . .
Figure 23.9 Sketches of two structures found in metamorphic rocks. (a) Kink bands in ultramylonite from the Cranberry Gneiss, Mesoproterozoic from the banks of the New River east of Mouth-of-Wilson, Virginia. Scale bar is 2 cm. (b) Sketch of mylonite zone forming "horse-tail shear" in Striped Rock Granite, Proterozoic, near Independence, Virginia. Scale bar is 2 cm.
(Both sketches from Raymond, 1984c)

Faults, in contrast, are fractures, along which there has been significant movement parallel to the plane of the structure. Brittle, brittle-ductile, and ductile faults occur in metamorphic rocks (J. G. Ramsay, 1980). The former tend to be relatively clean breaks characterized by polished and scratched surfaces (slickensides) *or* cataclastic zones marked by breccia, tectonic melanges, or related rock masses. Tectonic melanges are tectonically deformed *bodies* of rock "mappable at a scale of 1:24,000 or smaller and characterized both

by the lack of internal continuity of contacts or strata and by the inclusion of fragments and blocks of all sizes, both exotic and native, embedded in a fragmented matrix of finer-grained material" (figure 23.8c) (Raymond, 1984a). Native blocks are a part of the original unit that was fragmented, whereas exotic blocks were formed elsewhere and were incorporated into the melange. Ductile shear zones are zones of dislocation a few millimeters to hundreds of meters wide, composed of rocks called **mylonites** that are recrystallized and/or neocrystallized under the influence of a shearing stress (figure 23.8d and 23.9b).[39]

Veins are tabular joint or fault fillings composed of one or more minerals (figure 23.8b). The distinction between veins and dikes is somewhat arbitrary, but the term vein tends to be applied to tabular bodies that are monomineralic, bimineralic, or contain ore minerals, whereas the term dike is used for igneous rock types crystallized from magmas or magmatic vapor phases. Veins form where fluids penetrate a fracture and precipitate minerals.

Textures

Textures are a function of grain size, grain shape, intergrain relationships, grain distribution, and grain orientation (Spry, 1969, p. 5). Somewhat different sets of textural terms are commonly used in field and handspecimen work than are used in thin-section petrography. In this text, those sets of terms are integrated.

Texture Types

Metamorphic textures may be divided into five major types—foliated textures, granoblastic textures, diablastic textures, cataclastic textures, and relict textures.[40] The word relict assumes the same meaning here that it has with regard to structures; it refers to textures retained from a protolith. In general, the names of relict textures are preceded by the prefix *blast-* or *blasto-* (e.g., blastoporphyritic) to indicate that the texture is metamorphic, but retains the relict character.[41] "Blast" as a suffix indicates a new metamorphic texture (e.g., porphyroblastic).

Each of the major texture types is characterized by distinctive grain shapes or grain orientations, or both. **Foliated textures** are textures characterized by an alignment of mineral grains in such a way as to give the rock the appearance of or the tendency for splitting into layers or flat pieces (figure 23.10a). Commonly, the minerals in foliated rocks are predominantly acicular or tabular. **Granoblastic textures** are those characterized by more or less equidimensional mineral grains (figure 23.10b). As used here, the term granoblastic refers to a general category of texture and implies nothing about grain boundary shape, grain size variation, or preferred orientation or the lack thereof.[42] **Diablastic textures** are those in which tabular or acicular minerals are intergrown in a nonfoliated, interlocking, locally radiating manner (figure 23.10c) (Harker, 1932).[43]

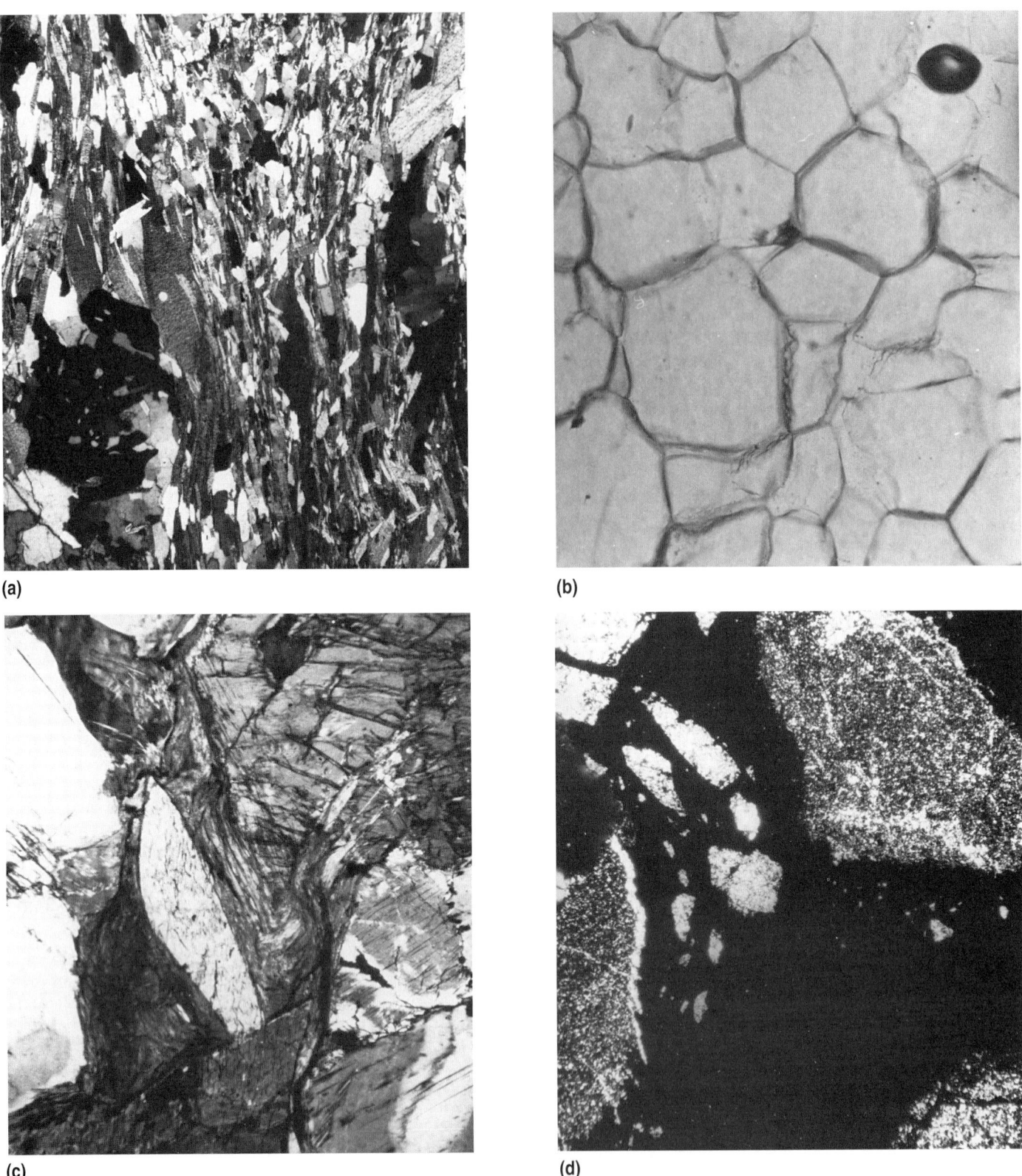

(a)

(b)

(c)

(d)

Figure 23.10 Photomicrographs showing major types of metamorphic textures. (a) Foliated (lepidoblastic) texture in pelitic schist of the Ashe Metamorphic Suite, Neoproterozoic (?), northwest North Carolina. (XN). (b) Granoblastic (equigranular-mosaic) texture in metadunite, Corundrum Hill, western North Carolina. (PL). (c) Diablastic texture in talc-anthophyllite-chlorite diablastite, Greer Hollow ultramafic body, Todd, North Carolina. (XN). (d) Cataclastic texture in fault breccia in Rome Formation, Cambrian, Mountain City, Tennessee. (XN). Long dimension of photos is 6.5 mm in (a), (c), and (d), 1.27 mm in (b).

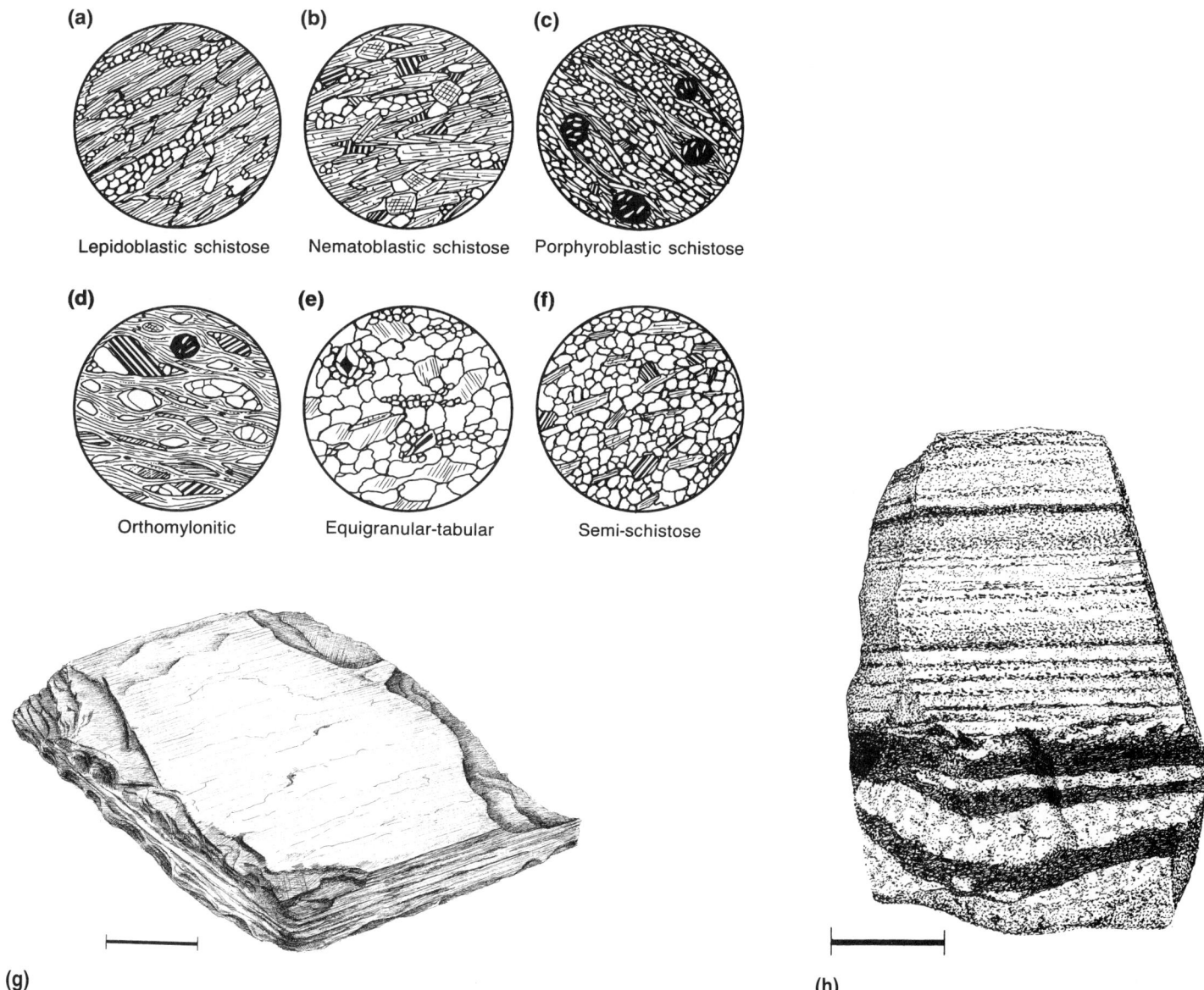

(a) Lepidoblastic schistose

(b) Nematoblastic schistose

(c) Porphyroblastic schistose

(d) Orthomylonitic

(e) Equigranular-tabular

(f) Semi-schistose

(g)

(h)

Figure 23.11 Sketches of selected types of foliated metamorphic textures. (a)–(f) are thin-section views; (g) and (h) are handspecimen views (scale bars = 2 cm). (a) Lepidoblastic schistose texture in chlorite-albite-quartz-white mica schist. (b) Nematoblastic schistose texture in plagioclase-hornblende schist. (c) Porphyroblastic schistose texture in garnet-biotite-plagioclase-white mica-quartz schist. (d) Orthomylonitic texture in plagioclase-quartz-white mica- chlorite orthomylonite. (e) Equigranular tabular texture in metadunite. (f) Semischistose texture in white mica-biotite-plagioclase-quartz semi-schist. (g) Slaty texture (and continuous "slaty" cleavage) in slate, from the Mettawee Formation, Pawlet, Vermont. (h) Gneissose texture in magnetite-quartz-feldspar gneiss, from the Valley Springs Group, Llano Co., Texas.

[(g) and (h) from Raymond, 1984c]

Cataclastic textures are nonfoliated textures characterized by fractured rock materials and mineral grains (figure 23.10d). Each of these major textural types may be subdivided into two or more individual types (see table 23.2 and figures 23.11 and 23.12).

Foliated textures characterize metamorphic rocks and therefore assume the most importance. The two major categories of foliated textures are strongly foliated and weakly foliated. Strongly foliated rocks include (1) rocks dominated mineralogically by platy, bladed, or acicular minerals, (2) rocks deformed to the extent that minerals, such as quartz, which typically show little tendency to be elongate parallel to foliation, are very elongate and define a well-developed cleavage, and (3) rocks characterized by mineral segregations, microlithons, or spaced cleavages. In strongly foliated rocks, continuous cleavage, spaced crenulation cleavage, or smooth,

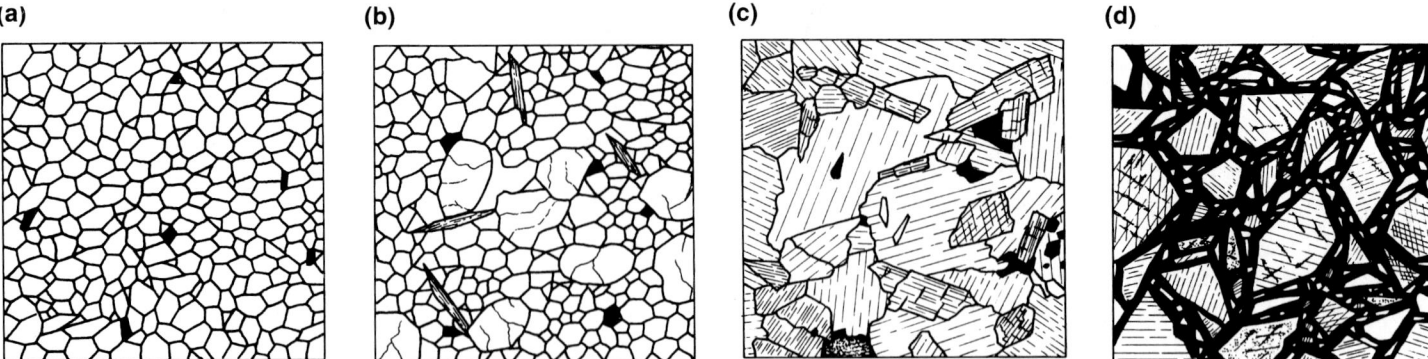

(a)   **(b)**   **(c)**   **(d)**

....
Figure 23.12   Sketches of selected nonfoliated metamorphic textures. (a)–(d) are thin-section views; (e) and (f) are handspecimen views. Scale bars are 2 cm. (a) Granoblastic-polygonal texture in metadunite (olivine granoblastite). (b) Heterogranoblastic texture in metadunite (olivine granoblastite). (c) Diablastic texture in tremolite-chlorite diablastite, Ashe Metamorphic Suite, northwestern North Carolina. (d) Cataclastic texture (schematic) in fault breccia; with Mn-oxide cement. (e) Hypidioblastic-heterogranular texture in quartz-plagioclase-hornblende granoblastite, Winding Stair Gap, southwestern North Carolina. (f) Porphyroblastic, allotrioblastic-heterogranoblastic texture in metadunite (olivine granoblastite) from Day Book Mine, Spruce Pine District, North Carolina.

[(e) and (f) from Raymond, 1984c]

rough, or anastamosing disjunctive-spaced cleavage are well developed. Weakly foliated rocks are those in which (1) linear, but not planar, arrangements of bladed to acicular grains dominate the texture, (2) equant to subequant grains, such as quartz and feldspar, comprise most of the rock, or (3) platy to bladed minerals, such as phyllosilicates, are present but are only weakly aligned. In such rocks, cleavage is stylolitic, anastamosing, or rough disjunctive-spaced cleavage and the microlithons between the cleavage surfaces lack a preferred orientation of grains.

Recrystallization, Neocrystallization, Nucleation, and Crystal Growth: An Overview

The development of metamorphic textures is similar in many ways to the development of igneous textures. *Neocrystallization* involves nucleation and crystal growth, as is the case with igneous textures. In metamorphic rocks, however, diffusion (migration) of chemical species towards the growing crystal occurs in a crystalline rock, rather than in a melt. In addition, *recrystallization* of the phases that are already present may occur, either simultaneously with neocrystallization or independent of it. In any case, recrystallization also involves nucleation and growth of crystals.

The metamorphic reactions and processes (e.g., neocrystallization and crystal growth) that give rise to new or reoriented crystals, that is, to new textures, are attempts by the rock system to attain equilibrium under the new metamorphic conditions. Those conditions are generally different from the conditions under which the rock formed. At equilibrium, the rock will have attained the lowest possible energy state under the extant conditions.

All metamorphic reactions and processes, including nucleation, diffusion, and growth, are activated by energy, or more specifically by differences in energy states (Vernon, 1976, p. 75). Chemical reactions (neocrystallization processes) are attempts to reduce the thermodynamic energy (called the Gibbs Free Energy) of the system.[44] Similarly, recrystallization involves progress towards reduction of the structural *free energy* of the system, especially that energy associated with the boundaries of each phase, the *surface free energy*. Surface free energy results from the fact that atoms at the surface of a crystal, while bonded internally to other atoms, are not bonded on the side of the atom facing the surface. Thus, that side *and* the surface have a higher energy. The more surface there is, the more surface energy there will be (all other factors being equal). Because a few large grains have less surface area than numerous small grains (in an equal volume of rock), larger grains are favored at equilibrium, because they have less surface free energy. Both nucleation and recrystallization aid the rock (the system) in its progress towards equilibrium, that is, towards a lower state of free energy.

Nucleation and Growth

Recall that nucleation is the process in which a relatively few atoms of the right type cluster together to form the rudiments of a new crystal structure. Most nucleation in metamorphic

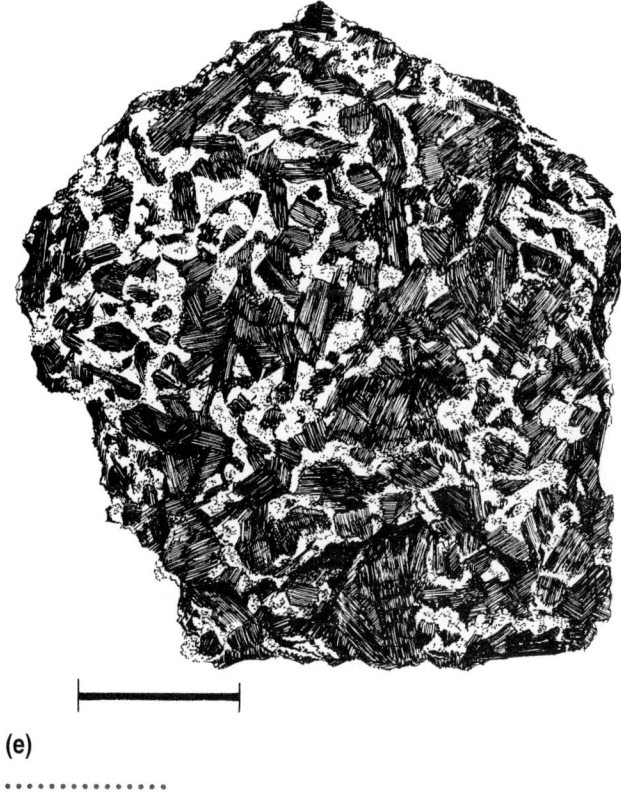

(e)

Figure 23.12 Continued

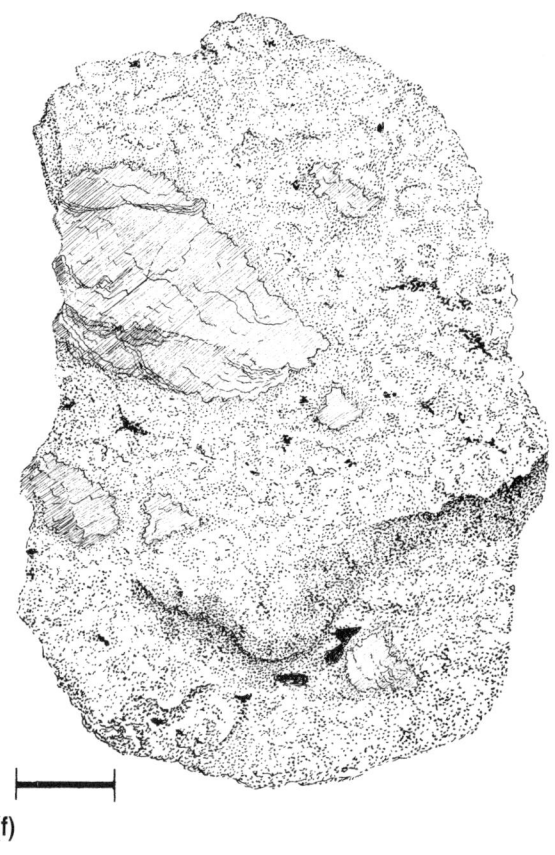

(f)

rocks is heterogeneous, because crystal lattices already exist upon which new minerals may be constructed. Nucleation occurs where (1) the appropriate atoms are available to form the nucleus, and (2) the appropriate conditions exist for the persistence of the nucleii. The latter condition prevails where the energy state, based on a given set of P-T conditions, is such that the free energy of the nucleus is favored over that of unbonded atoms.[45] The availability of ions depends either on their initial presence or on their ability to migrate to the nucleation site (i.e., on the process of diffusion).

Once the nucleus is formed, growth may occur. Growth will occur as long as it is favored by the energy state, that is, by large differences in free energy, and by the availability of ions. Grains that are strained or have relatively large surface areas (i.e., small grains) tend to recrystallize to reduce their free energy. New minerals will form where a new phase is more stable than a preexisting mineral that formed under different thermal conditions. The availability of ions depends on their presence in the surrounding rock and their ability to migrate to the growth site. Again, diffusion is important.

Diffusion

Clearly, diffusion is a critical process in the development of metamorphic textures. Both nucleation and crystal growth depend on it. **Diffusion** may be defined as a process in which chemical species migrate, in a solvent phase, under the influence of a chemical potential gradient between two sites (Jensen, 1965).[46] The chemical potential gradient rep-

resents a difference in chemistry, as well as temperature and/or pressure, between two parts of the system. In small volumes of rock, differences in chemistry exist where different mineral phases exist side by side or near one another. Differences in pressure exist because pressures are high where grains are in contact, but are generally lower in adjoining voids. In general, differences in temperature do not exist at the small mesoscopic scale. The solvent phase in metamorphic rocks may be a crystal or a pore fluid.

Within crystals, diffusion occurs where atoms jump from one lattice position to another (Condit, 1985; Nicolas and Poirier, 1976, p. 52). In rocks, diffusion occurs along crystal boundaries and along linear and planar defects within crystals (e.g. Spry, 1969, p. 14). The presence of a fluid phase on crystal boundaries greatly facilitates the diffusion process, because ions migrate more easily through a fluid.

Recrystallization

Recrystallization, which depends on diffusion, is more easily understood than neocrystallization, because in recrystallization, chemical changes do not accompany the textural changes. The rocks involved are generally monomineralic. Considering recrystallization only, a rock undergoing metamorphism may experience changes in grain size, grain shape, and grain orientation.

In situations of LP-contact metamorphism, stress will not be a significant factor, and recrystallization may be considered to result primarily from the influence of temperature.

Table 23.2 Metamorphic Rock Textures

A. Textural Terms Relating to Intergrain Relationships, Grain Shapes and Sizes, Grain Orientations, and Grain Distributions

Foliated Texture Aligned minerals (i.e., a preferred orientation of minerals) give a layered, flaky, or lineated appearance.

Strongly Foliated Texture Rock exhibits well-developed continuous or finely spaced cleavage.

Slaty Texture Very fine-grained texture (grains < 0.1 mm) characterized by abundant phyllosilicates strongly aligned in a planar or subplanar orientation. In handspecimen, this texture is aphanitic and specimens break into flat, smooth pieces.

Phyllitic Texture Very fine-grained to fine-grained texture (grains < 0.5 mm) characterized by the presence of crenulation cleavage, microfolds, or kink bands.

Schistose Texture Fine-grained to very coarse-grained texture (grains > 0.1 mm) characterized by subparallel arrangement of acicular, bladed, and/or tabular minerals (especially phyllosilicates and amphiboles).

Lepidoblastic texture Platy or sheet-structured minerals predominate.

Nematoblastic texture Acicular or elongate prismatic minerals predominate.

Gneissose Texture Fine-grained to very coarse-grained texture with more or less continuous bands (at the handspecimen scale) of contrasting mineralogy.

Porphyroblastic-Foliated Texture Very fine-grained to very coarse-grained rock with larger crystals (porphyroblasts) in a finer-grained, foliated matrix. (The term *foliated* should generally be replaced with the appropriate specific term, e.g., *porphyroblastic schistose* or *porphyroblastic gneissose texture*.) The porphyroblasts are commonly mineralogically distinct from the matrix.

Nodularblastic-Schistose Texture A texture consisting of nodular clusters of small grains of one or two minerals in a matrix of a different composition.

Mylonitic Texture (sensu lato) Foliated to porphyroclastic-foliated texture with fine-grained to very fine-grained matrix or fabric materials that typically show evidence of crystal-plastic deformation and syntectonic recrystallization or recovery features (i.e., unstrained, sutured grains).

Porphyroclastic texture A weakly to strongly foliated texture characterized by a bimodal distribution of grain sizes; with larger, typically deformed, protolith grains (the porphyroclast) in a finer-grained matrix of materials derived, in total or in part, from the porphyroclast. Matrix may be strain free, but typically is not.

Protomylonitic texture Mylonitic texture with > 50% porphyroclasts.

Orthomylonitic texture Mylonitic texture with 10–50% porphyroclasts.

Ultramylonitic texture Mylonitic texture with < 10% porphyroclasts. (Phyllonitic texture, as used in older literature, is more or less equivalent to ultramylonitic texture and is used especially where a crenulation cleavage is present.)

Foliated Cataclastic Texture (see *Cataclastic Textures* below).

Weakly Foliated Texture Texture characterized by lineations or weakly developed or widely spaced cleavages.

Semi-slaty Texture Very fine-grained to fine-grained weakly foliated texture.

Semi-schistose Texture Fine-grained to very coarse-grained weakly foliated texture.

Lepidoblastic-semi-schistose texture Fabric dominated by poorly aligned phyllosilicates or other platy minerals.

Nematoblastic-semi-schistose texture Fabric dominated by poorly aligned elongate prismatic or acicular minerals.

Sheaf texture Texture containing grain clusters consisting of arrays of diverging platy to acicular grains.

Comb texture Texture characterized by parallel to subparallel grains arranged perpendicular to a surface or set of surfaces upon which the grains nucleated.

Equigranular-Tabular Texture (granoblastic elongate) Texture characterized by aligned, elongate to tabular, subequant grains.

Porphyroclastic Texture A weakly to strongly foliated texture characterized by a bimodal distribution of grain sizes, with larger, deformed grains in a finer-grained matrix of materials derived, in total or in part, from the porphyroclast. Matrix may be strain free, but typically is not.

Diablastic Texture (*decussate texture*) Composed of non-aligned, radiating to randomly oriented acicular to platy grains.

SHEAF TEXTURE Texture characterized by grain clusters consisting of arrays of diverging platy to acicular grains.

SPHERULOBLASTIC TEXTURE (spherulitic texture, rosette texture) Texture characterized by clusters of radiating acicular to bladed minerals.

Table 23.2 Continued

FIBROBLASTIC TEXTURE Diablastic texture characterized by acicular minerals of approximately the same size.

Granoblastic Texture Equant to subequant grains dominate a granular aggregate.

HOMOGRANULAR TEXTURE Grains are all approximately the same size.

Granoblastic-Polygonal Texture (mosaic texture, equigranular-mosaic texture, equigranular texture, granulitic texture, granular texture) A texture characterized by polygonally shaped crystals of equal or nearly equal size with straight to slightly curved grain boundaries. Rocks with this texture are typically almost monomineralic. In monomineralic rocks, grain boundaries typically meet at 120° angles and grains are anhedral.

Protogranular texture (granular texture) Medium- to coarse-grained, granoblastic-polygonal texture typical of some ultramafic rocks. In ultramafic rocks, spinel forms vermicular intergrowths in pyroxenes. Note that this name has a genetic implication; i.e. that the texture is an original (versus second-stage) metamorphic texture.

Equigranular-mosaic texture (equigranular texture, granulitic texture) Fine-to coarse-grained, granoblastic-polygonal texture typical of some ultramafic rocks, granulites, and hornfelses.

Granoblastic-Polysutured Texture Texture characterized by polygonally shaped crystals of equal or nearly equal size with lobate or serrate margins.

HETEROGRANULAR TEXTURE Grains are of notably different sizes.

Heterogranoblastic Texture Texture characterized by equant or subequant minerals of varied sizes.

Nodularblastic Texture A texture consisting of nodular clusters of small grains of one or two minerals in a matrix of a different composition.

Clastic Texture Nonfoliated to foliated texture in which broken grains or rock fragments are enclosed in a matrix of more finely broken materials.

MORTAR TEXTURE (cataclastic texture) Nonfoliated texture with larger broken grains in a matrix of smaller, broken grain fragments of the same composition.

FOLIATED CATACLASTIC TEXTURE Foliated texture consisting of broken grains of rock or mineral materials in elongate segregations (meso-to microlithons) arranged in preferred orientations and separated by anastamosing to subparallel shear surfaces or microfoliated laminae.

VITRICLASTOBLASTIC TEXTURE Nonfoliated texture with broken fragments of rock or mineral in a matrix of frictionally generated glass.

Relict Textures Primary epiclastic, pyroclastic, or crystalline textures preserved in a metamorphic rock.

B. Textural Terms Relating to Nonpenetrative, Nonplanar, and Intragrain Textures

Porphyroblastic Texture A texture in which there is a bimodal distribution of grain sizes (equivalent to porphyritic texture in igneous rocks; large grains are called porphyroblasts). In general, the smaller grains surrounding the porphyroblast are of varied minerals, some or all of which are different from the composition of the porphyroblast.

Allotrioblastic (Texture) This term, which should be used as a prefix to other textural terms, indicates that grains are dominantly anhedral (e.g., allotrioblastic-granoblastic-polygonal texture).

Hypidioblastic (Texture) This term, which should be used as a prefix to other textural terms, indicates that grains are dominantly subhedral.

Idioblastic (Texture) This term, which should be used as a prefix to other textural terms, indicates that grains are dominantly euhedral.

Augen Texture A texture characterized by eye-shaped, larger grains or grain clusters enclosed in a finer-grained matrix.

Coronitic Texture A texture in which larger grains have rims of a generally finer-grained mineral, presumed to be produced by a reaction between the core mineral and the surrounding material.

Poikiloblastic Texture A texture in which a larger grain encloses many smaller grains.

Symplectic Texture A texture in which there is an intimate, commonly wormlike, intergrowth between two minerals.

Helicitic Texture A texture consisting of bands of small relict inclusions, typically folded or arranged in spiral patterns, enclosed in poikilitic grains or porphyroblasts.

Sources: Spry (1969), Bell and Etheridge (1973), Collerson (1974), Mercier and Nicolas (1975), J. E. N. Pike and Schwarzman (1976), Basu (1977b), Bates and Jackson (1980, 1987), Raymond (1984c), D. U. Wise et al. (1984), Chester, Friedman, and Logan (1985), and Bard (1986).

Under such conditions, major reorientation of grains is unlikely, so grain size and grain shape changes are the expected metamorphic effects. Metamorphism will involve breaking and reforming of bonds, heterogeneous nucleation, diffusion of ions to the newly nucleated sites, and grain growth. These changes allow the rock system to progress towards a state of lower free energy. Rocks with large amounts of grain surface area are the most susceptible to recrystallization. These would include rocks with fine grain size, irregular grain shapes, or large porosities. Recrystallization promotes a reduction in surface free energy by reducing the amount of surface.

We can observe the expected effects of recrystallization in rocks. The two effects anticipated, an increase in grain size and a smoothing of grain boundaries, are reported by Joesten (1983) from an ideal natural occurrence of LP-contact metamorphism in the Christmas Mountains of Texas. Here, a gabbro intrudes a chert-nodule-bearing limestone. At positions progressively closer to the contact, where temperatures increase correspondingly, the texture of the chert changes from a "mosaic texture" (granoblastic-polysutured texture) to a "granoblastic polygonal microstructure" (equigranular-mosaic texture). The average grain size increases from 0.0075 mm at 101.7 m from the contact to 1.06 mm at 1.8 m from the contact (figure 23.13).

These same effects are reproduced in experiments and predicted on the basis of theoretical considerations.[47] In monomineralic rocks composed of equant grains, the grain boundary between any two grains will be expected to have the same relative surface energy as the boundary between any other two grains. As a result, under equilibrium conditions, grains will be equidistant from one another, will meet at a triple point, and will have interfacial angles of approximately 120° (figure 23.14a). This is the lowest-energy configuration for the grain boundaries. The resulting texture is a granoblastic polygonal texture (figure 23.14b). Once such a configuration is developed, further reduction of surface free energy is attained by an increase in grain size.

Where a second mineral is introduced into the aggregate, surface free energies will vary, as will the resulting interfacial angles (figure 23.14c) (Vernon, 1976). The resulting textures will also be equigranular if equant grains are dominant (figure 23.14d). In rocks in which acicular or platy minerals are dominant, the textures will be diablastic.

The addition of deviatoric stress to the variables affecting the system creates additional variations in the textures. In general, the result will be the development of preferred orientations.[48] Two types of preferred orientations exist (Vernon, 1976, p. 24). **Dimensional preferred orientations** or **shape preferred orientations** (SPO), are those in which inequant grains have a tendency towards parallel alignment. Such preferred orientations are generally quite visible to the unaided eye and are referred to as foliation or lineation. **Lattice preferred orientations** (LPO) are those that result from special orientations of optical and crystallographic axes. Commonly,

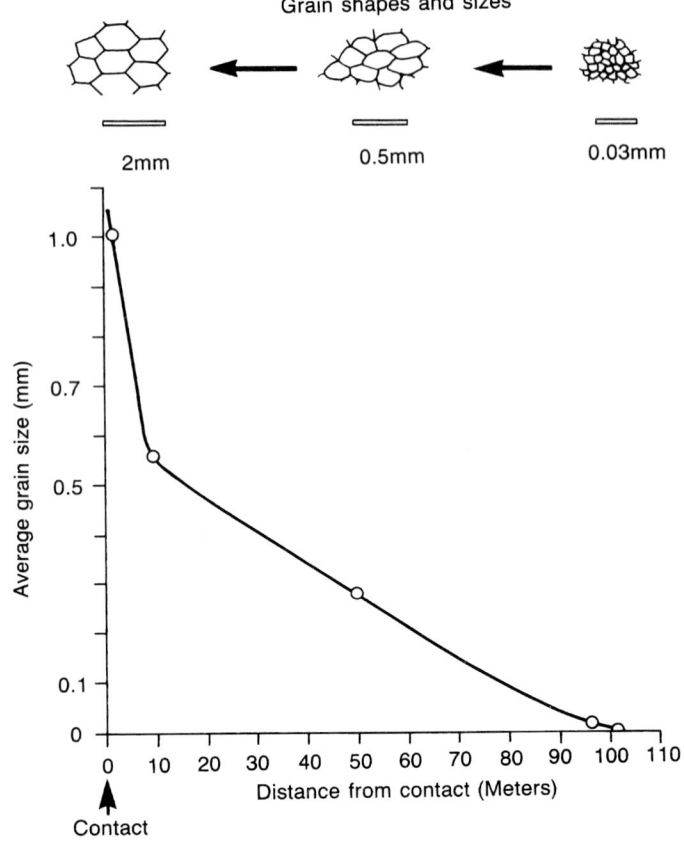

Figure 23.13
Graph showing grain size in metachert plotted against distance from contact with gabbro.
(Data from Joesten, 1983)

such preferred orientations are not visible to the unaided eye and occur in apparently randomly oriented grain assemblages.

In monomineralic or nearly monomineralic rocks, recrystallization under the influence of a deviatoric stress results in the following progressive sequence of textural changes and characteristics.[49]

1. Grains develop deformation features, such as deformation bands, and, in some cases, serrated boundaries (figure 23.15b).
2. Polygonization of grains occurs—a process in which larger, strained grains are reorganized into a number of smaller, strain-free grains, typically with serrated or irregular boundaries—beginning along grain boundaries and deformed regions in the host grain (figure 23.15c). This results in a porphyroclastic texture.
3. Coarsening of grain size and the straightening of grain boundaries follows the elimination of older, deformed grains, resulting in a granoblastic-polysutured texture.
4. A granoblastic polygonal texture develops, as enlarged grains form interfacial angles of 120° (figure 23.15d).

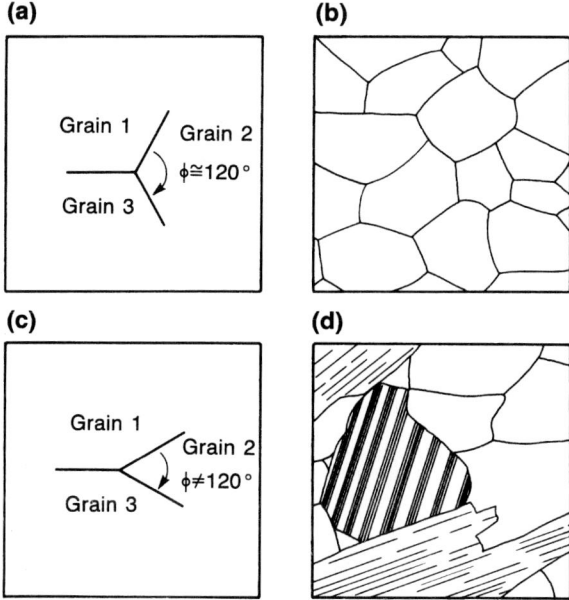

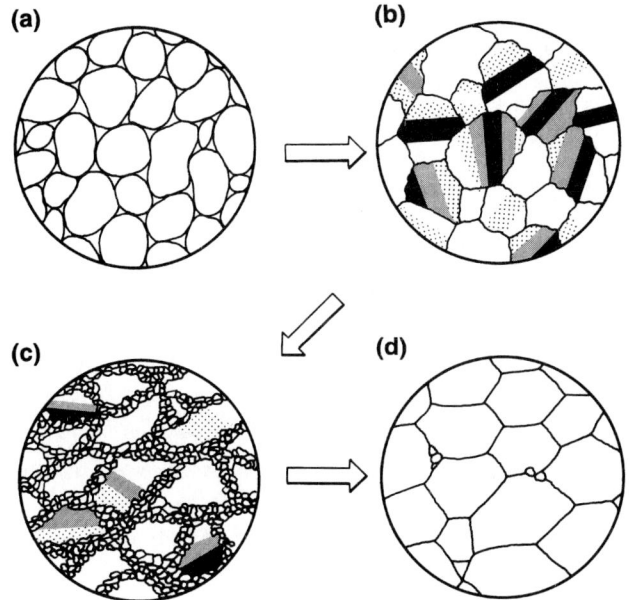

Figure 23.14 Equilibrium configurations for grain boundaries in rocks with granoblastic polygonal texture. (a) Grain boundaries and interfacial angles for three grain boundaries at equilibrium in a monomineralic rock. (b) Sketch of granoblastic polygonal texture with 120° grain-boundary interfacial angles. (c) Interfacial angles for three grain boundaries at equilibrium in a polymineralic rock, $\phi \neq 120°$. (d) Equilibrium texture in polymineralic rock in which equant minerals predominate.

Figure 23.15 Sketches showing idealized sequence of textural changes in a monomineralic rock under the influence of a deviatoric stress during metamorphism. (a) Original grains (in this case, of quartz sand). (b) Stage 1, development of deformation bands. (c) Stage 2, polygonization of grain margins. (d) Stage 3, coarsening of grain size with accompanying straightening of some grain boundaries, is followed by stage 4 in which granoblastic polygonal texture results from continued grain size increases and straightening of grain boundaries.

The textural changes may vary, depending on the pressure, temperature, strain rate, and the nature of the stress.[50] In particular, lattice preferred orientations will vary substantially under different conditions. In some cases, stage (4) is followed by the development of a porphyroblastic-granoblastic-polygonal texture, with porphyroblasts 1 cm or more in length (C. J. L. Wilson, 1973). Further grain growth yields a *coarse* granoblastic-polysutured texture. In cases of high strain rate, mylonitic, rather than granoblastic, textures are developed (C. Simpson, 1983).

The reduction in grain size during the polygonization stage, as well as during mylonitization, is a result of the high strain energies imposed by the deviatoric stress. These energies overcome the tendency to reduce surface area, because the smaller unstrained grains represent a *lower* free energy state than the larger strained grains.

All of the processes described above may be affected by phase changes and the presence of additional phases. C. J. L. Wilson (1973) found, for example, that the presence of additional phases inhibited the development of larger grains. In contrast, the presence of only a few nucleii of a certain phase may result in the development of a porphyroblastic texture with porphyroblasts of phases different from those of the matrix.

Metamorphic Differentiation

The development of gneissic structure and gneissose texture, like polygonization, is a process that seems at first glance to operate in opposition to the direction of lower free energy. In general, one would expect that metamorphism would tend to homogenize rocks, i.e., to make them uniform in composition and mechanical properties. This is so because homogenization eliminates chemical potential gradients resulting from differences in composition. Thus, a homogeneous rock should be one of lower free energy. In spite of this logic, compositional layering is very common in metamorphic rocks (figure 23.16). In many cases, the layering may represent relict primary features, but not in all.

Metamorphic differentiation is the process or processes that lead to the development of banded or lenticular segregations of minerals from an initially homogeneous rock. The process of metamorphic differentiation has been discussed by a number of petrologists.[51] To date, several explanations have been proposed. (1) The bands represent reaction zones developed between chemically incompatible rock types. (2) The bands represent synmetamorphic dikes or veins, in some cases formed by anatexis (Sawyer and Robin, 1986). (3) The bands develop due to preferential nucleation of

Figure 23.16 Banded (gneissic) structure in Cranberry Gneiss near Roan Mountain, Tennessee.

particular phases in preexisting structural zones (Bramwell, 1985). (4) The bands form via some combination of shearing, solution, and precipitation (M. B. Stephens, Glasson, and Keays, 1979). All of the above models may be possible, with each pertaining to particular cases. In the first two, continued metamorphism would presumably homogenize the rocks. In (3) and (4) (which may represent the more widely applicable models and the only models of metamorphic differentiation *sensu stricto*), the bands are produced and *enhanced* by the metamorphic process. In such cases, high strain may overcome the influence of the chemical potential gradient resulting from chemical differences. In other cases, the rates of solution and precipitation may vary and be enhanced in phyllosilicate layers by the growing physical differences (e.g., porosity differences) between layers (M. B. Stephens, Glasson, and Keays, 1979). In still other cases, physical separation of minerals with various mechanical properties may accompany deformation at high strain rates. Because all of these processes seem viable, each case must be assessed on its own merits.

SUMMARY

Metamorphism is the textural, structural, and mineralogical response of rocks subjected to the agents of pressure, temperature, deviatoric stress, and fluid phase under conditions between those of diagenesis and those of anatexis. Those conditions generally fall in the range $T = 100-750°$ C and $P = 0.1$ Gpa to 1.5 Gpa (1–15 kb). A fluid phase is commonly a critical part of the system during metamorphism and, though typically dominated by H_2O, may contain significant quantities of or be dominated by CO_2, CH_4, or other components. Each of the agents of metamorphism (pressure, temperature, chemically active fluid, deviatoric stress) may dominate during a given metamorphic event, but most commonly, two or more of these agents act in concert. Either local or regional forms of metamorphism may result.

Metamorphic rocks are commonly distinctive because of their structures and textures. In outcrop, the association of folds, faults, veins, boudins, and bands, with foliation, the dominant fabric element, is characteristic. Rock cleavage—which straddles the boundary between nonpenetrative structures and generally penetrative fabric elements (i.e. the textures)—is the most common feature of metamorphic rocks. Continuous cleavages are closely spaced (<0.01 mm), pervasive at the handspecimen scale, or both, and are distinguished from spaced cleavages that are characterized by a wider spacing (>0.01 mm). The dominant process in cleavage formation is syntectonic recrystallization, but mechanical rotation of phyllosilicates and pressure solution occur locally. Metamorphic layering may be a relict structure or it may result from intrusion, veining, transposition of bedding, or metamorphic differentiation.

Textures are a function of grain size, grain shape, grain distribution, grain orientation, and intergrain relationships. The major texture categories are (1) foliated textures, which are characterized by mineral grains showing a dimensional preferred orientation, (2) granoblastic textures, characterized by subequant to equant grains, (3) diablastic textures, in which elongate to tabular grains are radially to randomly arranged, (4) cataclastic textures, generally nonfoliated textures characterized by broken grains or rock fragments, and (5) relict textures, those textures inherited from the protoliths. The development of these various textures through recrystallization, neocrystallization, or both processes, involves nucleation and crystal growth. Both of the latter depend on diffusion, the migration of chemical species through grains or along their boundaries.

EXPLANATORY NOTES

1. This definition follows that of F. J. Turner (1948, p. 3) and is also generally consistent with that of Barth (1962, p. 232), A. R. Philpotts (1990), and other petrologists.
2. Refer to figure 5.1 and the accompanying discussion for a description of these states of a system.
3. See standard texts in structural geology such as G. H. Davis (1984), Suppe (1985), Hatcher (1990), or Twiss and Moores (1992) for more detailed discussions of stress here and in the text that follows.
4. See Cho, Liou, and Bird (1988) and Becker et al. (1989) for a description of very low P metamorphism (at < 0.05 Gpa).
5. In some cases, the term foliation has been applied to layered rocks that lack grain alignment. Such uses of the term should be avoided unless it can be shown that the layering is a metamorphic feature, rather than one inherited from an igneous or sedimentary protolith.
6. For example, Etheridge and Hobbs (1974), P. F. Williams, Means, and Hobbs (1977), and Fernandez (1987).
7. Compare J. C. Maxwell (1962), P. F. Williams, Collins, and Wiltshire (1969), and J. C. Moore and Geigle (1974) with Geiser (1975), Groshong (1976), T. H. Bell (1978), Beutner (1978), D. R. Gray (1978), W. J. Gregg (1985), J. H. Lee et al. (1986), Ishii (1988), and S. J. Sutton (1989). See D. S. Wood (1974) for a review. Also see Lundberg and J. C. Moore (1986).

8. For more information on the melting of ultramafic compositions, see Bowen and Shairer (1935) and Wyllie (1971c), as well as reviews in Wyllie (1971a), Yoder (1976), S. A. Morse (1980), and chapters 5 and 6 of this text.

9. Examples of unusual thermal conditions of metamorphism may be found in contact metamorphic environments discussed in chapter 26. For additional references to unusual conditions see references in note 16. See Bayly (1968), Winkler (1979), and F. J. Turner (1981) for additional discussions of the limits of metamorphism.

10. Wyllie (1971a) and standard geophysics texts such as A. H. Cook (1973) and Stacey (1977) provide discussions of these phenomena.

11. Frictional heating is discussed by Scholz (1980), Turcotte and Schubert (1982, pp. 189–190), and Molnar, Chen, and Padovani (1983).

12. Note that the same kind of fluid (i.e. a supercritical fluid phase) evolves during the origin of granitoid pegmatites (see chapter 12).

13. Also see Mariner and Wiley (1976), Sorey (1985), Alt et al. (1986), Hulen and Nielson (1986), and A. J. R. White (1986).

14. For additional information on fluids in metamorphism see Trommsdorff and Skippen (1986), Bickle and McKenzie (1987), Brady (1988), Ferry (1987; 1988), S. M. Peacock (1989), Ferry and Dipple (1991), the reviews of Rumble (1989) and Torgerson (1990), and the references therein.

15. Touret (1970), Roedder (1972, 1984, ch. 12, 13), Greenwood (1976, p. 198ff.), Fyfe, Price, and Thompson (1978, ch. 6), Ferry and Burt (1982), H. P. Taylor (1983), Touret and Dietvorst (1983), M. L. Crawford and Hollister (1986), Newton (1986).

16. J. D. Lowell and Guilbert (1970), Beane and Titley (1981), and Guilbert and Park (1986, ch. 5) review and discuss alteration around ore bodies associated with plutons. P. B. Larsen and Taylor (1986) discuss alteration surrounding a caldera.

17. Harker (1932), F. J. Turner (1948, ch. 1; 1981, pp. 3–5), Miyashiro (1973a, p. 22ff.), Mason (1978, p. 4), Winkler (1979, p. 1ff.), Best (1982, p. 348ff.), Suk (1983, ch. 1), Hyndman (1985, ch.13).

18. Shock metamorphism is discussed by Boon and Albritton (1936), Shoemaker (1960), Chao, Shoemaker, and Madsen (1960), Chao et al. (1962), Bunch and Cohen (1964), N. M. Short (1966), Currie (1967), B. M. French and Short (1968), and Spry (1969, pp. 247–249). Metamorphic core complexes are discussed by Coney (1979) and G. H. Davis (1980) and in the various papers in Crittenden, Coney, and Davis (1980). Examples of local dynamic metamorphism in high-temperature peridotites and related rocks are described by Nicolas and Poirier (1976, ch. 10). Dynamic metamorphism associated with fault zones is described by Beach (1980) and R. Kerrich et al. (1980). Associated metamorphic and/or deformational effects along discrete fault zones and small-scale shear zones are described by Brabb, Maddock, and Wallace (1966), Mosher (1980), and M. J. Watts and Williams (1983). Also refer to Spry (1969) and A. J. Barker (1990) for discussions of these various types of metamorphism and the resulting textures.

19. Mylonites, rocks produced by ductile deformation, are defined and described elsewhere in this text (ch. 30). Examples of regional shear zones, typically characterized by mylonites, are described by Theodore (1970), Roper and Justus (1973), Bak, Korstgard, and Sorenson (1975), Hatcher et al. (1979), Beach (1980), and C. Simpson (1985). Also see J. G. Ramsay and Graham (1970), Nicolas et al. (1977), and J. G. Ramsay (1980) . Melanges and associated broken and dismembered formations are discussed by K. J. Hsu (1968, 1969, 1974), Berkland et al. (1972), and Raymond (1975, 1984a), as well as in the various papers in Raymond (1984c) and Horton and Rast (1989). Also see papers in J. C. Moore (1986).

20. Bates and Jackson (1980).

21. Discussions of static metamorphism (typically under the heading "burial metamorphism") are provided by Coombs (1960, 1961), Dickinson et al. (1969), and Ernst (1971a,b,c), and in standard metamorphic texts such as F. J. Turner (1981) and Yardley (1989).

22. H. Ramberg (1952, p. 229). For examples, see Holser (1950), Floyd (1975), Joesten (1977), S. E. Swanson (1981), and Bowman, O'Neil, and Essene (1985).

23. For discussions of these processes see H. H. Read (1944, 1948), Ramberg (1952), Sederholm (1967), and Beloussov (1980, pp. 306–314). Guilbert and Park (1986, ch. 5) discuss alteration. Granitization was discussed in chapter 11. For more information on that process, see that chapter and the references therein.

24. The origins of the structures mentioned here are described in the text below and are discussed in more detail in various books on structural geology, such as J. G. Ramsay (1967), Davis (1982), G. Wilson (1982), Suppe (1985), Dennis (1987), and Twiss and Moores (1992).

25. Sander (1930, in Turner and Weiss, 1963; Sander, 1970) introduced the terminology now used to define the structural character of metamorphic rocks. He used the German word *Gefuge*, translated as "fabric," to refer to the geometrical character of rocks discussed here (Paterson and Weiss, 1961; Turner and Weiss, 1963; Sander, 1970; Bates and Jackson, 1987).

26. Additional tectonite types have been defined. See Turner and Weiss (1963), Sander (1970), Schwerdtner, Bennett, and Janes (1977), and Shelly (1989), and see Raymond (1987) for a review.

27. Scanning electron microscopy (SEM) has now shown that even those cleavages dominated by slickensided shear surfaces consist of small domains of aligned phyllosilicates (J. C. Moore et al., 1986). Rocks with such cleavages, nevertheless, are characterized by brittle shear fractures and are fundamentally different in structure than mylonitic S-tectonites. The SF-tectonites exhibiting such structures are foliated cataclasites (Chester, Friedman, and Logan, 1985; Kano and Sato, 1988).

28. The definition of *relict* or *palimpsest* offered here is essentially the same as that of Bates and Jackson (1987). Their definition follows that used in older works (Grout, 1932).

29. Some geologists (G. Wilson, 1982) continue to use this type of binary classification of secondary cleavage types.

30. The volume of literature and the diversity of ideas on cleavage formation is considerable. A detailed treatment is beyond the scope of this text. The interested reader should consult J. C. Maxwell (1962), J. G. Ramsay (1967, p. 177ff.),

P. F. Williams, Collins, and Wiltshire (1969), J. C. Moore and Geigle (1974), D. S. Wood (1974), Geiser (1975), Means (1975), Alvarez, Engelder, and Lowrie (1976), D. R. Gray (1976, 1978, 1979b, 1981), Groshong (1976), T. E. Tullis (1976), Beutner, Jancin, and Simon (1977), Tobisch et al. (1977), P. F. Williams, Means, and Hobbs (1977), Alvarez, Engelder, and Geiser (1978), T. H. Bell (1978a), Beutner (1978), Etheridge and Wilkie (1979), Lebedeva (1979), M. B. Stephens, Glasson, and Keays (1979), Borradaile, Bayly, and Powell (1982), D. S. Cowan (1982), Onasch (1983), W. J. Gregg (1985), B. E. Hobbs (1985), Ogawa and Miyata (1985), Rosenfeld (1985); J. H. Lee et al. (1986), Ishii (1988), S. M. Agar, Prior, and Behrmann (1989), Lash (1989), S. J. Sutton (1989), Bhagat and Marshak (1990), Kanagawa (1991), the papers on cleavage in Stauffer (1983), and the papers referred to in the reference lists in these various reports, for more information on conflicting views on cleavage origins. The work of Borradaile, Bayley, and Powell (1982) provides a good introduction to cleavage studies for those who wish to expand their understanding.

31. The origin of stylolitic cleavage is also discussed by Stockdale (1943), Heald (1955), Logan and Semeniuk (1976), Alvarez, Engelder, and Lowrie (1976), Alvarez and Engelder (1982), Borradaile (1982), and Engelder and Alvarez (1982). The origin of crenulation cleavage is also discussed by Cosgrove (1976), D. R. Gray (1976, 1978, 1979b) and T. H. Bell and Rubenach (1980). For more information and excellent photographs, see Borradaile, Bayley, and Powell (1982b). Pressure solution, an important process in the formation of stylolitic and other cleavages is discussed by Renton et al. (1969), A. Beach (1974, 1979), A. Beach and King (1978), and P.-Y. F. Robin (1978).

32. See the references to papers on the development of slaty cleavage in note 30, such as Geiser (1975), Maxwell (1962), and Lee et al. (1986).

33. Phaneritic metamorphic rocks with handspecimen-scale layers are called gneiss and those with megalayers are here called megagneiss.

34. Gneisses, as defined here, are characterized by a layered or banded structure and like texture. The two main types of phaneritic, strongly foliated textures are the schistose and gneissose textures. In much of the literature and as used by many geologists, definitions of gneissose texture and the corresponding rock type, gneiss, and the distinctions between schistose and gneissose textures and between schists and gneisses, are at best ambiguous. Harker (1932, p. 61) acknowledged the problem in describing the term gneiss as "vague and unsatisfactory."

Three uses of the term gneiss are currently applied by geologists. In one use, gneiss is a phaneritic, metamorphic rock of any composition characterized by alternating bands of contrasting mineral composition, at least some of which are characterized by the preferred orientation of the included minerals (Raymond, 1984c; R. S. Mitchell, 1985; cf. Bergman, 1784, in Tomkeieff, 1983). Here, both bands and preferred orientation are considered to be definitive characteristics. A second definition, based on both common practice and the definition of Werner (1787, in Tomkeieff, 1983), is that gneiss is a phaneritic, metamorphic rock of

granitoid composition with a preferred orientation of minerals. In this definition, the granitoid composition and preferred orientation are central, but banding is not required. A third definition is that a gneiss is a phaneritic, metamorphic rock of any composition that has a preferred orientation of some minerals, but that lacks continuous cleavage (L. Acker, pers. comm., 1987). In the latter case, gneiss is contrasted with schist, which has continuous cleavage. The word foliation has been judiciously avoided in these definitions because some geologists consider *foliation* to be synonymous with continuous cleavage, whereas others use it, as I do in this book, to refer to a visible, dimensional, preferred orientation.

Note that two of the contrasting definitions are based on definitions first presented in the eighteenth century. The controversy is long-standing. The fact is, there is a continuum of rock types extending from phyllosilicate-rich rocks with continuous cleavage to quartz- and feldspar-rich, banded rocks with spaced cleavage. The boundary between the two rock types, schist and gneiss, and between their corresponding textures, is arbitrarily chosen. Some definitions proffered in the last half century have added to the confusion. For example, C. M. Rice (1941) defines gneiss as "a foliated or banded crystalline rock in which granular minerals, or lenticles and bands in which they predominate, alternate with schistose minerals, or lenticles and bands in which they predominate. . . . It is most commonly of the same composition as granite." A similar definition is given by Bates and Jackson (1987). There are several problems with this definition: (1) linking the composition to granite supports the definition of Werner; (2) the suggestion that banding may occur supports the definition of Bergman; and (3) the suggestion that lenticles can occur in gneiss or schistose materials does not provide for a distinction between gneiss and schist, inasmuch as the latter is typically composed of lenticles of granular minerals separated by layers or lenticles of platy and/or acicular minerals.

In the past, many geologists have applied the terms schist and gneiss loosely. Rocks with abundant mica were called schist. Those not dominated by mica, especially those with considerable amounts of quartz and feldspar or those containing augen, were called gneiss—regardless of the presence or absence of continuous cleavage, banding, or granitoid composition. Precision of language and clarity of communication require that such applications of these terms be abandoned.

35. Metamorphic layers, cleavage, and relict bedding are all crudely planar features found in metamorphic rocks. Following the lead of Sander (1930, in Turner and Weiss, 1973), petrologists and structural geologists refer to all planar features in metamorphic rocks as S-surfaces. Bedding is designated S_0 and successively younger S-surfaces are designated $S_1, S_2, \ldots S_n$.

36. The interested reader should consult modern structural geology texts (see note 3) for discussion of the differences between brittle and ductile deformation. In a very general sense, ductile deformation involves flow and recrystallization, whereas brittle deformation involves breaking of materials.

37. G. Wilson (1982, ch. 9) reviews the history of the use and misuse of the term boudin. See J. G. Ramsay (1967) for a detailed discussion and analysis of boudins, related "chocolate tablet" structure, and the process of boudinage.

38. See G. Wilson (1982) for a discussion of the origin of mullions.

39. J. G. Ramsay (1980) and J. G. Ramsay and Huber (1987, Session 26) discuss brittle and ductile shear zones. Also see Nicolas et al. (1977), Sibson (1980), L. Anderson et al. (1983), C. Simpson (1983), and Blenkinsop and Rutter (1986). Chapter 30 focuses on mylonites. Refer to that chapter for details and references to the subjects of ductile deformation, cataclasis, and mylonite formation.

40. Additional information on textures is available in Spry (1969), Nicolas and Poirier (1976), Wenk (1985), Bard (1986), A. J. Barker (1990), and Yardley, MacKenzie, and Gilbert (1990).

41. Becke (1903, in Grout, 1932, p. 353) introduced the use of the root *blast* ("to sprout") to refer to metamorphic textures.

42. The term granoblastic has been used in several ways. As defined here, the use is much like Harker's (1932) original use. He did, however, specify that granoblastic textures were mosaic textures, a term that likewise has several meanings (Spry, 1969, p. 186). Heinrich (1956) also refers to granoblastic texture as a mosaic texture. Spry (1969) specifies that the texture be equidimensional-xenoblastic (composed of equant grains lacking crystal faces). Bard (1986, p. 182) defines the texture as one in which there is no preferred orientation of minerals. Because determination of the presence of preferred orientation (or demonstration of its absence) in rocks with equant minerals like quartz or olivine typically requires time-consuming petrofabric analysis, such a stricture renders the term useless for normal petrographic work and is rejected here.

43. Such textures have been loosely assigned to the foliated category by many petrologists simply because they are composed of inequant mineral grains. In spite of that, the textural arrangement of grains neither promotes the breaking of rocks with this texture into flakes, tabular pieces, or folia; nor does it give the rock the appearance of having aligned grains. Therefore, the use of the term foliated for such textures and the use of the word schist for rocks exhibiting such textures should be abandoned.

44. The Free Energy of a system can be considered in a general way to be the capacity of the system to do work. Additional discussion and quantitative treatment of this topic may be found in standard texts on thermodynamics, physical chemistry, or geochemistry (e.g. Krauskopf, 1967). Also see Ehrlich et al. (1972).

45. See Fyfe et al. (1958), Spry (1969), Vernon (1976, ch. 3), Gottstein and Mecking (1985), Rubie and Thompson (1985), Bard (1986), and Joesten and Fisher (1988) for more thorough discussions of nucleation and crystal growth.

46. Diffusion is discussed in more detail by Fyfe et al. (1958, p. 60ff.), Spry (1969), H. W. Green (1970), G. W. Fisher (1973, 1978), Vernon (1976), Walther and Wood (1984), Joesten (1985), Lasaga (1986), Wheeler (1987), and Joesten and Fisher (1988).

47. See Spry (1969), Nicolas and Poirier (1976, p. 163ff.), Vernon (1976, ch. 5), Ricoult (1979), and Mercier (1985).

48. The works of Nicolas and Poirier (1976) and in H. R. Wenk (1985) focus on problems related to the development of preferred orientations. Also see N. L. Carter, Christie, and Griggs (1964), H. W. Green (1967), Ave Lallemant and Carter (1970), J. Tullis, Christie, and Griggs (1973), C. J. L. Wilson (1973), Poirier and Nicolas (1975), T. H. Bell (1979), Etheridge and Wilkie (1979), Lister and Hobbs (1980), Zeuch (1983), Zeuch and Green (1984a,b), Toriumi and Karato (1985), Dell'Angelo and Tullis (1986), Shelley (1989), Ji and Mainprice (1990), Ildefonse, Lardeaux, and Cason (1990), H. R. Wenk and Pannetier (1990), and the references therein.

49. A. G. Sylvester and Christie (1968) describe such a sequence in deformed quartz arenites in the Inyo Mountains of California, C. J. L. Wilson (1973) describes a similar sequence in the same kind of rocks from the Mount Isa area of Australia, and Masuda (1982) details like changes in quartz schists from Shikoku, Japan. A detailed discussion of the causes of these changes is beyond the scope of this text. For more information, see the works listed in note 22 and in Means (1980).

50. For example, see Raleigh (1968), Ave Lallemant and Carter (1970), Lister and Hobbs (1980), Zeuch and Green (1984a), Ave Lallemant (1985), Mercier (1985), G. P. Price (1985), and Blumenfeld, Mainprice, and Bouchez (1986).

51. Suk (1983) and Hyndman (1985) summarize explanations and examples of metamorphic differentiation, and F. J. Turner (1948) and H. Ramberg (1952) devote chapters to the subject. Also see Dietrich (1963), Vidale (1974), Vernon (1976), Gray (1977), Dick and Sinton (1979), M. B. Stephens, Glasson, and Keays (1979), C. Simpson (1983), Bramwell (1985), and Sawyer and Robin (1986).

PROBLEMS

23.1 Examine figure 23.4. Consider a rock with the assemblage quartz + biotite + andalusite + muscovite + garnet + plagioclase, with veins of laumontite cutting the grain boundaries between andalusite and quartz. Describe the metamorphic history and explain whether the metmorphism indicated is prograde or retrograde.

23.2 Compare the texture in figures 23.12b with that in figure 23.11e. Note that the two rocks mineralogically are almost identical. Provide a general explanation, based on your present knowledge, of how a dunite like that in figure 23.12b might be converted to one with a texture like that in figure 23.11e. Comment on processes and their effects.

23.3 Joesten and Fisher (1988) provide information on wollastonite grain sizes in the Christmas Mountains contact aureole for which chert grain sizes are plotted in figure 23.13. The values for matrix wollastonite are 0.011 mm at 101.7 m, 0.053 mm at 46.5 m, and about 0.25 mm at 20 m from the contact. Plot these data on a copy of figure 23.13 and compare and contrast the plotted curve with the chert grain size data. (*Note:* The contact aureole also contains large wollastonite porphyroblasts with diameters of more than 20 mm. See Joesten (1983) and Joesten and Fisher (1988) for a discussion of how the various textures develop.)

24

Metamorphic Conditions, Mineralogies, Protoliths, Facies, and Facies Series

INTRODUCTION

Metamorphic rocks may exhibit a wide range of chemistry and minerals. The mineral composition reflects the conditions of metamorphism. Rock chemistry controls the mineral composition and, as mentioned in chapter 23, it also controls the upper and lower limits for metamorphic conditions. This chapter summarizes the common minerals of metamorphic rocks, notes the possible range of mineralogies and chemistries, discusses the significance of minerals and chemistry, and links minerals and rock chemistry through the concept of metamorphic facies.

The concept of metamorphic facies connects the observed mineralogical composition of metamorphic rocks to their conditions of metamorphism. A series of facies in a mountain belt reflects the tectonic setting of the metamorphism.

MINERALOGY, PROTOLITHS, AND ROCK CHEMISTRY

Petrographic observations are a powerful tool for deciphering the basic aspects of metamorphic rock history. In particular, the minerals present in a metamorphic rock can be used to identify the P-T conditions under which the rock formed. Although many of the minerals present in metamorphic rocks also occur in igneous and sedimentary rocks, additional minerals particularly characteristic of metamorphic rocks comprise a significant part of some rocks and distinguish the metamorphic rocks from rocks of other classes.

Minerals are also important, in part, because they reveal something of the chemistry of the protolith. Any kind of rock can be metamorphosed. Consequently, the range of protoliths for metamorphic rocks encompasses all rock types and rock chemistries that exist in nature. In order to make the discussion of metamorphic rocks manageable, the range of rock chemistries is divided into several groups—ultrabasic (silicate), basic, carbonate/nonsilicate, aluminous, siliceous-alkali-calcic, and silicic rocks. Ultrabasic (silicate) rocks are those with silica contents of less than 45%. Basic rocks have silica contents between 45 and 52%. The terms carbonate and nonsilicate are self-explanatory. Here, aluminous rocks are those siliceous rocks that are relatively rich in alumina, the term being akin to the term peraluminous used for igneous rocks. Siliceous-alkali-calcic rocks are rocks with moderate to abundant amounts of lime *and* the alkalis. Siliceous rocks are those that are very rich in silica (> 90% SiO_2). Each chemical group is characterized by a particular set of minerals (table 24.1). Differences in mineral content among the various

Table 24.1   Selected Common Minerals of Metamorphic Rock Chemical Groups

Ultrabasic Rocks	Basic Rocks	Carbonate Rocks	Aluminous Rocks	Siliceous-Alkali-Calcic Rocks	Siliceous Rocks
Olivines	Augite	Calcite	Quartz	Quartz	Quartz
Augite	Omphacite	Dolomite	White micas	Plagioclases	Plagioclases
Diopside	Jadeites	Aragonite	Biotite	Alkali feldspar	Alkali feldspar
Orthopyroxenes	Orthopyroxenes	Olivine	Chlorites	Chlorites	Biotite
Tremolite	Glaucophane	Diopside	Plagioclases	Biotite	White micas
Anthophyllite	Hornblende	Tremolite	Alkali feldspar	White micas	Chlorites
Serpentines	Actinolite	Wollastonite	Pyrophyllite	Sillimanite	Garnets
Chlorites	Epidotes	Talc	Sillimanite	Kyanite	Sillimanite
Talc	Lawsonite	Phlogopite	Kyanite	Andalusite	Kyanite
Phlogopite	Plagioclases	Periclase	Andalusite	Garnets	Andalusite
Chromite	Biotite	Idocrase	Staurolite	Cordierite	Cordierite
Magnetite	Zeolites	Graphite	Garnets	Jadeites	Aegirine
	Quartz	Garnets	Calcite	Lawsonite	Crossite
	Calcite	Pyrite	Chloritoid	Epidotes	Stilpnomelane
	Sphene	Pyrrhotite	Cordierite	Pumpellyite	Hematite
	Garnets		Tourmaline	Zeolites	Magnetite
	Magnetite		Kaolinite	Glaucophane	
	Ilmenite		Magnetite	Calcite	
			Ilmenite	Magnetite	

groups of metamorphic rocks emphasize the point made by Eskola (1915, in Turner, 1958) that the mineralogy is *controlled* by the chemistry, an indication that metamorphic rocks form in response to processes that are controlled by physical-chemical laws.

Each chemical group includes major metamorphic rock types that represent various protoliths. The ultrabasic group encompasses the metamorphosed silicate-ferromagnesian mineral-rich rocks, including peridotites, pyroxenites, and dunites. The basic group includes metamorphosed basalts and gabbros and related rocks. The carbonate/nonsilicate group includes the marbles, most of which have sedimentary protoliths (e.g., limestone or dolostone), and various other metamorphic rocks derived from evaporites or other less common rocks. The aluminous category refers to aluminum-rich rocks typically characterized by abundant micas and containing aluminum silicates such as kyanite or andalusite. These rocks are principally derived from shales and are referred to as

pelitic (e.g., pelitic schists), but may also be derived from peraluminous igneous rocks. Most intermediate to siliceous igneous rocks give rise to quartz-feldspar rocks that may be assigned to the siliceous-alkali-calcic group. Feldspathic and lithic sandstones are also protoliths for the metamorphic rocks of this group. Silicic rocks include metamorphosed quartz arenites, quartz wackes, cherts, siliceous sinter, and "silexites."

The chemistries of the various protoliths are represented by the chemistries presented in chapters 3 through 22. Although metamorphic rock chemistries may have been altered by metasomatism, they nevertheless fall in the general range provided by the limits of sedimentary and igneous rock compositions. The general fields for the chemistries of the various protoliths are presented in a triangular plot in figure 24.1. Note that the boundaries of some fields overlap, in part because alumina is not accounted for by the compositions of the corners.

Figure 24.1   Triangular diagram showing fields representing the chemistry of the six general chemical groups of metamorphic rocks. Values are in weight percent. A = aluminous rocks; B = basic rocks; C = carbonate and nonsilicate rocks; SAC = siliceous-alkali-calcic rocks; S = siliceous rocks; U = ultrabasic (ferromagnesian) silicate rocks.

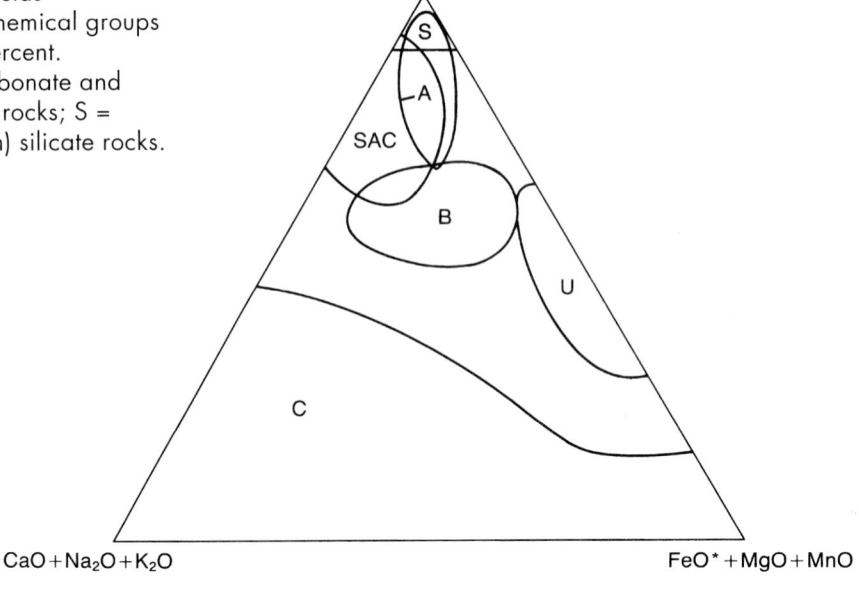

CLASSIFICATIONS OF METAMORPHIC ROCKS

Unlike igneous and sedimentary rock names, many metamorphic rock names are based entirely on rock texture. Others are based only on mineral content. This makes metamorphic rock classification somewhat different from igneous and sedimentary rock classification. In several texts, metamorphic rock names are simply presented on a list.[1]

Published classifications are based on texture, mineralogy, chemistry, or combinations of these parameters.[2] It is not uncommon to find classifications in which some names are based on mineralogy and others are based on texture, because

that is the *common practice* among geologists.[3] Thus, schist, a name based solely on the texture of the rock appears in the same classification with marble, a name based primarily on composition.

Textural Classifications

Several petrologists have adopted metamorphic rock classifications based primarily on textural criteria. These are designated here as textural classifications. Spock (1962), W. T. Huang (1962), Best (1982), and Raymond (1984c) have adopted this approach. Spock (1962, p. 241) divides metamorphic rocks into two main categories—those with parallel structure and those lacking parallel structure (figure 24.2a).

(a)

Rocks with visible parallel structure (foliates and banded rocks)	
Slate	Schist (continued)
Mylonite (in part)	Tremolite
Phyllite	Actinolite
Schist	Staurolite
Muscovite	Graphitic
Chlorite	Gneiss
Talc	Granitic, diorite, etc.
Biotite	Hornblende
Quartz-mica	Biotite
Garnetiferous	Banded
Hornblende	Augen
Rocks apparently lacking parallel structure	
Quartzite	Soapstone
Marble	Amphibolite
Dolomitic	Granulite
Serpentine, etc.	Eclogite
Hornfels	

(b)

Strongly foliated rocks	Weakly foliated rocks	Non-foliated to weakly foliated rocks
Slate	Gneiss	Granofels
Phyllite	Migmatite	Amphibolite
Schist	Mylonite	Serpentinite
		Greenstone
		Greisen
		Hornfels
		Quartzite
		Marble
		Argillite
		Skarn

Figure 24.2   Two classifications of metamorphic rocks based on texture. a) Binary classification of Spock (slightly modified from Spock, 1962). b) Threefold classification of Best (1982).

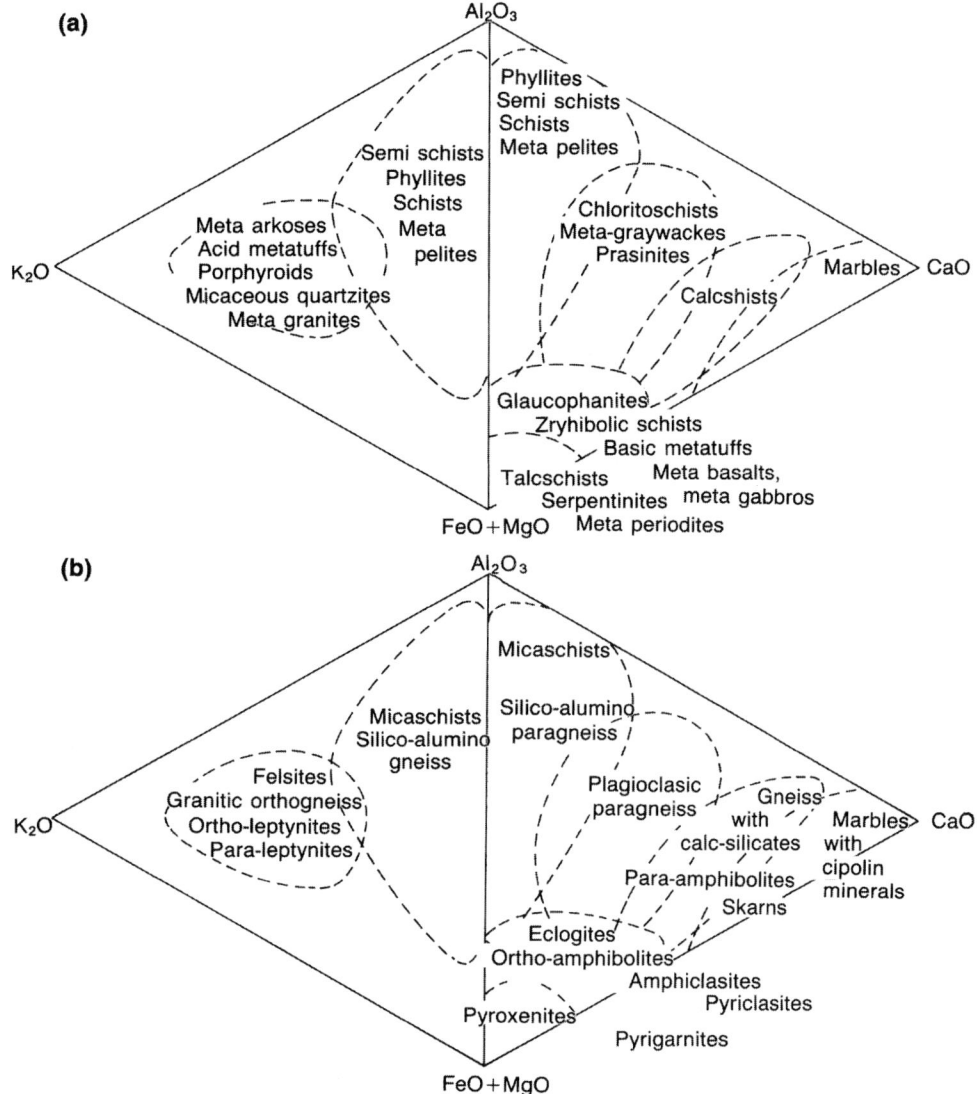

Figure 24.3 Bard's chemical classification of metamorphic rocks: a) for low-grade rocks, b) for medium- to high-grade rocks.

(From Bard, 1986)

Best (1982, p. 390) adopts three categories—strongly foliated rocks, weakly foliated rocks, and nonfoliated to weakly foliated rocks (figure 24.2b). W. T. Huang (1962, pp. 384–85) also uses three categories—cataclastic, nonfoliated, and foliated. He includes mylonites as cataclastic rocks, a practice that has been abandoned, because of the current understanding that mylonitic rocks are rocks with textures resulting primarily from recrystallization.

Raymond (1984c, pp. 128–29) adopts two main categories—crystalline and clastic rocks—which are further subdivided. Clastic rocks are subdivided into foliated and nonfoliated types. Crystalline rocks are subdivided into strongly foliated, weakly foliated, granoblastic, and diablastic types. The classification of Raymond (1984c) has been adopted here, but has been modified in an attempt to produce a more rational (albiet less conventional) classification (table 24.2).

Other Classifications

Two other classifications, one based primarily on chemistry and one based primarily on mineralogy, are presented here for the purpose of demonstrating that nontextural classifications are possible and have been proposed. These are the classifications presented in Winkler (1979) and Bard (1986). Bard's (1986) classification is a double triangle, chemical classification in which four endmember components are used as parameters of classification (figure 24.3). These are K_2O, Al_2O_3, CaO, and $FeO + MgO$. Two versions are presented, one for low-grade rocks and one for higher-grade rocks. Many of the names used by Bard are unusual (e.g., amphiclasite, zryhibolic schist) or archaic (e.g., prasinite, leptynite) and no definitions or explanations are provided.

499

Table 24.2 Classification of Metamorphic Rocks

Texture and Composition	Root Name	Examples of Names[1]
Crystalline Rocks		
STRONGLY FOLIATED		
Slaty	Slate	Black Slate
Phyllitic	Phyllite	Quartz-chlorite Phyllite
Schistose	Schist	Biotite-quartz-white mica Schist
Serpentine-rich		Serpentine Schist (Serpentinite)
Hornblende-rich		Hornblende Schist (Amphibolite)
Calcite-rich		Calc-schist
Gneissose	Gneiss	Biotite-quartz-plagioclase Gneiss
Mylonitic	Mylonite	Quartz-chlorite Mylonite
Protomylonitic	Protomylonite	Biotite-quartz Protomylonite
Orthomylonitic	Orthomylonite	Chlorite-quartz Orthomylonite
Ultramylonitic	Ultramylonite	Muscovite-quartz Ultramylonite
WEAKLY FOLIATED		
Semi-slaty	Semi-slate *or*	Maroon Semi-slate
	Argillite	Black Argillite
Semi-schistose	Semischist	Biotite-garnet-sillimanite-
	or	plagioclase-quartz Semischist
	prefix *meta-*	Porphyroclastic Meta-dunite
	followed by a	Muscovite Meta-arkose
	protolith name	Epidote Metabasalt
		Serpentine Semi-schist
		(Serpentinite)
		Hornblende Semi-schist
		(Amphibolite)
		Calcite Marble Semi-schist

The classification presented by Winkler (1979, pp. 340–344) was developed by Austrian petrographers. It, too, has a double triangle shape (figure 24.4). The classification is quantitative and the classification parameters are the minerals quartz, carbonate, micas, and feldspars. Textural terms are used as a secondary parameter for naming rocks. Among the drawbacks of this classification are (1) it employs commonly used textural names for compositional categories, and (2) it is ambiguous in specifying that textures may override the names based on the mineralogical categories.

CONDITIONS OF METAMORPHISM AND PETROGENETIC GRIDS

Metamorphism is usually described in terms of the general field of P-T space in which it occurs (figure 23.2). The limits of this P-T space are based on experimentally determined conditions, such as the melting of wet granite or the P-T conditions for crystallization of certain phases. Within these limits, it is possible and desirable to divide P-T space into

Table 24.2 Classification of Metamorphic Rocks

Texture and Composition	Root Name	Examples of Names[1]
DIABLASTIC	Diablastite	Biotite Diablastite
		Serpentine Diablastite (Serpentinite)
		Hornblende Diablastite (Amphibolite)
GRANOBLASTIC	Granoblastite	Garnet-quartz-plagioclase Granoblastite
Varieties include:	Marble- granoblastite	Calcite Marble-granoblastite
		Dolomite Marble-granoblastite
	Tactite (Skarn) (Ca-silicate-rich rock)	Garnet-epidote Tactite
	Metaquartzitic Granoblastite	Muscovite Metaquartzitic Granoblastite
	Eclogite (Pyrope-omphacite rock)	Phlogopite Eclogite
Clastic Rocks		
FOLIATED		
Foliated cataclastic	Quasimylonite	Polymict Quasimylonite
		Illite-quartz Quasimylonite
NON-FOLIATED		
Cataclastic	Cataclasite	Plagioclase-quartz Cataclasite
Mortar	Breccia	Rhyolite Breccia
	or	
	Cataclasite	Rhyolite Cataclasite
Vitriclastoblastic	Pseudotachylite	Quartz-plagioclase Pseudotachylite

[1]Minerals are placed before the root name in order of increasing abundance, with the most abundant mineral adjacent to the root name.

subregions that reflect more restricted or specific conditions of metamorphism. Similarly, subdivision of P-X (X = composition) or T-X space is possible.

In practice, a metamorphic petrologist observes mineral assemblages in rocks, not P-T conditions. In order to link those assemblages to specific conditions of metamorphism, it is essential to know the P-T conditions of stability (and metastability) of any given assemblage that has been observed. Experimental laboratory analyses of phase relations provide that link. Once the stability of a mineral or assemblage is known. The mineral or assemblage can be assigned properly to a region in P-T space. Numerous phase studies

have been completed (see appendix C). Thus, P-T space has been subdivided into regions characterized by particular minerals or mineral assemblages, each corresponding to a particular, limited set of P-T conditions. Bowen (1940) referred to graphs of subdivided P-T space, with regions characterized by particular phases or phase assemblages, as **petrogenetic grids.**

Conditions of Metamorphism

In order to determine the limits of metamorphic conditions, it is necessary to specify exactly what physical-chemical indicators mark those limits. It was noted in chapter 23 that the

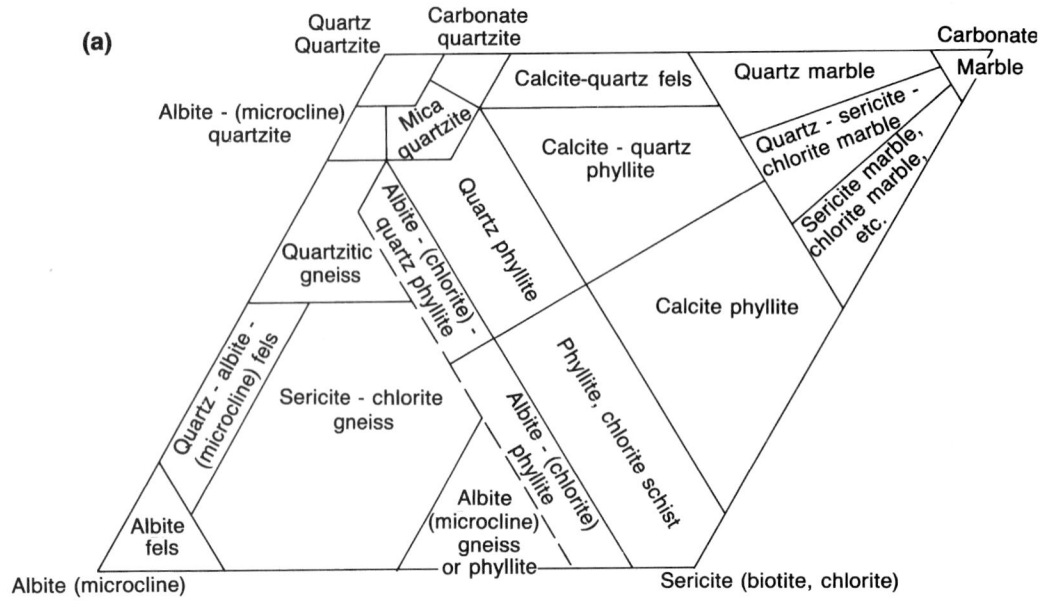

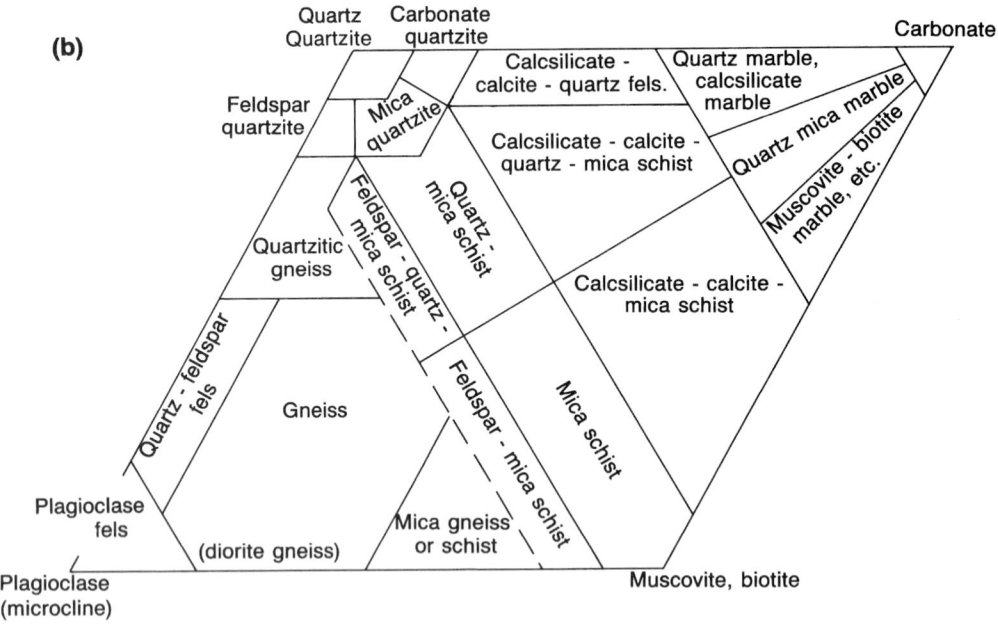

Figure 24.4   The Austrian mineralogical classification of metamorphic rocks: a) for low-grade rocks and b) for high-grade rocks.
(From Winkler, 1979)

temperature limits of metamorphism are approximately 100° C and 600–2000°C. At the lower limit, the beginning of metamorphism is marked by the first appearance of a mineral, mineral assemblage, or texture that is not stable under the surface to near-surface conditions of weathering and diagenesis. Knowledge of the stability limits of that mineral, as-

semblage, or texture is based on experimental analysis and field measurements. At the upper limit, the end of metamorphism, marked by the beginning of melting, is also determined by experiment. Similarly, pressure and fluid phase limits require experimental or theoretical delineation.

The Beginnings of Metamorphism

To recognize the beginnings of metamorphism, one needs to know the nature of weathering and diagenesis (see chapter 16) and to recognize the occurrence of minerals and textures not associated with those processes. Recall that weathering commonly produces such minerals as calcite, the clays, and metallic oxides like hematite, goethite, or todorokite.[4] Similar minerals, plus additional phases like dolomite and phillipsite, are diagenetic products.

What minerals mark the inception of metamorphism? Some zeolites, such as phillipsite, form in sediments on the ocean floor[5] and are, therefore, diagenetic; but other zeolites such as wairakite, heulandite, and laumontite do not appear under diagenetic conditions. Therefore, we reason that the appearance of these minerals, as well as white mica, certain chlorites, and other minerals that appear *only* in rocks buried beyond the levels of weathering or heated beyond the ambient temperature of the surface zones of the Earth, must mark the beginning of metamorphism.

By definition, the conditions at the beginning of metamorphism must exceed the surface conditions, which lie in the range $T = <0–60°C$ and $10^5 Pa–0.1\ Gpa\ (0.001–1\ kb)$.[6] Though some studies have been completed, experimental syntheses of the minerals that form under the lowest P-T conditions of metamorphism are fraught with difficulties because of the slow reaction rates under such low-energy conditions.[7] For example, Liou (1971c) showed that laumontite formed from stilbite at about 150°C and 0.1 Gpa (1 kb) and Thompson (1970) demonstrated that kaolinite and quartz react to form pyrophyllite at 0.1 Gpa (1 kb) and 325°C. Yet, because experimental data are inadequate, we rely considerably on observations of *where* the minerals in question occur in nature as a basis for inferring *what* the conditions of formation were.[8]

Such observations have been made in a number of tectonic settings. For example, Alt et al. (1986) and Becker et al. (1989) discovered that Na-zeolite, laumontite, talc, and mixed-layer chlorite-smectite formed at shallow levels in the oceanic crust, representing conditions with lower limits of about 100°C and 0.02 Gpa (0.2 kb). Similar data are derived from sedimentary basins (C. E. Weaver, 1960, 1984, ch. 8; B. M. French, 1973; Hower et al., 1976) and the Salton Sea geothermal field (Muffler and White, 1969; Cho, Liou, and Bird, 1988). Together, experiments and observations all suggest lower limits of metamorphism of about 50–150°C and 0.02–0.15 Gpa (0.2–1.5 kb).

Diagenesis occurs over a range of low temperatures and low pressures (see chapter 16). Review of the P-T conditions of diagenesis reveals that they overlap with those of the lower limits of metamorphism. This is possible because of variations in rock and fluid phase compositions and in temperature. Diagenetic minerals can form at depths of burial that, given different bulk rock or fluid phase compositions or temperature, would yield metamorphic minerals. Consequently, the lower limit of metamorphism, as shown in figures 24.5a and 23.2, is a broad boundary zone.

The Upper Limit of Metamorphism

The upper limit of metamorphism is more diffuse than the lower limit, but for different reasons. Recall that the upper limit is marked by the beginning of melting. Experimental analyses of the conditions of beginning melting for various bulk rock compositions provide well-defined P-T values for that anatexis.[9] Yet, because the bulk rock chemistry has such a profound influence on the temperature of melting, the temperature range for the upper limit of metamorphism is 600–2000°C. The solidus for wet granite and the liquidus for dry dunite bracket that range of conditions. At a specific temperature between those bracketing values, a quartz-feldspar rock might melt entirely, a pelitic or amphibole schist might melt partially, and a basalt might recrystallize into an olivine-pyroxene-plagioclase granoblastite without melting at all.

In terms of pressure, upper mantle pressures of about 1.2–1.5 Gpa (12–15 kb) provide the upper limit *normally considered* in discussions of metamorphism (figure 24.5). (Considering that the mantle and core rocks are, in fact, metamorphic, the extreme pressure in the center of the core provides the actual upper limit.) As is the case with temperature, pressure-controlled melting is a function of bulk rock and fluid phase compositions.

The presence or absence of a fluid phase provides an important control on melting and hence on the upper limit of metamorphism. For example, the melting temperature of granitic rock may be depressed by as much as 450°C by the addition of a fluid.[10] Similarly, the presence of a fluid phase depresses the melting temperature of ultrabasic rocks by values of up to 450°C.[11]

Clearly, then, the upper limit, like the lower limit of metamorphism, cannot be defined as a single, fixed value. Rather, the limit may be specifically defined for particular bulk rock compositions at specified P-T conditions, and a fluid phase composition. Consequently, that boundary too, as shown on figure 24.5a, is a broad boundary zone.

Petrogenetic Grids, Geothermometry, and Geobarometry

Once the P-T limits of metamorphism are established, either in a general way for rocks of variable bulk composition or more specifically for rocks of particular bulk compositions, the petrogenetic grid is outlined and may be subdivided (figure 24.5a). It may also be expanded into the third dimension by adding an additional axis representing the composition of the fluid phase. In any case, subdivision is based on experimentally determined fields of mineral stability.

A number of mineral reaction curves are particularly important in examining the metamorphism of various terranes of metamorphic rock. For example, in terranes metamorphosed under conditions of static metamorphism, the reaction

$$NaAlSi_3O_8 \iff NaAlSi_2O_6 + SiO_2$$

albite $\qquad\qquad$ jadeite+ quartz $\quad$ (24.1)

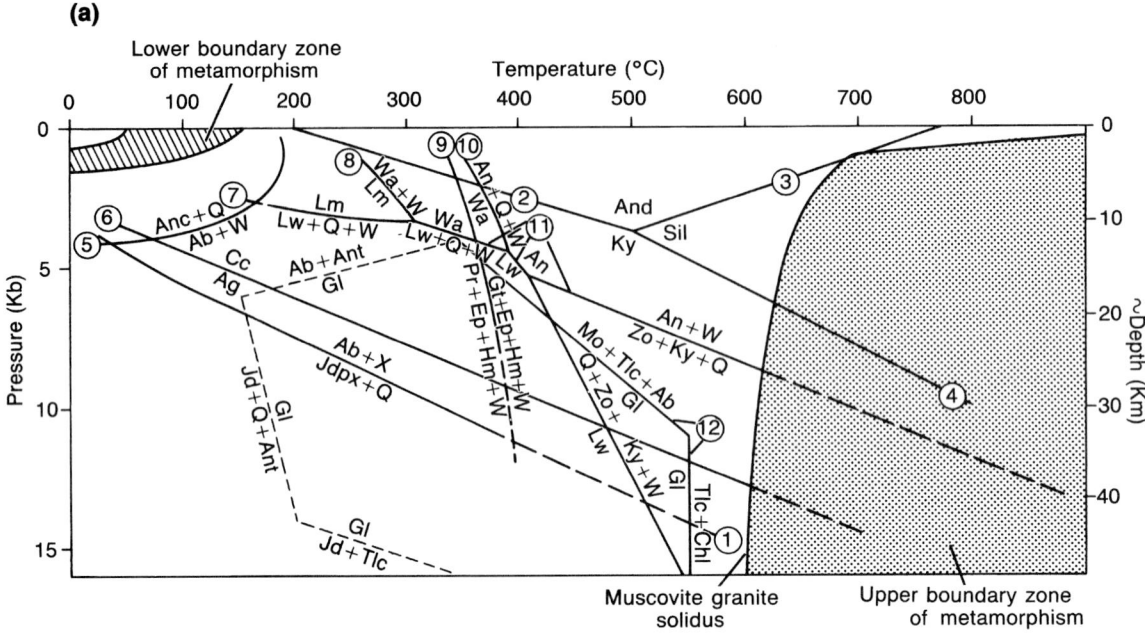

Figure 24.5   Partial petrogenetic grid for metaclastic rocks. (a) Grid showing the locations of various experimentally determined (solid lines) and calculated (short dashed lines) univariant mineral reaction curves. Long dashes show projected extensions. Curves represent the following reactions and are from the noted works. (1) Albite (Ab) + X (necessary elements) = Jadeitic pyroxene (Jdpx) + quartz (Q). Jdpx = $Jd_{82}Ac_{14}Di_4$ (Newton and Smith, 1967). (2) Andalusite (And) = kyanite (Ky) (Holdaway, 1971). (3) Andalusite = sillimanite (Sil) (Holdaway, 1971). (4) Kyanite = sillimanite (Holdaway, 1971). (5) Analcite (Anc) + quartz = albite + water (W) (Campbell and Fyfe, 1965). (6) Calcite (Cc) = aragonite (Ag) (Jamieson, 1953; Clark, 1957). (7) Laumontite (Lm) = lawsonite (Lw) + 2 quartz + 2 water (Liou, 1971a). (8) Laumontite = wairakite (Wa) + 2 water, and wairakite = lawsonite + 2 quartz (Liou, 1971b). (9) Prehnite (solid solution)(Pr) + epidote (solid solution)(Ep) + hematite-magnetite buffer (HM) + water = garnet (solid solution)(Gt) + epidote (solid solution) + HM + water

explains the appearance of jadeitic pyroxene in metawackes and marks the boundary between moderate and high pressure conditions. The curve for this reaction, which was studied by F. Birch and LeCompte (1961) and Newton and Smith (1967), is shown in figure 24.5b. If only this curve were depicted, the diagram would be a phase diagram showing that albite is stable at low pressures, whereas the combination of jadeite plus quartz is stable at high pressures. This phase diagram becomes a part of the petrogenetic grid.

The petrogenetic grid in figure 24.5a also shows curves representing the three reactions

$$\begin{array}{lcl} Al_2SiO_5 & \Longleftrightarrow & Al_2SiO_5 \\ \text{andalusite} & & \text{kyanite} \end{array} \quad (24.2)$$

$$\begin{array}{lcl} Al_2SiO_5 & \Longleftrightarrow & Al_2SiO_5 \\ \text{andalusite} & & \text{sillimanite} \end{array} \quad (24.3)$$

$$\begin{array}{lcl} Al_2SiO_5 & \Longleftrightarrow & Al_2SiO_5 \\ \text{kyanite} & & \text{sillimanite} \end{array} \quad (24.4)$$

These reactions are particularly important in pelitic schists (metamorphosed shales) and other aluminous rocks. Because of several peculiarities involving these apparently simple reactions, the exact positions of the reaction curves in the petrogenetic grid have been somewhat difficult to establish.[12] A reasonably

satisfying set of curves is now available and they meet at an invariant point that lies at 0.387 Gpa (3.87 kb) and 511°C (Hemingway et al., 1991). P-T space is divided into three distinct phase fields by the three reaction curves. At low pressures, andalusite is the stable aluminum silicate. At relatively low temperatures and relatively higher pressures, kyanite is stable. Sillimanite is the stable phase at high temperature.

The experimental studies on which petrogenetic grids are based and the petrogenetic grids themselves provide a framework for estimating *approximate* conditions of metamorphism for particular rocks. These P-T estimates are based on distinctive mineral assemblages known as *critical mineral assemblages* (discussed below). Because reaction curves are generally bivariant, assessments of the P-T-X conditions of metamorphism based on petrogenetic grids are often quite general. For example, if a rock found in a study area contained the assemblage

kyanite + jadeitic pyroxene + lawsonite + glaucophane + white mica + quartz

the P-T conditions of formation indicated on a petrogenetic grid such as that shown in figure 24.5a would be those shown in the shaded area in figure 24.5b. The area of stability for the rock, within the petrogenetic grid, is determined by finding that part of the grid in figure 24.5a in which *all* of the listed phases are stable. Albeit this limits the possible P-T conditions, the range of

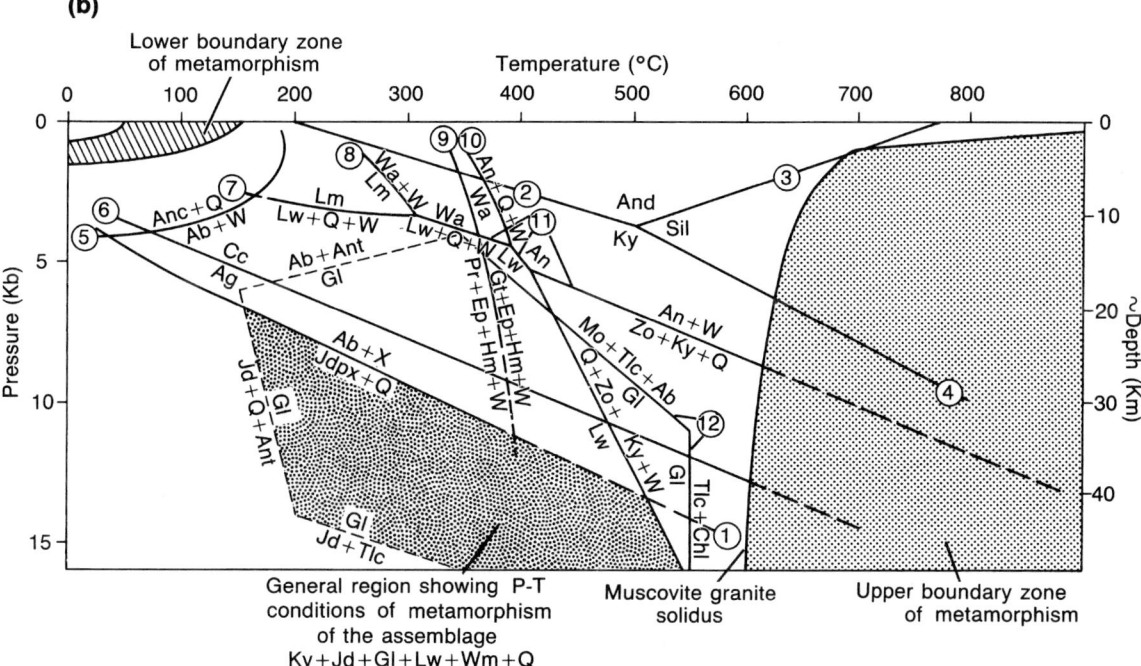

(b)

Figure 24.5 Continued

(Liou, Kim, and Marayama 1983). (10) Wairakite = anorthite (An) + quartz + water (Liou, 1970). (11) Lawsonite = anorthite + water, zoisite + kyanite + quartz = anorthite + water, and lawsonite = zoisite (Zo) + kyanite + quartz + water (Newton and Kennedy, 1963). (12) Upper stability limit of glaucophane; glaucophane (Gl) = montmorillonoid clay (Mo) + talc (Tlc)? + albite, and glaucophane = talc? + Chlorite (Chl) (Maresch, 1977). Lower stability limits of glaucophane (short dashes) calculated by Muir Wood (1980) Jd = jadeite; Ant = antigorite. b. Petrogenetic grid (P,T space identical to (a)) showing general region of stability for the assemblage Ky + Jdpx + glaucophane + lawsonite + white mica + quartz, shaded for emphasis. Note the wide range of conditions (*T* > 300°C and *P* > 10 kb) under which this six-phase assemblage will crystallize.

conditions under which the rock may have formed is still considerable (*T* = 150°–550°C, *P* = 0.6–2 Gpa, or 6–20 kb).

Because one of the goals of studying metamorphic terranes is to determine the *specific* P-T-X conditions of metamorphism, a method of further refining P, T, and X estimates is needed. The composition of the fluid phase (X_{fluid}) may be restricted by the mineralogy present in the rock. Yet, with regard to pressure and temperature, many minerals are stable over a wide range of conditions. Furthermore, where diagnostic (critical) mineral assemblages are absent, that is, where the phase assemblages in the rocks are stable over temperature ranges of more than 200°C and pressure ranges of several kilobars, petrogenetic grids provide little help in determining the exact P-T conditions. Quantitative studies used to determine more exact T and P conditions than those provided by the petrogenetic grid are referred to as **geothermometry** and **geobarometry**, respectively.[13] Obviously, what is needed for geothermometry and geobarometry are indicators of very specific conditions. Such specificity is realized via study of the P-T conditions associated with the particular chemistries of certain minerals that comprise mineral assemblages.

The assemblage biotite + garnet, which occurs over a wide range of conditions in a variety of metamorphic rocks, has been widely used as a geothermometer (A. B. Thompson, 1976a; Ferry and Spear, 1978; Indares and Martignole,

1985a,b).[14] The basis for the geothermometer is the idea that, if pressure is accounted for, the temperature of metamorphism will control the ratios of Fe to Mg in the coexisting mineral phases. The ratios will be different in the individual minerals. Thus, considering the endmember reaction

$$Fe_3Al_2Si_3O_{12} + KMg_3AlSi_3O_{10}(OH)_2 \Longleftrightarrow$$
$$\text{garnet} \quad + \quad \text{biotite}$$

$$Mg_3Al_2Si_3O_{12} + KFe_3AlSi_3O_{10}(OH)_2$$
$$\text{garnet} \quad + \quad \text{biotite}$$

the distribution of iron and magnesium in the actual minerals can be expressed by a constant distribution coefficient K_D, given by

$$K_D = \frac{(Fe/Mg)_{biotite}}{(Fe/Mg)_{garnet}}$$

Chemical analysis of the respective Fe and Mg values for garnets and biotites presumed to be in equilibrium in a rock are then used to determine the specific temperature, either via a thermodynamic calculation (using this equilibrium constant) or by a comparison with experimentally studied phases.

Other geothermometers have been developed for a range of rock compositions and conditions of metamorphism.[15] Thus, a wide variety of rocks may now be used for

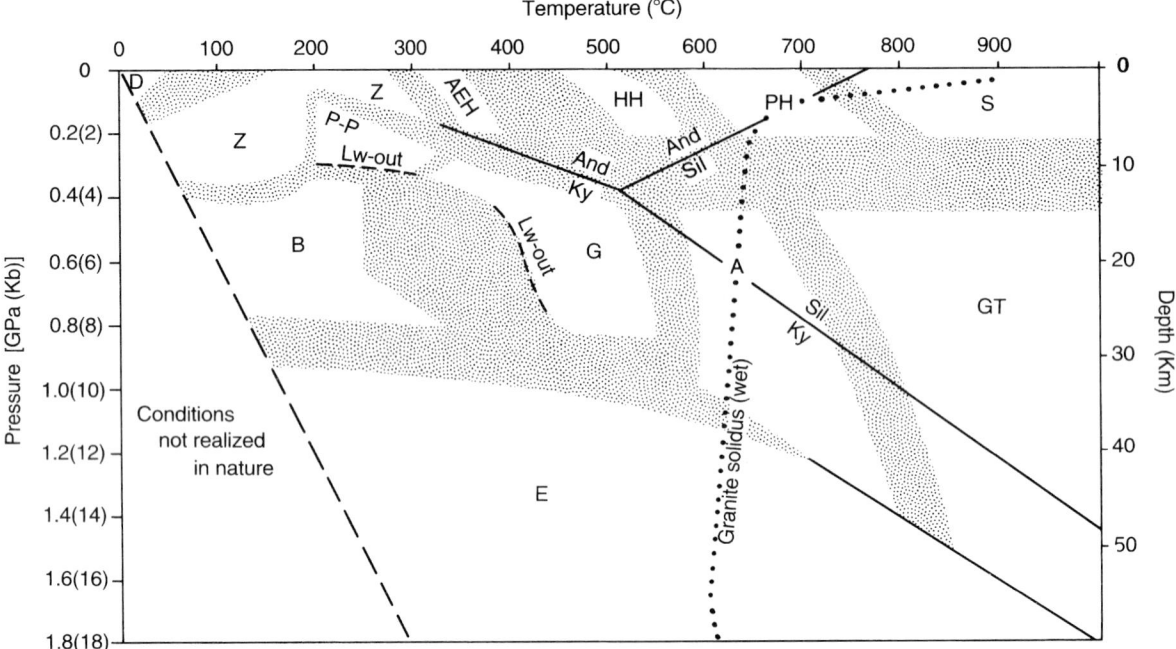

Figure 24.6 Metamorphic facies diagram, showing the general stability fields of the eleven metamorphic facies used in this text. Facies are as follows: AEH = albite-epidote hornfels, A = amphibolite, B = blueschist, E = eclogite, G = greenschist, GT = granulite, HH = hornblende hornfels, PP = prehnite-pumpellyite, PH = pyroxene hornfels, S = sanidinite, and Z = zeolite. D = zone of diagenesis. See appendix C for a compilation of specific reactions on which the boundaries are based. The facies names are those used by F. J. Turner (1981).

analyses of temperatures of metamorphism. In a like manner, several mineral assemblages may be used for geobarometry, to evaluate the pressures of metamorphism.[16]

THE FACIES CONCEPT

Facies and the Petrogenetic Grid

A **metamorphic facies** is defined as a set of rocks representing the full range of possible rock chemistries, with each rock characterized by an equilibrium assemblage of minerals that reflects a specific, but limited, range of metamorphic conditions (cf. Eskola, 1915, 1920, in Turner, 1958).[17] Again, the conditions of metamorphism and the chemistry of the rock control the mineralogy of each metamorphic rock. To illustrate this concept and clarify its meaning, consider the following subset of three rocks, all metamorphosed under the same conditions (3.5 kb, 300°C). The three rocks—a *marble*, a crystalline metamorphic rock composed of calcite, dolomite, and minor quartz; a *metabasite* (a metabasalt) composed of chlorite, albite, actinolite, epidote, and minor quartz; and a *pelitic schist* (metamorphosed shale) composed of quartz, chlorite, muscovite, and albite—are quite different mineralogically. Yet, because they were metamorphosed under the *same conditions*, their respective mineral assemblages each represent the same region on a petrogenetic grid. Similarly, rocks of any other composition metamorphosed

under these same conditions also exhibit mineral assemblages that represent the same area on the petrogenetic grid. Thus, although the mineral assemblages are not the same, each represents the same general conditions of metamorphism. All of these rocks belong to the same metamorphic facies.

Three rocks of the same chemical composition as the three listed earlier but metamorphosed under *different conditions* (e.g., 4 kb and 550°C) are found to have different minerals. Under these conditions, a metabasite would consist of andesine + hornblende ± quartz, the pelitic schist might consist of quartz + sillimanite + white micas + biotite + garnet + oligoclase, and the marble might contain calcite + dolomite + tremolite + talc. Again, the minerals in the three rocks are quite different, yet they represent the same general P-T conditions and, therefore, the same facies.

Eskola,[18] having noted these relationships between mineralogy, bulk chemistry, and metamorphic conditions, advanced the idea of the *metamorphic facies*. He emphasized that under a given set of conditions, the minerals of a metamorphic rock are solely a function of its chemistry. To each facies, Eskola assigned a name. Over the years, his facies scheme has been modified by many petrologists, in the interest of developing a more usable or more complete subdivision of P-T space.[19]

Here, P-T space, the petrogenetic grid, is divided into 11 regions, each of which corresponds to a facies (figure 24.6).[20] The low-P, high-T facies—the Albite-Epidote Hornfels Facies, Hornblende Hornfels Facies, Pyroxene

Table 24.3 Selected Rock Types Typical of Various Facies

Facies	Protoliths			
	Shale	Basalt	Dunite	Limestone
Zeolite	Zeolitic shale	Zeolitic greenstone	Lizardite serpentinite	Calcite marble-granoblastite
Prehnite-Pumpellyite	Chlorite phyllite	Pumpellyite greenstone	Lizardite serpentinite	Calcite marble-granoblastite
Blueschist	Glaucophane phyllite	Glaucophane schist	Lizardite serpentinite	Aragonite marble-granoblastite
Eclogite	Kyanite granoblastite	Eclogite	Dunite	Wollastonite granoblastite
Greenschist	Slate	Actinolite greenstone	Antigorite serpentinite	Calcite marble-granoblastite
Amphibolite	Kyanite-mica schist	Amphibolite	Anthophyllite-talc schist	Tremolite-calcite marble-granoblastite
Granulite	Sillimanite-orthoclase granoblastite	Garnet-pyroxene granoblastite	Dunite	Wollastonite-diopside marble-granoblastite
Albite-Epidote Hornfels	Chloritoid hornfels	Epidote hornfels	Antigorite diablastite	Calcite marble-granoblastite
Hornblende Hornfels	Andalusite hornfels	Hornblende diablastite	Talc dunite	Diopside marble-granoblastite
Pyroxene Hornfels	Cordierite hornfels	Augite hornfels	Dunite	Wollastonite-diopside marble granoblastite
Sanidinite	Cordierite-mullite hornfels	Augite-hypersthene hornfels	Dunite	Akermanite-spurrite marble-granoblastite

Hornfels Facies, and Sanidinite Facies—characterize LP-contact metamorphism. The Greenschist, Amphibolite, and Granulite Facies are typical of dynamothermal metamorphism. The Zeolite, Prehnite-Pumpellyite, and Blueschist Facies are found in regions of static metamorphism. Note that facies names are just that, names, they do not indicate minerals that are always present in every rock of the facies. Also note that the boundaries between the facies shown in figure 24.6 are diffuse. This is because, for each of the various chemical groups of rocks, specific reactions are used to define each boundary between two facies, but the locations of the individual reaction curves do not coincide. For example, the reaction marking the boundary between the Greenschist and Amphibolite facies for metabasites will not be exactly the same as that marking the same facies boundary for aluminous rocks (see appendix C).

With the petrogenetic grid subdivided into named areas, petrographic observations may be used to assign rocks to a specific facies. For example, the first subset of rocks described, metamorphosed under the lower-grade conditions, belongs to the Greenschist Facies. The higher-grade subset belongs to the Amphibolite Facies. Such assignments facilitate communication about specific metamorphic conditions. Every protolith yields a rock composed of a particular mineral assemblage that may be assigned to a facies. That assemblage will vary, depending on the facies, and will contrast with the assemblages indicative of other protoliths under similar or different conditions (table 24.3).

Critical Minerals

Observed minerals and mineral assemblages are useful for mapping and distinguishing rocks of one facies from those of another. Minerals and mineral assemblages for which the

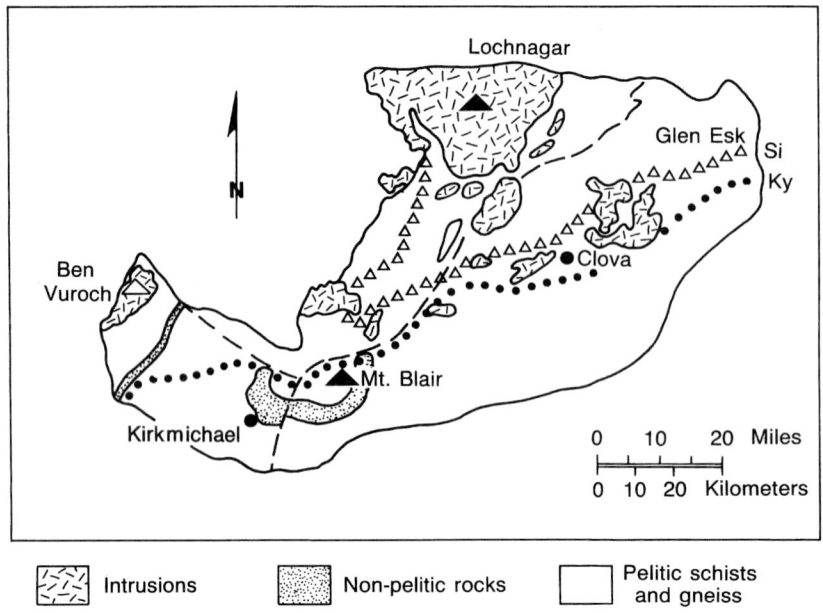

| Intrusions | Non-pelitic rocks | Pelitic schists and gneiss |

Figure 24.7 Generalized geologic and metamorphic map of a part of the Grampian Highlands of the type area of Barrovian metamorphism, Scottish Highlands. Ky (shown by the dotted line) is the kyanite isograd and marks the first appearance of kyanite in the pelitic rocks along south to north traverses. Si (shown by the line of triangles) marks the southwestern boundary of the regional zone of sillimanite-bearing rocks.

(From E. L. McLellan, "Metamorphic reactions in the kyanite and sillimanite zones of the Barrovian type area" in Journal of Petrology, *26:789–818, 1985. Copyright © 1985 Oxford University Press, Oxford England. Reprinted by permission.)*

reaction curves and stability fields are generally known and which may be used to distinguish between one facies or zone and another are called **critical minerals** and **critical mineral assemblages,** respectively. For example, in terranes containing quartz-mica schists, assemblages with the critical mineral andalusite are recognized as lower-pressure assemblages than are similar schists bearing the critical mineral kyanite. In practice, petrologists map the occurrences of such minerals or mineral assemblages in order to show changing conditions of metamorphism across a region (figure 24.7). In mapping or subsequent petrographic study, the locations of the first appearance of a critical mineral are marked by a line. This map line was originally considered to correspond to a reaction curve in the petrogenetic grid. Thus, the line marked "Si" in figure 24.7 would have been considered to correspond to a reaction curve such as curve 4 in figure 24.5a. Because each point along the map line was considered to represent the *same* reaction, and therefore the *same* P-T condition or "grade" of metamorphism, the line was called an **isograd** (*iso* = same, *grad* = grade)(Tilley, 1924).

Winkler (1974, 1976, ch. 7) rightly pointed out that, in fact, the first appearance of a mineral at different locations in an area may result from different reactions. For example, kyanite might form via the reaction

$$\text{Mg-chlorite} + \text{staurolite} + \text{quartz} + \text{muscovite} \longleftrightarrow \\ \text{kyanite} + \text{biotite} + \text{water}$$

or via the reaction

$$\text{Mg-chlorite} + \text{quartz} + \text{muscovite} \longleftrightarrow \text{kyanite} + \\ \text{biotite} + \text{water.}[21]$$

It is likely that these two reactions occur at somewhat different P-T conditions, and the absence of staurolite from the second reaction suggests a different bulk composition for the two parent rocks represented by the assemblages on the left. Thus, the first appearance of a mineral along an isograd may actually represent somewhat different P-T-X conditions from place to place. To account for this fact and to distinguish more well-defined isograds, Winkler proposed that isograds based on specific reactions be called *isoreaction-grads* (Winkler, 1976, p. 66). He later changed this term to *reaction iso-grads* (Winkler, 1979, p. 66).

The minerals (and mineral assemblages) most useful for mapping isograds—the critical minerals (or critical mineral assemblages)—are those that are distinctive and that have a limited range of conditions over which they are stable. Minerals like quartz and biotite, which are present in a wide variety of rocks metamorphosed over a wide range of conditions, have limited use in this regard. Nevertheless, the *first appearance* of biotite has been used as an isograd. Minerals such as andalusite, sillimanite, kyanite, jadeitic pyroxene, and laumontite and mineral assemblages such as lawsonite + jadeitic pyroxene + glaucophane, which have a more restricted range of stability, assume more importance in mapping metamorphic changes across a region.

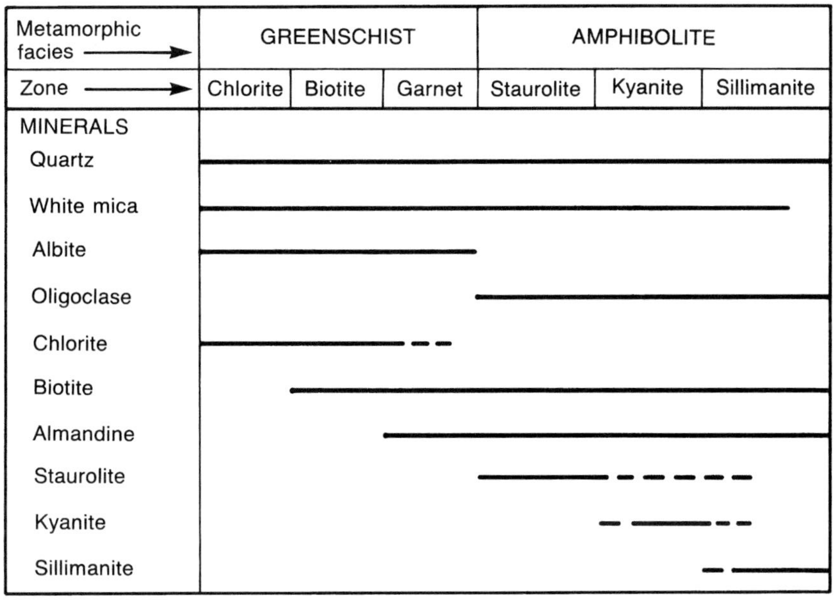

Metamorphic facies →	GREENSCHIST			AMPHIBOLITE		
Zone →	Chlorite	Biotite	Garnet	Staurolite	Kyanite	Sillimanite
MINERALS						
Quartz						
White mica						
Albite						
Oligoclase						
Chlorite						
Biotite						
Almandine						
Staurolite						
Kyanite						
Sillimanite						

Figure 24.8 Mineral-zone diagram showing the occurrence of various mineral phases in pelitic (aluminous) rocks across the zones of the classical Barrovian Facies Series in the Highlands of Scotland. Solid lines show the common occurrence of the phase; dashed lines indicate sporadic occurrences of the phase in rocks of restricted bulk composition.
(Based on data from Barrow, 1893; Harker, 1932; Chinner, 1960; 1978; Harte and Hudson, 1979; McLellan, 1985a, 1985b)

FACIES SERIES

Because early studies in metamorphism were done in northwestern Europe and the eastern United States, two areas that were actually linked as one prior to the plate tectonically driven separation of the continents,[22] the particular pattern of metamorphic zones present in the above areas was considered to be normal. That pattern of zones, each of which is bounded by isograds, was first discovered by Barrow (1893, 1912) in a study of the Scottish Highlands and was later described by Harker (1932).[23] From lower to higher grade, it consists of chlorite, biotite, and garnet zones (of the Greenschist Facies) followed by staurolite, kyanite, and sillimanite zones (of the Amphibolite Facies)(figures 24.7 and 24.8). These zones were recognized in pelitic rocks on the basis of the first appearance of the named critical minerals. The rocks types and major mineral assemblages of each zone are as follows.

Chlorite Zone (slates, phyllites, and schists)
    quartz + albite + white mica + chlorite
Biotite Zone (phyllites and schists)
    quartz + albite + white mica + chlorite + biotite
Almandine (Garnet) Zone (phyllites and schists)
    quartz + albite + white mica ± chlorite +
    biotite + garnet
Staurolite Zone (schists)
    quartz + oligoclase + white mica + biotite
    + garnet + staurolite
Kyanite Zone (schists)
    quartz + oligoclase + white mica + biotite + garnet ±
    staurolite + kyanite
Sillimanite Zone (schists, gneisses, and granoblastites)
    quartz + oligoclase ± K-rich alkali feldspar ± white mica
    + biotite ± kyanite + sillimanite

Notice that once a mineral appears at an isograd, it may persist through several zones. Chlorite, albite, staurolite, kyanite, and white mica eventually disappear. Notice also that the rock types change from slates and phyllites, which are fine-grained rocks, to schists and gneisses, which are coarser-grained rocks.

Miyashiro (1961), in a classic work on regional metamorphic belts, pointed out that different metamorphic belts (1) have different mineralogies, (2) are characterized by different isograds, and (3) have had different histories. These differences are reflected by the different metamorphic facies present in the belts. In metamorphic belts in which there were high temperatures (e.g., in northern New England, of North America), the series of facies present across the belt is Zeolite, Prehnite-Pumpellyite, Greenschist, Amphibolite (curve BFS, figure 24.9). Here, andalusite occurs (rather than kyanite), as the aluminum silicate of medium-temperature rocks. Cordierite, sillimanite, and biotite are also present.[24] In contrast, in the Franciscan Complex of California, the geothermal gradient was low (curve FFS, figure 24.9). The facies represented there are the Zeolite, Prehnite-Pumpellyite, and Blueschist Facies. Pelitic rocks contain minerals such as stilpnomelane, glaucophane, lawsonite, quartz, and white mica.[25] In summary, metamorphic belts in different places

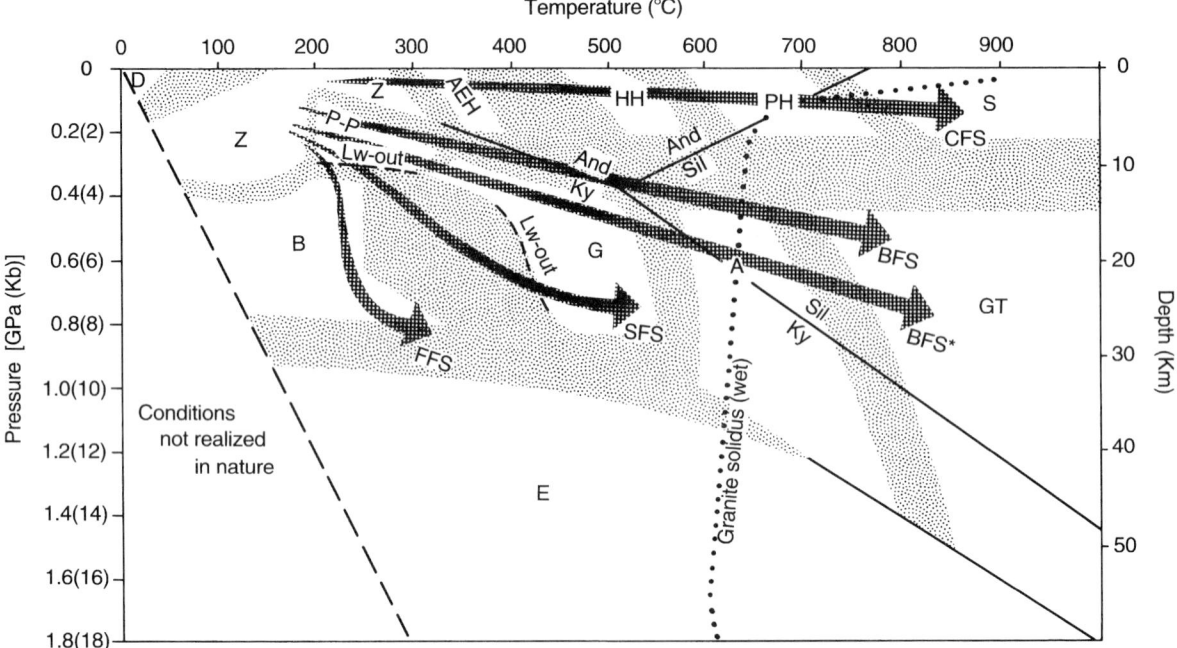

Figure 24.9 Metamorphic facies diagram showing geothermal gradient paths for the five facies series discussed in the text. CFS = Contact Facies Series, BFS = Buchan Facies Series, BFS* = Barrovian Facies Series, SFS = Sanbagawa Facies Series FFS = Franciscan Facies Series. Facies are as follows: AEH = albite-epidote hornfels, A = amphibolite, B = blueschist, E = eclogite, G = greenschist, GT = granulite, HH = hornblende hornfels, PP = prehnite-pumpellyite, PH = pyroxene hornfels, S = sanidinite, and Z = Zeolite. Note that the Blueschist Facies and the Greenschist Facies overlap in a wide zone at pressures above about 3.8 kb. Aluminum-silicate triple point and reaction curves are from Holdaway (1971; modified by Hemingway et al., 1991). The minimum melting curve (solidus) for muscovite granite is from Huang and Wyllie (1981).

were formed under different conditions and are characterized by different minerals and sequences of facies. Miyashiro (1961) called the progression of facies across a metamorphic belt a **metamorphic facies series.**

Five types of facies series were initially proposed by Miyashiro (1961). These were the andalusite-sillimanite type, the low-pressure intermediate group, the kyanite-sillimanite type, the high-pressure intermediate group, and the jadeite-glaucophane type. Later, Miyashiro (1973a, pp. 71–86) dropped the intermediate groups and adopted the names low-pressure baric type, medium-pressure baric type, and high-pressure baric type. In this book, following Miyashiro's original study (1961), five types are recognized.

1. A very low-pressure, andalusite-sillimanite type, represented by the rocks of contact metamorphic zones, is called the **Contact Facies Series.**
2. A low-pressure andalusite-sillimanite type, represented by the rocks of the Buchan area of northeastern Scotland, is designated the **Buchan Facies Series.**
3. A medium-pressure, high-temperature, kyanite-sillimanite type, represented by rocks of the Scottish Highland described by Barrow (1893), is named the **Barrovian Facies Series.**

4. A high-pressure, moderate-temperature type, represented by rocks of the Sanbagawa Metamorphic Belt of Japan, is here designated the **Sanbagawa Facies Series.**[26]
5. A high-pressure, very low temperature, jadeite-glaucophane type, represented by the rocks of the Franciscan Complex of California, is here named the **Franciscan Facies Series.**

Each of the series is characterized by a particular sequence of metamorphic facies (table 24.4). Just as facies are characterized by particular minerals, each facies series is represented by a specific sequence of mineral assemblages that occurs in low- to high-grade rocks. As is always the case, the mineral composition of any particular rock is a function of the bulk chemistry of that rock.

Each facies series represents a particular line through the petrogenetic grid and this line mimics a geothermal gradient (figure 24.9). Based on the experimentally determined stabilities of the phase assemblages present in the rocks, these *apparent geothermal gradients*[27] are approximately as follows:

Contact-type Facies Series	> 80°C/km
Buchan-type Facies Series	40–80°C/km
Barrovian-type Facies Series	20–40°C/km

Table 24.4 Facies and Critical Minerals in the Five Facies Series

Facies Series	Facies Included	Critical Minerals[1,2]
Contact	Zeolite → Albite-Epidote Hornfels → Hornblende Hornfels → Pyroxene Hornfels → Sanidinite	Analcite, *Wairakite*, Albite + epidote + chlorite + chloritoid, Hornblende + tremolite + diopside, Andalusite, Cordierite, Sillimanite, *Sanidine, Mullite, Monticellite, Tillyite, Spurrite*
Buchan	Zeolite → Prehnite-Pumpellyite → Greenschist → Amphibolite → Granulite	Analcite, Laumontite, Prehnite + pumpellyite, Albite + epidote + chlorite, *Andalusite + chloritoid, Cordierite* + garnet + biotite, Sillimanite + Orthoclase, *Hypersthene + orthoclase +* garnet + quartz + plagioclase
Barrovian	Zeolite → Prehnite-Pumpellyite → Greenschist → Amphibolite → Granulite	Analcite, Laumontite, Prehnite + pumpellyite, Epidote + albite + chlorite ± actinolite, Epidote + hornblende + garnet, *Kyanite*, Sillimanite, *Hypersthene + orthoclase* + garnet + quartz + plagioclase
Sanbagawa	Zeolite → Prehnite-Pumpellyite → Blueschist → Greenschist → Amphibolite	Analcite, Laumontite, *Heulandite*, Prehnite + pumpellyite, *Lawsonite* + glaucophane ± stilpnomelane, *Epidote + glaucophane* or crossite, Epidote + hornblende + garnet, Kyanite
Franciscan	Zeolite → Prehnite-pumpellyite → Blueschist → Eclogite	Analcite, *Laumontite, Heulandite*, Prehnite + pumpellyite, Ferrocarpholite, *Lawsonite* + glaucophane ± stilpnomelane, *Lawsonite* ± aragonite, *Jadeitic pyroxene, Omphacite + pyrope*

Sources: Zen (1960), Miyashiro (1961, 1973a), Hietenan (1967), Zwart (1969), Ernst et al. (1970), Chopin and Schreyer (1983), Liou, Maruyama, and Cho (1985) and observations of the author.

[1] Not all critical minerals will appear in every metamorphic belt.

[2] Particularly distinctive minerals of a facies series are italicized.

Sanbagawa-type Facies Series	10–20°C/km
Franciscan-type Facies Series	< 10°C/km

Facies series with high apparent geothermal gradients typify volcanically active regions. Those with low apparent geothermal gradients represent areas of subduction and rapid burial, where heat flow from the interior of the Earth is neither aided by movement of magmas nor enhanced by convection.

Complete facies series are rarely present in an orogenic belt. For example, in the Grampian region of the Scottish Highlands, the Zeolite Facies is apparently missing, the first sign of metamorphism being the development of phase as-semblages of the Greenschist Facies.[28] England and Thompson (1984) and D. Robinson (1987) suggest that, in many cases, the incompleteness of a facies series is a function of the tectonically controlled thermal conditions. For example, in extensional settings (e.g., back-arc spreading centers), high heat flow results from magmatic and associated hydrothermal activity at low pressures (i.e., there is a high geothermal gradient). However, spreading removes the rocks relatively rapidly from the region of high heat flow, restricting the complete development of a facies series (Robinson, 1987). In such regions, upper Amphibolite and Granulite

(a)

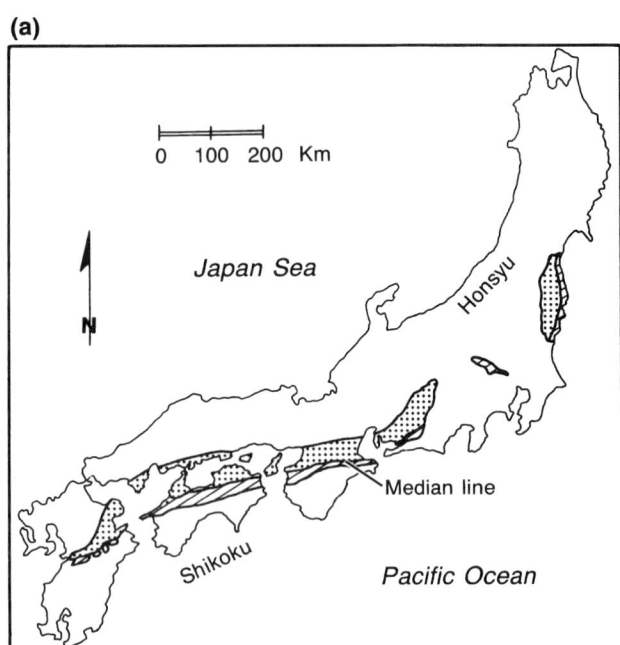

(b)

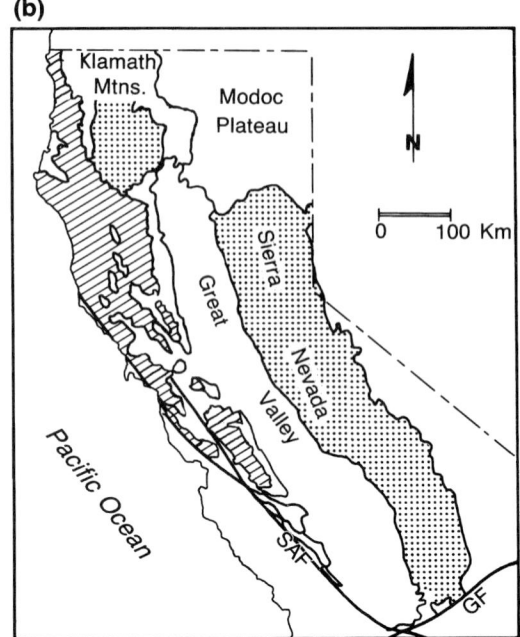

(c)

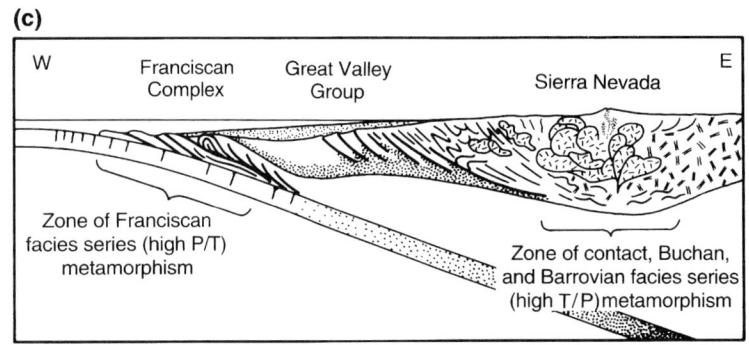

Figure 24.10   Schematic maps and a section showing paired metamorphic belts in two parts of the circum-Pacific region. a) Map showing Mesozoic low- and high-temperature belts in southeastern Japan. Diagonal lines represent Sanbagawa Metamorphic Belt (low to high *P*, low *T*) and stippled pattern marks location of Ryoke-Abukuma Metamorphic Belt (high *T*, low *P*). b) Map showing late Mesozoic metamorphic belts in California. Diagonal lines mark the extent of the low temperature, low to high pressure Franciscan belt. Stipples mark the inferred extent of the high-temperature belt (based on data from Durrell, 1940; Hietenan, 1951, 1973, 1976, 1977; L. D. Clark, 1954, 1970; Eric, Stromquist, and Swinney, 1955; E. H. Bailey, Irwin, and Jones, 1964; Creely, 1965; Davis et al., 1965; Bateman and Wahrhaftig, 1966; G. A. Davis, 1966; Kistler, 1966; Blake, Irwin, and Coleman, 1967, 1969; Lanphere, Irwin, and Hotz, 1968; Ernst, 1971b; Kistler et al., 1971; B. M. Page, 1981; Dickinson, 1981; Irwin, 1985; Mortimer, 1985; Donato, 1987; and observations by the author). c) Diagrammatic section across California for the late Mesozoic Period, showing paired metamorphic belts developing in association with an east-dipping subduction zone.

((a) From A. Miyashiro, "Evolution of metamorphic belts" in Journal of Petrology, *2:277-311, 1961. Copyright © 1961 Oxford University Press, Oxford England. Reprinted by permission.)*

Facies rocks are usually absent. Slow reaction rates, the effects of various fluid phases, or the absence of appropriate P-T conditions may also result in the absence of a particular facies from a series.

In the circum-Pacific region, mountain systems commonly exhibit pairs of metamorphic belts (figure 24.10) (Miyashiro, 1961). Miyashiro named such zones **paired metamorphic belts.** On the continent side of a pair, the facies series of an "inner metamorphic belt" is typically of the Buchan or Barrovian type. Within the inner belt, granitoid igneous rocks, reflecting *in situ* igneous activity, are common. The inner metamorphic belt represents a volcanic arc, plus its basement and its roots. The "outer metamorphic belt," on the ocean side of the pair, is characterized by facies series of the Sanbagawa or Franciscan type. Mafic and ultramafic rocks, including ophiolitic rocks, are the usual igneous rocks found in these outer metamorphic belts. In many cases, these igneous rocks formed elsewhere (e.g., at a mid-ocean ridge) and have been tectonically emplaced into the belt. The outer metamorphic belt represents an accretionary complex, a crustal expression of subduction. Together, the paired metamorphic belts mark the site and direction of subduction, with the high-T, lower-P belt marking the position of the volcanic arc, the low-T, high-P belt marking the subduction zone and plate boundary; and the pair indicating subduction in the direction from outer towards the inner metamorphic belt (figure 24.10).

SUMMARY

Metamorphic rocks may be grouped into six broad chemical groups—the silicic, siliceous-alkali-calcic, aluminous, carbonate/nonsilicate, basic, and ultrabasic groups—each of which encompasses the chemistry of various sedimentary and/or igneous protoliths. Mineralogically, each of the groups is distinctive. Because the chemistry of the protolith and the conditions of metamorphism control the equilibrium mineral assemblage, the observed assemblages can be used as indicators of the chemistry of the protolith.

Metamorphic rock classification is commonly founded on a textural subdivision of the rocks. Nevertheless, mineralogical and chemical classifications have been proposed. In this text, the classification adopted is based on texture and has two major categories—the crystalline textures, including foliated, weakly foliated, diablastic, and granoblastic types, and the clastic textures, including foliated and nonfoliated types.

One of the major goals of metamorphic petrology is to discover the general and specific conditions under which metamorphism has occurred. Laboratory analyses provide data for a composite phase diagram referred to as a petrogenetic grid. Petrogenetic grids are used for comparative purposes and to estimate the general conditions under which particular metamorphic mineral assemblages formed. Specific P-T conditions are assessed using geothermometry and geobarometry, analyses based on chemical variations within the individual minerals of a metamorphic mineral assemblage.

A set of all rocks, each with a distinctive mineral assemblage, formed under a particular set of conditions is called a metamorphic facies. Eleven metamorphic facies are recognized here. Linear groups of adjacent facies comprise metamorphic facies series and each facies series represents an apparent geothermal gradient. The Franciscan Facies Series represents the lowest apparent geothermal gradient, whereas these gradients are increasingly higher in the Sanbagawa, Barrovian, Buchan, and Contact Facies series. Orogenic belts, formed via subduction-related events, may exhibit paired metamorphic belts, with outer, high-pressure Franciscan or Sanbagawa facies series oceanward of inner, high-temperature Barrovian or Buchan facies series.

EXPLANATORY NOTES

1. Among those who have simply listed metamorphic rock names are Grout (1932), Miyashiro (1973a), and Ehlers and Blatt (1982).
2. Winkler (1979, p. 340ff.) adopts a quantitative mineralogical classification suggested by Austrian petrographers, whereas Bard (1986) adopts a chemical classification. Most other authors have used texture as a primary basis of classification (W. T. Huang, 1962; Spock, 1962; Best, 1982; D. L. Williams et al., 1982; and Raymond, 1984c) or have mixed texture, genesis, mineralogy, and/or chemistry in some combination (J. F. Kemp, 1929; Harker, 1932, 1939; Mason, 1978; F. J. Turner, 1981; Hyndman, 1985). Fry (1984) promotes the use of mineralogically based names first, then texturally based names where mineralogically based names are inadequate or inappropriate.
3. For example, F. J. Turner (1981) and Raymond (1984c).
4. For more information on weathering, see Reiche (1950), Ollier (1969b), and Ritter (1986, chs. 3,4). The stabilities of various oxides are discussed in texts such as Garrels and Christ (1965) and Krauskopf (1967, 1979). The common manganese oxide stains and veins in rocks are composed of minerals such as birnessite, romanechite, todorokite, and hollandite (R. M. Potter and Rossman, 1979).
5. For example, see reviews in R. L. Hay (1966) and Boles (1981) and see Alt et al. (1986).
6. This range of conditions is based on typical surface conditions, both on the continents and beneath the water in the ocean basins. Different petrologists suggest various criteria for defining the beginning of metamorphism. For example, B. M. French (1973) indicates that the appearance of minerals that consistently show a secondary textural relationship and are associated with axial plane cleavage and fractures marks the beginning of metamorphism. The definition of Mason (1978) is like that used in this text. Winkler (1979, p. 11) suggests that the beginning of metamorphism is marked by the first appearance of a mineral assemblage that cannot form in a sedimentary environment. One result of the use of these various criteria is that the P-T conditions at the *beginning* of metamorphism cited by various authors differ, ranging from those cited here up to temperatures of 250–350°C and pressures of 0.6 Gpa (6 kb) (see B. Bayly, 1968; B. M. French, 1973; Winkler, 1979; Weaver, 1984).
7. For example, see A. S. Campbell and Fyfe (1965), A. B. Thompson (1970b), Liou (1971a,b), Nitsch (1971),Velde (1973), Ivanov and Gurevich (1975), and Donahoe and Liou (1985). Also see the review of Zen and Thompson (1974).

8. Examples include those reported by Muffler and White (1969), Blake, Irwin, and Coleman (1967, 1969), Dickinson et al. (1969), Ernst et al. (1970), B. M. French (1973), Floran and Papike (1978), M. Frey (1978), S. D. Weaver et al. (1984), Alt et al. (1986), and Becker et al. (1989). Also see R. L. Hay (1966, 1977) and Surdam (1977).

9. Refer back to chapter 6 for information and references on anatexis. See Bowen and Anderson (1914), S. A. Morse (1980, p. 130), and Ohtani and Kumazawa (1981) for information on the melting of forsterite ($T = 1890°C$ at 1 bar to 2000°C at 20 kb).

10. Luth (1969), Huang and Wyllie (1975, 1981), C. R. Stern and Wyllie (1981).

11. For example, see D. H. Green (1973a), Kushiro (1973a), Eggler (1978), and Ribe (1985).

12. Richardson et al. (1969), Holdaway (1971), Robie and Hemingway (1984), Grambling and Williams (1985), Salje (1986), Kerrick (1990), Hemingway et al. (1991).

13. Discussions of geothermometry and geobarometry may be found in Essene (1982) and Newton (1983). For additional examples and discussions, see the papers cited in the text. D. M. Carmichael (1978) earlier proposed that pressure zones ("bathozones") be defined on the basis of mineral assemblages, the first appearance of which marks a surface called a *bathograd*. The development of geobarometry enhances the capability of metamorphic petrologists to define depth zones specifically, thereby making the delineation of bathograds more widely applicable.

14. Also see Goldman and Albee (1977), Baltatzis (1979), Ghent and Stout (1981), Schreurs (1985), and Hoisch (1990).

15. Geothermometers include such mineral assemblages as garnet-clinopyroxene (Banno, 1970; Raheim and Green, 1974; Ellis and Green, 1979; Krogh, 1988; T. H. Green and Adam, 1991), garnet-hornblende (P. R. A. Wells, 1979; C. M. Graham and Powell, 1984; Ghent and Stout, 1986), hornblende-plagioclase-bearing assemblages (Plyusnina, 1982; Blundy and Holland, 1990), garnet-staurolite-Al_2SiO_5-quartz-H_2O (Hodges and Spear, 1982), two feldspars (Barth, 1951; Stormer, 1975; Fuhrman and Lindsley, 1988; but see W. L. Brown and Parsons, 1985, who argue that this geothermometer is based on an erroneous assumption), plagioclase-muscovite- Al_2SiO_5-quartz-H_2O (Cheney and Guidotti, 1979), muscovite-biotite (Hoisch, 1989), and calcite-dolomite (R. Powell, Condliffe, and Condliffe, 1984). For other geothermometers and additional discussion, see Ghent (1976), Ghent and Stout (1981), N. L. Green and Usdansky (1986), Kawasaki (1987), Pownceby, Wall, and O'Neill (1987), Triboulet and Bassias (1988), Sen and Jones (1989), Grambling (1990), Spear et al. (1990), Berman and Koziol (1991), and Witt-Eickschen and Seck (1991).

16. Geobarometric analyses have been attempted on a range of mineral assemblages, including sphalerite-pyrrhotite-pyrite (Hutcheon, 1978), olivine-clinopyroxene (G. E. Adams and Bishop, 1986), garnet-rutile-ilmenite-plagioclase-quartz (Bohlen and Liotta, 1986), garnet-rutile-ilmenite-Al_2SiO_5-quartz (Bohlen, Wall, and Boettcher, 1983), garnet-plagioclase-biotite-muscovite (Ghent and Stout, 1981), and biotite-muscovite-chlorite-quartz (R. Powell and Evans, 1983). For additional discussions of geobarometry, see Anovitz and Essene (1987), Bucher-Nurminen (1987),

Holdaway, Dutrow, and Hinton, (1988), Koziol and Newton (1988), McKenna and Hodges (1988), Kohn and Spear (1989; 1990), Hoisch (1990), and Mukhopadhyay (1991).

17. Although the wording is different, this definition follows the original definitions of Eskola (1915; 1920, in Turner, 1958) in part. The definition embraces two stipulations, noted by Turner (1958, p. 18): (1) rocks of each facies form in response to the same physical conditions, (2) equilibrium is attained at the time of formation of the constituent mineral assemblage of each facies. Eskola (1920, 1939; in Turner, 1958) also stipulated that for each specific rock composition, *the same set of minerals* is produced by the physical conditions for each facies (i.e., each set of minerals in a bulk composition corresponds to *one* facies). This means that a basalt metamorphosed to the assemblage pumpelleyite + actinolite + chlorite + albite + quartz in one place, but to the assemblage epidote + actinolite + chlorite + albite + quartz in another, should be assigned to different facies in each of the two places. Some petrologists have moved toward this (Liou, Maruyama, and Cho, 1985). Petrographic studies show that numerous mineralogical assemblages exist in each bulk composition over the range of possible metamorphic conditions. Thus, following Eskola, the number of facies should be quite large, and it will increase as knowledge of mineralogical variability increases. Winkler (1974) correctly pointed out that, as a result, the number of facies (and "subfacies") has increased over time to the point that the facies concept is losing its utility. He chose to abandon the use of the concept (Winkler, 1974, 1976, 1979). As have others, I choose to lump like assemblages under single facies names.

18. The views of Eskola reported here are based on quotations in Turner (1958) and on the summaries by numerous authors, including Winkler (1979).

19. Published facies schemes vary from author to author. The reaction curves, on which facies boundaries are based, are selected on the basis of criteria each author thinks are important. Consequently, the facies boundaries of one author may lie in the middle of a facies field of another (e.g., compare the facies fields of F. J. Turner, 1981, p. 420, and those of Dobretsov and Sobolev, 1972, p. 199).

   As noted above, Winkler (1979, ch. 6) reasoned that because the facies concept has been made obsolete by the wealth of petrographic information now available, it should be abandoned. That information, he argues, shows that within areas like those designated here as individual facies on the P-T grid, there are, in fact, numerous groups of equilibrium mineral assemblages (i.e., numerous subsets of rocks), each of which represents a restricted range of P-T conditions and chemistry. Thus, *sensu stricto*, there are numerous metamorphic facies. For certain bulk compositions, this may be the case. For others it is not. However, I concur with Winkler (1974) in the belief that the proliferation of facies and subfacies titles is cumbersome. It inhibits and confuses communication because the facies names are established in the literature and reflect general sets of P-T conditions to most geologists. If the concept of metamorphic facies is to remain a useful tool for the metamorphic petrologist, Eskola's original idea (1920, in Turner, 1958) that each specific mineral assemblage must represent a new

514

facies must be abandoned in favor of the kind of broader definition adopted in this text, in which certain *similar assemblages* that represent a limited range of metamorphic conditions are lumped together under the title of a facies. I here retain the facies concept, as modified, because of its utility in designating areas on the petrogenetic grid and because its use is widespread. As Winkler (1979) did, I consider specific mineral assemblages within each facies to mark *zones* (e.g., the staurolite zone).

20. The facies names used here are those used by F. J. Turner (1981, pp. 202–210). F. J. Turner (1981, p. 202) and others apply the designation Eclogite Facies to rocks thought to have been recrystallized under conditions of metamorphism of high pressure and moderate to high temperature. The validity of the concept of an Eclogite Facies is discussed in chapter 29.

21. See J. B. Thompson and Norton (1968), D. M. Carmichael (1970), and Winkler (1979, p. 227).

22. See P. M. Hurley (1968), Harold Williams and Max (1980), and standard texts on tectonics (Condie, 1982; Windley, 1977, p. 148ff.) for discussions of this tectonic history and references to the important body of literature supporting this contention.

23. Additional works on this area are numerous. For more information, see Chinner (1960, 1966, 1978, 1980), Harte and Hudson (1979), McLellan (1985), and the works cited therein.

24. Miyashiro (1961, 1973a) emphasizes that andalusite characterizes low-pressure facies series, notes that sillimanite occurs at higher temperatures in the facies series, and indicates that cordierite is a common phase in both the higher-temperature parts of the andalusite zone and in the sillimanite zone.

25. Work on the pelitic rocks of the Franciscan Complex is limited. The Franciscan terrain and several other metamorphic belts developed in areas of low temperature are characterized by metasandstones (siliceous-alkali-calcic rocks) and metabasites (see Coombs, 1960; E. H. Bailey, Irwin, and Jones, 1964; McKee, 1962; Ernst, 1964, 1965, 1971b; Ernst et al., 1970; Brothers, 1974; Roeske, 1986). In the Franciscan Complex, siliceous rocks, carbonate rocks, ultramafic rocks, and pelitic rocks also occur, but in some low-T, high-P terranes, these rocks are characteristic (Seki, 1958; Ernst et al., 1970; Black, 1977; Liou, 1981a; Cloos, 1983; Matthews and Schliestedt, 1984; E. H. Brown, 1986; Mottana, 1986; Okay, 1986).

26. Miyashiro (1961) initially cited areas of the Sanbagawa Metamorphic Belt as examples of his jadeite-glaucophane-type facies series. However, the definitive assemblage jadeite + lawsonite + quartz apparently does not occur in Sanbagawa schists (Hirajima, 1983, in Banno, 1986). Instead, assemblages containing epidote ± Na-amphibole ± actinolite, which form at somewhat higher temperatures, occur (Ernst et al., 1970; Miyashiro, 1973a; Toriumi, 1975; E. H. Brown, 1977; Liou, Maruyama, and Cho, 1985; Maruyama, Cho, and Liou, 1986). Thus, the Sanbagawa terrane represents a high-pressure intermediate type of facies series, rather than the high-pressure (jadeite-glaucophane) type.

27. Continuous geothermal gradients did not exist in most mountain systems, the thermal histories of individual belts within the mountain range reflect the disparate individual tectonic histories of those belts. Consequently, the apparent geothermal gradients, in many cases, are an artifact of the tectonic history, rather than an actual record of the thermal history of the mountain system as a whole. Within individual belts, it may be possible to assess the actual geothermal gradient.

28. See Barrow (1893, 1912), Harker (1932), Chinner (1960, 1966), and Harte and Hudson (1979).

PROBLEMS

24.1. Assign each of the following rock groups to a facies series based on its mineral assemblages.
(a) Group A
Rock 1 Quartz-albite-laumontite-muscovite-chlorite
Rock 2 Quartz-albite-epidote-muscovite-chlorite
Rock 3 Quartz-plagioclase-muscovite-biotite-andalusite-staurolite
Rock 4 Quartz-plagioclase-alkali feldspar-biotite sillimanite-garnet
(b) Group B
Rock 1 Quartz-albite-laumontite-muscovite-chlorite
Rock 2 Quartz-albite-prehnite-muscovite-chlorite
Rock 3 Quartz-albite-muscovite-chlorite-stilpnomelane-lawsonite
Rock 4 Quartz-white mica-lawsonite-jadeitic pyroxene glaucophane
(c) Group C
Rock 1 Quartz-albite-wairakite-muscovite-chlorite
Rock 2 Quartz-albite-epidote-muscovite-chlorite
Rock 3 Quartz-plagioclase-muscovite-biotite-cordierite-garnet
Rock 4 Quartz-plagioclase-cordierite-orthopyroxene (hypersthene)

25

Metamorphic Phase Diagrams

INTRODUCTION

Phase diagrams are as useful for the study of metamorphic rocks as they are for the study of igneous rocks. The typical diagram used in metamorphic studies differs from that used in igneous studies, however, because melts are not present in most metamorphic rocks and are not included on the diagrams. Inasmuch as metamorphic rocks, by definition, remain solid as they undergo changes, metamorphic petrologists are concerned with changes of one mineral assemblage to another. The nature of these reactions, the progress of the reactions, and the controls on the reactions are important in understanding the metamorphic history. At the highest grades of metamorphism, where melting occurs, igneous and metamorphic processes overlap.

Recall that the Phase Rule is

$$P = C - F + 2$$

where P is the number of phases, F is the number of degrees of freedom, and C is the number of components. As a generalization, minerals are considered to be phases in natural metamorphic systems at equilibrium.[1] Because there are commonly at least two degrees of freedom (P and T), the number of phases (minerals) will be less than or equal to the number of components. In short, P = C. This relationship was first noted by Goldschmidt (1911, in Mason,

1978) and is referred to as **Goldschmidt's Mineralogical Phase Rule.**

Metamorphic rocks, like other rocks, may be characterized chemically by 10 or 11 major oxides. Substitutions (e.g., Mn for Fe, Ti for Al) reduce from 10 or 11 to 6 or 7, the number of components that must be considered. Thus, metamorphic rocks can be treated as if they have six or seven components and, following Goldschmidt's Mineralogical Phase Rule, metamorphic phase assemblages will consist of about six or seven phases. In plotting such metamorphic phase assemblages, one encounters the same difficulty as in plotting multiphase igneous assemblages. A maximum of only four coexisting phases may be plotted conveniently in three-dimensional space. To deal with this problem, in metamorphic petrology, systems are considered to be saturated in certain phases; that is, *certain phases are assumed to be present in all assemblages*. For example, in pelitic rocks quartz is always present,[2] so that SiO_2 is in excess and does not affect the number of degrees of freedom (F). Consequently, in plotting the phase assemblages, we can consider the system saturated in SiO_2 and we can delete one phase (quartz) and its corresponding component (SiO_2) from the diagram without affecting the system or the Phase Rule calculations. In general, three-phase *partial assemblages* are plotted on triangular phase diagrams, with the additional phases (those present in all assemblages) listed to the side of the triangle.

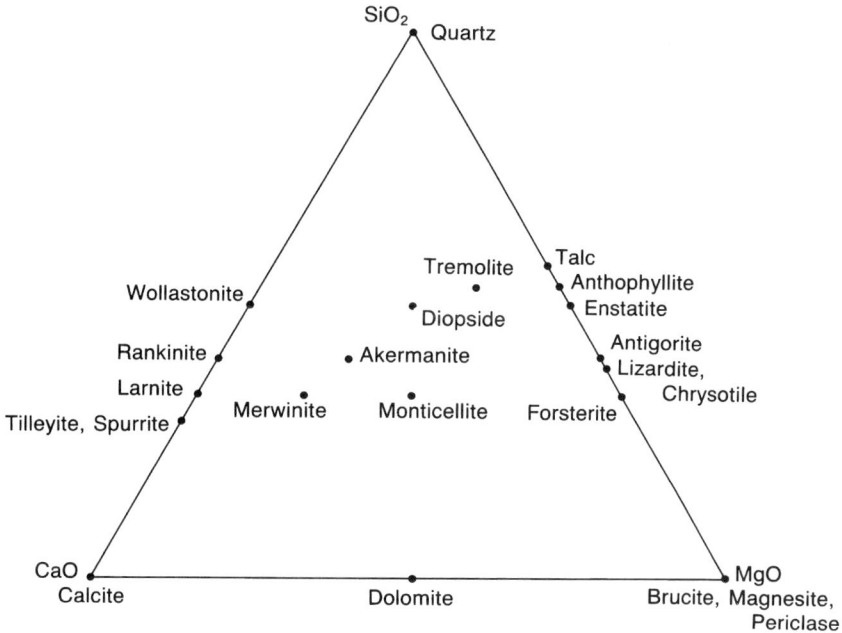

Figure 25.1 Diagram representing the system CaO-MgO-SiO_2-H_2O-CO_2. Here only the CaO-MgO-SiO_2 plane of the system is presented, with dots indicating the positions of minerals in the system that plot on or project onto the plane. Table 25.1 lists the chemical formulae for each of the phases shown.

(Based on Bowen, 1940, and B. W. Evans, 1977)

THE SYSTEM SiO_2-CaO-MgO-H_2O-CO_2 AS A MODEL

One of the easiest phase diagrams to understand is that for a simplified form of the system SiO_2-CaO-MgO-H_2O-CO_2.[3] In order to reduce the complexity of the initial analysis, we temporarily ignore the components of the fluid phase—water and carbon dioxide. This can be done *if* these components are externally controlled, that is, if they are **externally buffered** by a large influx of fluid from outside of the system (outside of the rock being considered in the analysis).[4] Ignoring the fluid phase allows one to plot the remaining three components at the corners of a composition triangle (figure 25.1). Each corner of the triangle represents 100 mole percent of the component at that corner. Points located on lines or within the triangle represent combinations of the components at the corners, the position of the point reflecting the molecular percentage of the various components. Reading the triangle is directly analogous to reading a triangular rock classification chart, except that instead of percentages of minerals, we read moles of components.

Rock, mineral, or chemical compositions may be plotted on triangular phase diagrams as molecular ratios. Thus, in figure 25.1, wollastonite, which has the formula $CaSiO_3$ (table 25.1) and consists of one mole of CaO and one mole of SiO_2, plots exactly one-half of the way between CaO and SiO_2. It contains no MgO and plots on the 0% MgO line. Similarly, diopside plots at a point above the center of the tri-

angle, as it consists of one mole of CaO, one mole of MgO, and two moles of SiO_2 (25% CaO, 25% MgO, 50% SiO_2). Dolomite is plotted exactly halfway between the CaO and MgO corners. It contains one mole of MgO per mole of CaO. Of course, dolomite also contains CO_2, but here one may consider that CO_2 is available in abundance; it is externally buffered and is available wherever it is needed to form a mineral. Actually, CO_2 can be plotted at the apex of a tetrahedron (figure 25.2). The position of dolomite, $CaMg(CO_3)_2$, is projected from the CO_2 apex onto the CaO-MgO-SiO_2 (CMS) plane of the tetrahedron, and lies at a point exactly halfway between CaO and MgO. Note that silica is not present in dolomite, with the result that dolomite is plotted on the 0% SiO_2 line.

To plot a rock analysis on the three-component CaO-MgO-SiO_2 (CMS) diagram, the number of moles (in the rock) of each of the three endmember components of the diagram is determined by dividing the three respective oxides in the chemical analysis by their corresponding molecular weights. For example, the weight percent of silica in the analysis is divided by 60.09 g/mol. Similarly, the lime and magnesia are divided by their molecular weights. The three values are then summed to 100% (moles SiO_2 + moles CaO + moles MgO = 100%) and mole percentages are calculated. The analysis is plotted by locating the point that corresponds to the molecular *percentages* of the three components. The procedure is identical to that used in plotting points on triangular variation diagrams in igneous petrology.

Table 25.1   Chemical Formulae of Minerals Plotted in the CMS System

Mineral	Formula	Mineral	Formula
Quartz	SiO_2	Wollastonite	$CaSiO_3$
Calcite	$CaCO_3$	Rankinite	$Ca_3Si_2O_7$
Dolomite	$(Ca,Mg)CO_3$	Larnite	Ca_2SiO_4
Magnesite	$MgCO_3$	Tillyite	$Ca_3Si_2O_7 \cdot 2CaCO_3$
Brucite	$Mg(OH)_2$	Spurrite	$2Ca_2SiO_4 \cdot CaCO_3$
Periclase	MgO	Merwinite	$Ca_3Mg[SiO_4]_2$
Forsterite	Mg_2SiO_4	Akermanite	$Ca_2MgSi_2O_7$
Lizardite	$Mg_3Si_2O_5(OH)_4$	Diopside	$CaMgSi_2O_6$
Chrysotile	$Mg_3Si_2O_5(OH)_4$	Tremolite	$Ca_2Mg_5Si_8O_{22}(OH)_2$
Antigorite	$Mg_3Si_2O_5(OH)_4$	Anthophyllite	$Mg_7Si_8O_{22}(OH)_2$
Talc	$Mg_6Si_8O_{20}(OH)_4$	Enstatite	$MgSiO_3$

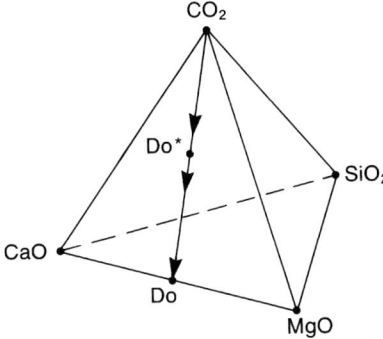

Figure 25.2   Sketch of the system CaO-MgO-SiO_2-CO_2. The point marked Do* is a plot of the composition of dolomite. For consideration in the CMS diagram, Do* is projected to the point marked Do on the basal plane of the system CaO-MgO-SiO_2-CO_2.

The minerals commonly found in the rocks that plot in the system CaO-MgO-SiO_2-H_2O-CO_2 are shown in figure 25.1 and their formulae are listed in table 25.1. Minerals such as tremolite, talc, and dolomite do not plot directly on the CMS plane of the diagram because they contain H_2O or CO_2. Therefore, they are *projected* onto the CMS plane (as was demonstrated for dolomite in figure 25.2). The listed minerals are those that are common in carbonate rocks (marbles) and in metamorphosed ultramafic rocks (e.g., metadunite, metaperidotite), the rock types for which this diagram is used.

Mineral Assemblages, Reactions, and Facies

Equilibrium mineral assemblages representing a facies or subfacies may be plotted on the triangular phase diagram. The stable coexistence of any mineral pair is indicated on

the diagram by a *tie line*, the line connecting two stable phases. Each three-phase assemblage is plotted as a triangle, the corners of which are connected by tie lines. Because the diagram has only three corners (three components), only three phases from any one assemblage can be plotted as an equilibrium assemblage.[5] Each point within the diagram represents a system.

A stable assemblage in limestones and diagenetically altered limestones (at low P-T conditions), for example, is calcite + dolomite + quartz. This assemblage is shown by the three corners of the triangle (Qz, Cc, Do) on the phase diagram in figure 25.3. The bulk composition of a representative rock is shown by a star (this is the system under consideration). The mineral assemblage calcite + dolomite + quartz represents a stable, three-phase assemblage for the divariant field of the petrogenetic grid in which the phase diagram is plotted. That this field is divariant is evident from application of the Phase Rule ($F = C - P + 2 \Rightarrow F = 3 - 3 + 2 \Rightarrow F = 2$).

The univariant curve in the petrogenetic grid represents the reaction

3 dolomite + 4 quartz + H_2O = talc + 3 calcite + 3 CO_2

Dolomite and quartz are stable together to the left of the curve. Under the higher-temperature conditions on the right, talc and calcite are stable together. Note that in the phase diagram to the right of the curve, dolomite and quartz are no longer connected by a tie line, indicating that they are no longer stable together. Note also that the triangular phase diagram is subdivided into smaller triangles, the sides of which are tie lines connecting stable phases. The small triangles indicate stable three-phase assemblages (quartz + calcite + talc, calcite + dolomite + talc, dolomite + magnesite + talc) in the Amphibolite Facies.

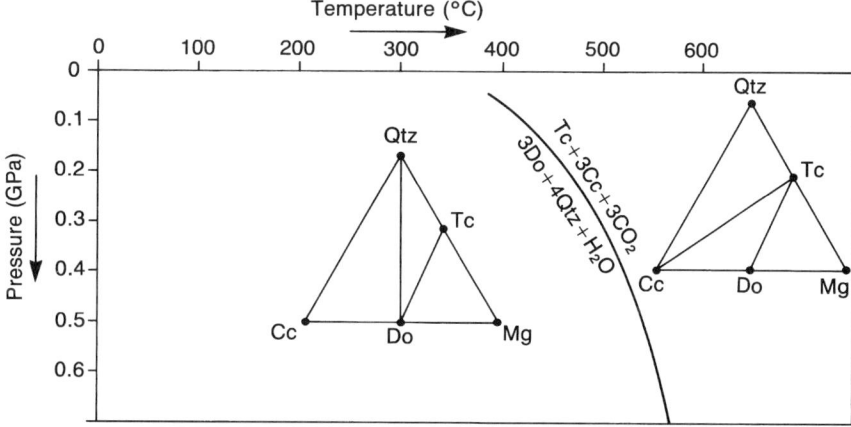

Figure 25.3 Part of the petrogenetic grid for carbonate rocks showing the *nonterminal* reaction curve and phase diagrams for the reaction 3 dolomite + 4 quartz + H$_2$O = talc + 3 calcite + 3 CO$_2$ (see appendix C for the complete grid). The reaction curve here, studied by T. M. Gordon and Greenwood (1970), Metz and Puhan (1971), and Eggert and Kerrick (1981), is located on the basis of the work of Metz and Puhan (1971). The star represents one possible bulk composition for which the reaction shown would result in a nonterminal reaction.

Reactions and the Use of Triangular Phase Diagrams

Three-phase assemblages represented by the smaller triangles within a triangular phase diagram, in general, represent mineral assemblages stable over some range of P-T-X conditions. That range is represented by the area between reaction curves in the petrogenetic grid. If the assemblage changes as a result of a change in pressure, temperature, or composition, the change will be represented in one of two ways, as a terminal or a nonterminal reaction. In **terminal reactions,** one or more phases cease to be possible on one or the other side of the reaction. In such cases, minerals appear or disappear from the phase diagram. In **nonterminal reactions,** some mineral pairs become unstable, while others become stable. On the phase diagram, there is a *tie line flip,* but mineral phases neither appear nor disappear from the diagram.

The reaction examined above (see figure 25.3) is representative of a nonterminal reaction. The first phase diagram shows that quartz, calcite, and dolomite are stable together. They are connected by tie lines. The second phase diagram shows that—for the bulk composition shown—calcite, dolomite, and talc are stable together. As a result of the reaction, the tie line connecting dolomite and quartz is replaced by one connecting calcite and talc. Because one of the rules of using triangular phase diagrams is that under equilibrium conditions *tie lines do not cross,* the dolomite-quartz line is removed and replaced by the calcite-talc tie line. In considering the change from one phase diagram to the other, it appears that the tie line has "flipped" in position. (Note also that the two solid phases on the left side of the reaction are the two that are connected by a tie line and are stable together at low T, whereas the two solid phases on the right side of the equation are the two that are connected by a tie line and are stable together at higher T.)

As may be induced from the above discussion, in working with triangular phase diagrams, two general rules are maintained. First, the phase diagrams must be subdivided into three-phase fields (i.e., divided into triangular subareas). Second, tie lines should not cross. If they do cross, the assemblage involved represents (1) a disequilibrium assemblage, (2) an equilibrium assemblage stable over a range of P-T conditions, (3) a condition resulting from projection of phases onto a plane (in which case the tie lines only appear to cross), or (4) an assemblage in which one or more phases partition the components used to define the diagram (F. J. Turner, 1968, p. 178).

As noted, terminal reactions involve the appearance or disappearance of a phase. In diagramming such reactions, tie lines are added or subtracted to maintain triangular subdivision of the interior of the phase diagram, but no tie line flip occurs. As an example, consider the reaction

calcite + quartz = wollastonite + CO$_2$

In this reaction, two phases connected by a tie line (calcite and quartz) react to form a new phase that plots *between* them on the same line (figure 25.4). Note that to retain a triangular subdivision within the phase diagram, an additional tie line must be added to the diagram after wollastonite is plotted. Terminal reactions involving the appearance or disappearance of a phase *within* one of the small triangles of the phase diagram can also occur. Both terminal and nonterminal reactions that involve univariant curves are **discontinuous reactions.**

Triangular phase diagrams containing solid solution phases may contain assemblages in which only two phases, at least one of which is a solid solution phase, coexist stably. In these cases, the systems are trivariant ($F = C–P + 2 => F = 3 – 2 + 2 => F = 3$). Such is the case in the AFM diagrams discussed

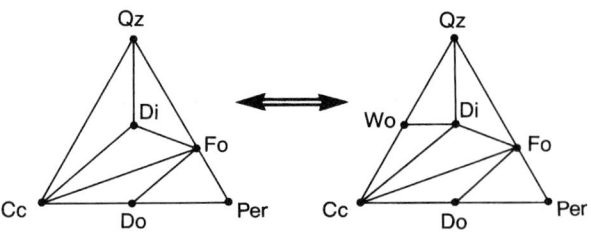

Figure 25.4 CMS phase diagrams for the *terminal* reaction quartz + calcite = wollastonite + CO_2. Note that the only changes are the addition of Wo and a tie line in the triangle on the right. Cc = calcite, Di = diopside, Do = dolomite, Fo = forsterite, Mg = magnesite, Qz = quartz, and Wo = wollastonite.
(After Bowen, 1940)

below, where two-phase assemblages are represented by groups of tie lines rather than by a single line and the solid solution phases are shown as areas or bars rather than points (figure 25.5). Both terminal and nonterminal reactions, including discontinuous reactions, may be depicted on such diagrams. However, some reactions involving solid solutions are continuous reactions. To illustrate such a reaction, recall that iron and magnesium may be partitioned between phases such as garnet and biotite (see chapter 24). During metamorphism, as the temperature changes, the ratios of iron to magnesium in the two phases change; that is, the compositions of the garnet and biotite vary. As a consequence, the position, but not the general configuration, of the tie lines will vary as the reactants and products change composition (see figure 25.5). Reactions such as this, in which the compositions of coexisting phases change gradually over the course of the reaction, are called **continuous reactions.**

Continuous reactions and metamorphic rock histories involving a number of solid solution phases stable over a wide range of conditions may be examined using the concept of **reaction space** (J. B. Thompson, 1982a, 1982b, 1991; Schneiderman, 1990).[6] This method of analysis involves defining, for a given bulk composition, an imaginary volume, the axes of which are reactions and the faces of which are the mineralogical limits beyond which a reaction cannot proceed (figure 25.6). The limits result from the depletion of a reacting phase or some similar chemical control. By examining the chemical data, minerals, and textures of a rock, it is possible to define one *specific reaction path* within this volume that was followed by the rock during metamorphism. Quantification of the progress of the reactions involved, including changing modal mineralogy, chemistry, or fluid volume during metamorphism, is possible using a function called the *reaction progress variable* (Ferry, 1983a,b; Ridley, 1986).[7]

Facies and Phase Diagrams

Triangular phase diagrams can be used to depict stable mineral assemblages representing a facies. Observation of the rocks from a region *covering a wide range of compositions*

reveals a set of mineral assemblages, each corresponding to one of the bulk rock compositions. Using these mineral assemblages, a complete phase diagram or a set of diagrams can be constructed that will show the possible, stable, three-phase assemblages for the region in which the rocks occur. Particular metamorphic grades (i.e., zones, subfacies, and facies) are represented by the specific arrangement of tie lines that plot on each diagram.[8] Different assemblages in rocks of the same bulk composition from adjoining regions represent different conditions of metamorphism and hence, different facies, subfacies, zones, or grades of metamorphism.

A series of triangular phase diagrams can be used to depict the mineralogical and, therefore, the facies changes in a metamorphic belt. Such a series of diagrams may show graphically the facies changes over an entire facies series. Recall that the P-T conditions indicated by facies series reflect the tectonic setting and may reflect the geothermal gradient.

THE ACF DIAGRAM

The ACF diagram (figure 25.7) is a triangular diagram used to plot mineral assemblages in metabasites and impure carbonate rocks (Eskola, 1939, in Winkler, 1976). It is not a true phase diagram in a Phase Rule sense; rather, it is a pseudophase diagram, because the diagram combines components, notably FeO and MgO, that are not completely interchangeable (Guidotti, 1982).[9] Nevertheless, for some purposes, the ACF diagram can be and is used as a phase diagram. Each of the three corners of the diagram represents moles of a particular major element oxide or group of oxides. The oxides SiO_2, Al_2O_3, Fe_2O_3, FeO, MnO, MgO, and CaO are accounted for in the diagram, but in order to plot seven oxides on a three-component diagram, adjustments must be made. First, all assemblages are assumed to be saturated with silica, that is, quartz must be present in all assemblages for which the diagram is used and it is not plotted. Next, Fe^{2+}, Mg, and Mn are assumed to substitute freely for one another, because they occupy the same sites in mineral structures.[10] The F-corner represents moles of the oxides of these three cations. Similarly, Al and Fe^{3+} are assumed to substitute freely for one another and they are lumped together. The sum $Na_2O + K_2O$ is subtracted from the sum $Al_2O_3 + Fe_2O_3$, as the former combine with Al to make feldspar (this manipulation allows the diagram to function as a projection from feldspar). The molecular value remaining after this subtraction is plotted at the A corner. The C corner represents the moles of CaO. In summary, the three corners of the diagram are represented by the molecular values

$$A = Al_2O_3 + Fe_2O_3 - (Na_2O + K_2O)$$

$$C = CaO$$

$$F = FeO + MgO + MnO$$

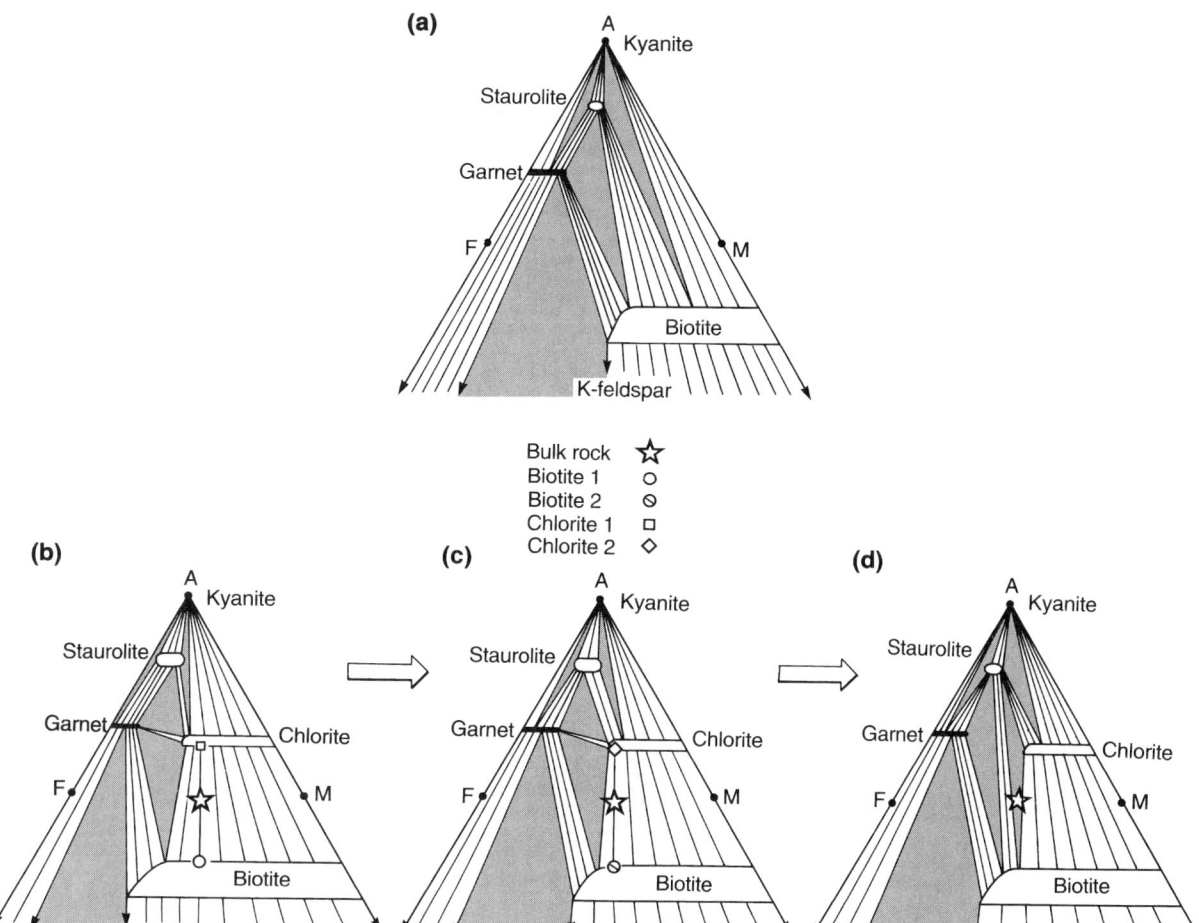

Figure 25.5 Phase diagrams showing two-phase assemblages plotted on a triangular diagram and tie line shifts due to continuous reactions. (a) AFM diagram showing areas in which two, rather than three phases form assemblages plotted on the diagram. The groups of subparallel tie lines show two-phase fields. The ends of each tie line connect a mineral of one particular chemistry (within the range of compositions possible for that phase) with another mineral of a particular chemistry (within the range of compositions possible for the second phase). The shaded areas represent three-phase fields. (b)–(d) Hypothetical AFM diagrams showing a change in chemistry of two phases during a continuous reaction [(b)→(c)] and a subsequent change from a two-phase assemblage on the diagram to a three-phase assemblage on the diagram [(c)→(d)]. The bulk composition of the rock is shown as a star. During the continuous reaction, the two phases, chlorite 1 and biotite 1, initially in equilibrium and connected by a tie line, change composition to become chlorite 2 and biotite 2, connected by a different tie line. Continued metamorphism [(c)→(d)] causes a shift in tie lines that is large enough that the bulk composition, which at lower grades was within a two-phase region, is encompassed by a three-phase region.
(Based in part on J. B. Thompson, 1957; Guidotti, 1974; and Abbott, 1979b).

Figure 25.6 Schematic reaction-space polyhedron showing the direction of a reaction during the metamorphism of a pelitic rock . The axes, labelled A, C, and D, represent particular reactions (not shown). The arrow represents the reaction path followed by the rock, which is a path away from phases that are increasing (B, K, S, M) and towards those being depleted (C, G, P). B = biotite, C = chlorite, G = garnet, K = kyanite, M = muscovite, P = plagioclase, and S = staurolite.

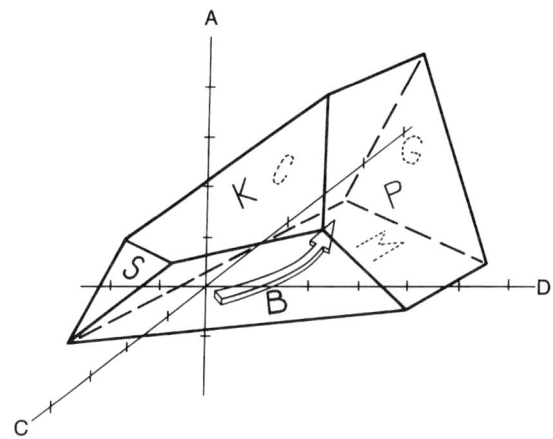

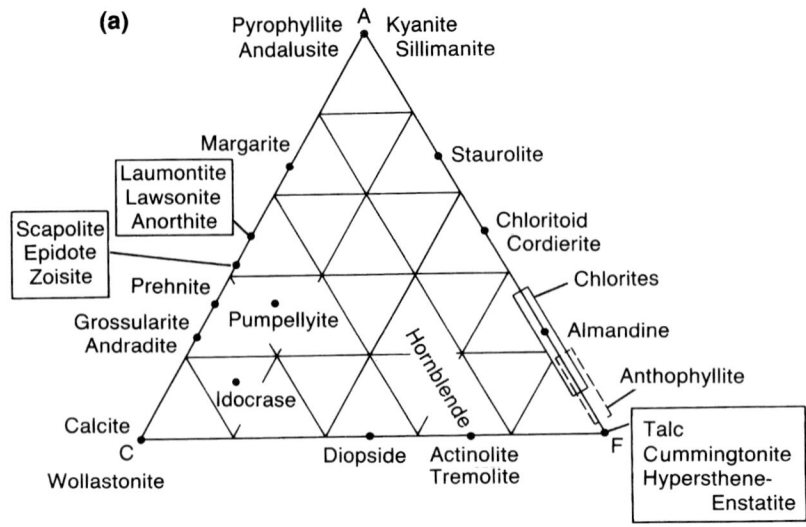

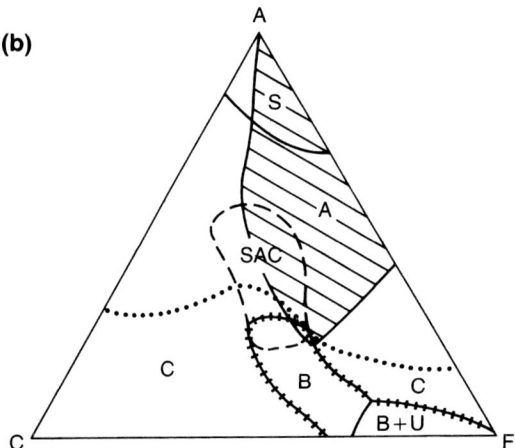

Figure 25.7 ACF diagram. (a) Diagram showing composition plots of minerals that occur in the ACF compositional triangle (from Winkler, 1979). (b) ACF diagram showing the approximate compositional limits of the six chemical groups of metamorphic rocks defined in chapter 24. A = aluminous rocks, B = basic rocks, B + U = basic and ultrabasic rocks, C = carbonate rocks and other chemical precipitates, S = siliceous rocks, and SAC = siliceous alkali-calcic rocks.

Before plotting data on the diagram, these values must be normalized by making A + C + F = 100%. To plot the analysis of a rock, further adjustments must be made to the analysis to account for minor or additional minerals that may appear in the assemblage and affect the plot, but that cannot be plotted on the diagram.[11]

Some typical minerals that plot on the ACF diagram are shown in figure 25.7a. Minerals that have a substantial range of composition (solid solution phases) plot as areas rather than points. Figure 25.7b shows where the bulk rock chemistries of various common rock types will plot. From the positions of the bulk rock chemistries, one can easily determine which minerals might be expected in the various metamorphic rocks. For example, calcite, dolomite, wollastonite, anorthite, and grossular are among the minerals expected in metamorphosed limestones (carbonate rocks). In contrast, metabasites might contain diopside, hornblende, actinolite, glaucophane, hypersthene, olivine, epidote, laumontite, pumpellyite, or anorthite, whereas andalusite, kyanite, sillimanite, cordierite, and anorthite are minerals that might appear in an aluminous rock such as a metapelite.

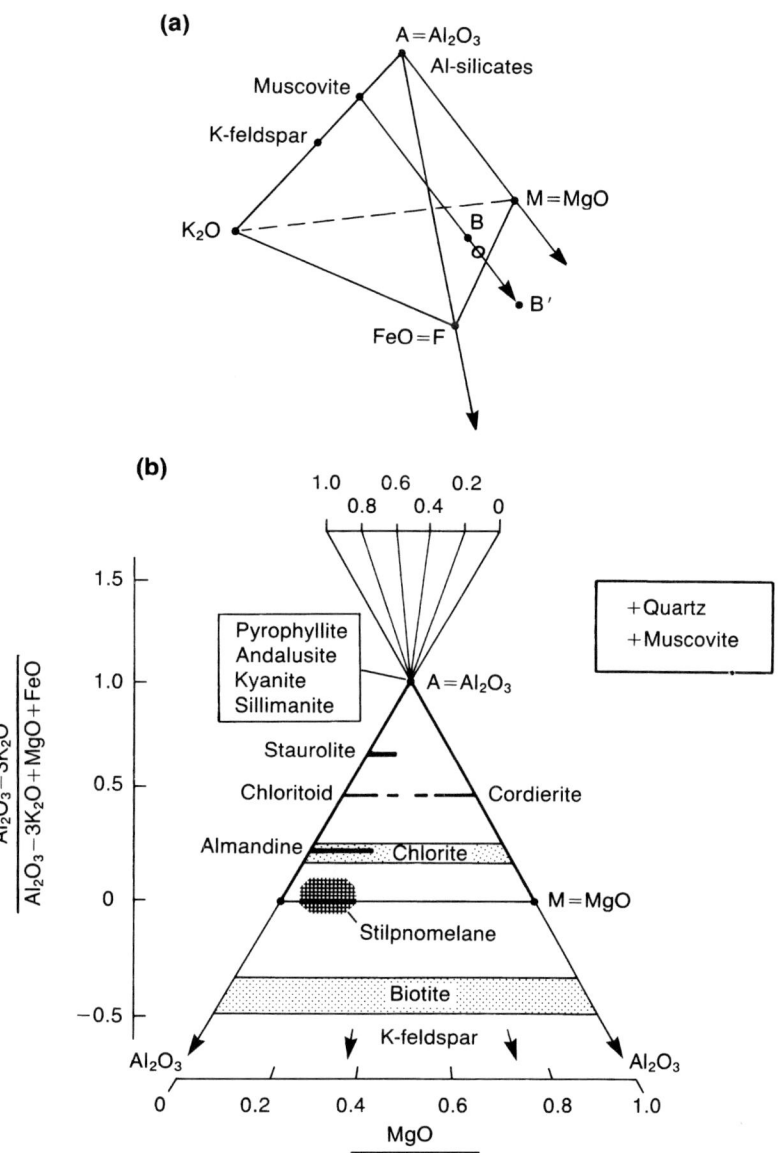

Figure 25.8 AFM diagram. (a) AKFM tetrahedron showing positions of muscovite and K-feldspar. Mineral positions on the AFM plane are projected from muscovite (+ quartz). For example, a biotite of composition B that plots within the tetrahedron is projected through the side (circled point of intersection) onto the AFM plane at B'. Minerals lacking K_2O plot directly on the AFM plane. (b) The AFM plane and its extension showing the compositional points, bars, or fields of minerals that plot in the AFM diagram. Scales for plotting molecular ratios of compositional data are shown at the top, bottom, and left side of the diagram.
(Modified from J. B. Thompson, 1957)

THE AFM DIAGRAM

The AFM diagram is a true phase diagram. Developed by J. B. Thompson (1957), it is used to depict the mineral assemblages present in metapelites, some metasandstones, metaigneous rocks of generally similar chemistry, and any other rock containing both quartz and muscovite.[12] In this diagram, FeO and MgO are treated as separate components (as they should be). The diagram is derived from a four-component plot, the AKFM tetrahedron (figure 25.8a).

Six major element components are accounted for in the system SiO_2-Al_2O_3-FeO-MgO-K_2O-H_2O, which serves as the foundation for the AFM diagram. Adjustments may be made for the additional components TiO_2, Fe_2O_3, Na_2O, and P_2O_5.[13] The diagram is used only for systems saturated in silica; i.e. quartz is always present and need not be depicted other than to be listed adjacent to the diagram. Similarly, H_2O is considered to be available where needed, either because the system will be saturated with it or because the external environment can provide adequate water, if it is needed to form a particular phase. Adjustments made to account for the presence of the "additional" components listed leaves only the four remaining components—Al_2O_3, FeO, MgO, and K_2O—to be plotted.

Any aluminous or ferromagnesian mineral or rock composition may be plotted in the AFM diagram. Mineral or rock compositions that lack K_2O (e.g., garnet) plot directly on the basal AFM plane. Projecting from the composition of muscovite (point M, figure 25.8a), one may plot any point lying within the volume of the tetrahedron onto the basal AFM plane or its projection to infinity. Minerals like biotite (point B) that are K-rich project beyond the FeO-MgO line (outside of the AFM triangle). Inasmuch as muscovite is used as a projection point, the system is considered to contain excess muscovite, that is, it can only be used for assemblages containing muscovite. Muscovite is listed with quartz adjacent to the diagram (figure 25.8b). Actual plotting of compositions in the AFM diagram is done using calculated coordinate values derived from the chemical analysis. The coordinate axes, shown in Figure 25.8b, are the molecular values MgO/(MgO+FeO) and $(Al_2O_3 - 3K_2O)/(Al_2O_3 - 3K_2O + MgO + FeO)$.

The positions at which common mineral compositions are plotted in the AFM diagram are shown in figure 25.8b. Pelitic rock compositions may plot anywhere within the diagram, but most commonly fall in the lower center. Similarly, sandstone compositions are variable. Quartz arenites typically plot near the A corner, because they consist of quartz and aluminous clays, whereas lithic wackes and arenites have compositions that plot nearer the base (the M–F line), because they contain Fe-Mg-bearing minerals and volcanic rock fragments.

THE CFM DIAGRAM

The CFM phase diagram is useful for plotting the phase changes in metabasites (Abbott, 1982, 1984). In this diagram (figure 25.9), a number of major components are taken into account, including SiO_2, Al_2O_3, Fe_2O_3, FeO, MgO, CaO, Na_2O, and K_2O, but the two-dimensional plane used for graphical display of phase assemblages is the CFM plane. That plane has corners with the following molecular values:

$$C = CaO + Na_2O + K_2O - Al_2O_3$$

$$F = FeO - Fe_2O_3$$

$$M = MgO$$

Additions and subtractions are made to adjust for the presence of the additional oxides in various minerals. The system is considered to be saturated with quartz, plagioclase, water, and magnetite, and may also contain alkali feldspar.

The compositions of common minerals found in metabasites are plotted on the CFM diagram in figure 25.9. Bulk rock compositions for basalts, andesites, gabbros, and diorites plot near, but commonly above, the F–M line.

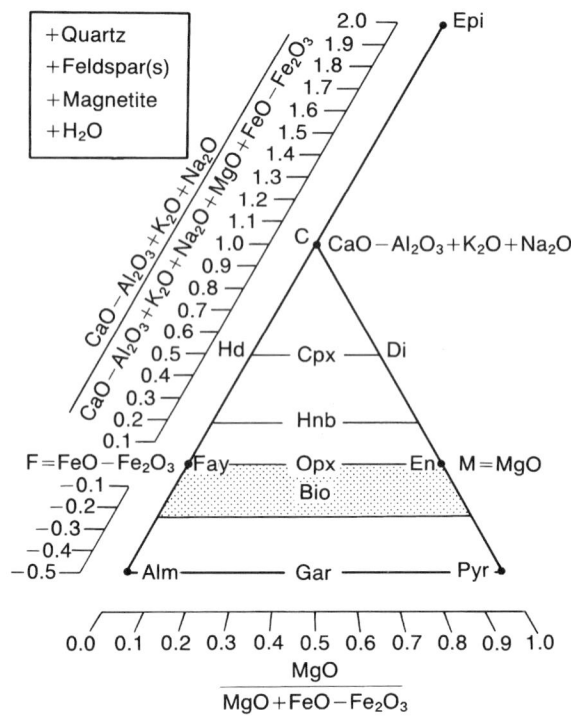

Figure 25.9 CFM diagram showing the endmember compositions for C, F, and M, the plotting scale, and the compositional ranges (lines and shaded area) of various minerals, as projected from quartz, H_2O, the feldspars, and magnetite onto the CFM plane. Epi = epidote, Cpx = clinopyroxene (Hd = hedenbergite = $CaFeSi_2O_6$, Di = diopside = $CaMgSi_2O_6$), Hnb = hornblende, Opx = orthopyroxene (En = enstatite = $Mg_2Si_2O_6$), Fay = fayalite, Bio = biotite (shaded area), Gar = garnet (Alm = almandine, Pyr = pyrope).
(Modified from Abbott, 1982)

OTHER DIAGRAMS

A number of additional diagrams have been used to present the phase assemblages of particular types of rock for which the diagrams above were either not appropriate or not yet available. These include the A'KF, AKN, and ACF[3] diagrams (figure 25.10).[14] The A'KF diagram allows the plotting of muscovite and biotite, as well as K-rich alkali feldspar, but has the same flaw as the ACF diagram in having FeO and MgO combined at the F corner. The AKN diagram, based on the system Na_2O-K_2O-Al_2O_3-SiO_2-H_2O, like the AFM diagram, is useful for showing phase changes in metashales and metasandstones, as well as in felsic metaigneous rocks. The rocks plotted on this diagram should be poor in ferromagnesian minerals. The ACF[3] diagram was developed for use with low- to moderate-temperature, low- to high-pressure metamorphosed shales, sandstones, and mafic igneous rocks.

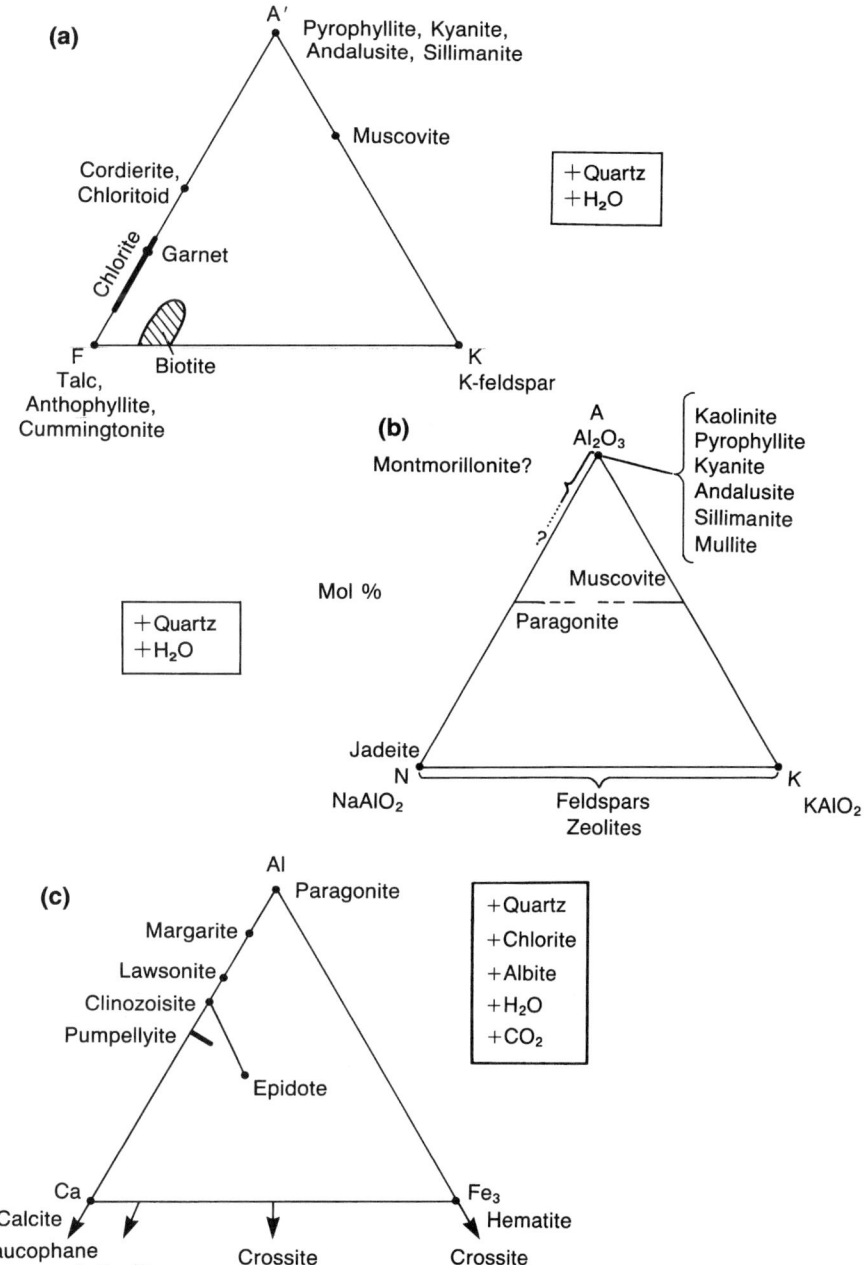

Figure 25.10 A'KF, AKN, ACF³ diagrams showing compositional ranges of minerals that plot in the diagrams. (a) A'KF diagram (after Eskola, 1939, in Winkler, 1965). (b) AKN diagram (after J. B. Thompson and A. B. Thompson, 1976). (c) ACF³ diagram (after E. H. Brown, 1977). Minerals and components in box represent phases from which projection is made. For details on corner compositions, refer to the cited references.

The A'KF, AKN, and ACF³ diagrams, as well as those representing other systems, provide some versatility in showing phase changes for different kinds of rocks. Each diagram is used in the same way, with terminal, nonterminal, and continuous reactions marking the changes from one metamorphic grade to another. In the chapters that follow, where various facies series are described, selected diagrams are used to show distinctive phase assemblages for various grades of metamorphism.

A method of constructing other (geometrical) phase diagrams from data on phase compositions was developed by Shreinemaker (in Zen, 1966; Yardley, 1989). This method allows the development of diagrammatic petrogenetic grids for specific systems in which univariant reaction curves meet at one or more invariant points.[15]

525

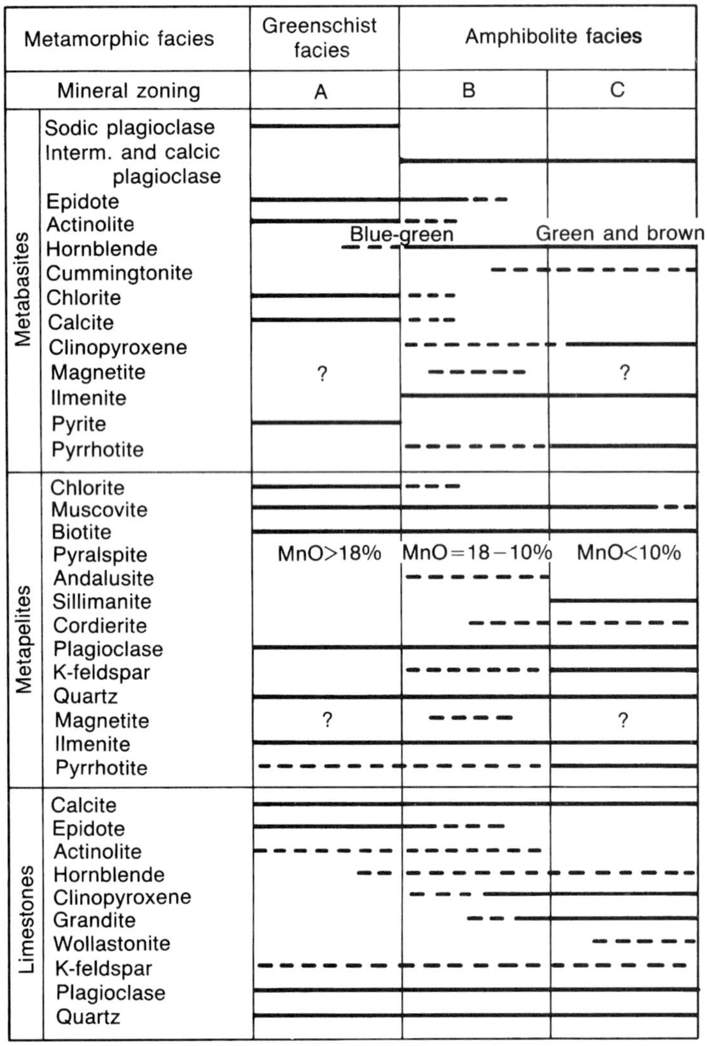

Metamorphic facies	Greenschist facies	Amphibolite facies	
Mineral zoning	A	B	C

Metabasites

Mineral	A	B	C
Sodic plagioclase			
Interm. and calcic plagioclase			
Epidote			
Actinolite			
Hornblende	Blue-green	Green and brown	
Cummingtonite			
Chlorite			
Calcite			
Clinopyroxene			
Magnetite	?		?
Ilmenite			
Pyrite			
Pyrrhotite			

Metapelites

Mineral	A	B	C
Chlorite			
Muscovite			
Biotite			
Pyralspite	MnO>18%	MnO=18–10%	MnO<10%
Andalusite			
Sillimanite			
Cordierite			
Plagioclase			
K-feldspar			
Quartz			
Magnetite	?		?
Ilmenite			
Pyrrhotite			

Limestones

Mineral	A	B	C
Calcite			
Epidote			
Actinolite			
Hornblende			
Clinopyroxene			
Grandite			
Wollastonite			
K-feldspar			
Plagioclase			
Quartz			

Figure 25.11 Mineral-facies chart for the Central Abukuma Plateau of Japan.
(From Miyashiro, 1973a)

MINERAL-FACIES CHARTS

The changes in mineral assemblages through a facies series or across several grades of metamorphism are clearly revealed by mineral-facies charts. Ernst (1965; 1971a, 1971b, 1971c, 1973b, 1977a) and Miyashiro (1973) have popularized the use of such charts, which consist of a rectangular plot with facies or grades of metamorphism listed on the horizontal axis and individual minerals, lumped by major bulk rock composition, listed on the vertical axis (figure 25.11). Lines extending parallel to the horizontal axis show the range of facies or grades over which each mineral is stable. Dashed lines reveal limited conditions of stability.

In the example in figure 25.11, note that two aspects of the changing mineralogy are easily observed. First, the points of appearance or disappearance (by terminal reactions) of critical minerals and mineral assemblages are obvious. Second, the correlation between specific mineralogical changes and the facies or grade boundaries based upon them is evident. In addition, the mineralogical variations between the various bulk rock chemistries may be compared with ease. The principal detriment of such diagrams is that stable phase assemblages are not shown. Thus, such diagrams cannot be used as a substitute for phase diagrams, but rather should be used in conjuction with them.

SUMMARY

Several types of phase diagrams are used in metamorphic petrology. In each case, the diagram is selected on the basis of its usefulness in portraying the mineralogical changes in a particular rock type. All of these diagrams show assemblages of solid phases, rather than melt compositions. Changes from one mineral assemblage to another are depicted on the diagrams as terminal reactions (addition or subtraction of a phase from the diagram) or nonterminal reactions (tie-line flips on the diagram). Univariant curves depict discontinuous reactions, whereas some reactions involving solid solutions are continuous. Reaction paths involving solid solutions may be depicted in reaction space and quantitatively defined using a reaction progress variable.

The most commonly used phase diagrams include the AFM and the CMS diagrams. The former is used for pelitic (aluminous) and quartz-feldspar (siliceous-alkalic-calcic) rocks and the latter is useful in the study of carbonate and ultramafic rocks. The CFM diagram is used for basic rocks. In addition, the pseudo-phase diagrams, A'KF and ACF, are used for pelitic and mafic plus carbonate rocks, respectively.

Mineral-facies charts are useful for portraying general changes in mineralogy across a facies or facies series. Such charts are not phase diagrams, however, and cannot be used in place of phase diagrams, because they do not depict stable phase assemblages.

EXPLANATORY NOTES

1. Mineral grains are not strictly phases in many cases, as they contain inclusions and exsolution lamellae, which, in fact, are mechanically separable.
2. Quartz may be replaced in pelitic rocks by other phases under extreme conditions. In the Sanidinite Facies, the silica phase may be tridymite, and in the Eclogite Facies, it may be coesite.

3. The metamorphic relations in the system SiO_2-CaO-MgO-H_2O-CO_2, related systems, and particular examples have been studied by several workers. Important works include those of N. L. Bowen (1940), R. I. Harker and Tuttle (1956), H. J. Greenwood (1967), Turner (1968), Metz and Trommsdorf (1968), Trommsdorf and Evans (1972), B. W. Evans and Trommsdorf (1974), Skippen (1974), Slaughter, Kerrick, and Wall (1975), Joesten (1976), B. W. Evans (1977), Kase and Metz (1980), G. Franz and Spear (1983), and S. B. Tanner, Kerrick, and Lasaga (1985). Winkler (1967, 1974, 1976, 1979) and F. J.Turner (1981) provide good reviews of the subject. Also see chapters 26 and 31 of this text for additional discussions.

4. Systems are *internally buffered* if the mineral assemblage controls the fluid phase composition. This occurs if reactions produce or consume CO_2 or H_2O in amounts that significantly affect the fluid phase composition. Under such circumstances, externally generated fluids are relatively insignificant in volume.

5. Any number of phases may be plotted by projection onto the plane, but information is lost in doing so and the diagram cannot be as easily used to represent equilibrium assemblages.

6. A detailed treatment of this subject is beyond the scope of this text. For additional examples, see J. B. Thompson, Laird, and Thompson (1982) and Poli (1991).

7. Detailed treatment of this topic is beyond the scope of this text.

8. To construct a complete phase diagram from field and petrographic data, it is important to collect samples of rock covering a wide range of bulk compositions. By doing so, one is able to obtain a wide range of three-phase mineralogies that will allow complete definition of the tie lines of the phase diagram. This point has been emphasized by J. B. Thompson (1957) and Guidotti (1982).

9. In this chapter, the term *pseudo-phase diagram* refers to phase diagrams that look like typical triangular metamorphic phase diagrams but do not meet the requirements of the Phase Rule, because of the way that oxides are combined to make the components used as definitive endmembers for the apices of the triangle. Guidotti (1982) discusses this problem.

10. Though it was assumed in older works that Fe^{2+} and Mg^{2+} substitute freely for one another in various minerals in rocks, more recent studies reveal that there is an uneven distribution of these two ions between coexisting ferromagnesian minerals. As noted in chapter 24, uneven distributions are referred to as partitioning of the element. See chapter 24 for a brief discussion and for references to important works on the subject.

11. For details on the specific procedures involved in ACF calculations and more thorough discussions of the use of this and other diagrams, see the quotations of Eskola's (1939) work and other discussions in older metamorphic petrology texts such as Winkler (1965, 1967, 1974, 1976, 1979), F. J. Turner (1968, 1981), Miyashiro (1973a), and Mason (1978).

12. More detailed discussions of the nature and use of AFM diagrams are provided in metamorphic petrology texts such as those cited in note 11. Also see Best (1982, p. 403ff.) and Yardley (1989, ch.3).

13. See J. B. Thompson (1957) and Mason (1978).

14. The A'KF diagram was introduced by Eskola (see F. J. Turner, 1968, p. 175 ff.) and is described in metamorphic petrology texts such as Turner (1968), Miyashiro (1973a), and Winkler (1976). The AKN diagram (a name applied here) was introduced by J. B. Thompson (1961, in A. B. Thompson, 1974). The AKN diagram is derived from the system $NaAlO_2$-$KAlO_2$-Al_2O_3-SiO_2-H_2O, with points projected from SiO_2 and H_2O onto the Al_2SiO_5-$KAlSi_3O_8$-$NaAlSi_3O_8$ plane. A. B. Thompson (1974) and J. B. Thompson and Thompson (1976) discuss its use. The ACF[3] diagram was designed by E. H. Brown (1977a). See E. H. Brown (1974) and E. H. Brown et al. (1981) for additional discussions relating to the sorts of rocks for which this diagram was devised. H. J. Greenwood (1975) provides a calculation for the production of any desired projection of phase relations in any system.

15. See Zen (1966) and Yardley (1989, appendix) for details on how to construct Schreinemaker diagrams.

PROBLEMS

25.1. Using the chemical analyses of ultramafic rock 3 in table 10.2 and carbonate rock 3 in table 21.1, (a) calculate the coordinates for and plot the positions of these bulk compositions on a CMS diagram. (b) Using the low-T phase diagram shown in figure 25.3, determine what mineralogy would be present in these rocks. (c) Similarly, determine the stable minerals for these rocks for the higher-T phase diagram in figure 25.3.

25.2. Using the chemical analysis of pelitic rock 6 in table 18.3, (a) calculate the coordinates for and plot the position of this bulk composition on an AFM diagram. (b) Using a chemical analysis of an almandine garnet from a mineralogy book, calculate the coordinates for and plot the position of its composition on the AFM diagram.

26

Contact Metamorphism

INTRODUCTION

Contact metamorphism, as the name implies, occurs locally, at and near the contacts between intrusions and the surrounding country rock. As might be expected in such a setting, the metamorphism is dominantly controlled by the heat introduced by the intrusion. The effects of increased temperature are most pronounced where intrusions occur at shallow levels in the crust. There, contrasts in temperature between country rock and intrusion are at a maximum. With increasing depth of intrusion, the temperature contrast generally decreases, as do the contact effects, except in relatively rare cases where intrusion occurs in higher-P/T facies series.

As will become evident, the fluid phase is also an important agent of contact metamorphism. It transports heat and has a profound influence on the chemistry and mineral composition of the rocks with which it comes in contact. Fluids are particularly important (1) in the metamorphism of carbonate rocks, where metamorphism yields CO_2, (2) in rocks undergoing metamorphism in hydrothermally active areas, and (3) in rocks that yield H_2O upon metamorphism, such as those with abundant phyllosilicates. Especially along the mid-ocean ridges and in other active hydrothermal areas, H_2O-rich fluids are very important agents of metamorphism.

Contact metamorphism produces fine-grained granoblastites and diablastites, commonly called **hornfelses** (figure 26.1).[1] In addition to a variety of common minerals, such as quartz, feldspars, and epidote, hornfelses locally contain unique phases. Such minerals as spurrite and tilleyite (calcium silicates) may form in carbonate rocks; mullite, an aluminum silicate, forms in pelitic rocks; and minerals more commonly associated with igneous rocks, such as sanidine, form in rocks of appropriate chemistry. The high temperatures of metamorphism, in their extreme, may even cause local melting in the contact zone, yielding glass.

Examination of the effects of contact metamorphism allows us to gain insight into metamorphic processes without the complicating influences of high pressure and deviatoric stress. Typically, contact metamorphism occurs at shallower levels of the crust, where the pressure is relatively low (< 0.4 Gpa = 4 kb). At those shallow levels, the deviatoric stresses characteristic of the deeper levels of mountain belts are generally absent and the contact metamorphic rocks lack foliation. Locally, however, high-pressure contact rocks do exist, for example, in forearc regions, shields, and root zones of mountain belts.

Contact Facies Series rocks are mineralogically similar and in some cases identical to those of the Buchan Facies Series (Pattison and Tracy, 1991), but contact

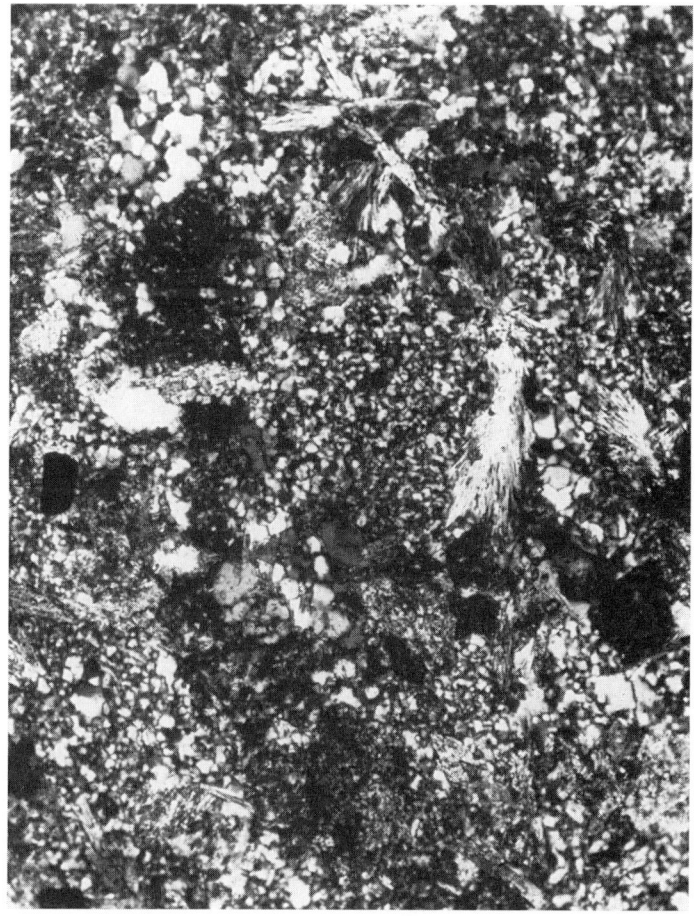

Figure 26.1 Photomicrograph of epidote-rich hornfels. Tioga Lake, Sierra Nevada, California. Note dominant granoblastic texture. (PL). Long dimension of photo is 0.33 mm.

rocks are distinguished by their general lack of foliation. They differ from rocks of the other facies series in their mineralogy, textures, and local distribution.

FACIES AND FACIES SERIES

Contact metamorphic rocks are found in **aureoles**, zones of metamorphic rock surrounding and associated with plutons (figure 26.2), as well as in roof pendants within plutons and in xenoliths in plutons and lava flows. In this text, the hydrothermal metamorphism associated with igneous activity along the mid-ocean ridges is discussed with contact metamorphism because (1) the metamorphic effects are restricted to distances of a few kilometers perpendicular to strike and vertically within the rock column, (2) the effects are produced primarily as a result of the igneous and associated hydrothermal activity, and (3) deviatoric stress does not play a dominant role in fabric development in this environment.

Observation of the occurrences of contact metamorphic rocks and examination of petrogenetic grids (see appendix C) reveals that Zeolite, Prehnite-Pumpellyite, Albite-Epidote Hornfels, Hornblende Hornfels, Pyroxene Hornfels, and Sanidinite facies constitute the *Contact Metamorphic Facies Series*. At progressively higher temperatures, in pelitic and siliceous-alkali-calcic rocks, minerals indicative of these facies include analcite, stilbite, wairakite, pyrophyllite, cordierite, andalusite, sillimanite, K-feldspar,

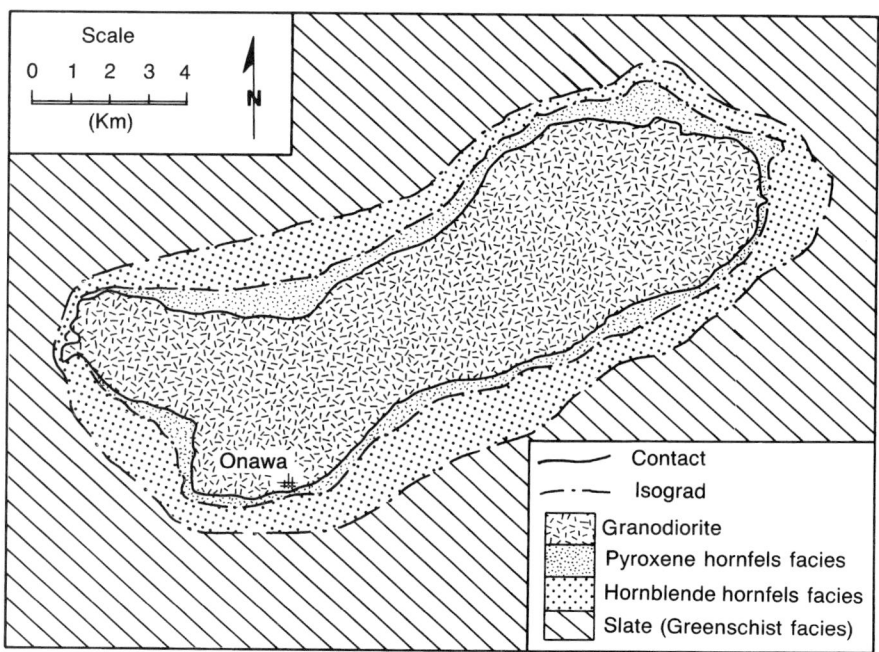

Figure 26.2 Map of the contact aureole of the Onawa Pluton, Maine, showing crudely concentric metamorphic zones. *(Modified from Philbrick, 1936; J. M. Moore, 1960)*

Scale

0 1 2 3 4

(Km)

N

Onawa

——— Contact
— · — Isograd
Granodiorite
Pyroxene hornfels facies
Hornblende hornfels facies
Slate (Greenschist facies)

orthopyroxene, sanidine, and mullite. Wairakite, albite, actinolite, epidote, hornblende, the pyroxenes, and olivine occur in corresponding basic rocks. In carbonate rocks, minerals such as talc, tremolite, diopside, forsterite, grossularite, wollastonite, and spurrite may develop.

In any given aureole, a complete sequence of facies will not likely occur. Both low-temperature facies and high-temperature facies are often missing. Because many intrusions occur in previously metamorphosed rocks, occurrence of the lower-grade facies is precluded by the resistance of preexisting, partially dehydrated, metamorphic country rocks to retrograde metamorphism. At the highest grades, Sanidinite Facies rocks can only develop where very low pressures, particular bulk rock compositions, or the absence of a fluid phase inhibit melting. The upper parts of both the Pyroxene Hornfels Facies and the Sanidinite Facies overlap the zone of melting, where P-T conditions and a fluid phase cause melting in many bulk compositions (see figure 24.6 and appendix C).

A typical example of a partial facies series is provided by the contact aureole of the Devonian Onawa pluton of Maine (figure 26.2). (Philbrick, 1936; J. M. Moore, 1960)[2] The pluton is an elongate, composite mass of granitoid rock that was intruded into slate country rock previously metamorphosed under conditions of the lower Greenschist Facies. The country rocks contain the assemblage Fe-Ti oxide + white mica + chlorite + quartz. The first evidence of contact metamorphism is the appearance of spots in the slates as far as 2 km from the pluton margin. The spots were cordierite porphyroblasts (now largely replaced by phyllosilicates) and are part of the assemblage biotite + andalusite + cordierite + white mica + quartz + albite (figure 26.3). This assemblage is representative of the Hornblende Hornfels Facies.

The outer zone of spotted slates surrounds a second zone. This second zone is composed of porphyroblastic granoblastites (hornfelses) with the same mineral assemblage as the rocks of the outer zone, but lacking their continuous (slaty) cleavage. The second zone surrounds a third zone, adjacent to the pluton, composed of coarser-grained granoblastites with the assemblage biotite + sillimanite + cordierite + alkali feldspar + quartz (figure 26.3). This assemblage indicates the Pyroxene Hornfels Facies. Local, small-scale dikes and fine-mesoscopic, millimeter-size patches of quartz-alkali feldspar rock indicate that partial melting has occurred locally. In summary, only two facies are recognized in the Onawa aureole, the Hornblende Hornfels Facies and the Pyroxene Hornfels Facies.

CONDITIONS OF CONTACT METAMORPHISM

The conditions of contact metamorphism, with rare exception, are those of low to moderate pressure and low to high temperature. Both isotopic studies and comparisons of

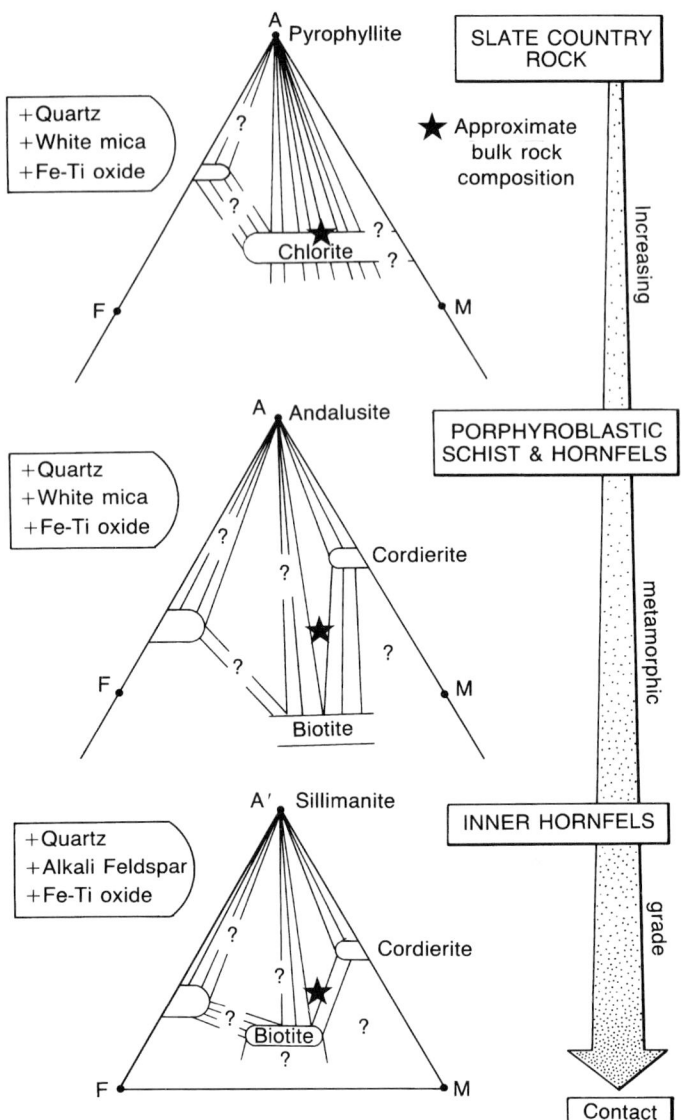

Figure 26.3   AFM diagrams showing the metamorphic mineral assemblages of the contact aureole and metamorphic terrane surrounding the Onawa Pluton, Maine. Sodic plagioclase is present in some lower-grade assemblages.

observed mineral assemblages with experimentally determined phase equilibria yield information about the P-T conditions. Table 26.1 lists the maximum P-T conditions estimated for several contact aureoles. Note that pressures are generally less than 0.4 Gpa (4 kb). The common presence of andalusite in contact aureoles is a good mineralogical indicator that the pressure in the middle grade of most contact aureoles was lower than that of the aluminum silicate triple point, which occurs at 0.387 Gpa (3.87 kb) (Hemingway et al., 1991).

Temperatures of metamorphism vary widely. Among the controlling factors are (1) the temperature of the

Table 26.1 Estimated Maximum P-T Conditions for Selected Contact Aureoles

Locality	T (°C)	P (kb)	References
DSDP Hole 504B	380	0.4	Alt et al. (1986)
Brewster Co., Tex.	470	0.3	Droddy and Butler (1979)
Mid-Atlantic Ridge 6° N	>500	0.5	Bonatti et al. (1975)
Blakes Ferry, Ala.	510	7.0	R. G. Gibson and Speer (1986)
Hope Valley, Calif.	540	2.0	Kerrick, Crawford, and Randazzo (1973), Ferry (1989)
Marysville, Mont.	600	1.0	J. M. Rice (1977), Lattanzi, Rye, and Rice (1980)
Notch Peak, Utah	600	2.0	Hover Granath, Papike, and Labotka (1983)
Duluth Complex, Minn.	620	1.5	Labotka (1983), Labotka, White, and Papike (1984)
Tioga Pass, Calif.	670	2.0	Kerrick (1970)
Bergell Intrusion, Italy	700	3.5	Trommsdorf and Evans (1972, 1977a), Wenk, Wenk, and Wallace (1974)
Mathematician Ridge	700	0.6	Stakes and Vanko (1986)
Liberty Hill, S. C.	725	4.5	Speer (1981, 1987)
Onawa, Maine	725?	3.6	J. M. Moore (1960), Pattison and Tracy (1991)
Connecticut Valley	745	<0.1	April (1980)
Lilesville, N. C.	750	3.5	N. H. Evans and Speer (1984)
Mount Royal, Quebec	750	0.5	Williams-Jones (1981)
King Island, Tasmania	800	0.7	Hing and Kwak (1979)
Ronda, Spain	800	4.3	Loomis (1972a)
Morton Pass, Wyo.	933	3.0	Russ-Nabelek (1989)
Christmas Mtns., Tex.	1035	0.3	Joesten (1974, 1976)
Bushveld Complex, S. Africa	>1200	1–2	Wallmach, Hatton, and Droop (1989)

magma, which is a function of its chemistry, (2) the temperature of the country rock at the time of intrusion, (3) the conductivities of the solidifying magma and the country rock, (4) a factor called the diffusivity (of both the country rock and the intrusion), (5) the heat of crystallization of the magma, (6) the heat capacity (the rate of change in the energy of reaction with change in temperature), (7) fluid transport, the heating or cooling by influx of water, and (8) contributions from other sources, such as radioactive decay (Jaeger, 1957, 1959; F. J. Turner, 1981; Ferry, 1983b).[3] In a simplified case, considering only the temperatures of the intrusion (T_i) and the country rock (T_{cr}), the temperature at the contact (T_c) will be $T_c = 1/2(T_i + T_{cr})$. In fact, however, the effects of the additional factors will increase or reduce that value. These factors and the temperature of the contact will also affect the *width* of the contact aureole.

PROCESSES IN CONTACT METAMORPHISM

The brief description of the Onawa aureole (above) provides a glimpse of some of the kinds of processes that operate during contact metamorphism. In particular, the growth of porphyroblasts and progressive recrystallization and neocrystallization were indicated. Local melting accompanied metamorphism at the highest grades.

In addition to the above processes, processes involving a fluid phase may be quite important. The fluid phase may result from decarbonation and dehydration reactions, which are common in metamorphism, and may cause melting or metasomatism in rocks invaded by it.

The Fluid Phase

Evidence that a fluid phase exists in metamorphic rocks was presented in chapter 23. The fluid phase may begin as (1) meteoric water, (2) interstitial water or brine in sedimentary and igneous protoliths, (3) juvenile water or fluids derived from intrusions, or (4) as water or other volatile species chemically bound in volatile-bearing mineral phases, such as clays or carbonate minerals.[4] Particularly important in the evolution of the fluid phase are decarbonation and dehydration reactions.

If we examine the phase assemblages present in the Onawa aureole, we find that dehydration is clearly revealed. The Fe-Ti oxide + white mica + chlorite + quartz assemblage of the country rocks is replaced in the outer zone of the aureole by one containing andalusite and cordierite. Chlorite [$(Mg,Fe,Al)_6(Al,Si)_4O_{10}(OH)_8$], a hydrous phase, disappears from the assemblage, and is replaced by the assemblage biotite + cordierite + andalusite, two minerals of which are anhydrous phases [andalusite = Al_2SiO_5 and cordierite = $(Mg,Fe)_2Al_4Si_5O_{18}$]. Closer to the pluton, white mica [$(K, Na)Al_2(Si_3Al O_{10})(OH)_2$], a hydrous phase, is replaced by alkali feldspar [$(K,Na)AlSi_3O_8$], an anhydrous phase. This pattern is typical of prograde metamorphism, in which the rocks become increasingly anhydrous at progressively higher grades of metamorphism as fluids are driven off. Thus, progressive metamorphism generates a fluid phase that apparently migrates away from the zones of highest-grade metamorphism.

A similar situation pertains in rocks involving carbonate phases. For example, a reaction that occurs during the contact metamorphism of carbonate rocks is

$$calcite + quartz <=> wollastonite + CO_2 \qquad (26.1)$$

$$CaCO_3 + SiO_2 <=> CaSiO_3 + \qquad CO_2$$

In this reaction, the carbonate phase calcite combines with quartz to yield wollastonite and the volatile species CO_2.[5] As is the case with dehydration, progressively higher grades of metamorphism yield increasingly CO_2-poor rocks. CO_2 is driven from the minerals during progressive metamorphism and becomes a part of a fluid phase.

In some cases, fluids evolved from the rocks themselves are overwhelmed by fluid derived from an intrusion or that involved in hydrothermal activity. Below we will examine a case where an intrusion induces Si-Al-Fe metasomatism by introducing these elements via a magmatic fluid phase. In contrast, along mid-ocean ridges, heated ocean water in great volumes moves through newly formed oceanic crust, altering the rock chemistry (Rona et al., 1983).

The *composition* of the fluid phase controls the mineral assemblage developed under given sets of metamorphic conditions and vice versa (Ferry and Burt, 1982, A. B. Thompson, 1983).[6] Thus, it is important to know the nature of the fluid phase present during metamorphism. That fluid phase

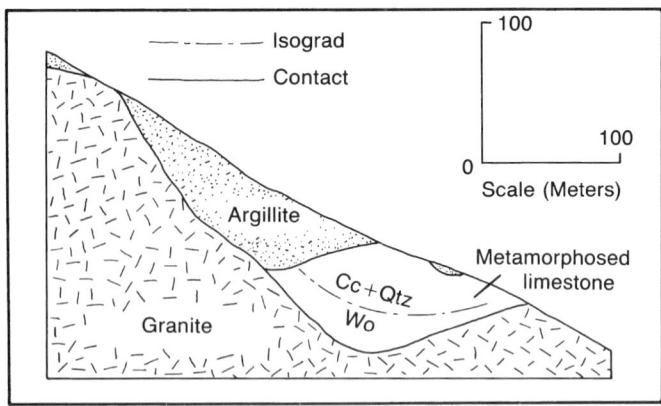

Figure 26.4 Cross section of the Victory Mine contact aureole near Salmo, British Columbia. The contact geology is based on drilling and surface mapping. Only the wollastonite isograd in the carbonate rock is shown.
(Simplified from H. J. Greenwood, 1967)

may be dominated by H_2O, but in carbonate rocks and at higher grades of metamorphism, the mole fraction of H_2O (X_{H_2O}) may decrease to values of 0.1 or less.[7] In carbonate rocks, the ratio of H_2O to CO_2 is especially important. Other components of the fluid phase may include Cl, F, B, S, Na, C and H (CH_4) (Russ-Nabelek, 1989; Sisson and Hollister, 1990; Labotka, 1991).

The influence of varying mole fractions of CO_2 and H_2O in metamorphism of carbonate rocks was demonstrated by H. J. Greenwood (1967) in his work on a contact aureole near Salmo, British Columbia (figure 26.4). Here, during metamorphism, water and carbon dioxide formed a fluid phase dominated by these two components, and the fluid phase controlled the development of wollastonite in the contact aureole. The important phase relations are best depicted on a T-X diagram constructed for a fixed pressure (figure 26.5). At low mole fractions of CO_2, wollastonite is stable at lower temperatures. At higher values of X_{CO_2}, wollastonite requires higher temperatures to form. A change in pressure simply raises or lowers the temperature of the reaction.

In general, a series of such $T-X_{CO_2-H_2O}$ curves are applicable to metamorphism of carbonate and ultramafic rocks. Where more than one curve exists, the curves will tend to intersect at invariant points (figure 26.6), beyond which the stable phase assemblage will differ. More generally, $T-X_{fluid}$ curves may be used for any fluid phase in any composition of rock.

Recrystallization and Neocrystallization

Recrystallization and neocrystallization are evident in the Onawa aureole, just as was dehydration. In terms of recrystallization, it was noted that there was an increase in grain size in the rocks from the outer to the inner zones of the aureole.

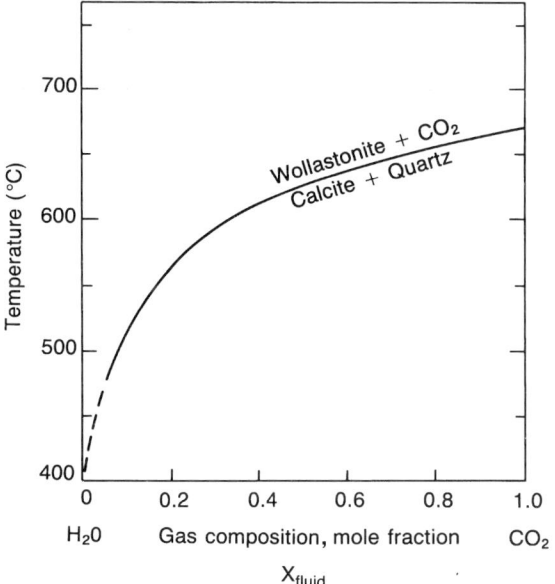

Figure 26.5 T-X_{fluid} diagram showing the stability relations of calcite, quartz, and wollastonite at one kilobar (P_f = 1 kb). Note the strong downward bend of the reaction curve at high values of X_{H_2O}. An increase in pressure raises the position of the curve to higher temperatures but does not change its general shape.

(Simplified from H. J. Greenwood, 1967)

Recall that the same effect, quantified by Joesten (1983), was described in chapter 23. In general, rocks become progressively more coarse-grained with increasing grades of metamorphism, though polygonization and grain-size reduction may occur as an intermediate stage in the coarsening process.

Neocrystallization in the Onawa aureole involved the development of biotite, andalusite, cordierite, sillimanite, and alkali feldspar. The appearance of each new phase or phase assemblage marks a reaction in which there is a chemical and structural readjustment in the rocks. In general, such readjustments occur as the rock equilibrates with new P-T conditions and responds geochemically to the fluid phase present at the time.

The first indication of contact metamorphism in the Onawa contact aureole, the development of porphyroblasts, is a common form of neocrystallization in the outer zones of contact aureoles. Of course, porphyroblasts also develop in regionally metamorphosed rocks, wherever conditions are appropriate. In general, porphyroblast development involves the formation of a few nucleii and the subsequent growth of those nucleii into large crystals.[8] Nucleation is possible where the necessary phases are present to yield chemical species needed to form the nucleus of a new phase stabilized by the

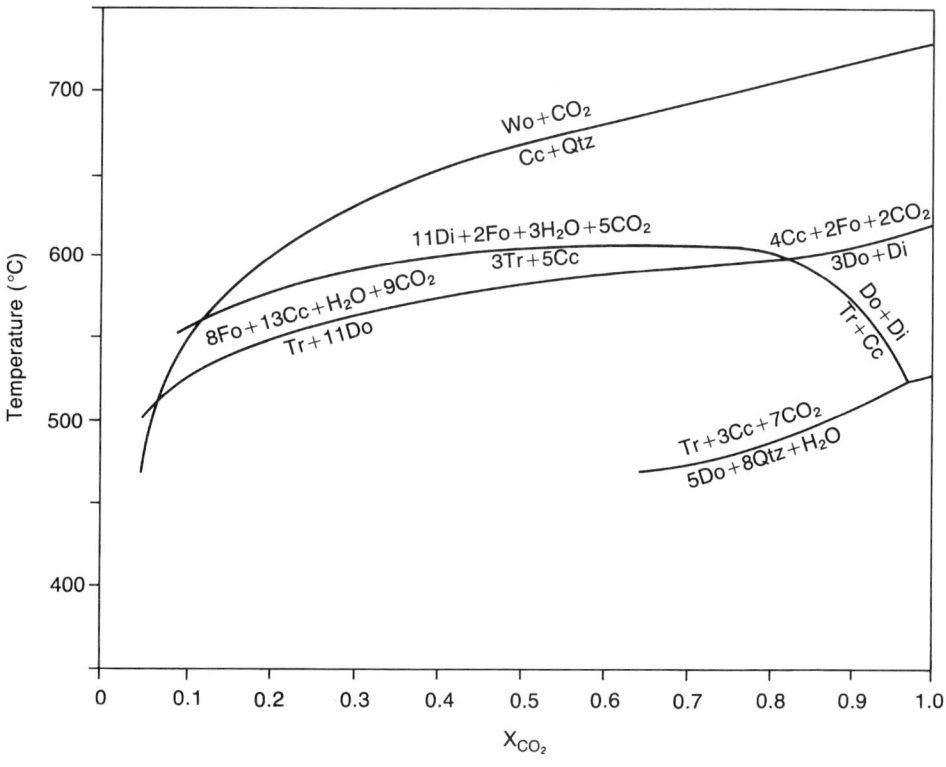

Figure 26.6 T-X phase diagram showing intersecting stability curves for some reactions important in the metamorphism of carbonate rocks. Reactions involve the phases calcite (Cc), diopside (Di), dolomite (Do), forsterite (Fo), quartz (Qtz), tremolite (Tr), wollastonite (Wo), and the components of the fluid phase P = 2 kb (0.2 Gpa).

(Simplified from Hover Granath, Papike, and Labotka, 1983)

imposition of different P, T, or X_{fluid} conditions. Diffusion of species through the surrounding rock allows both nucleation and growth.

In zones of high shear stress, nucleation and growth of porphyroblasts are prohibited by the effects of that stress, especially effects such as dissolution and solution transfer that result from chemical potential gradients created by deformation of crystals (T. H. Bell, Rubenach, and Fleming, 1986). If shear stress is moderate, porphyroblast nucleii may form at local sites of lower stress. Once nucleated, porphyroblasts grow. The increased size of the porphyroblast may then create **pressure shadows,** zones of lower stress and equant grain growth adjacent to the porphyroblast that facilitate further porphyroblast growth (see figure 23.11c) (D. J. Prior, 1987). In contact zones, shearing stress is typically moderate to nil and porphyroblasts commonly lack pressure shadows. Where they are present, they indicate that shearing stresses accompanied contact metamorphism.

MINERALOGICAL CHANGES DURING CONTACT METAMORPHISM

Mineralogical changes during contact metamorphism are a function of bulk rock chemistry, P-T conditions, and the nature of the fluid phase. Below, each of the major categories of bulk rock is discussed in terms of the mineralogical changes that occur during contact metamorphism.

Aluminous Rocks (Pelitic Rocks)

The Onawa aureole is representative of contact aureoles in pelitic rocks. Yet, as is typical, it does not exhibit all of the facies of contact metamorphism. To see the range of mineralogies that may develop, a composite array of facies diagrams (figure 26.7) must be compiled from observed occurrences of the Contact Metamorphic Facies Series or calculated using model systems such as the so-called KFMASH system (Pattison and Tracy, 1991).[9] The phase diagram most useful for portraying assemblages in aluminous rocks is the AFM diagram. *Note:* Only a few representative diagrams are presented here, and the addition or subtraction of each phase during the progressive metamorphism of any assemblage of rocks yields a new topology.

At the lowest grades of metamorphism, in the Zeolite Facies, various combinations of quartz, chlorite, alkali feldspar, calcite, kaolinite and various mixed-layer clays (smectite/illite, vermiculite/chlorite), and illite, as well as a number of zeolites and iron oxides, characterize the rocks.[10]

With progressive metamorphism, the clays are replaced by white mica, chloritoid may appear in iron-rich rocks, kaolinite is replaced by pyrophyllite, and biotite appears (see figure 26.7). The reaction

$$\text{kaolinite} + 4 \text{ quartz} <==> 2 \text{ pyrophyllite} + 2 \text{ H}_2\text{O} \quad (26.2)$$

(Velde, 1969) is one of the important reactions marking the boundary between the Zeolite and Albite-Epidote Hornfels Facies.[11] Assemblages such as

quartz–white mica–chlorite–biotite–chloritoid–albite–epidote

and

quartz–chlorite–white mica–alkali feldspar–albite–epidote

are typical of the Albite-Epidote Hornfels Facies. The presence of biotite in this facies indicates that a reaction such as

$$\text{chlorite} + \text{alkali feldspar} <==> 2 \text{ biotite} + 2\text{H}_2\text{O} \quad (26.3)$$

has occurred. The assemblages listed above are essentially identical to those of the adjacent, higher-pressure Greenschist Facies. This is no doubt one of the reasons that the Albite-Epidote Hornfels Facies is less distinct and less commonly recognized than the higher-temperature Hornblende Hornfels Facies, which is characterized by more distinctive mineral assemblages. The similarity also raises the question of whether or not the Albite-Epidote Hornfels Facies should be recognized as a separate and distinct facies.

The Albite-Epidote Hornfels Facies is succeeded at higher temperatures by the Hornblende Hornfels Facies. This facies, at the middle grades of contact metamorphism, exhibits some of the most characteristic contact metamorphic phase assemblages (see figure 26.7). These include

quartz–white mica–andalusite–cordierite–biotite

and

quartz–alkali feldspar–cordierite–plagioclase–magnetite.[12]

Staurolite and garnet may also occur in this facies, as may Mg-amphiboles.[13] Important reactions leading to Hornblende Hornfels Facies assemblages include

$$\text{pyrophyllite} <==> \text{andalusite} + 3 \text{ quartz} + \text{H}_2\text{O} \quad (26.4)$$

(Kerrick, 1968: Haas and Holdaway, 1973),

$$\text{Fe-chlorite} + \text{quartz} + \text{magnetite} <==> \text{garnet (almandine)} + \text{H}_2\text{O} \quad (26.5)$$

(L. C. Hsu, 1968), and

$$2 \text{ muscovite} + 2 \text{ chlorite} + 4 \text{ quartz} <==> 2 \text{ cordierite} + 2 \text{ biotite} + 7 \text{ H}_2\text{O} \quad (26.6)$$

(Fawcett and Yoder, 1966; N. H. Evans and Speer, 1984; Pattison, 1987).

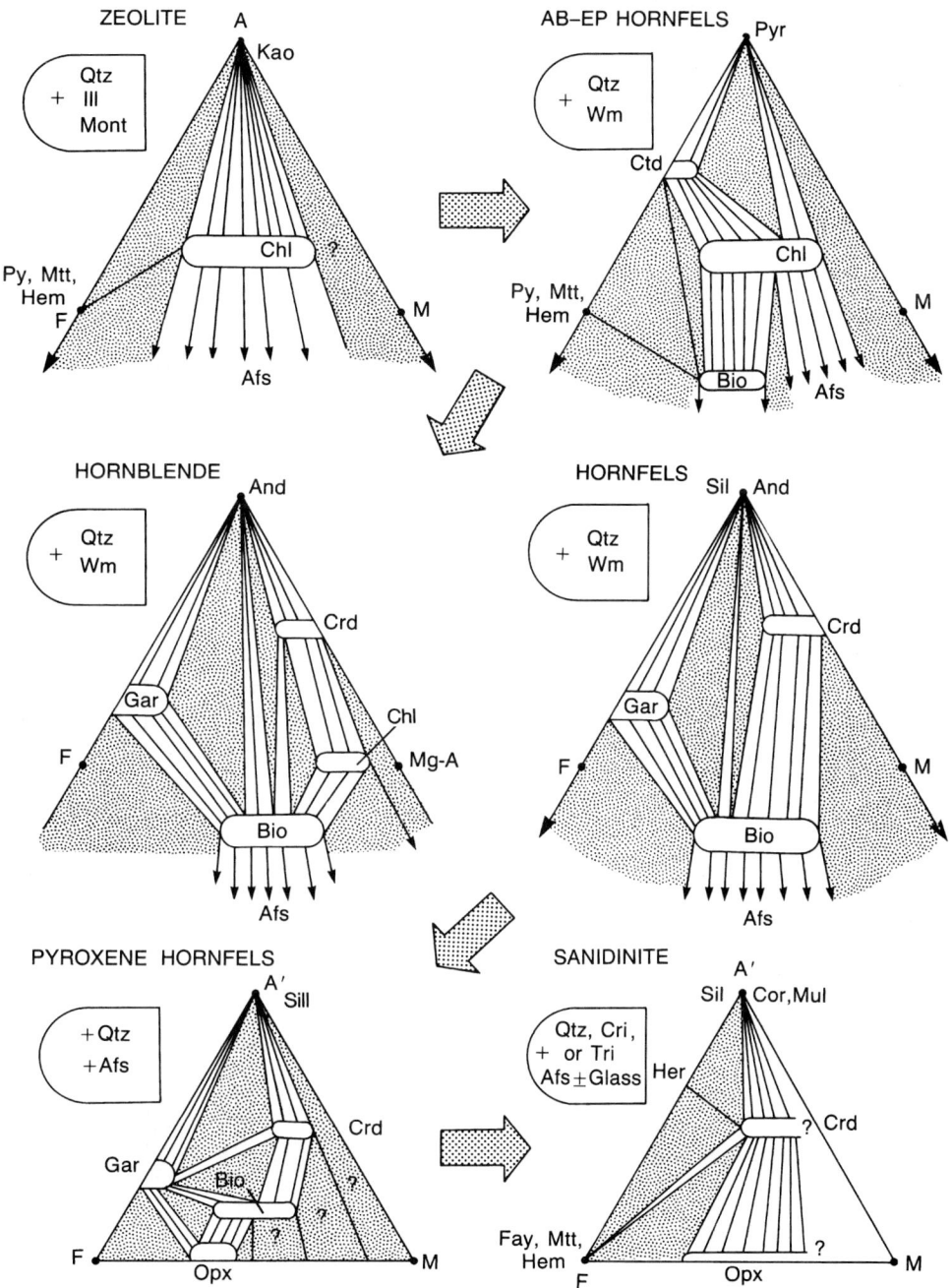

Figure 26.7 AFM and A′FM phase diagrams, showing typical phase assemblages for the facies of contact metamorphism. These diagrams are applicable to pelitic rocks and aluminous siliceous-alkali-calcic (quartz-feldspar) rocks. Afs = alkali feldspar, And = andalusite, Bio = biotite, Chl = chlorite, Cor = corundum, Crd = cordierite, Ctd = chloritoid, Cum = cummingtonite, Fay = fayalite, Gar = garnet, Hem = hematite, Her = hercinite, Ill = illite, Kao = kaolinite, Mg-A = magnesium amphibole, Mont = montmorillonite, Mtt = magnetite, Mul = mullite, Opx = orthopyroxene, Py = pyrite, Pyr = pyrophyllite, Qtz = quartz, Sil = sillimanite, Tri = tridymite, Wm = white mica.

(Based on sources cited in the text. A′FM plot after E. W. Reinhardt, 1968)

Where intermediate to mafic magmas intrude aluminous country rocks and incorporate fragments of those rocks as xenoliths, Pyroxene Hornfels and Sanidinite facies rocks are developed. Occurrences of the former are not uncommon in contact aureoles, but those of the latter are rare. In these two facies, the hydrous phases white mica and biotite may yield to the anhydrous phases alkali feldspar and orthopyroxene. Muscovite breaks down via the reaction

$$\text{muscovite} + \text{quartz} \Longleftrightarrow \text{andalusite (or sillimanite)} + \text{alkali feldspar} + H_2O \quad (26.7)$$

(Evans, 1965).[14] This reaction is one of the important reactions marking the boundary between the Hornblende Hornfels and Pyroxene Hornfels facies. White mica has also been found to break down, under disequilibrium conditions that are probably typical of xenolithic Sanidinite Facies metamorphism, via the coupled reactions

$$\text{white mica} \Longleftrightarrow \text{alkali feldspar} + \text{biotite} + \text{mullite} \quad (26.8)$$

and

$$\text{white mica} \Longleftrightarrow \text{alkali feldspar} + \text{biotite} + \text{corundum} + \text{hercynite} \quad (26.9)$$

(Brearley, 1986).
Biotite may break down via reactions such as

$$4 \text{ biotite} + 18 \text{ quartz} + 12 \text{ andalusite} \Longleftrightarrow$$

$$6 \text{ cordierite} + 4 \text{ alkali feldspar} + H_2O \quad (26.10)$$

(Holdaway and Lee, 1977; Pattison, 1987) or

$$\text{biotite} + \text{quartz} \Longleftrightarrow \text{sanidine} + \text{magnetite} \pm \text{hematite} \quad (26.11)$$

(Eugster and Wones, 1962; Wones and Eugster, 1965). Both reactions occur generally within the P-T conditions of the Pyroxene Hornfels Facies, but the latter reaction is important at low pressures in defining the boundary between Pyroxene Hornfels and Sanidinite facies. Typical assemblages of the Pyroxene Hornfels Facies[15] include

quartz–K-feldspar–biotite–cordierite–garnet–plagioclase,

quartz–K-feldspar–andalusite–cordierite–biotite , and

quartz–alkali feldspar–plagioclase–biotite–garnet–cordierite–hypersthene.

Note the absence of white mica in these assemblages.

In the Sanidinite Facies,[16] garnet may break down to form Fe-cordierite, fayalite, and hercynite, via the reaction

$$5 \text{ almandine} \Longleftrightarrow 2 \text{ Fe-cordierite} + 5 \text{ fayalite} + \text{hercynite} \quad (26.12)$$

(L. C. Hsu, 1968). Other reactions important within this facies are

$$\text{orthoclase} \Longleftrightarrow \text{sanidine} \quad (26.13)$$

(J. V. Smith, P. H. Ribbe, and D. B. Stewart, 1965, in Hyndman, 1972) and

$$\text{quartz} \Longleftrightarrow \text{tridymite} \quad (26.14)$$

(Tuttle and England, 1955). Each yields phases characteristic of the Sanidinite Facies. Thus, assemblages such as

tridymite–sanidine–Fe-cordierite–hercynite–mullite

are particularly definitive of the facies (see figure 26.7). Other assemblages include

sillimanite–mullite–cordierite–glass,

mullite–cordierite–cristobalite, and

corundum–hercynite–ilmenite–magnetite.[17]

Reviewing all of the reactions listed above, we can see that most represent either dehydration reactions or polymorphic changes. The parageneses of the Zeolite Facies are quite hydrous, whereas those of the Sanidinite Facies are rather dry. Clearly, dehydration is the norm in progressive contact metamorphism. In some cases, the fluid phase generated by the dehydration (and related) reactions migrates away from the region of metamorphism, whereas in others it promotes melting of the less refractory part of the rock, yielding a glass phase.

Silicic and Siliceous-Alkalic-Calcic Rocks

The silica-rich rocks typically have a restricted bulk composition, with high silica, moderate to low alumina, and minor amounts of other components. Because of the limited range of chemistry, mineral assemblages are also limited. Where either alumina is relatively abundant or calcium and the alkalis are significant, the mineralogical changes will be like those of the pelitic rocks and siliceous-alkali-calcic rocks, respectively.

The siliceous-alkali-calcic (SAC) rocks are quartz- and feldspar-rich and are sometimes referred to as quartzofeldspathic. Protoliths include sandstones and felsic to intermediate igneous rocks. Some of these rocks are aluminous and those that are bear mineralogical similarities to the pelitic rocks. The major differences are that the SAC rocks are characterized by calcium-rich phases and generally lack the aluminum silicates (i.e., andalusite, sillimanite). Those that are more calcium-, iron-, and magnesium-rich are gradational in composition with the basic rocks. In general, the SAC rocks differ from the basic rocks in containing abundant quartz and lacking abundant Ca-amphiboles (and clinopyroxenes).

SAC rocks may be represented on either AFM or ACF diagrams. The AFM diagrams in figure 26.7 are applicable to *aluminous* SAC rocks. Mineral assemblages of those with appreciable calcium may be shown more clearly on the ACF pseudo-phase diagram (figure 26.8). The dotted polygon outlined on the Zeolite Facies diagram in figure 26.8 shows the general range of compositions for SAC rocks. A few unusual

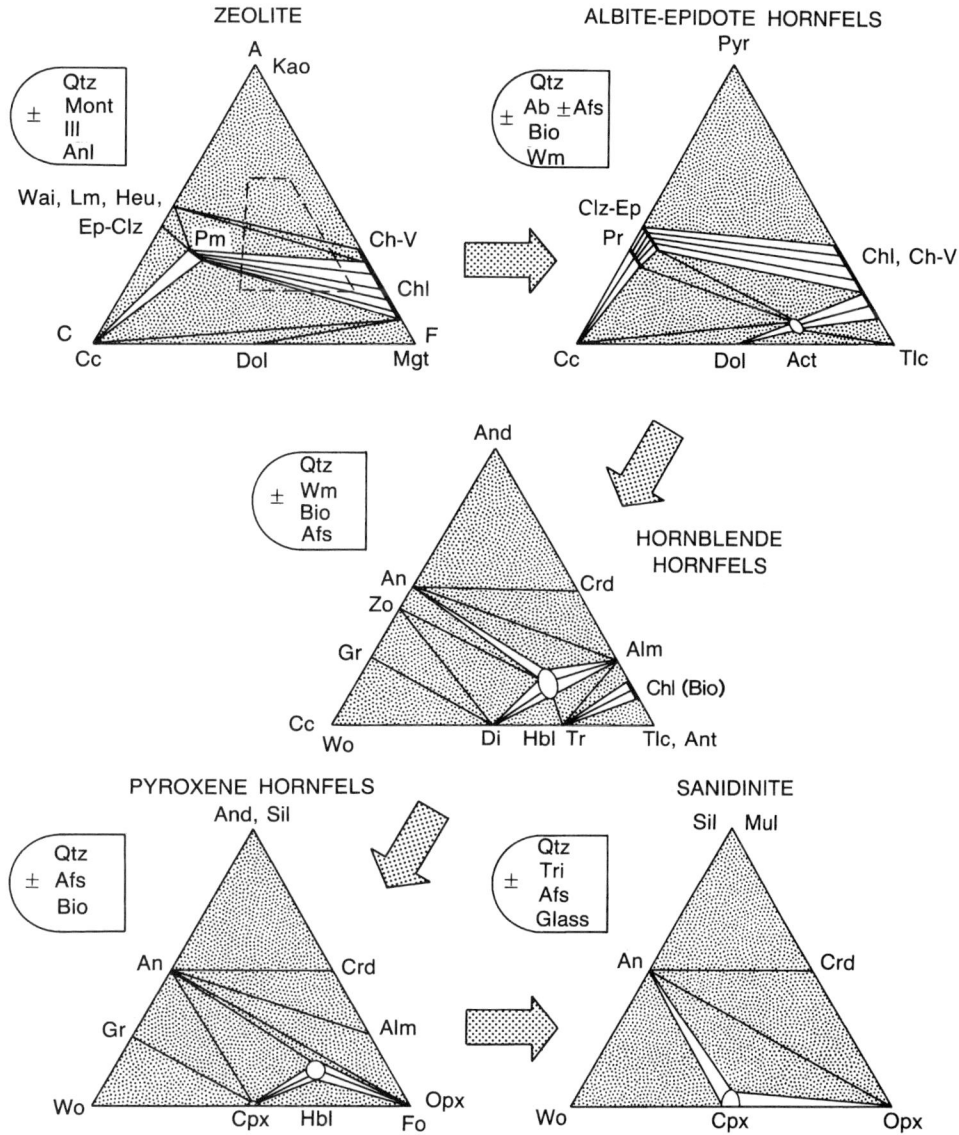

Figure 26.8 ACF diagrams depicting typical mineral assemblages for siliceous-alkali-calcic rocks, intermediate igneous rocks, and basic igneous rocks in the various facies of contact metamorphism. Abbreviations as in figure 26.7 and as follows: Ab = albite, Act = actinolite, Alm = almandine (garnet), An = anorthite, Anl = analcite, Ant = anthophyllite, Ch-V = chlorite/vermiculite, Clz = clinozoisite,

Cpx = clinopyroxene, Di = diopside, Dol = dolomite, Ep = epidote, Fo = forsterite, Gr = grossularite (garnet), Hbl = hornblende, Heu = heulandite, Lm = laumontite, Mgt = magnesite, Pr = prehnite, Pm = pumpellyite, Tlc = talc, Tr = tremolite, Wai = wairakite, Wo = wollastonite, Zo = zoisite.
(Based on sources cited in text)

compositions may lie outside of this field, but sandstones and siliceous igneous rocks generally plot on the left and upper parts of the polygon, whereas intermediate igneous rocks plot near the bottom and bottom right. Basic rocks (discussed below) plot in the bottom center and below the bottom line of the polygon. Ultrabasic rocks (also discussed below) plot in the lower right corner of the ACF triangle. *Note:* Only a few representative diagrams are presented here; numerous additional topologies exist between the diagrams shown.

At the lowest grades of metamorphism, in the Zeolite Facies, a number of phase assemblages characterize rocks metamorphosed under various conditions.[18] Figure 26.8 shows a common group of assemblages. Metasandstones typically contain an assemblage that includes quartz, analcite, clay minerals, heulandite, and chlorite or chlorite mixed-layer minerals (chlorite/vermiculite, chlorite/smectite). More calcic sandstones may also include pumpellyite, calcite, or both. At lower temperatures, xanthophyllite may occur (Droddy and Butler, 1979), whereas higher temperatures may yield laumontite, prehnite, or wairakite (Liou, Maruyama, and Cho, 1987; Inoue and Utada, 1991).

Marking the transition from the Zeolite Facies to the Albite-Epidote Hornfels Facies are the conversion of kaolinite to pyrophyllite (see equation 26.2), the conversion of analcite to albite, and the disappearance of Ca-zeolites. Albite forms via the reaction

$$\text{analcite} + \text{quartz} \Longleftrightarrow \text{albite} + H_2O \qquad (26.15)$$

(Liou, 1971a; A. B. Thompson, 1971).

Typical assemblages in SAC rocks of the Albite-Epidote Hornfels Facies include

quartz–albite–epidote–chlorite–white mica–biotite,
quartz–albite–chlorite–white mica–
epidote–prehnite–sphene, and
albite–quartz–biotite–calcite–chlorite –white mica

(Loney et al., 1975).[19] Talc and actinolite may form in more mafic sandstones (S. D. McDowell and Elders, 1980). In the lower-temperature part of the facies, pumpellyite is present. In the highest-temperature part of the facies, pumpellyite, prehnite, or both minerals may be absent. The Prehnite-Pumpellyite Facies is indicated where prehnite and pumpellyite occur together.

The appearance of andalusite (see equation 26.4), garnet, and hornblende mark the transition to the Hornblende Hornfels Facies (figure 26.8).[20] Plagioclase is more calcic than in lower-grade facies and zoisite may be present. In the higher-temperature part of the facies, cordierite occurs and calcite is replaced by wollastonite (see equation 26.1). Typical mineral assemblages mimic those of pelitic rocks, except that plagioclase and quartz are more abundant than they are in pelitic rocks. A characteristic assemblage is

quartz–plagioclase–cordierite–andalusite–biotite.

In rocks of intermediate composition, the assemblage

plagioclase–quartz–hornblende–garnet–epidote

is representative.

The Pyroxene Hornfels Facies, in its lower-temperature part, contains aluminous SAC rocks that are mineralogically like their pelitic counterparts, except that quartz and the feldspars are more abundant.[21] White mica is absent. A typical assemblage is

quartz–plagioclase–alkali feldspar–
cordierite–andalusite–biotite.

In rocks of intermediate composition, epidote-group minerals are absent and clinopyroxene is present. At higher temperatures, andalusite rather than sillimanite is present in aluminous rocks, orthopyroxene is an important phase in more basic assemblages, and hornblende does not occur. A typical assemblage might consist of

quartz–plagioclase–alkali feldspar–cordierite–
orthopyroxene.

SAC rocks may melt under Pyroxene Hornfels Facies conditions. For example, in the Sierra Nevada of California, a trachybasalt intruded a biotite-hornblende granodiorite and partially melted it (Dodge and Calk, 1978).[22] The resulting rocks contain the assemblage

quartz–plagioclase–alkali feldspar–
glass–orthopyroxene–sphene.

The phase assemblages of the Sanidinite Facies are less complex than those of the lower-temperature facies (figure 26.8).[23] SAC rocks either contain an aluminum silicate (sillimanite or mullite) or they contain a pyroxene. In either case, they contain quartz (reverted from tridymite), plagioclase, and may contain cordierite. In addition, these rocks may have alkali feldspar and glass. The glass-bearing rocks are sometimes called *buchites*. Characteristic assemblages include

mullite–cordierite–glass,
quartz–alkali feldspar–plagioclase–titanomagnetite–glass,
and
sanidine–clinopyroxene–hematite–glass.

Basic Rocks

Basic rocks—the metamorphic equivalents of gabbros, basalts, and related rocks—are characterized by the phase assemblages depicted by tie lines that cross the lower, right-hand third (the F-rich corner) of the ACF triangle (figure 26.8).[24] Contact/hydrothermally metamorphosed basic rocks are widely distributed among oceanic crustal rocks[25] and contact-metamorphosed mafic rocks exist elsewhere as well.

In the Zeolite Facies, typical assemblages include

chlorite–chlorite/smectite–analcite–heulandite–quartz–calcite–
smectite,
chlorite–albite–laumontite–epidote–pumpellyite–quartz–sphene,
and
chlorite–wairakite–albite–pumpellyite –quartz– calcite.

A wide variety of additional minerals may occur, including K-feldspar, thomsonite, chabazite, stilbite, prehnite, mixed-layer minerals (e.g., chlorite/vermiculite), various smectites, talc, anhydrite, pyrite, hematite, and magnetite. Reactions marking the upper boundary region of the Zeolite Facies, as defined here, include the following:

$$5 \text{ prehnite} + \text{chlorite} + 2 \text{ quartz} \Longleftrightarrow 4 \text{ epidote} +$$
$$\text{actinolite} + 6 H_2O \qquad (26.16)$$
$$\text{and}$$

$$\text{wairakite} \Longleftrightarrow \text{anorthite} + 2 \text{ quartz} + 2 H_2O \qquad (26.17)$$

These reactions mark the appearance of actinolite and the upper stability limit of the common zeolites, respectively (Liou, Maruyama, and Cho, 1985; Liou, 1970).[26]

The typical assemblage of the Albite-Epidote Hornfels Facies is

chlorite–albite–epidote–actinolite–quartz–sphene,

but assemblages containing prehnite *or* pumpellyite do occur within this facies. Calcite, iron oxides, and iron sulfides are also common. The upper boundary of the facies and corresponding lower boundary of the Hornblende Hornfels Facies is a broad zone, the lower-temperature limit of which is marked mineralogically by the appearance of two plagioclases (albite coexisting with oligoclase) and the coexistence of two Ca-amphiboles (an actinolite and a hornblende) (Maruyama, Liou, and Suzuki, 1982; Maruyama, Suzuki, and Liou, 1983).

The Hornblende Hornfels Facies thus begins with reactions such as

$$\text{actinolite} + \text{albite} <==> \text{hornblende} + \text{oligoclase} + \text{quartz} \quad (26.18)$$

and

$$\text{epidote} + \text{actinolite} + \text{chlorite} <==> \text{hornblende} + H_2O \quad (26.19)$$

that yield hornblende and oligoclase at temperatures just below 400°C (Spear, 1981; Maruyama, Suzuki, and Liou, 1983). At slightly higher temperatures, albite disappears. The appearance of garnets is defined by reactions such as

$$\text{prehnite} <==> 2 \text{ zoisite} + 2 \text{ grossularite} + 3 \text{ quartz} + 4 H_2O \quad (26.20)$$

and

$$\text{Fe-chlorite} + \text{quartz} + \text{magnetite} <==> \text{almandine} + H_2O \quad (26.21)$$

(L. C. Hsu, 1968; Liou, 1971; Helgeson et al. 1978). In the middle of the facies, diopside is produced via the reaction

$$\text{dolomite} + 2 \text{ quartz} <==> \text{diopside} + 2 CO_2 \quad (26.22)$$

(F. J. Turner, 1981, pp. 163–64). As calcium is consumed in the reactions above, sphene is converted to ilmenite (Moody, Meyer, and Jenkins, 1983). Together these various reactions yield parageneses such as

$$\text{actinolite–hornblende–albite–oligoclase–chlorite–epidote–sphene}$$

in lower-grade Hornblende Hornfels Facies metabasites, and

$$\text{hornblende–andesine–clinopyroxene–quartz –ilmenite, or}$$

$$\text{anthophyllite–cummingtonite–cordierite–plagioclase–quartz–ilmenite}$$

at higher grades. Biotite and garnet occur in some assemblages. Note that the latter assemblages contain no actinolite and only one plagioclase.

The transition from the Hornblende Hornfels Facies to the Pyroxene Hornfels Facies is marked by the disappearance of zoisite and Mg-chlorites. A zoisite-out reaction is

$$6 \text{ zoisite} <==> 6 \text{ anorthite} + 2 \text{ grossular} + \text{corundum} + 3 H_2O \quad (26.23)$$

Boettcher (1970). Mg-chlorite may be eliminated by reactions such as

$$5 \text{ chlorite} <==> \text{cordierite} + 3 \text{ spinel} + 10 \text{ forsterite} + 20 H_2O \quad (26.24)$$

(Helgeson et al., 1978) or

$$\text{chlorite} + 2 \text{ quartz} <==> \text{garnet} + \text{orthopyroxene} + 4 H_2O. \quad (26.25)$$

Representative assemblages in metabasites of the Pyroxene Hornfels Facies are

$$\text{orthopyroxene–clinopyroxene–hornblende–biotite–Fe-Ti oxide, and orthopyroxene–clinopyroxene–Ca-plagioclase–hornblende–ilmenite}$$

(Russ-Nabelek, 1989). Quartz or olivine may also occur.

The Sanidinite Facies and the higher-pressure parts of the Pyroxene Hornfels Facies are characterized by the absence of amphibole. A representative assemblage in a metabasite at this grade is

$$\text{Ca-plagioclase–augite–olivine–ilmenite–magnetite}$$

(I. D. Muir and Tilley, 1957). The pyroxenes are more Fe-rich and the clinopyroxenes are more aluminous. Garnet is absent from basic rocks of the Sanidinite Facies.

The Origin of Spilites

Spilite is a term used to refer to sodic basic rocks composed of assemblages such as albite–chlorite–clinopyroxene–epidote–quartz–calcite–sphene. **Keratophyre** and **quartz keratophyre** are related rocks of more siliceous character. The origin of these rocks has long been controversial, with some workers arguing that they represent the products of direct crystallization from a magma (e.g. Amstutz and Patwardhan, 1974) and others favoring some form of low-grade metamorphic origin (e.g. Battey, 1974). The former interpretation is based primarily on textural data, whereas the latter is based primarily on phase equilibria studies. The compositions of the clinopyroxenes and their textures appear to be igneous, but it has been argued that many of the other phases have a secondary appearance.

The question of the origin of spilites is raised here because the mineralogy of the spilites (and keratophyres) is much like that of metabasites of low metamorphic grade (compare the above assemblage to those listed on page 538 for the Zeolite Facies). In fact, spilite assemblages lacking pyroxene and containing zeolites or pumpellyite are identical to the Zeolite Facies assemblages. Do identical assemblages form both by low-grade metamorphism at low P and T and via crystallization of magma at high T and low P?

A number of lines of evidence suggest that the answer to the above question is no. First, we have seen that at higher T and low P the phase assemblages that are stable are those containing hornblende and a plagioclase more calcic than albite; the albite–chlorite–epidote assemblage is not stable. Second, experimental phase-stability studies of rocks of spilite-like composition indicate that the assemblage clinopyroxene–chlorite (as well as other assemblages containing chlorite) is not stable under magmatic conditions (Yoder, 1967; Liou, Kuniyoshi, and Ito, 1974; Moody, Meyer, and Jenkins, 1983). Furthermore, no spilite is known to have been erupted in historic times and the only contemporary volcanic rocks that are similar are hydrothermally metamorphosed oceanic basalts (Humphris and Thompson, 1978). Experimental studies of seawater-basalt interaction indicate that spilitization is a viable hydrothermal-metamorphic process (Wedepohl, 1988). Together these data make a strong case for a metamorphic origin for the spilites. Inasmuch as many spilites contain > 4% Na_2O, it is likely that sodium metasomatism, probably on the seafloor, was an important process in the development of the spilites and keratophyres.

Carbonate Rocks

Because they commonly develop into marbles and skarns containing interesting and valuable minerals, limestones and dolostones in contact aureoles have received considerable study.[27] The formation of the interesting minerals, however, requires that the carbonate rock be impure. Calcite, dolomite, or magnesite alone are stable, at slightly elevated pressures and in the absence of H_2O, to temperatures in excess of 700° C.[28] Consequently, contact metamorphism of a pure limestone or dolostone may yield no more than a coarsening of grain size in the carbonate minerals of the rock.

Most carbonate rocks are not pure calcite or dolomite. Many contain silica in the form of chert or sand grains, and calcite-dolomite-quartz mixtures are also common. Clay minerals in carbonate rocks provide aluminum to the system. Organic carbon, another common impurity, is converted to graphite upon metamorphism. Iron, potassium, and sodium are among the other elements that may occur, become involved in reactions, or control the stability of phases during metamorphism (see Skippen and Trommsdorff, 1986).

Bowen (1940) presented the first comprehensive analysis of progressive metamorphism of siliceous limestones and dolostones, which focussed on the anhydrous system $CaO\text{-}MgO\text{-}SiO_2\text{-}CO_2$. He proposed that through a series of decarbonation reactions, siliceous dolostones and limestones yield successive assemblages marked by ten index minerals: tremolite (which, of course, is a hydrous phase), forsterite, diopside, periclase, wollastonite, monticellite, akermanite, spurrite, merwinite, and larnite.[29] The compositions of these phases and their positions on the $CaO\text{-}MgO\text{-}SiO_2$ (CMS) plane are shown in figure 26.9. Figure 26.10 shows a petrogenetic grid with coordinates of T and $P \approx P_{CO_2}$ on which the positions of selected, experimentally determined and calcu-

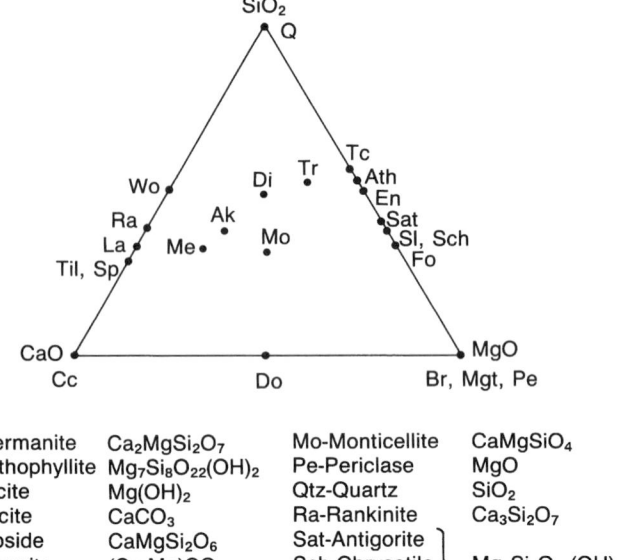

AK-Akermanite	$Ca_2MgSi_2O_7$
Ath-Anthophyllite	$Mg_7Si_8O_{22}(OH)_2$
Br-Brucite	$Mg(OH)_2$
Cc-Calcite	$CaCO_3$
Di-Diopside	$CaMgSi_2O_6$
Do-Dolomite	$(Ca,Mg)CO_3$
En-Enstatite	$Mg_2Si_2O_6$
Fo-Forsterite	Mg_2SiO_4
La-Larnite	Ca_2SiO_4
Me-Merwinite	$Ca_3Mg(SiO_4)_2$
Mgt-Magnesite	$MgCO_3$
Mo-Monticellite	$CaMgSiO_4$
Pe-Periclase	MgO
Qtz-Quartz	SiO_2
Ra-Rankinite	$Ca_3Si_2O_7$
Sat-Antigorite	
Sch-Chrysotile }	$Mg_6Si_4O_{10}(OH)_8$
Sl-Lizardite	
Sp-Spurrite	$Ca_5(SiO_4)_2CO_3$
Tc-Talc	$Mg_3Si_4O_{10}(OH)_2$
Til-Tilleyite	$Ca_5Si_2O_7(CO_3)_2$
Tr-Tremolite	$Ca_2Mg_5Si_8O_{22}(OH)_2$
Wo-Wollastonite	$CaSiO_3$

Figure 26.9 The $CaO\text{-}MgO\text{-}SiO_2$ plane of the system $CaO\text{-}MgO\text{-}SiO_2\text{-}H_2O\text{-}CO_2$ with plotted positions of various calc-silicate minerals.

lated curves relevant to such decarbonation reactions are plotted. Between the curves, phase diagrams depict the assemblages (tie lines connect the various phases, the compositions and plotted positions of which are shown in figure 26.9). Note that in each of the reactions, CO_2 is produced (and may become a part of the fluid phase).

The Importance of the Fluid Phase

In many aureoles, tremolite and talc are phases that develop at the lowest grades of metamorphism.[30] Both are hydrous phases. Neither is produced by the reactions shown in figure 26.10. During progressive metamorphism, these minerals develop through interaction with H_2O and yield H_2O to the fluid phase, which commonly also derives H_2O from the intrusion that causes the contact metamorphism. Given the reactions depicted in figure 26.10 and those involving talc and tremolite, it is clear that progressive metamorphism yields both CO_2 and H_2O. These components are completely miscible and form a single fluid phase. Thus, it is necessary to consider phases in the system $CaO\text{-}MgO\text{-}SiO_2\text{-}CO_2\text{-}H_2O$ if we are to correctly discuss phase assemblages of carbonate rocks in the facies of contact metamorphism.

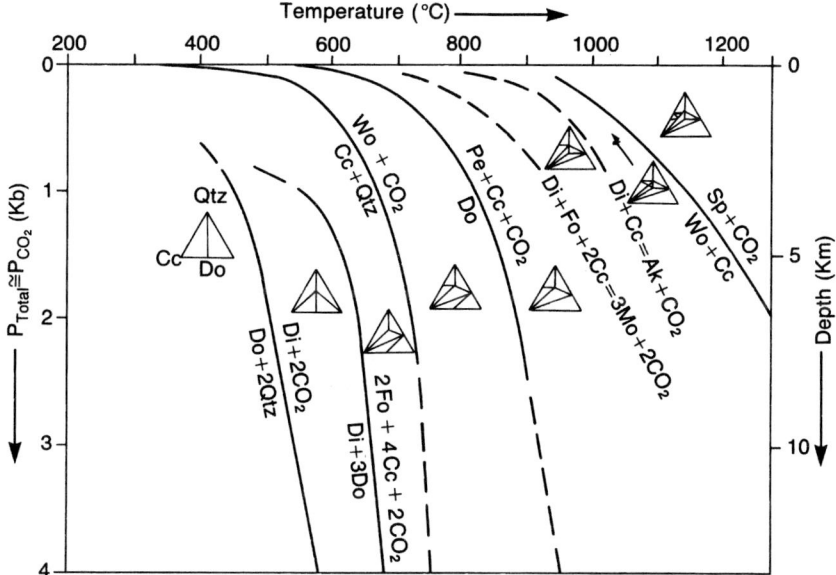

Figure 26.10 Petrogenetic grid for carbonate rocks in the system CaO-MgO-SiO_2-CO_2. $P_{total} \approx P_{CO_2}$. Selected reaction curves are shown; phase diagrams show the phase assemblages that are stable between the curves. Positions of minerals in the CaO-MgO-SiO_2 plane are as shown in figure 26.9. All curves are experimentally determined, with slight modifications based on thermodynamic calculations. *(Sources as in table 26.2 and appendix C; also see Tracy and Frost, 1991)*

The inclusion of two (or more) fluid components in the system means that (1) by the Phase Rule, most reactions will be bivariant (at the least), and (2) the effect of various ratios of CO_2 to H_2O (and other components) in the fluid phase, that is, the effect of the mole fraction of CO_2 (X_{CO_2}) or H_2O (X_{H_2O}) on the stability fields of minerals in carbonate rocks must be considered.[31] That effect, discussed briefly in chapter 25, is shown schematically in figure 26.11. In this T-X phase diagram, pressure is fixed. At very low values of X_{CO_2}, the sequence of key minerals that will develop with increasing temperature during the metamorphism of a siliceous dolomitic marble is talc → tremolite → antigorite → wollastonite → brucite → periclase (path a-a, figure 26.11). At intermediate values of X_{CO_2}, the sequence will be talc → tremolite → forsterite → wollastonite (b-b'); and at very high values of X_{CO_2}, the sequence will be diopside → forsterite → wollastonite (c-c'). The pressure and the sequence of increasing temperatures are the same in each case. Only the composition of the fluid phase varies. Thus, a particular key mineral or phase assemblage cannot be used as a P-T indicator *unless* the composition of the fluid phase is known and is taken into

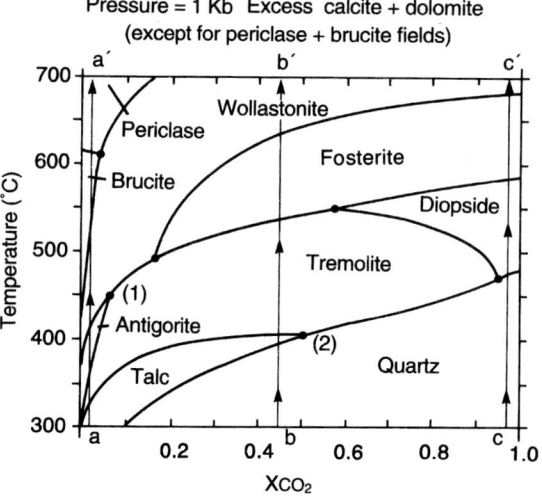

Figure 26.11 T-X_{fluid} phase diagram showing three equivalent thermal paths (a–a', b–b', c–c'), at different X_{CO_2} values, which result in three different mineral sequences. *(Modified from Bucher-Nurminen, 1982)*

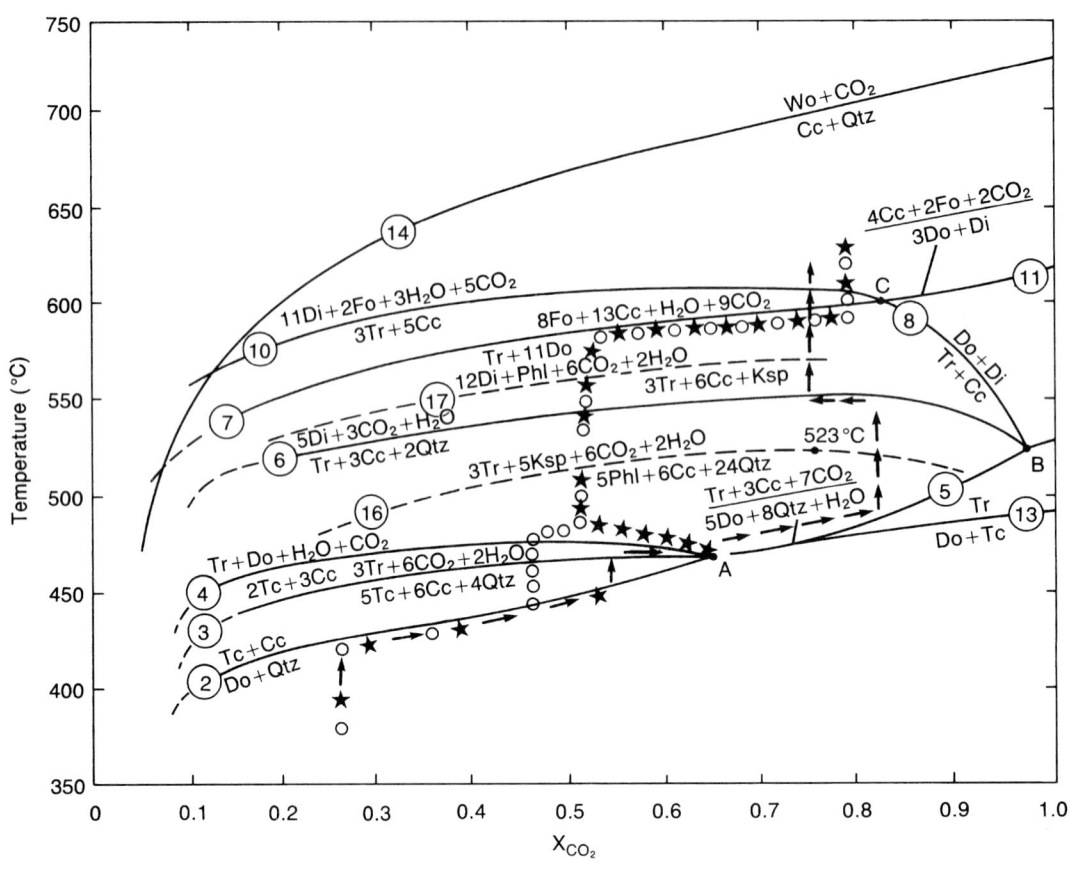

Figure 26.12

T-X$_{fluid}$ phase diagram showing suggested paths of metamorphism in T-X space (stars and arrows) for the Notch Peak Aureole, Utah. Notice the changing composition of the fluid phase (as indicated by left-to-right shifts in paths). A simplified version of the top part of this diagram is shown in figure 26.6. Abbreviations not listed in previous figures: Ksp = K-rich alkali feldspar, Phl = phlogopite. P = 2 kb (0.2 Gpa).

(From Hover Granath, Papike, and Labotka, 1983)

account. One additional complexity that exists is that the fluid phase may change composition during the progress of the metamorphic event, because of additions or subtractions of CO_2 and H_2O (figure 26.12).[32] For example, decarbonation reactions may increase the mole fraction of CO_2 in the fluid phase: In figure 26.11, this would force the reaction path to the right.

Because the fluid phase is transient, we must seek information about its former composition through direct or indirect methods. One approach is through fluid inclusion studies, which can give direct information about the fluid phase.[33] A second, but indirect, method is based on mass balance calculations of the volume of fluid generated during the length of the metamorphic event (Labotka, White, and Papike, 1984; Ferry, 1989; Labotka, 1991). A third method involves examining the phase assemblages in the aureole in order to discover phase assemblages that indicate the locations of invariant points in the T-X diagrams (where the various reaction curves intersect). For example, at a given pressure, if the sequence of key minerals in the aureole was identical to the sequence a–a′ of figure 26.11, it would be possible to conclude that the

composition of the fluid phase was $X_{CO_2} < 0.1$ and $X_{H_2O} > 0.9$ i.e. to the left of the invariant point (1). Because the positions of invariant points and the configurations of tie lines change with changing pressure, the pressure must be known if we are to use this technique. If the pressure can be determined independently, however, discovery of the sequence of key minerals and assemblages that represent invariant assemblages allows us to restrict the possible fluid phase composition to a particular range. As another example, an equilibrium assemblage of talc–tremolite–calcite–dolomite–quartz would indicate conditions at invariant point 2, indicating a fluid phase composition of $X_{CO_2} = X_{H_2O} = 0.5$. Here again, an independent determination of pressure may be necessary, although in some instances the invariant assemblages are stable only over a restricted range of pressures.

Metamorphism of Dolomitic, Argillaceous, and Siliceous Carbonate Rocks

Taking into account X_{H_2O} and X_{CO_2}, we can now define phase assemblages for the various facies of contact metamorphism of impure carbonate rocks. Typical examples are

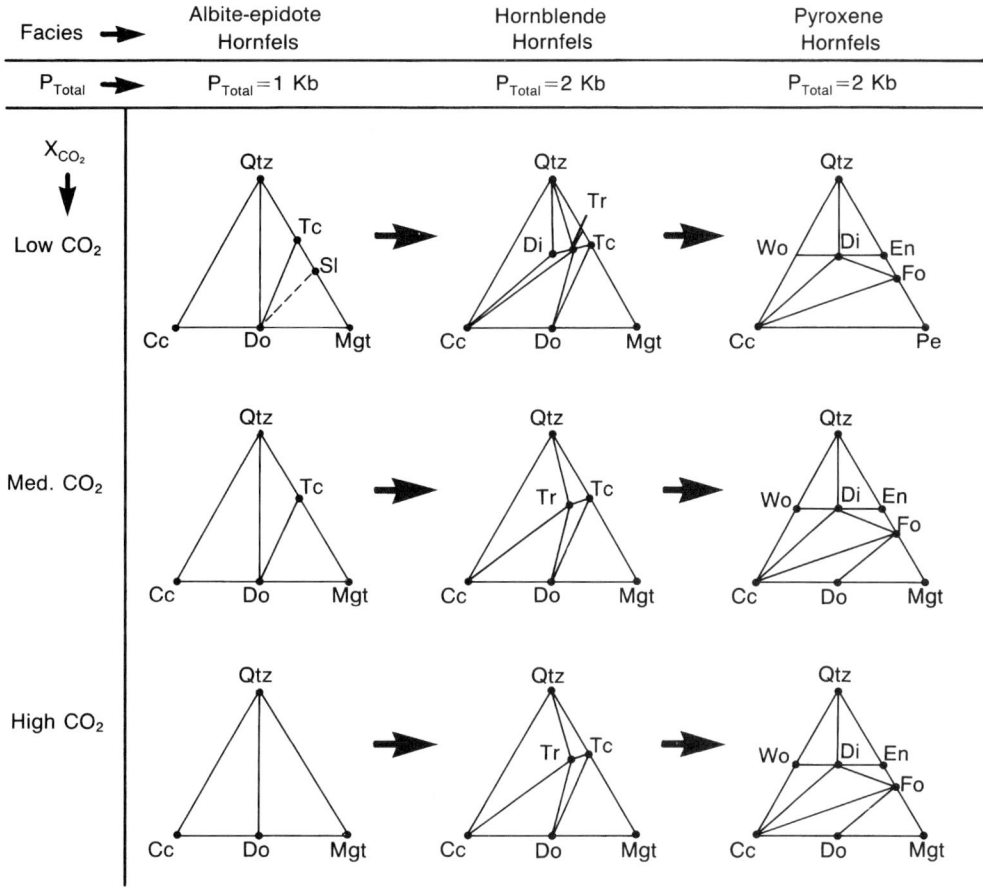

Figure 26.13 Phase assemblages in carbonate rocks (plotted on the CMS phase diagram) for various facies of contact metamorphism at low, medium, and high values of X_{CO_2} (based on various sources cited in the text). Refer to figure 26.9 for mineral compositions and plotted positions.

depicted in figure 26.13. Assemblages for low, medium, and high values of X_{H_2O} and X_{CO_2} are presented. Several important reactions are listed in table 26.2. As was the case in the figures for other bulk compositions of rock, only representative phase diagrams are shown. Additional diagrams between those shown are necessary to develop a complete sequence of diagrams for the range of possible conditions.

The important phases that develop from these dolomite- and calcite-rich rocks include talc, tremolite, diopside, forsterite, wollastonite, grossularite, phlogopite, and a host of other, uncommon minerals. The sequence in which they appear depends on the chemistry of both the rock and the fluid phase, as well as on the P-T conditions. Not all carbonate rocks containing aluminous, siliceous, or magnesian phases were initially impure. Some relatively pure limestones or other carbonate rocks have undergone *metasomatism*, resulting in the formation of Mg-, Al-, and Si-bearing phases.

Because the chemistry of carbonate rocks differs significantly from that of most other rocks (see figure 25.1), we can readily observe the effects of metasomatism in carbonate rocks

that have experienced this process. For example, consider the metamorphic aureole at Crestmore, California (figure 26.14) described by Burnham, (1954, 1959). Here, quartz diorite and porphyritic quartz monzonite have intruded a relatively pure Mg-bearing limestone unit, causing recrystallization, isochemical neocrystallization, and metasomatically induced neocrystallization. The igneous rocks (which locally exhibit a contaminated contact unit that resulted from assimilation of carbonate country rocks) are surrounded by an aureole of variable width (< 3 cm–> 15 m) consisting of four parts. The outermost zone, here referred to as the *marble zone*, consists of calcite marble and brucite-calcite marble.[34] In this zone, isochemical recrystallization, resulting in a coarsening of grain size in calcite, and neocrystallization, yielding brucite, produced these two main rock types. The marble zone is succeeded inwardly by the *monticellite zone*, consisting of rocks composed of calcite and monticellite in association with one or more of the various minerals clinohumite, forsterite, melilite, spurrite, tilleyite, and merwinite. An *idocrase zone* occurs interior to the monticellite zone. The idocrase zone contains rocks composed

543

Table 26.2   Selected Reactions Important in the Metamorphism of Carbonate Rocks

$\cdots\cdots\cdots\cdots\cdots\cdots\cdots\cdots\cdots\cdots\cdots\cdots\cdots\cdots\cdots\cdots\cdots\cdots\cdots\cdots\cdots\cdots\cdots\cdots\cdots\cdots\cdots\cdots\cdots$

3 magnesite	+	4 quartz	+	H_2O	=	talc	+	3 CO_2				
3 $MgCO_3$	+	4 SiO_2	+	H_2O	=	$Mg_3Si_4O_{10}(OH)_2$	+	3 CO_2				
3 dolomite	+	4 quartz	+	H_2O	=	talc	+	3 calcite	+	3 CO_2		
3 $CaMg(CO_3)_2$	+	4 SiO_2	+	H_2O	=	$Mg_3Si_4O_{10}(OH)_2$	+	3 $CaCO_3$	+	3 CO_2		
2 dolomite	+	talc	+	4 quartz	=	tremolite	+	4 CO_2				
2 $CaMg(CO_3)_2$	+	$Mg_3Si_4O_{10}(OH)_2$	+	4 SiO_2	=	$Ca_2Mg_5Si_8O_{22}(OH)_2$	+	4 CO_2				
5 talc	+	6 calcite	+	4 quartz	=	3 tremolite	+	6 CO_2	+	2 H_2O		
5 $Mg_3Si_4O_{10}(OH)_2$	+	6 $CaCO_3$	+	4 SiO_2	=	3 $Ca_2Mg_5Si_8O_{22}(OH)_2$	+	6 CO_2	+	2 H_2O		
tremolite	+	3 calcite	+	2 quartz	=	5 diopside	+	3 CO_2	+	H_2O		
$Mg_5Si_8O_{22}(OH)_2$	+	3 $CaCO_3$	+	2 SiO_2	=	5 $CaMgSi_2O_6$	+	3 CO_2	+	H_2O		
dolomite	+	2 quartz	=	diopside	+	2 CO_2						
$CaMg(CO_3)_2$	+	2 SiO_2	=	$CaMgSi_2O_6$	+	2 CO_2						
diopside	+	3 dolomite	=	2 forsterite	+	4 calcite	+	2 CO_2				
$CaMgSi_2O_6$	+	3 $CaMg(CO_3)_2$	=	2 Mg_2SiO_4	+	4 $CaCO_3$	+	2 CO_2				
3 tremolite	+	5 calcite	=	2 forsterite	+	11 diopside	+	5 CO_2	+	3 H_2O		
3 $Ca_2Mg_2Si_8O_{22}(OH)_2$	+	5 $CaCO_3$	=	2 Mg_2SiO_4	+	11 $CaMgSi_2O_6$	+	5 CO_2	+	3 H_2O		
calcite	+	quartz	=	wollastonite	+	CO_2						
$CaCO_3$	+	SiO_2	=	$CaSiO_3$	+	CO_2						
dolomite	=	periclase	+	calcite	+	CO_2						
$CaMg(CO_3)_2$	=	MgO	+	$CaCO_3$	+	CO_2						
2 calcite	+	forsterite	+	diopside	=	3 monticellite	+	2 CO_2				
2 $CaCO_3$	+	Mg_2SiO_4	+	$CaMgSi_2O_6$	=	3 $CaMgSiO_4$	+	2 CO_2				
diopside	+	calcite	=	akermanite	+	CO_2						
$CaMgSi_2O_6$	+	$CaCO_3$	=	$Ca_2MgSi_2O_7$	+	CO_2						
3 calcite	+	2 wollastonite	=	tilleyite	+	CO_2						
3 $CaCO_3$	+	2 $CaSiO_3$	=	$Ca_5Si_2O_7(CO_3)_2$	+	CO_2						
3 calcite	+	2 wollastonite	=	spurrite	+	2 CO_2						
3 $CaCO_3$ + 2 $CaSiO_3$	=	$Ca_5(SiO_4)_2CO_3$	+	2 CO_2								
2 monticellite	+	spurrite	=	2 merwinite	+	calcite						
2 $CaMgSiO_4$	+	$Ca_5(SiO_4)_2CO_3$	=	2 $Ca_3Mg(SiO_4)_2$	+	$CaCO_3$						

Table 26.2 Continued

tilleyite	+	4 wollastonite	=	3 rankinite	+	2 CO_2
$Ca_5Si_2O_7(CO_3)_2$	+	4 $CaSiO_3$	=	3 $Ca_3Si_2O_7$	+	2 CO_2
spurrite	+	rankinite	=	4 larnite	+	CO_2
$Ca_5(SiO_4)_2CO_3$	+	$Ca_3Si_2O_7$	=	4 Ca_2SiO_4	+	CO_2
spurrite	=	calcite	+	2 larnite		
$Ca_5(SiO_4)_2CO_3$	=	$CaCO_3$	+	2 Ca_2SiO_4		
rankinite	=	larnite	+	wollastonite		
$Ca_3Si_2O_7$	=	Ca_2SiO_4	+	$CaSiO_3$		

Sources: Bowen (1940), R. I. Harker and Tuttle (1955), Burnham (1959), Metz and Winkler (1963), Metz and Trommsdorff (1968), Metz (1970), Metz and Puhan (1970), Skippen (1971, 1974), Puhan and Hoffer (1973), Joesten (1974), Slaughter, Kerrick, and Wall (1975), Zharikov, Schmulovich, and Vulatov (1977), Winkler (1979), Eggert and Kerrick (1981).

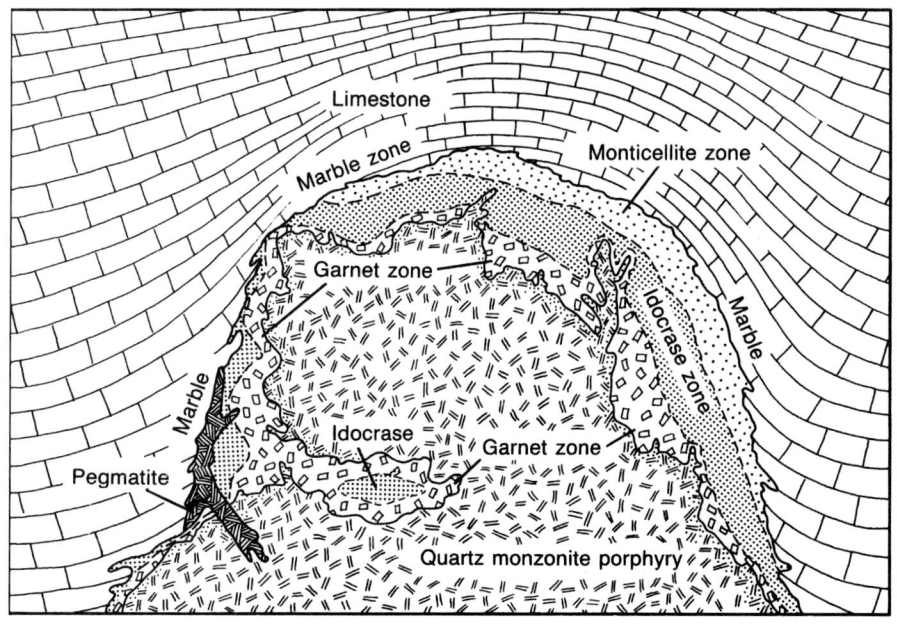

Figure 26.14 Idealized cross section through the quartz monzonite and contact aureole at Crestmore, California, showing the various zones of contact metamorphism.
(Slightly modified from Burnham, 1959)

of idocrase in association with such minerals as calcite, diopside, wollastonite, phlogopite, monticellite, and xanthophyllite. Closest to the intrusion is the *garnet zone*, where diopside-wollastonite-grossularite rocks, containing minor calcite and quartz, are the dominant rock types.

Examination of the key minerals indicates that metasomatism has occurred. The progressive sequence of key minerals and their chemistries is as follows:

calcite — $CaCO_3$
calcite + brucite — $CaCO_3 + Mg(OH)_2$
monticellite — $CaMgSiO_4$
idocrase — $Ca_{10}Mg_2Al_4Si_9O_{34}(OH)_4$

grossularite - wollastonite - diopside — $Ca_3Al_2Si_3O_{12}$ - $CaSiO_3$ $CaMgSi_2O_6$.

Notice that there is a progressive increase in the ratio Si/Ca towards the contact and a similar increase in Al. Chemical analyses of the rocks confirm these trends and also indicate a slight enrichment in Fe^{3+} (figure 26.15). Inasmuch as the original rock was a Mg-bearing limestone, the first two assemblages indicate isochemical metamorphism (the water was probably present initially or is a retrograde addition). The latter three assemblages reflect introduction of silica and alumina (i.e., they reflect metasomatism). Although less obvious in less pure rocks, similar metasomatism may occur during contact metamorphism in rocks of any composition.

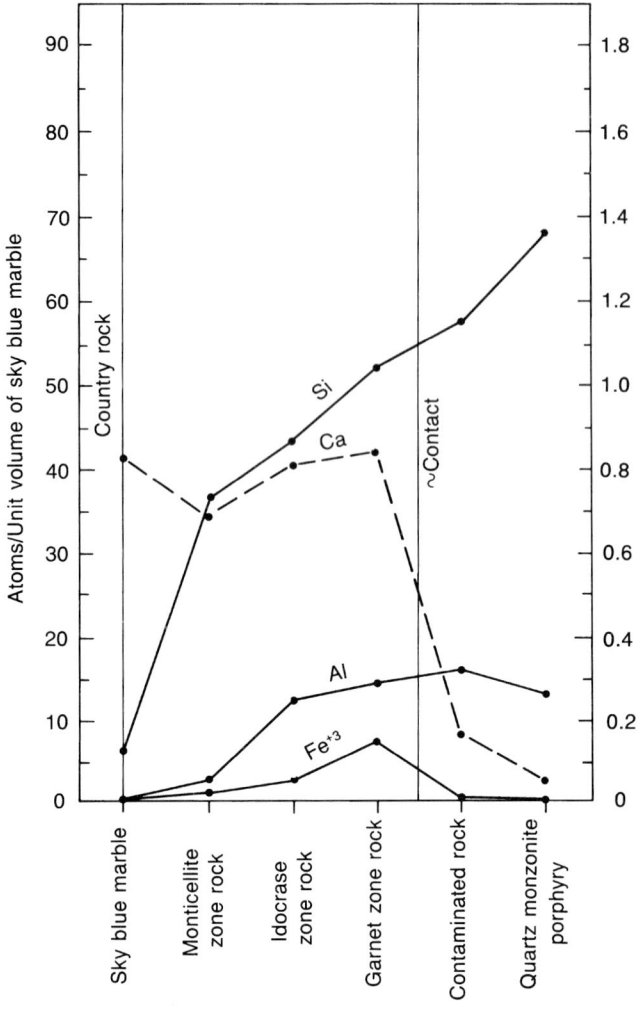

Figure 26.15 Chemical plot showing the variation in abundance of selected elements across the contact aureole at Crestmore, California.
(Simplified from Burnham, 1959)

Ultramafic Rocks

In considering contact metamorphism of ultramafic rocks, the system CaO-MgO-SiO$_2$-CO$_2$-H$_2$O is pertinent, but for these rocks, we must focus on the Mg-rich side of the CMS triangle (B. W. Evans, 1977). The following idealized sequence of phase assemblages, based principally on assemblages reported from several localities,[35] represents a progression from lower to higher grade.

Zeolite Facies

talc–chrysotile–lizardite–dolomite–chlorite–magnetite
talc–chrysotile–lizardite–tremolite–chlorite–magnetite

Albite-Epidote-Hornfels Facies

talc–chrysotile–lizardite–tremolite–chlorite–magnetite
talc–antigorite–tremolite–chlorite–magnetite

Hornblende Hornfels Facies

talc–antigorite–tremolite–chlorite–magnetite
talc–forsterite–tremolite–chlorite–chromite

Pyroxene Hornfels Facies

forsterite–tremolite–anthophyllite–chlorite–chromite
forsterite–tremolite–enstatite–chlorite–chromite
forsterite–hornblende–enstatite–spinel–chromite
forsterite–diopside–enstatite–spinel.

These assemblages represent a limited range of bulk composition. Other bulk compositions will yield somewhat different assemblages in which, for example, diopside and/or forsterite will appear at lower-temperature conditions. Note that hornblende does not represent the Hornblende Hornfels Facies in these bulk compositions, but rather forms in the Pyroxene Hornfels Facies. In the absence of H$_2$O (i.e., at very low activites of water) or at very low pressures, the temperatures at which the neocrystallization reactions occur are substantially lower (see figure 26.10).

Carbonate phases are generally absent from the assemblages listed above. In those assemblages in which they are absent, the composition of the fluid phase may be essentially X_{H_2O} = 1.0. Under such circumstances, brucite is a common phase in Mg-rich rocks. Carbonate phases *are* present in some rocks (S. E. Swanson, 1981), and in such rocks the phase relations will be controlled by the composition of the fluid phase (i.e., by the ratio X_{CO_2}/X_{H_2O})(B. W. Evans and Trommsdorff, 1974).

Metamorphic reactions in ultramafic rocks are discussed more fully in chapter 31. Nevertheless, reactions of note include

$$17 \text{ chrysotile} <==> \text{antigorite} + 3 \text{ brucite} \qquad (26.26)$$

$$\text{antigorite} + 20 \text{ brucite} <==> 34 \text{ forsterite} + 51 \text{ H}_2\text{O} \qquad (26.27)$$

and

$$9 \text{ talc} + 4 \text{ forsterite} <==> 5 \text{ anthophyllite} + 4 \text{ H}_2\text{O} \qquad (26.28)$$

(H. J. Greenwood, 1963; 1971; B. W. Evans, 1977).[36] These define the approximate lower limits of the Albite-Epidote Hornfels Facies, the Hornblende Hornfels Facies, and the Pyroxene Hornfels Facies, respectively.

SUMMARY

Contact metamorphic rocks are typically nonfoliated granoblastic rocks called hornfelses. They form from elevated temperatures in aureoles that surround igneous intrusions. The temperatures at some contacts may have exceeded 1000° C, but in all aureoles, they diminished to regional values over relatively short distances. Because of this decline in temperature, the aureoles are typically narrow, seldom measuring more than 2 km across and commonly measuring only a few to hundreds of meters across. Pressures of metamorphism seldom exceeded 0.4 Gpa (4 kb) and were typically less than 0.2 Gpa (2 kb), indicating that contact metamorphic rocks are characteristically associated with epizonal intrusions. Hydrothermal metamorphism in active geothermal regions, including the mid-ocean ridges, is a form of contact metamorphism characterized by large fluxes of H_2O-rich fluids.

The facies recognized in contact metamorphic terranes include the Zeolite Facies, the Prehnite-Pumpellyite Facies, the Albite-Epidote Hornfels Facies, the Hornblende Hornfels Facies, the Pyroxene Hornfels Facies, and the Sanidinite Facies. Together, these form a Contact Metamorphic Facies Series. Across this facies series, aluminum silicates change from kaolinite to pyrophyllite to andalusite to sillimanite to mullite or corundum. Partial melting may yield glass in the Sanidinite Facies. Chloritoid and cordierite are also characteristic phases in pelitic rocks. In metabasites, actinolite is the stable inosilicate at the lower grades of metamorphism, whereas hornblende and the pyroxenes (diopside plus orthopyroxenes) characterize the medium and higher grades, respectively. In impure carbonate rocks, a variety of minerals is produced, including talc and tremolite at the lower grades; diopside, forsterite, periclase, and wollastonite at the intermediate grades; and monticellite, merwinite, spurrite, tilleyite, akermanite, rankinite, and larnite at the highest grades (high T, low P). In all of these rock types, but especially in the carbonate rocks, careful attention must be given to understanding the fluid phase that existed during metamorphism, if the conditions of metamorphism are to be understood. The fluid phase may, in some cases, promote metasomatism, and in many cases controls the mineral composition of the rocks.

EXPLANATORY NOTES

1. The term hornfels, a genetic term for granoblastic to diablastic rocks formed in contact metamorphism, is well ingrained in the literature and practice of geology. It is a different kind of term (being based on origin) than schist or slate, which are based on texture, or marble, which is based on composition. The term usually implies a fine-grained, nonfoliated texture, but is also used for coarse-grained rocks or those that may have a foliation. Furthermore, while some geologists envision hornfelses as dark-shaded rocks, others apply the term to any contact rocks, whether dark or light. Consequently, I recommend that this term be abandoned over time. **Contact granoblastite** and **contact diablastite** are suitable substitutes for *hornfels*, where there is a need to imply genesis with the rock name. As descriptive terms, granoblastite and diablastite are superior.

2. Additional descriptions of this contact aureole, more readily accessible, may be found in F. J. Turner (1968, pp. 6–8) and Best (1982, pp. 415–417).

3. Also see Ferry (1980a) and Nisbet and Fowler (1982), Carslaw and Jaeger (1959) give a quantitative treatment of the conduction of heat in solids. A treatment of the thermodynamics of metamorphic reactions is beyond the scope of this text, but the interested student may find a good introduction in Krauskopf (1967) and in basic texts on thermodynamics for geologists, such as R. Kern and Weisbrod (1967).

4. For example, see Burnham (1959), Lachenbruch et al. (1976), Fyfe, Price, and Thompson (1978, p. 33, chs. 3, 6), Kanaris-Sotiriou and Angus (1979), Ferry and Burt (1982), J. M. Rice and Ferry (1982), Tracy et al. (1983), Walther (1983), Labotka, White, and Papike (1984), Bowman, O'Neil, and Essene (1985), Truesdell and Janik (1986), Labotka et al. (1988), McKibben, Andes, and Williams (1988), Wedepohl (1988), Novick and Labotka (1990), and Sisson and Hollister (1990). Also see various papers in Walther and Wood (1986) for detailed treatments of subjects related to metamorphic fluid compositions, analyses, and reactions.

5. Kridelbaugh (1973) discusses this reaction and its kinetics.

6. H. J. Greenwood (1962, 1967, 1975a), J. M. Rice and Ferry (1982), Speer (1982), Hover Granath, Papike, and Labotka (1983), Ferry (1983b,c), Trommsdorf and Skippen (1986).

7. Greenwood (1962, 1967), Metz (1970), Touret and Dietvorst (1983).

8. Additional discussions of porphyroblast development may be found in Rosenfeld (1968), Spry (1969), Misch (1971), van den Eeckhout and Konert (1983), T. H. Bell, Rubenach, and Fleming (1986), D. J. Prior (1987), Vernon (1988), T. H. Bell and Johnson (1989), W. D. Carlson (1989), J. Reinhardt and Rubenach (1989), and S. E. Johnson (1990).

9. KFMASH is an abbreviation for the system K_2O-FeO-MgO-Al_2O_3-SiO_2-H_2O, which can generally be represented on AFM diagrams. Calculations using Shreinemakers' technique (see ch. 25, note 15) may be used to construct schematic petrogenetic grids for this system. Pattison and Tracy (1991) review contact metamorphism of pelitic rocks using the KFMASH system and divide the contact rocks into four contact facies series, corresponding with Contact- Buchan- (2 series), and Barrovian facies series. They note that low- to high-P conditions exist in various contact metamorphic aureoles, but their discussion emphasizes the fact that most contact metamorphism is low pressure metamorphism.

10. Zeolite Facies rocks are not common in metamorphic aureoles because intrusions commonly occur in previously metamorphosed rocks. Where pelitic sediments occur along mid-ocean ridges and on the cratons in hydrothermal areas, however, Zeolite Facies rocks are present. The phase diagrams presented here are based on data obtained from Muffler and White (1969), S. D. McDowell and Elders (1980), Barger and Beeson (1981), Evarts and Schiffman (1983), C. E.Weaver et al. (1984), Alt et al. (1986), and Liou, Maruyama, and Cho (1987). Additional data are from M. Frey (1978). Also see Shearer et al. (1988), Yau et al. (1988), Inoue and Utada (1991), and the summaries in Winkler (1979) and Hyndman (1985).

11. Helgeson et al. (1978) discuss this and many of the other reactions mentioned here and those important in subsequent discussions of metamorphism. Phase assemblages important in the Albite-Epidote Hornfels Facies are described by Muffler and White (1969), Verhoogen et al. (1970, p. 549ff.), Loney et al. (1975), Evarts and Schiffman (1983), C. E. Weaver et al. (1984), Liou, Maruyama, and Cho (1987), and Loney and Brew (1987) and are discussed by Winkler (1979), F. J. Turner (1981), and J. B. Thompson and Thompson (1976), among others. These works form the foundation for the discussion here. Also see M. Frey (1978, 1987c), Ashworth and Evirgen (1984), and Miyashiro and Shido (1985) for additional discussions of phase assemblages and mineralogical changes relevant to this facies.

12. Hornblende Hornfels Facies rocks are discussed widely in the literature. This discussion is based primarily on examples described in Philbrick (1936), Chapman (1950), Compton (1960), J. M. Moore (1960), Verhoogen et al. (1970, p. 549ff.), T. P. Loomis (1972a, 1976), Loney et al. (1975), Vaniman, Papike, and Labotka (1980), Speer (1982), Labotka (1983), Evans and Speer (1984), Labotka, White, and Papike (1984), Pattison and Harte (1985), Loney and Brew (1987), and Pattison (1987). Also see Reverdatto (1970), Miyashiro (1973a), Mason (1978), Winkler (1979), F. J. Turner (1981), and Best (1982) for general discussions, as well as E. W. Reinhardt (1968), J. B. Thompson and Thompson (1976), Holdaway (1978), April (1980), and Kerrick (1987) for treatments of particular relevant topics.

13. Mg-amphiboles are reported by Seki (1957, 1961 in Miyashiro, 1973a). Staurolite is reported, for example, by Compton (1960), T. P. Loomis (1972a), and Loney et al. (1975).

14. Also see Kerrick (1972), Chatterjee and Johannes (1974), and Schramke et al. (1987).

15. Rocks of the Pyroxene Hornfels Facies have been described by Tilley (1924), T. P. Loomis (1972a, 1976), Simmons, Lindsley, and Papike (1974), Dodge and Calk (1978), Vaniman, Papike, and Labotka (1980), Benimoff and Sclar (1984), N. H. Evans and Speer (1984), Droop and Charnley (1985), Pattison and Harte (1985), and Pattison (1987). The phase relations are reviewed by Miyashiro (1973a) and F. J. Turner (1981).

16. Sanidinite Facies pelitic rocks are described by Brauns (1911, in F. J. Turner, 1981), H. H. Thomas (1922), Agrell and Langley (1958), Searle (1962), and Brearley (1986), and their parageneses are reviewed by F. J. Turner (1981). These works provide the basis for the discussion here.

17. See Agrell and Langley (1958), Searle (1962), and F. J. Turner (1981, p. 283ff.). Also see Cosca et al. (1989) for some extremely unusual mineral assemblages.

18. The mineral assemblages of low T metamorphic rocks are reviewed by M. Frey (1987b) and Liou, Maruyama, and Cho (1987). For descriptions of specific Zeolite Facies siliceous-alkali-calcic rocks, see Muffler and White (1969), Droddy and Butler (1979), S. D. McDowell and Elders (1980), Barger and Beeson (1981), Brauckmann and Fuchtbauer (1983), and Evarts and Schiffman (1983), which together with the reviews in M. Frey (1987b) and Liou, Maruyama, and Cho (1987) provide the basis for the mineral associations and reactions discussed here. Some experimental temperature limits are reported by Barth-Wirsching and Höller (1989).

19. Additional works reporting Albite-Epidote Hornfels Facies assemblages include Verhoogen et al. (1970, p. 553ff.), Droddy and Butler (1979), S. D. McDowell and Elders (1980), Brauckmann and Fuchtbauer (1983), Evarts and Schiffman (1983), Cho and Liou (1986), and Inoue and Utada (1991). Also see the review in F. J. Turner (1981).

20. The character of Hornblende Hornfels Facies SAC rocks is reviewed by Miyashiro (1973a) and Turner (1981). See those works and studies of specific areas, including Harker and Marr (1893), Compton (1960), Verhoogen et al. (1970, p. 549ff.), Loney et al. (1975), Droddy and Butler (1979), R. G. Gibson and Speer (1986), and Loney and Brew (1987).

21. The mineral assemblages of Pyroxene Hornfels Facies SAC rocks are like those of the pelitic rocks. Few separate descriptions have been published, but see Le Maitre (1974), Loney et al. (1975), and Loney and Brew (1987). Also, refer to F. J. Turner (1981).

22. A similar case is described by Kitchen (1989).

23. H. H. Thomas (1922), Searle (1962), Turner (1981), Brearley (1986).

24. The phase assemblages reported and the discussion here are based on several works including H. H. Thomas (1922), Thomson (1935), G. A. MacDonald (1944), Muir and Tilley (1957, 1958), Agrell and Langley (1958), Seki, Ernst, and Onuki (1969, in Liou, 1971b), Spooner and Fyfe (1973), Chinner and Fox (1974), Hietanen (1974), Jolly (1974), Bonatti (1975), Floyd (1975), Kuniyoshi and Liou (1976b), Dimroth and Lichtblau (1979), S. E. Swanson and Schiffman (1979), Ernst, Liou, and Moore (1981), Spear (1981a,b), F. J. Turner (1981, ch. 8), Abbott (1982, 1984), Hynes (1982), J. B. Thompson, Laird, and Thompson (1982), Evarts and Schiffman (1983), Kristmannsdottir (1983), Maruyama, Suzuki, and Liou (1983), J. B. Moody, Meyer, and Jenkins (1983), Andrew (1984), Sivell and Waterhouse (1984a), Ishizuka (1985), Liou, Maruyama, and Cho (1985), Alt et al. (1986), Cho, Liou, and Maruyama (1986), Cho and Liou (1987), Ferry, Mutti, and Zuccala (1987), Liou, Maruyama, and Cho (1987), Becker et al. (1989), and Russ-Nabelek (1989).

25. For example, see Spooner and Fyfe (1973), Bonatti et al. (1975), Humphris and Thomson (1978), Dimroth and Lichtblau (1979), Evarts and Schiffman (1983), Kristmannsdottir (1983), Sivell and Waterhouse (1984a), Ishizuka (1985), Alt et al. (1986), JGR, v. 93, no. B5 (1988), Becker et al. (1989), Goodge (1989), Gillis and Robinson (1990), and the numerous works cited therein.

26. Liou, Maruyama, and Cho (1987) do not consider wairakite to be indicative of the Zeolite Facies, as they have adopted, for metabasites, the facies scheme of Liou, Maruyama, and Cho (1985). In that scheme, the Contact Metamorphic Facies Series would consist of the successive facies, Zeolite Facies–Prehnite-Actinolite Facies–Greenschist Facies–(Epidote Amphibolite Facies)–Amphibolite Facies. Liou, Maruyama, and Cho (1985) choose the reaction prehnite + chlorite + quartz = tremolite + zoisite + H_2O, which falls in the same general region of P-T space as the wairakite reaction, for the upper limit of their Prehnite-Actinolite Facies. The lower limit of that facies (i.e., the upper limit of their Zeolite Facies) occurs at a temperature that is about 150° C lower (see Appendix C). Thus, the reaction involving wairakite is not a critical reaction for their Zeolite Facies because it occurs at temperatures above the upper limit of the Zeolite Facies, *as they define it.*

27. References pertinent to the discussion here and on which it is based include A. Harker and Marr (1893), Bowen (1940), Tilley (1948, 1951a), Holser (1950), Burnham(1954, 1959), R. I. Harker and Tuttle (1955, 1956), R. L. Rose (1958), L. S. Walter (1963a, 1963b), H. J. Greenwood (1963, 1967a, 1967b, 1975a), Johannes and Metz (1968), Metz and Trommsdorff (1968), Johannes (1969), T. M. Gordon and Greenwood (1970), Metz (1970, 1976), Metz and Puhan (1970), Reverdatto (1970), Skippen (1971, 1974), Hoschek (1973), Kerrick, Crawford, and Randazzo (1973), Puhan and Hoffer (1973), Joesten (1974, 1976), B. W. Evans and Trommsdorff (1974), Slaughter, Kerrick, and Wall (1975), J. M. Rice (1977), Zharikov, Shmulovich, and Bulatov (1977), Winkler (1979), Lattanzi, Rye, and Rice (1980), Kase and Metz (1980), Bucher-Nurminen (1981, 1982), Eggert and Kerrick (1981), Hoersch (1981), F. J. Turner (1981), Williams-Jones (1981), Flowers and Helgeson (1983), Hover Granath, Papike, and Labotka (1983), Andrew (1984), S. B. Tanner, Kerrick, and Lasaga (1985), Skippen and Trommsdorff (1986), Labotka et al. (1988), Ferry (1989), Wallmach, Hatton, and Droop (1989), Jenkins and Clare (1990), Trommsdorf and Connolly (1990), and Tracy and Frost (1991).

28. R. I. Harker and Tuttle (1955, 1956), Johannes and Metz (1968), Winkler (1979).

29. Bowen suggested a jingle as an aid to remembering the sequence he proposed (though the jingle may be more difficult to remember than the list of mineral names): Tremble, for dire peril walks, / Monstrous acrimony spurning mercy's laws. Note that Bowen lists the appearance of periclase before wollastonite. Experimental work (R. I. Harker and Tuttle, 1955, 1956; H. J. Greenwood, 1967) reveals that wollastonite precedes periclase (under anhydrous conditions).

30. For example, Tilley (1948), Puhan and Hoffer (1973), Hoersch (1981), and Williams-Jones (1981).

31. The works of Korzhinskii (1959, in Skippen, 1974), Wyllie (1962), and H. J. Greenwood (1967) emphasized the necessity of using $T\text{-}X_{fluid}$ plots in analyzing carbonate phase assemblages in metamorphic rocks.

32. For example, see Puhan and Hoffer (1973), Williams-Jones (1981), Bucher-Nurminen (1982), and Flowers and Helgeson (1983). For an analogous situation in noncarbonate rocks see Labotka, White, and Papike (1984).

33. For example see Touret and Dietvorst (1983), Craw (1988), Nwe and Grundmann (1990), Sisson and Hollister (1990), Labotka (1991), and the references therein.

34. In Burnham (1954, 1959), the brucite-calcite marbles are called "predazzites."

35. See Pinsent and Hirst (1977), Trommsdorff and Evans (1972, 1977), B. W. Evans and Trommsdorff (1974), Springer (1974, 1980a, 1980b), B. R. Frost (1975), S. E. Swanson (1981), and the review of B. W. Evans (1977).

36. Also see Johannes (1969), B. W. Evans et al. (1976), Trommsdorf (1983).

PROBLEMS

26.1. Construct a series of ACF diagrams representing a theoretically possible step-by-step progression from the Zeolite Facies diagram to the Albite-Epidote Hornfels Facies diagram shown in figure 26.8.

26.2. Construct a series of ACF diagrams representing a theoretically possible step-by-step progression from the Pyroxene Hornfels Facies diagram to the Sanidinite Facies diagram shown in figure 26.8.

26.3. Write an equation for each of the topology changes shown in your answer to problem 26.2.

26.4. Select reaction curves for the appearance of talc and tremolite from appendix C, and using a copy of figure 26.10, plot the curves. Construct phase diagrams showing a consistent set of topologies for a contact-metamorphic sequence in which the minerals talc, tremolite, diopside, and forsterite appear, in that order.

26.5. Construct topologies in the system $CaO\text{-}MgO\text{-}SiO_2(\text{-}CO_2\text{-}H_2O)$ for each of the fields shown in figure 26.12.

26.6. Using the sequence of phase assemblages (representing low CO_2) listed in the section on ultramafic rocks (page 546), construct a consistent set of $CaO\text{-}MgO\text{-}SiO_2$ topologies representing a sequence of progressive metamorphism in which those assemblages were produced.

27

Regional Metamorphism Under Low to Medium P/T Conditions: Buchan and Barrovian Facies Series

INTRODUCTION

Mountain systems typically contain large belts of regionally metamorphosed rock. The more obvious of these regional belts, from which the term regional metamorphism was originally derived, are characterized by foliated metamorphic rocks developed under medium to high temperatures. The accompanying pressures vary from low to high values; thus, the ratio of P to T (P/T) is low to moderate and T/P is moderate to high. Geothermal gradients, which are likewise moderate to high, produce Buchan and Barrovian Facies series.

Because the pressures of Buchan and Barrovian Facies series are commonly higher than are those of Contact Facies Series, they may contain different sequences of critical minerals (see table 24.4).[1] Recall that Buchan Facies Series form under pressures, which, in the middle grades of metamorphism, are *lower* than that of the aluminum silicate triple point (i.e., pressures in the lower Amphibolite Facies are less than 0.387 Gpa or 3.87 kb). Consequently, the critical sequence of aluminum silicates is kaolinite → pyrophyllite → andalusite → sillimanite. The Barrovian Facies Series, in contrast, develops where pressures in the middle grades of metamorphism are *higher* than that of the aluminum silicate triple point (i.e., at intermediate temperatures, P > 3.87 kb). The resulting aluminum silicate mineral sequence is

kaolinite → pyrophyllite → kyanite → sillimanite. The presence in pelitic rocks of either andalusite or kyanite at the middle grades of metamorphism is one feature that distinguishes these facies series from one another.

In chapter 24, it was noted that Barrow (1893, 1912) described the sequence of mineral zones in the Scottish Highlands that later became known as the Barrovian Facies Series. Recall that this sequence was considered to be normal. Although the idea that Barrovian metamorphism is normal was no doubt overemphasized in the past, there is some tendency for metamorphic belts to evolve towards a Barrovian kind of sequence. Such a sequence represents a geothermal gradient that is somewhat average for the continents (about 20° C/km).[2] As colder metamorphic belts are heated up and hotter belts cool down, they tend to develop mineral assemblages characteristic of Barrovian Facies Series. The tendency to approach the average is one of the factors responsible for the paucity of low-temperature metamorphic belts in older mountain systems.[3] High-temperature, low-pressure belts may be more readily preserved, because high-temperature metamorphism induces dehydration, and later retrogression is inhibited by a lack of fluids. In this chapter, we first examine high T/P regional metamorphism—metamorphism that yields Buchan Facies Series—and then contrast that series with the Barrovian Facies Series.

BUCHAN FACIES SERIES

Buchan Facies Series, like Barrovian Facies Series, take their name from a region in the Scottish Highlands (H. H. Read, 1952).[4] The Buchan rocks lie to the east of the Barrovian rocks and were metamorphosed under lower pressures. Recall that the geothermal gradient of Buchan Facies Series rocks is 40°C–80°C/km. Thus, at depths of 10 km, temperatures are about 400–800° C and the pressure is 0.3 Gpa (3 kb); that is, pressures at the middle grades of metamorphism are *lower* than those of the aluminum silicate triple point.

In general, the geothermal gradients that give rise to the low pressures and high temperatures of Buchan Facies Series may be attributed to regional heating. *Where* that heating occurs may vary. In some cases, heating results from intrusion of groups of plutons at shallow to moderate depths (Lux, DeYoreo, and Guidotti, 1986; DeYoreo et al., 1989; R. B. Hanson and Barton, 1989). In northeastern North America, a thermal event generated by such a circumstance was caused by magmatic or orogenic events that resulted from plate collisions at a convergent margin.[5] Alternatively, heating associated with thinning of crust or lithospheric mantle may cause this type of metamorphism (Wickham and Oxburgh, 1985; Golberg and Leyleroup, 1990; Loosveld and Etheridge, 1990). Hence, low-pressure regional metamorphism may also develop in a rift zone at a divergent plate margin or along a major strike-slip fault. The possibilities require that a petrotectonic analysis of each Buchan-type metamorphic belt be made to assess the paleotectonic environment of metamorphism.

The appropriate paleotectonic environments for Buchan metamorphism apparently are common, and metamorphic belts with Buchan Facies Series are widely distributed. Miyashiro (1973a, ch. 7) describes a number of Buchan belts from various parts of the world, notably Spain and Japan. Other localities containing Buchan Facies Series include Maine, New Hampshire, Colorado, Oregon, Alaska, Australia, India, and Ireland.[6] Buchan Facies Series are also common in Precambrian shields, for example, in Canada and Finland (see Schreurs, 1985; Schreurs and Westra, 1986), as well as in Phanerozoic gneissic core complexes (Blumel and Schreyer, 1977). In the Precambrian, especially the Archean Eon, heat flow from the interior of the Earth may have been higher, resulting in steeper geothermal gradients and high-temperature, low-pressure metamorphism in the crust (Fyfe, 1974).

As noted above, the low pressures of metamorphism that produced the various Buchan Facies Series metamorphic belts yield the idealized critical mineral sequence in aluminous rocks: kaolinite → pyrophyllite → andalusite → sillimanite. In the type area in the Scottish Highlands, kaolinite has been reported as a possible retrograde phase, but it is andalusite that is distinctive of the series.[7] Here, as well as elsewhere, cordierite is also an important phase in the middle to high grades of Buchan Facies Series metamorphism.

The general sequence of phase assemblages in pelitic rocks at the type locality is as follows.

Hydrous Zone

muscovite–chlorite–quartz–albite–ilmenite±chloritoid

muscovite–chlorite–biotite–quartz–albite–ilmenite

Andalusite Zone

muscovite–biotite–quartz–oligoclase–garnet–andalusite–ilmenite

muscovite–biotite–quartz–oligoclase–andalusite–staurolite–cordierite–ilmenite

Sillimanite Zone

muscovite–biotite–quartz–oligoclase–andalusite–staurolite–cordierite–garnet–sillimanite–K-rich alkali feldspar

microcline–biotite–quartz–oligoclase–garnet–sillimanite–cordierite–magnetite[8]

The hydrous zone represents the Greenschist Facies, the Andalusite and lower Sillimanite zones represent the Amphibolite Facies, and the upper Sillimanite Zone represents the Granulite Facies (figure 27. 1).

Buchan Phase Assemblages and Reactions

The overall progression of mineralogical changes in the Buchan Facies Series at the type locality defines only the facies sequence Greenschist → Amphibolite → Granulite. As is typical, the Zeolite and Prehnite-Pumpellyite Facies are not represented. The complete facies series is revealed by the sequence of representative mineral assemblages listed in table 27.1. The low-grade assemblages are virtually identical to those of the Barrovian Facies Series described below. Similarly, Greenschist Facies rocks are mineralogically similar to their equivalents in Barrovian Facies Series. It is in the Amphibolite Facies, where andalusite and cordierite appear, that the Buchan Facies Series is distinguished from the higher-pressure Barrovian rocks. Although garnet and staurolite may appear in addition to cordierite in the Buchan Facies Series, the *relative order* of appearance of these minerals differs.

The various phase assemblages developed in each metamorphic zone of the Buchan Facies Series indicate various reactions. In pelitic rocks, at the lowest grade, the Zeolite Facies contains assemblages such as

kaolinite–illite–illite/smectite–chlorite–quartz–analcite

which is characterized by the clay minerals.[9] Alkali feldspar, smectites, iron oxides, calcite, dolomite, Ca-zeolites, and organic carbon may also occur.

The perceptive student will have noted the similarity between the above phase assemblage and those of low-grade contact and hydrothermal metamorphism. Because very

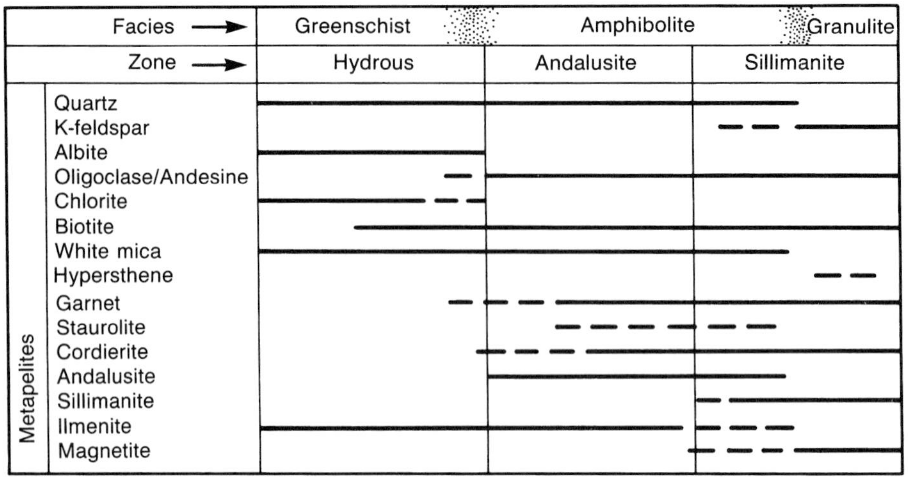

Facies →	Greenschist	Amphibolite	Granulite
Zone →	Hydrous	Andalusite	Sillimanite

Metapelites:
- Quartz
- K-feldspar
- Albite
- Oligoclase/Andesine
- Chlorite
- Biotite
- White mica
- Hypersthene
- Garnet
- Staurolite
- Cordierite
- Andalusite
- Sillimanite
- Ilmenite
- Magnetite

Figure 27.1 Mineral-facies chart for metapelites of the classic Buchan area of Buchan Facies Series metamorphism in the Scottish Highlands.

[Data sources include A. J. Baker (1985), A. J. Baker and Droop (1983), Chinner (1966), Harker (1932), Harte and Hudson (1979), N. F. C. Hudson (1980, 1985), Porteous (1973) and H. H. Read (1952).]

low-grade conditions for *all* facies series overlap, they contain identical or nearly identical phase assemblages. Thus, low-grade phase assemblages and reactions described in any of these facies series are generally pertinent to others.

At slightly higher-grade conditions, where assemblages of the Zeolite Facies are replaced by those of the Prehnite-Pumpellyite Facies, some minerals, such as K-rich alkali feldspar, are absent from many rocks, and new phases appear, such as "phengite" (greenish white mica), albite, and illite/chlorite or illite/smectite/chlorite mixed-layer minerals. These mineral changes indicate that metamorphic reactions have occurred. Smectites and K-rich alkali feldspar are among the first minerals that may disappear from aluminous rocks, and they do so via discontinuous reactions[10] such as

$$\text{smectite} + \text{K-rich alkali feldspar} <==> \text{illite} + \text{quartz} \pm \text{chlorite} \quad (27.1)$$

The amount of chlorite produced by such a reaction is small, as chlorite development is controlled by the limited number of Mg ions available from the smectite and pore fluid. Kaolinite also commonly disappears from pelitic assemblages before or during development of Prehnite-Pumpellyite Facies assemblages. The disappearance of kaolinite apparently does not result from the (perhaps anticipated) reaction

$$\text{kaolinite} + 4\,\text{quartz} <==> 2\,\text{pyrophyllite} + 2\,\text{H}_2\text{O} \quad (26.2)$$

which occurs at about 300° C. This is clear because (1) the reaction temperature is higher than that of the Zeolite/Prehnite-Pumpellyite Facies boundary (see figure 24.6); (2) in restricted bulk compositions, kaolinite is stable into the Prehnite-Pumpellyite Facies or lowermost Green-schist Facies, where it is replaced by pyrophyllite;[11] and (3) kaolinite typically disappears from pelitic rocks of metamorphic sequences within or at the top of the Zeolite Facies.[12] A reaction that would yield the appropriate change[13] is

$$\text{kaolinite} + \text{quartz} + \text{alkali feldspar} + \text{magnetite} + \text{calcite} + \text{Mg}^{++} + (\text{OH})^- <==>$$

$$\text{illite/montmorillonite/chlorite (mixed layer)} + \text{illite} + \text{CO}_2 + \text{H}_2\text{O}. \quad (27.2)$$

Such a reaction involves most of the phases present in the rock, except organic carbon, which yields methane (CH_4) to the fluid phase.

Greenschist Facies assemblages are perhaps the most widely distributed (or at least recognized) metamorphic assemblages on Earth. This is probably the case, because (1) the Greenschist Facies occurs in the lower grades of Buchan and Barrovian Facies series, and in the middle to upper grades of the Sanbagawa Facies Series, and (2) the P-T conditions of Greenschist Facies metamorphism are high enough to readily promote reactions, but low enough to be attained in many environments of metamorphism.

Typical assemblages in Greenschist Facies pelitic rocks include

white mica–chlorite–quartz–albite–magnetite,

white mica–chlorite–quartz–epidote–albite, and

white mica–chlorite–biotite–albite–quartz–ilmenite.

Stilpnomelane, alkali feldspar, pyrophyllite, chloritoid, sphene, and calcite are among the other phases that may appear in this facies. The appearance of the various new

Table 27.1 Typical Mineral Assemblages of Buchan Facies Series Rocks

Zeolite Facies

illite/smectite–kaolinite–chlorite/smectite–quartz–hematite (pelitic rock)

quartz–analcite–heulandite–chlorite–illite–hematite (SAC rock)

chlorite–prehnite–analcite–quartz–sphene (metabasite)

Prehnite-Pumpellyite Facies

illite/smectite–chlorite–quartz–albite–hematite (pelitic rock)

quartz–albite–white mica–chlorite–stilpnomelane–prehnite–pumpellyite–hematite (SAC rock)

chlorite–quartz–albite–white mica–pumpellyite–prehnite–stilpnomelane –sphene (metabasite)

Greenschist Facies

muscovite–chlorite–biotite–quartz–albite–magnetite (pelitic rock)

quartz–albite–muscovite–chlorite–K-feldspar–calcite–pyrite (SAC rock)

chlorite–actinolite–epidote–albite–quartz–sphene (metabasite)

Amphibolite Facies

muscovite–biotite–quartz–andalusite–cordierite–staurolite–oligoclase–ilmenite (pelitic rock)

quartz–oligoclase–muscovite–biotite–garnet–andalusite–ilmenite (SAC rock)

hornblende–plagioclase–garnet–biotite–quartz–sphene (metabasite)

Granulite Facies

K-feldspar–sillimanite–garnet–andesine–quartz–biotite–magnetite (pelite)

quartz–andesine–K-feldspar–biotite–garnet–magnetite (SAC rock)

hornblende–garnet–clinopyroxene–plagioclase–quartz–ilmenite–apatite–sphene (metabasite)

Sources: See notes 4 and 15; unpublished data from the author.

minerals signals several reactions. Changes in chlorite and white mica compositions involve continuous reactions. For example, biotite is produced by the reaction

$$\text{celadonite (Fe,Mg-bearing white mica) + chlorite} \Longleftrightarrow$$
$$\text{muscovite + biotite + quartz} + H_2O \qquad (27.3)$$

in which the composition of the white mica changes (Brown, 1971).[14] Biotite may also form from alkali feldspar and chlorite via the discontinuous reaction equation 26.3. Pyrophyllite is produced from kaolinite via reaction equation 26.2 or from the breakdown of smectites (Helgeson and others, 1978). Epidote is derived from Ca-zeolites, prehnite, or detrital plagioclase. Continuous reactions of the form

$$\text{chlorite + biotite}_1 + \text{quartz} \Longleftrightarrow \text{garnet + biotite}_2 + H_2O$$
$$(27.4)$$

give rise to garnet at higher grades.

The Greenschist-Amphibolite Facies boundary is a broad zone. The disappearance of albite marks the maximum upper limit of the Greenschist Facies. Both albite and pyrophyllite are absent from Amphibolite Facies rocks, whereas cordierite and the aluminum silicates andalusite (at lower grades) and sillimanite (at higher grades) characterize aluminous bulk compositions. Additional phases that may occur in pelitic rocks include, but are not restricted to, chloritoid, alkali feldspar, tourmaline, apatite, and sphene.

Reactions distinctive of Buchan Facies Series are those defining the appearance of andalusite and cordierite, which combined with the disappearance of albite, mark the transition to the Amphibolite Facies. Although the conversion of pyrophyllite to andalusite via reaction equation 26.4 may occur, it is probable that alternative reactions produce the andalusite. Reactions such as paragonite + quartz <==>

andalusite + albite (component in plagioclase) + H_2O
(27.5)

and

cordierite$_1$ + biotite$_1$ + muscovite <==>

cordierite$_2$ + biotite$_2$ + andalusite + quartz   (27.6)

may be important (Chatterjee, 1972; N. F. C. Hudson, 1980). Cordierite may form from muscovite and associated phases via reactions such as

chlorite$_1$ + biotite$_1$ + muscovite + quartz <==>

cordierite + chlorite$_2$ + biotite$_2$ (27.7)

(N. F. C. Hudson, 1980) or from assemblages that include chloritoid.

Pelitic rocks in the Granulite Facies are distinguished by the general absence of white mica, by the presence of alkali feldspar + sillimanite or orthopyroxene, and by the occurrence of the assemblage cordierite + orthopyroxene (figure 27.1, table 27.1). The characteristic assemblages develop through reactions such as

muscovite + quartz <==> sillimanite + orthoclase +
H_2O (27.8)

which describes the breakdown of muscovite (Evans, 1965; Althaus et al., 1970) and

biotite + quartz <==>

hypersthene + almandine + K-rich alkali feldspar +
H_2O (27.9)

which describes the breakdown of some biotite to form orthopyroxene (Winkler, 1979, p. 265). Additional cordierite may arise via reactions involving garnet or biotite, plus sillimanite and quartz (Holdaway and Lee, 1977; S. M. Lee and Holdaway, 1977).

The mineralogical changes described here for the pelitic rocks are paralleled by mineralogical changes in all other bulk rock compositions. Representative phase assemblages are presented for a variety of bulk compositions in table 27.1. In carbonate rocks, tremolite occurs with calcite, dolomite, and quartz in the Greenschist Facies, and is joined by diopside, forsterite, phlogopite, and wollastonite in the Amphibolite Facies. In the Granulite Facies, tremolite is absent, and phlogopite, quartz, and calcite may disappear. In metabasites, the lowest-grade rocks contain combinations of albite, epidote, prehnite, pumpellyite, heulandite or laumontite. At higher grades, chlorite, albite, and actinolite of the

Greenschist Facies are replaced by biotite, oligoclase, hornblende, cummingtonite, and augite in the Amphibolite Facies. Granulite Facies rocks contain two pyroxenes, brown hornblendes, and intermediate plagioclase (e.g., Bard, 1969).

Example: Buchan Metamorphism, Northern New England, U.S.A.

Perhaps the best-known Buchan Facies Series is that of northern New England, in the United States. As a result of detailed work by a number of petrologists, the petrology is rather well known.[15] A line representing the aluminum silicate triple point extends through New England—from Rhode Island, through central Massachusetts, across western New Hampshire, and into northeastern Vermont—marking a change from a Barrovian Facies Series on the southwest to a Buchan Facies Series on the northeast (figure 27.2) (J. B. Thompson and Norton, 1968).

In the Buchan Facies Series of northeastern New England, several Acadian (mid-Paleozoic) isograds have been mapped in the widely distributed pelitic rocks, including biotite, garnet, andalusite-staurolite, cordierite-staurolite, sillimanite, and K-feldspar–sillimanite isograds. Locally, muscovite coexists with sillimanite and K-feldspar in pelitic rocks of the uppermost zone; thus, the rocks containing these minerals belong to the Amphibolite Facies. Granulite Facies rocks are present only to the south, in New Hampshire, Massachusetts, and northern Connecticut (Osberg et al., 1989; Schumacher et al., 1989).[16] In carbonate rocks, biotite, amphibole, zoisite, and diopside isograds have been mapped (Ferry, 1976, 1982, 1983b,c). In northernmost Maine, Quebec, and New Brunswick, the Zeolite and Prehnite-Pumpellyite Facies are represented by analcite, prehnite-pumpellyite, and pumpellyite-epidote-actinolite zones in metaclastic and metavolcanic rocks (Pavlides, 1973).[17] In summary, Zeolite, Prehnite-Pumpellyite, Greenschist, Amphibolite, and Granulite Facies rocks are represented in central to northern New England. At least one late Carboniferous intrusion, the Sebago Batholith of western Maine, caused overprinting of earlier Acadian metamorphic rocks (Alienikoff, 1984; Hayward and Gaudette, 1984; Lux and Guidotti, 1985). Elsewhere, local areas containing low- to high-grade Taconic and Alleghenian metamorphic rocks adjoin the dominant Acadian regional metamorphic terrane (figure 27.2) (Murray and Skehan, 1979; Guidotti, 1985).

Radiometric dating, geothermometry, and geobarometry indicate that a diachronous, four-stage Acadian metamorphism (405–350 m.y.b.p.)[18] produced the Buchan rocks at variable temperatures and maximum pressures of 0.24–0.45Gpa (2.4–4.5 kb).[19] Temperatures ranged from 400° C at the biotite isograd (Ferry, 1984) to 670° C in the

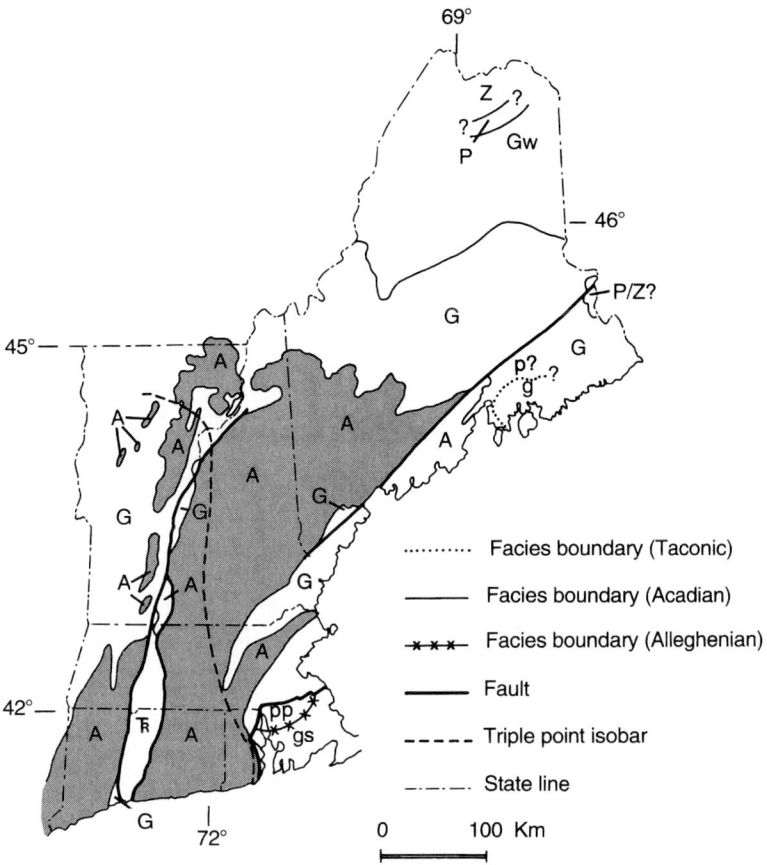

Figure 27.2 Generalized metamorphic map of New England showing zones of Paleozoic regional metamorphism, with Buchan Facies Series northeast of the triple-point isobar and Barrovian Facies Series southwest of the isobar. Acadian metamorphic facies are represented as follows: Z =Zeolite Facies, P = Prehnite-Pumpellyite Facies, Gw = weakly recrystallized Greenschist Facies, G = Greenschist Facies, and A = Amphibolite Facies. Taconic facies are g = Greenschist Facies and p = Prehnite-Pumpellyite Facies. Alleghenian-Hercynian metamorphic facies are pp = Prehnite-Pumpellyite Facies and gs = Greenschist Facies. Tr = unmetamorphosed Triassic rocks.

[Based principally on the syntheses of J. B. Thompson and Norton (1968) and Guidotti (1985); also see B. A. Morgan (1972), Richter and Roy (1974), and Murray and Skehan (1979).]

K-feldspar–sillimanite zone (Lux et al., 1986; DeYoreo et al., 1989).[20] Locally, contact-like geothermal gradients of 80–100° C/km may have been attained, especially near the numerous intrusions associated with the regional metamorphic event.[21]

A generalized phase-facies diagram for the northeastern New England part of the Northern Appalachian Orogen is presented as figure 27.3. Note the presence of andalusite and cordierite and the absence of kyanite in the middle grades of metamorphism, as compared to the Barrovian Facies Series rocks of the southern Appalachian Orogen described below. Local migmatites, formed by partial melting, are developed at the highest grades of metamorphism.[22]

In south-central Maine, an exceptional opportunity exists to examine mineralogical changes with increasing grade of metamorphism within single formations. The Wa-

terville and Vassalboro/Songerville Formations, which trend northeast-southwest, are crossed by isograds that trend approximately east-west (figure 27.4). The Waterville Formation is composed of metamorphosed shale, argillaceous sandstone, and argillaceous limestone, whereas the Vassalboro/Songerville Formation consists of metamorphic equivalents of argillaceous carbonate rocks, calcareous quartz wackes, and shales (Osberg, 1968, 1979; Ferry, 1983c; Ferry and Osberg, 1989). Metamorphic grade increases from chlorite-zone pelitic rocks and ankerite-zone carbonate rocks in the northeast to sillimanite-zone metapelites and diopside-zone metacarbonate rocks in the southwest (figures 27.4 and 27.5). As is typical in zones of increasing grade, there is an overall increase in grain size towards the southwest.

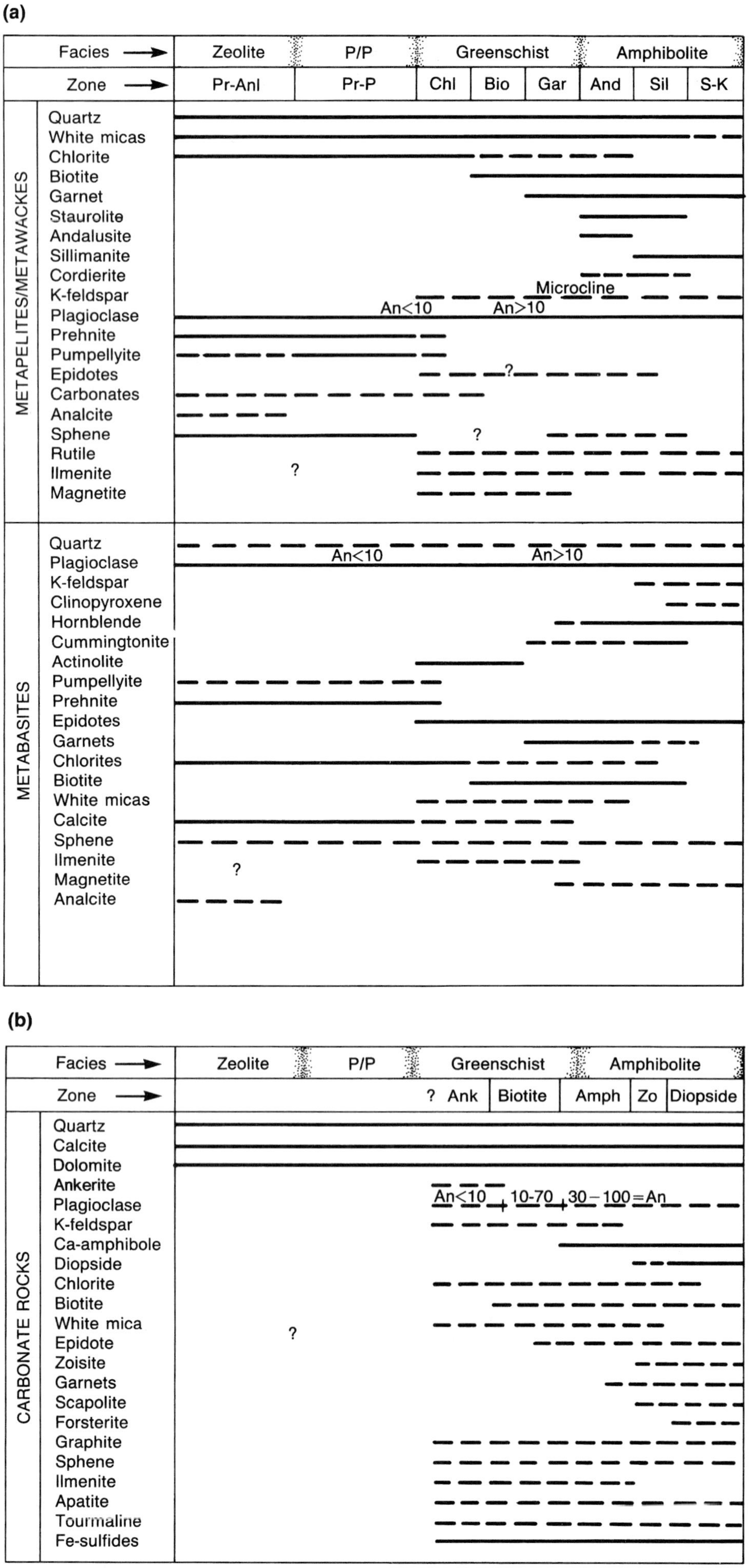

Figure 27.3 Mineral-facies chart for the middle Paleozoic, Buchan Facies Series of northeastern New England. (a) Metapelites, metawackes, and metabasites. (b) Carbonate rocks. Amph = amphibole, An = anorthite (component of plagioclase), And = andalusite, Ank = ankerite, Bio = biotite, Chl = chlorite, Gar = garnet, P/P = prehnite-pumpellyite (facies), Pr-Anl = prehnite-analcite, Pr-P = prehnite-pumpellyite (zone), S-K = sillimanite-alkali feldspar, Sil = sillimanite, Zo = zoisite.

[Sources are listed in explanatory note 15 and in the paragraph following that footnote number in the text. Unpublished observations by the author are also included.]

(a) Mineral-facies chart for metapelites/metawackes and metabasites.

Facies →	Zeolite	P/P	Greenschist			Amphibolite		
Zone →	Pr-Anl	Pr-P	Chl	Bio	Gar	And	Sil	S-K

METAPELITES/METAWACKES: Quartz, White micas, Chlorite, Biotite, Garnet, Staurolite, Andalusite, Sillimanite, Cordierite, K-feldspar (Microcline), Plagioclase (An<10, An>10), Prehnite, Pumpellyite, Epidotes, Carbonates, Analcite, Sphene, Rutile, Ilmenite, Magnetite.

METABASITES: Quartz, Plagioclase (An<10, An>10), K-feldspar, Clinopyroxene, Hornblende, Cummingtonite, Actinolite, Pumpellyite, Prehnite, Epidotes, Garnets, Chlorites, Biotite, White micas, Calcite, Sphene, Ilmenite, Magnetite, Analcite.

(b) Mineral-facies chart for carbonate rocks.

Facies →	Zeolite	P/P	Greenschist			Amphibolite		
Zone →			? Ank	Biotite	Amph	Zo	Diopside	

CARBONATE ROCKS: Quartz, Calcite, Dolomite, Ankerite, Plagioclase (An<10, 10-70, 30–100=An), K-feldspar, Ca-amphibole, Diopside, Chlorite, Biotite, White mica, Epidote, Zoisite, Garnets, Scapolite, Forsterite, Graphite, Sphene, Ilmenite, Apatite, Tourmaline, Fe-sulfides.

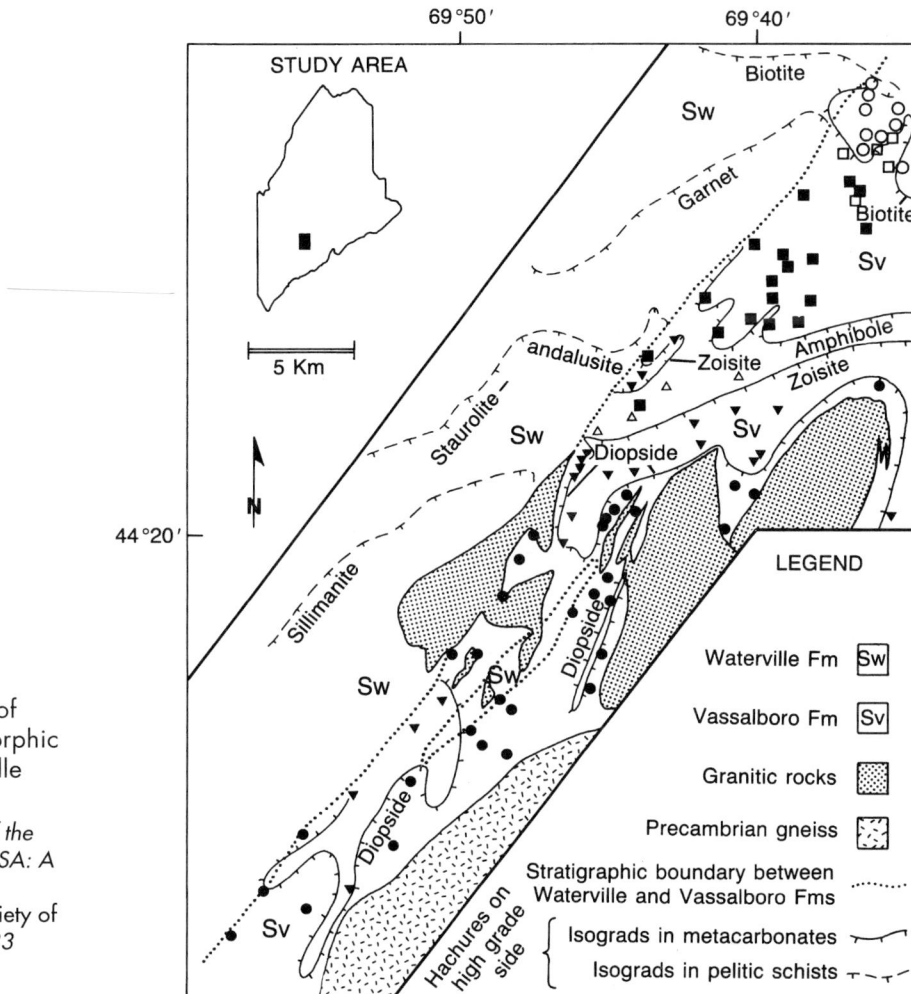

69°50' 69°40'

STUDY AREA

5 Km

44°20'

LEGEND

Waterville Fm [Sw]

Vassalboro Fm [Sv]

Granitic rocks

Precambrian gneiss

Stratigraphic boundary between
Waterville and Vassalboro Fms

Isograds in metacarbonates

Isograds in pelitic schists

Figure 27.4 Geological sketch map of part of southern Maine showing metamorphic isograds in the Vassalboro and Waterville formations.

(From J. M. Ferry, "Regional metamorphism of the Vassalboro Formation, south-central Maine, USA: A case study of the role of fluid in metamorphic petrogenesis" in Journal of the Geological Society of London, *140:551-76, 1983. Copyright © 1983 Geological Society of London. Reprinted by permission.)*

In the Vassalboro/Songerville Formation, the carbonate rocks are characterized by assemblages such as the following (after Ferry, 1983c):

Ankerite Zone

ankerite–quartz–albite–muscovite–calcite–chlorite–pyrite

Biotite Zone

ankerite–quartz–albite–muscovite–biotite–calcite–chlorite–pyrite

biotite–quartz–oligoclase/labradorite–muscovite–calcite–chlorite–pyrite

Amphibole Zone

Ca-amphibole–biotite–quartz–andesine/anorthite–calcite–chlorite–pyrrhotite

Zoisite Zone

zoisite–Ca-amphibole–quartz–andesine/anorthite–calcite–biotite–microcline–pyrrhotite

Diopside Zone

diopside–zoisite–Ca-amphibole–quartz–andesine/anorthite–calcite–biotite–pyrrhotite.

The minerals garnet, graphite, sphene, tourmaline, apatite, and scapolite are accessory minerals at various grades of metamorphism.

In the pelitic rocks, the mineral assemblages are typical of Buchan Facies Series rocks (figures 27.3 and 27.5). The following assemblages are representative (Osberg, 1968; Ferry, 1980a, 1982):

Chlorite Zone

muscovite–quartz–albite–chlorite–ankerite–calcite–magnetite

Biotite Zone

muscovite–quartz–plagioclase–biotite–chlorite–calcite–ilmenite

557

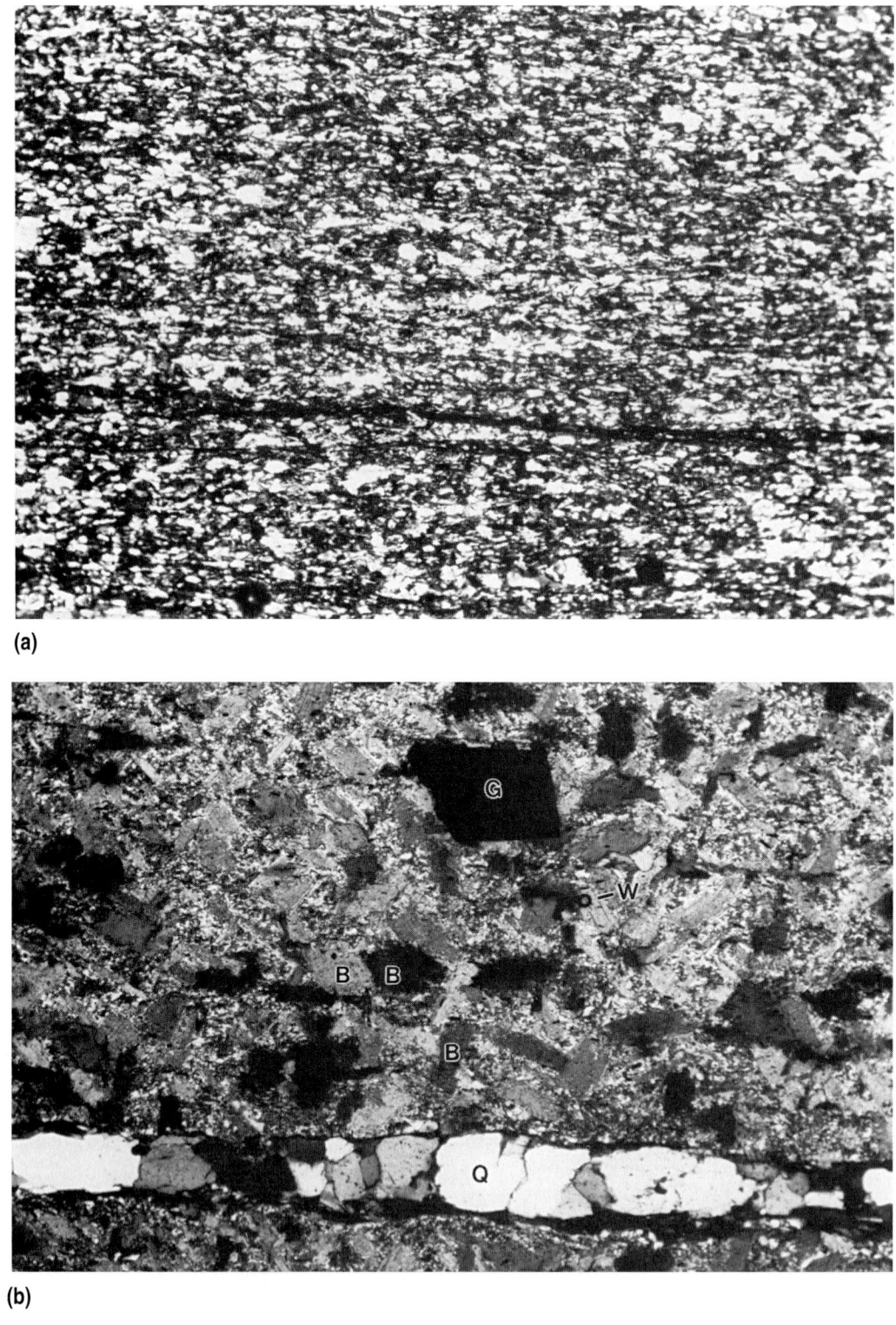

(a)

(b)

• • • • • • • • • • • • • •
Figure 27.5 Photomicrographs of pelitic schists from the Waterville Formation. (a) Chlorite Zone schist. (XN). Long dimension of photo is 1.27 mm. (b) Garnet Zone schist with quartz lens. (XN). (c) Andalusite Zone schist. (XN). (d) Sillimanite Zone schist. (XN). A = K-rich alkali feldspar,

B = biotite, CH = chlorite, CD = cordierite, G = garnet, O = opaque minerals, P = plagioclase feldspar, Q = quartz, SI = sillimanite, ST = staurolite, and W = white micas. In (b), (c), and (d) long dimension of photo is 3.25 mm.

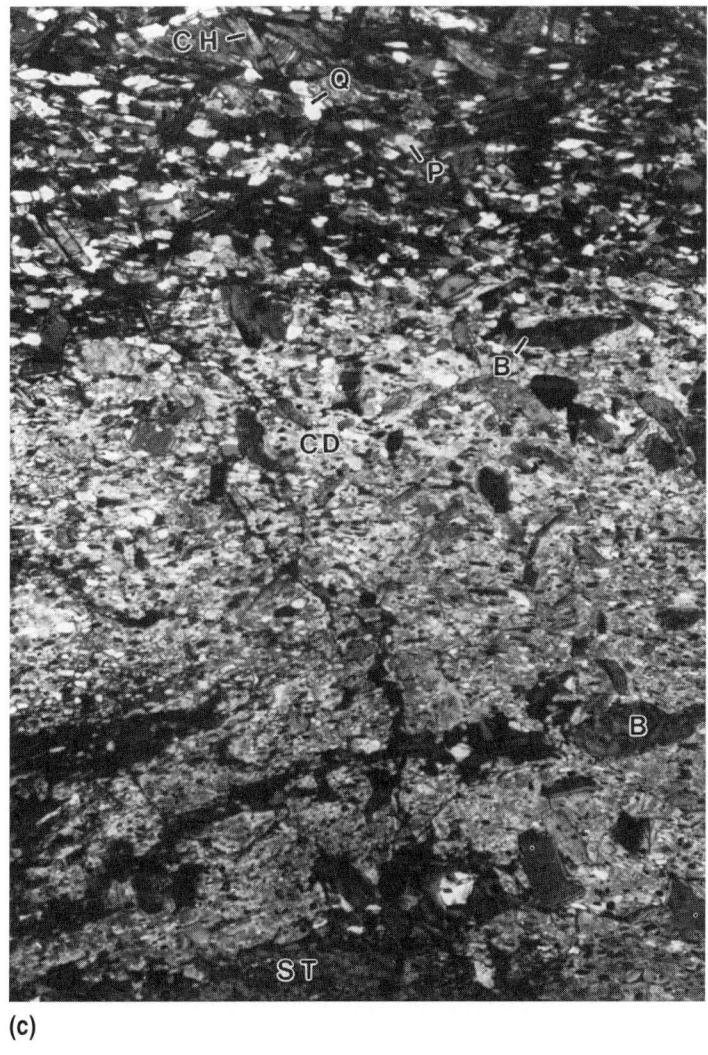

(c)

(d)

Figure 27.5 Continued

Garnet Zone

muscovite–quartz–plagioclase–biotite–garnet–chlorite–
calcite–ilmenite

Staurolite-Andalusite Zone

muscovite–quartz–plagioclase–biotite–garnet–staurolite–
andalusite–ilmenite

Sillimanite Zone

muscovite–quartz–plagioclase–biotite–garnet–sillimanite–
cordierite–microcline–ilmenite.

Pyrrhotite, pyrite, tourmaline, rutile, clinozoisite, and sphene
are among the accessory minerals that occur at various grades
of metamorphism.

The fact that all of the metamorphic rocks of each type
are derived from a single formation means that compositional
variations are minimized as a factor in controlling the mineral
assemblages. Similarly, the pressure was relatively uniform during the metamorphic event. As a consequence, temperature,
fluid phase composition, and element migration were responsible for the mineralogical differences that developed. The temperature was a primary controlling factor, but fluids were
extremely influential, as they transport heat, transport ions
(changing the rock chemistry), and drive mineral reactions
(Ferry, 1980b, 1982, 1983b,c). Fluid/rock ratios are estimated
to have been between 0.7 and 2.0 during metamorphism, and
fluid flow rates were about 10^{-1}mm/year (Ferry and Osberg,
1989).

Differences and Similarities Between Contact and Buchan Facies Series

A complete gradation exists between Contact and Buchan Facies series and the conditions of metamorphism overlap. The major differences that exist between rocks of Contact Facies Series and those of the Buchan Facies Series are that the latter are regional in distribution (rather than local), are characterized by foliated fabrics, and lack Sanidinite Facies at the highest grades. Buchan Facies Series are associated with orogenic belts that have experienced a regional influx of heat, rather than being locally associated with particular plutons. In contrast, in contact metamorphism, the heat producing the metamorphism is local and may clearly be related to an intrusion or a set of closely associated intrusions.

The second major difference—that Buchan rocks are characteristically foliated, whereas contact rocks are typically granoblastic to diablastic—is related to the tectonic regime in which such rocks form. In contact metamorphism, intrusions cause local metamorphism and may intrude a passive (nondeforming) country rock. As a consequence, many contact metamorphic rocks lack foliation altogether. Fabrics do develop where movements during intrusion and metamorphism generate deviatoric stresses that produce foliations or lineations in the rocks, but foliation is not a definitive characteristic of contact metamorphic rocks. In contrast, Buchan rocks develop in orogenic belts associated with zones of plate convergence, rifting, or transcurrent faulting. As a consequence, metamorphism may be either synkinematic (occurring at the same time as deformational movements) or slightly pre- or postkinematic. In general, either synkinematic metamorphism or prekinematic metamorphism followed by regional deformation will result in regionally developed foliations. It is these regionally developed foliations that characterize Buchan rocks.

TRIPLE-POINT ROCKS

Recall that Buchan Facies Series are metamorphosed at pressures lower than that of the aluminum silicate triple point (3.87 kb = 0.387 Gpa). Andalusite is a critical indicator of that kind of facies series. Barrovian Facies Series occur in rocks metamorphosed at pressures that, in the middle grades of metamorphism, are greater than the 0.387 Gpa aluminum silicate triple point. Kyanite is a critical mineral indicator of that kind of facies series.

As might be expected, in some occurrences of transitional facies series, rocks containing all three aluminum silicates (andalusite, kyanite, and sillimanite) are present. Such occurrences are reported in New England, Idaho, New Mexico, and Alaska.[23]

BARROVIAN FACIES SERIES

Occurrences

Barrovian Facies Series occur in a number of Phanerozoic orogenic belts, as well as in some of Precambrian age. Notable among the Phanerozoic belts are the Caledonides of northwestern Europe, including the classic region in the Scottish Highlands, and parts of the Appalachian Mountain System of eastern North America.[24] Other Phanerozoic belts with Barrovian rocks occur in western North America (e.g., Idaho, Colorado, British Columbia, Alaska), Venezuela, Spain, southern Europe and Asia (the Alpine-Himalayan Orogen), central Asia (the Ural Mountains), and Japan.[25]

Precambrian belts of Barrovian rocks occur in the Black Hills of South Dakota, the Rocky Mountains, and Labrador, Quebec, and Ontario (Canada).[26] Outside of North America, such belts are found in the former Soviet Union, India, and Sri Lanka.[27]

The Phanerozoic orogenic belts are clearly associated with convergent plate margins. Both Barrovian and Buchan Facies series develop at such margins. Those of Precambrian age may represent convergent margins as well, but their origins are debated (Windley, 1977; P. F. Hoffman, 1989). In convergent zones, regional heating due to the rise of plutons into the overlying plate (the plate above the subduction zone) is the general cause of metamorphism (Dewey and Bird, 1970; Miyashiro, 1982, p. 153), but migrating fluids may also transport heat.

The Barrovian Facies Series Revisited

The zones of metamorphism in the Scottish Highlands originally described by Barrow (1893, 1912) include six distinct mineral assemblages that occur in the rock types listed below:

Chlorite Zone

(slates, phyllites, and schists)

quartz–albite–white mica–chlorite–microcline ± calcite

Biotite Zone

(phyllites and schists)

quartz–albite–white mica–chlorite–biotite ± microcline ± calcite ± epidote

Almandine (Garnet) Zone

(phyllites and schists)

quartz–albite–white mica–biotite–garnet ± chlorite

Staurolite Zone

(schists)

quartz–oligoclase–white mica–biotite–garnet–staurolite

Kyanite Zone

(schists)

quartz–oligoclase–white mica–biotite–garnet–kyanite ± staurolite

Sillimanite Zone

(schists, gneisses, and granoblastites)

quartz–oligoclase–biotite–sillimanite ± kyanite ± K-rich alkali feldspar ± white mica.

The additional granulite facies assemblage

K-rich alkali feldspar–quartz–sillimanite–garnet–biotite–plagioclase–rutile–Fe-Ti oxide (?)

has been reported more recently (A. J. Baker and Droop, 1983; A. J. Baker, 1985). These assemblages in the pelitic rocks are paralleled by others in mafic rocks, in which there are increases in the calcium content of plagioclase, a change in the Ca-amphibole composition, and other mineralogical changes (figure 27.6).[28]

As was the case in occurrences of Contact and Buchan Facies series rocks, metamorphic belts exhibiting Barrovian Facies Series do not always exhibit *all* of the facies possible within the series. Clearly this is the case in the Scottish Highlands, where Zeolite and Prehnite-Pumpellyite Facies are not described as part of the sequence. To completely characterize the phase assemblages of a facies series, it is necessary to depict the phase assemblages of the entire facies series by constructing a composite series of phase diagrams and an associated set of petrogenetic grids (see appendix C). Such an illustration, showing phase topologies for *all* conditions of P T, and X_{fluid} in the facies series, can appear overwhelming due to the large number of phase diagrams required to present such a complete sequence.[29] A representative sequence of topologies, however, can reveal the general trends of phase changes that occur during progressive metamorphism, as shown by the selected phase diagrams for Barrovian Facies Series in figure 27.7.

Phase Assemblages and Reactions in Barrovian Facies Series

The representative phase diagrams for rocks of Barrovian Facies Series shown in the topologies in figure 27.7 serve four purposes. First, they allow us to obtain an impression of the overall mineralogical trends that characterize the facies series. Second, each phase diagram reveals typical assemblages for a part of the facies it represents. Third, the composite nature of

the illustration allows us to compare phase assemblages in rocks of different composition metamorphosed at the same general grade. For example, assemblages for aluminous and aluminum-rich siliceous and siliceous-alkali-calcic (SAC) rocks are shown on AFM diagrams, whereas more mafic SAC rocks plus intermediate to basic igneous rocks have mineral assemblages that are plotted on ACF pseudo-phase diagrams or CFM diagrams. Mineral assemblages of carbonate and ultramafic rocks are plotted on the CSM diagram. The diagrams for the carbonate rocks represent intermediate X_{CO_2} values, whereas those for ultramafic rocks represent low values of X_{CO_2} (high X_{H_2O}). Fourth, the sequences imply several reactions.

The reactions implied by the topologies in figure 27.7 are numerous. In most cases, several reactions must have occurred in order for adjoining topologies to develop. Experimentally determined and calculated reaction curves for many of the more important reactions are presented in appendix C.

Assemblages and Reactions in Pelitic Rocks

A survey of occurrences of pelitic rocks in Barrovian Facies Series reveals the important phase assemblages at various grades of metamorphism. These, in turn, suggest the kinds of reactions that are important during prograde metamorphism. At the lowest grade, in the Zeolite Facies, which forms under conditions just above those of diagenesis, assemblages are characterized by clay minerals (C. E. Weaver et al., 1984).[30] Assemblages may include

kaolinite–illite–illite/smectite–chlorite–quartz–analcite,

illite–illite/smectite–kaolinite–chlorite, and

illite–chlorite–kaolinite–K-rich alkali feldspar.[31]

In addition, smectites, iron oxides, calcite, and dolomite may occur, as may Ca-zeolites (in appropriate bulk compositions). Organic carbon is also generally present.

At slightly higher-grade conditions, assemblages of the Zeolite Facies are replaced by those of the Prehnite-Pumpellyite Facies. In this facies, some minerals present in lower-grade rocks, such as kaolinite and K-rich alkali feldspar, are absent from rocks of certain bulk compositions. New phases appear, including albite, phengitic white mica, and illite/chlorite or illite/smectite/chlorite mixed-layer minerals (figures 27.7 and 27.8). In addition, changes occur in the structures of some minerals, for example, in chlorite and illite (J. Hoffman and Hower, 1979). Notably, illite changes in crystallinity (C. E. Weaver, 1960; Kubler, 1967; Kisch, 1990).

The mineralogical changes imply several reactions. As was the case in Buchan Facies Series, K-rich alkali feldspar and smectites are among the first minerals to disappear from aluminous rocks. Discontinuous reactions such as equation 27.1 account for the replacement of smectite and feldspar by illite, chlorite, and quartz.[32] Kaolinite also is

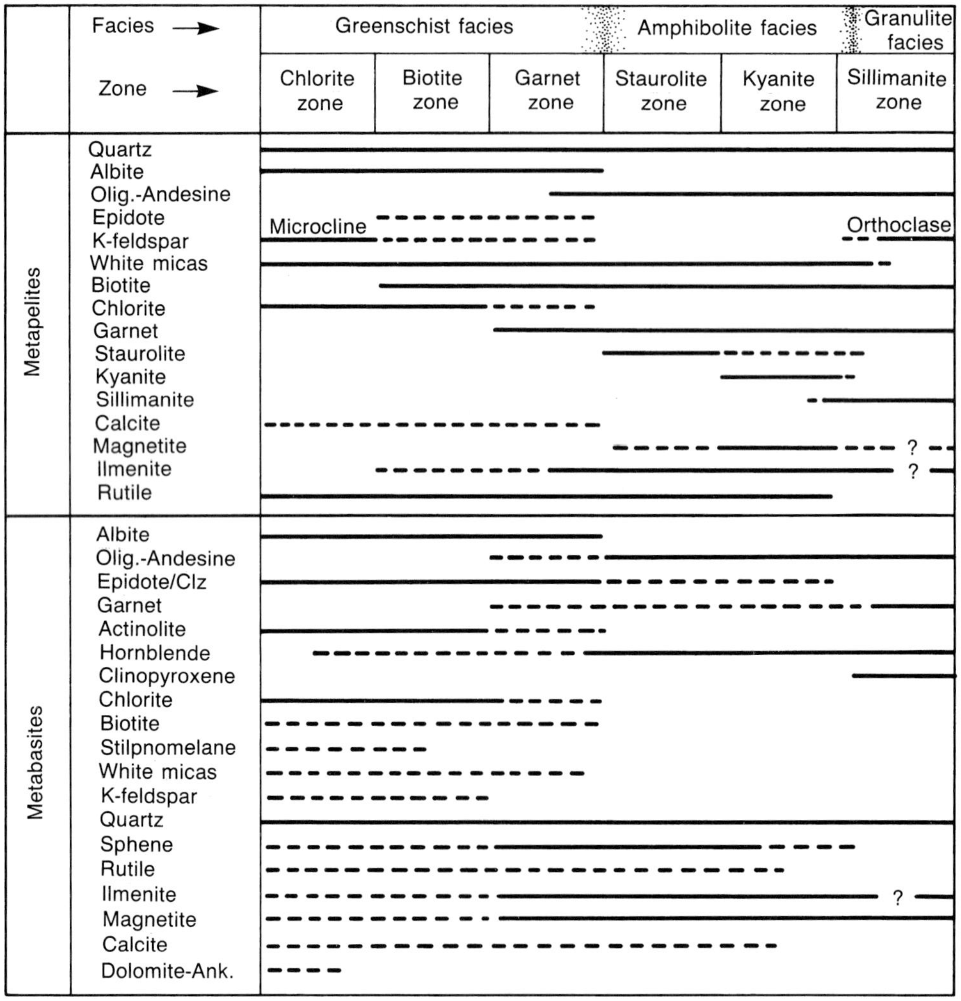

Facies →	Greenschist facies			Amphibolite facies		Granulite facies
Zone →	Chlorite zone	Biotite zone	Garnet zone	Staurolite zone	Kyanite zone	Sillimanite zone

Metapelites

Quartz
Albite
Olig.-Andesine
Epidote
K-feldspar — Microcline .... Orthoclase
White micas
Biotite
Chlorite
Garnet
Staurolite
Kyanite
Sillimanite
Calcite
Magnetite — ?
Ilmenite — ?
Rutile

Metabasites

Albite
Olig.-Andesine
Epidote/Clz
Garnet
Actinolite
Hornblende
Clinopyroxene
Chlorite
Biotite
Stilpnomelane
White micas
K-feldspar
Quartz
Sphene
Rutile
Ilmenite — ?
Magnetite
Calcite
Dolomite-Ank.

Figure 27.6 Mineral-facies chart for the type area of Barrovian Facies Series metamorphism in the Scottish Highlands. Ank = ankerite, Clz = clinozoisite, Olig = oligoclase.

[Data from Harker (1932), Chinner (1960, 1966), C. M. Graham (1974, 1985), Harte and Hudson (1979), Graham et al. (1983), A. J. Baker (1985), McLellan (1985a, 1985b), Moles (1985), and K. P. Watkins (1987).]

commonly absent from Prehnite-Pumpellyite Facies rocks. Although the reactions involved are unknown, kaolinite commonly diminishes or disappears as illite appears or increases in abundance, as chlorite increases in abundance, and as 2M white micas appear.[33] Illite/smectite mixed-layer minerals may also appear where kaolinite disappears from the rocks. The chemistry of the new phases suggests that Si, K, Na, Ca, Mg, and Fe must be available for these phases to form. Such elements may be derived from calcite, Mg-calcite, and dolomite; quartz, chalcedony, or opal; K-rich alkali feldspar; smectites; iron oxides; minor amounts of ferromagnesian minerals, especially detrital biotite and chlorite; and *pore fluids*. C. E. Weaver, Beck, and Pollard (1971) suggest that some of the necessary ions may be transported into the rock by migrating solutions, and several other workers suggest that Eh, pH, and the chemistry of the pore solutions may control the nature of the reactions that occur (Garrels and Christ, 1965; Grim, 1968; B. M. Sass, Rosenberg, and Kittrick, 1987).[34] A reaction such as equation 27.2, involving most of the phases in a typical rock, would account for the disappearance of kaolinite and would yield appropriate Prehnite-Pumpellyite Facies phases. Inasmuch as chlorite and illite may already exist in low-grade rocks or sediments, reactions such as equations 27.1 and 27.2 merely add to the abundance of these phases in the rock. Of course, as the pressure and temperature increase, continuous reactions may produce compositional and structural changes in these phases.[35]

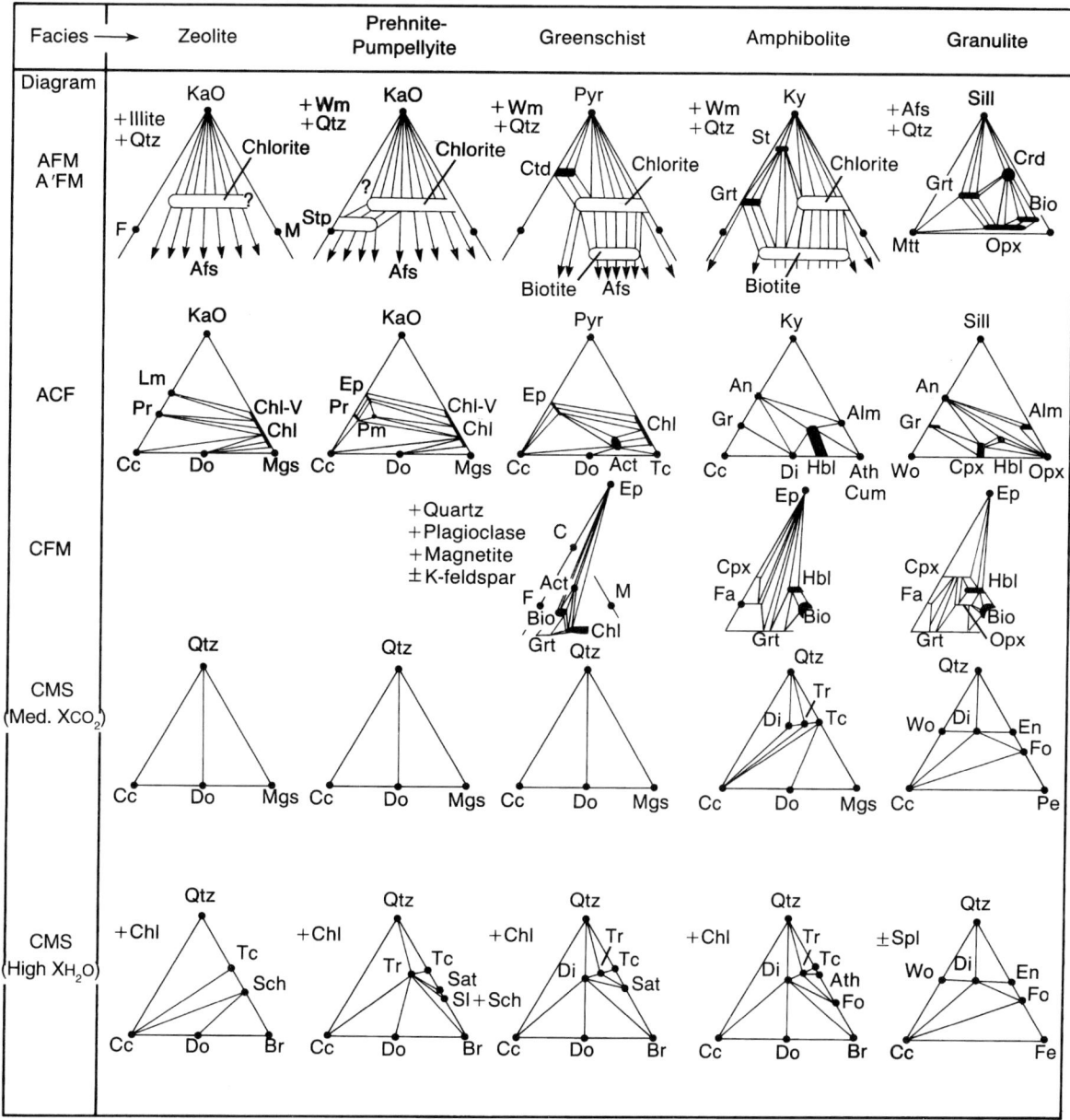

Figure 27.7 Representative phase diagrams for the various facies of the Barrovian Facies Series. AFM diagrams are useful for depicting phase assemblages in pelitic rocks and aluminous SAC rocks. ACF diagrams are useful for metabasites, metaigneous rocks of intermediate composition, and more iron- and aluminum-rich SAC rocks. CFM diagrams depict the phase assemblages of metabasic rocks. The CMS diagram is useful in representing phase assemblages in carbonate and ultramafic rocks.

[CFM diagrams are from Abbott (1982, 1984). Additional sources include Miyashiro (1973a), B. W. Evans (1977), Winkler (1979), F. J. Turner (1981), Evarts and Schiffman (1983), Absher and McSween (1985), various sources listed in the text and notes 24–66, and the observations of the author.]

An additional phase that may appear in pelitic rocks of the Prehnite-Pumpellyite Facies is stilpnomelane. Stilpnomelane may form either by modification of detrital biotite or via a reaction such as

chlorite + K-rich alkali feldspar + magnetite $\Longleftrightarrow$
stilpnomelane (27.10)

Stilpnomelane is generally restricted to iron-rich bulk rock compositions (e.g., see Floran and Papike, 1978).

As the P-T conditions increase, Greenschist Facies assemblages with new minerals form[36]. Typical assemblages in pelitic rocks (figure 27.9a) include

white mica (muscovite and paragonite)–chlorite–quartz–albite–magnetite,

white mica–chlorite–biotite–quartz–albite–magnetite,
white mica–chlorite–chloritoid–epidote–magnetite–
hematite, and
white mica–chlorite–garnet–biotite–quartz–albite–
oligoclase–magnetite.

Stilpnomelane, pyrophyllite, alkali feldspar, calcite, and sphene also occur in some rocks of this facies.

Reactions that produce the various new minerals include both discontinuous and continuous types. Changes in chlorite and white mica compositions involve continuous reactions, such as reaction equation 27.3, in which the composition of the white mica changes and biotite is produced (E. H. Brown, 1971).[37] Biotite may also form from chlorite and alkali feldspar via the discontinuous reaction equation 26.3 or via the continuous reaction

$$\text{microcline} + \text{chlorite} + \text{white mica}_1 <==>$$

$$\text{biotite} + \text{white mica}_2 + \text{quartz} + H_2O \qquad (27.11)$$

(Mather, 1970). Pyrophyllite forms from the breakdown of smectites or is produced from kaolinite via reaction equation 26.2 (Helgeson et al., 1978). Reactions such as

$$\text{chlorite} + 5\,\text{pyrophyllite} <==> 7\,\text{chloritoid} + 17\,\text{quartz} + 4\,H_2O \qquad (27.12)$$

may yield chloritoid (J. B. Thompson and Norton, 1968) and, at higher-grade conditions, continuous reactions such as

$$\text{chlorite} + \text{biotite}_1 + \text{quartz} <==> \text{garnet} + \text{biotite}_2 + H_2O \qquad (27.4)$$

generate garnet. Because of the variable composition of garnet, its first appearance in a regional metamorphic terrane probably occurs over a range of P-T conditions, depending on the composition of the rock and the fluid phase. In some cases, it may not appear until the lower part of the Amphibolite Facies.[38]

As is the case in the Buchan Facies Series, the Greenschist-Amphibolite Facies boundary is a broad zone. The disappearance of albite marks the maximum upper limit of the Greenschist Facies. Thus, albite, like pyrophyllite, is absent from Amphibolite Facies rocks. Staurolite, rather than chloritoid, occurs in the lower part of the Amphibolite Facies and the aluminum silicates kyanite (at lower grades) and sillimanite (at higher grades) characterize aluminous bulk compositions. Typical assemblages (figures 27.7 and 27.9b) include

white mica–biotite–quartz–plagioclase (oligoclase)–
garnet–magnetite,
white mica–biotite–chlorite–quartz–oligoclase–garnet–
staurolite–ilmenite,
white mica–biotite–quartz–oligoclase–garnet–kyanite–
ilmenite, and
white mica–biotite–quartz–oligoclase–garnet–sillimanite–
ilmenite.[39]

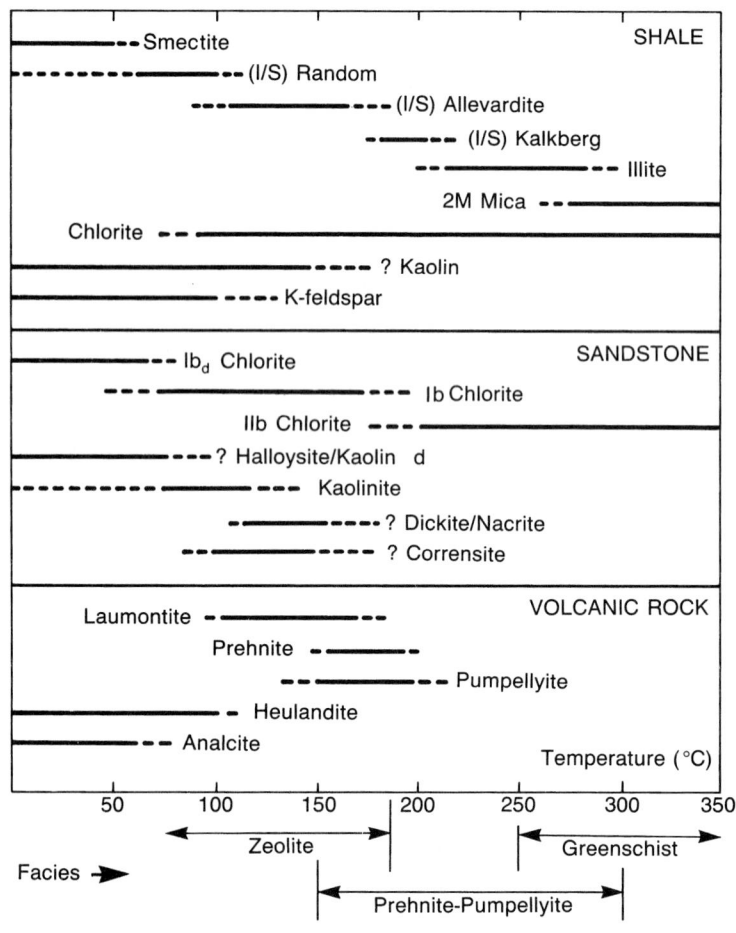

Figure 27.8 Progressive mineralogical changes in weakly metamorphosed rocks of the Montana disturbed belt. In the shale, note the changes from one mixed-layer illite-smectite (I/S) mineral to another. In the sandstones, different structural varieties of chlorite are stable under different conditions. Facies designations are assigned here on the basis of the mineralogical relations discussed in this chapter and P-T limits defined in chapter 24.
(Modified from J. Hoffman and Hower, 1979).

Additional phases that occur locally include, but are not restricted to, alkali feldspar, chloritoid, tourmaline, epidote, apatite, and sphene. Note that the first assemblage above is one that occurs over a wide range of conditions, the specific nature of which can only be determined through geothermobarometry.

Possible reactions leading to phase assemblages in pelitic rocks of the Amphibolite Facies are numerous.[40] A number of reactions have been proposed to yield the first appearance of staurolite. These involve various combinations of the phases chloritoid, biotite, chlorite, muscovite, quartz,

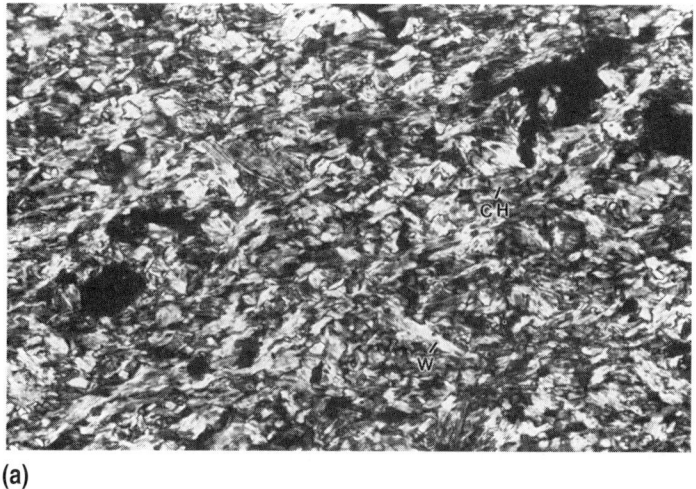

(a)

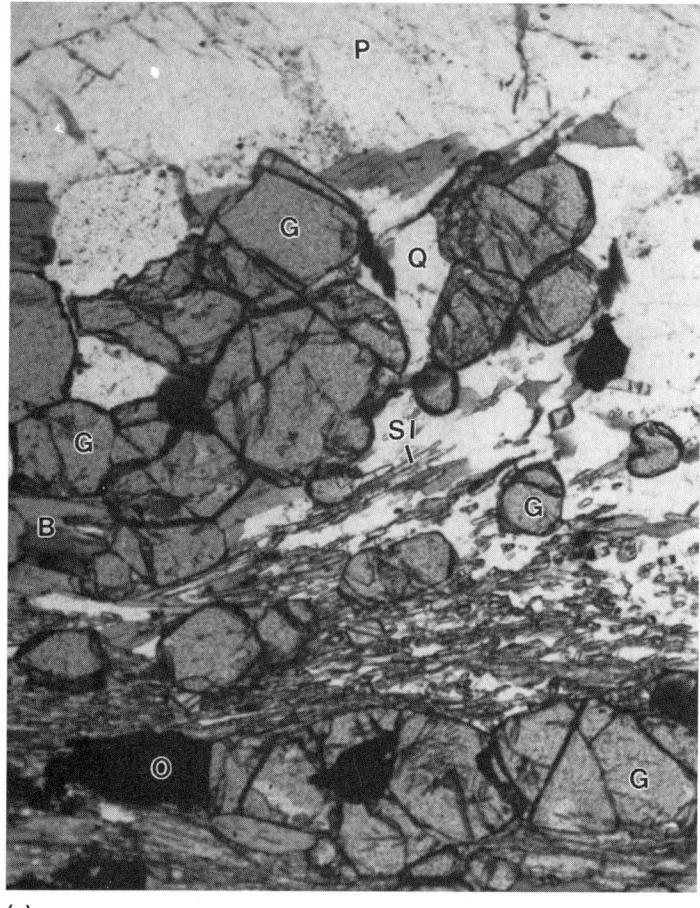

(c)

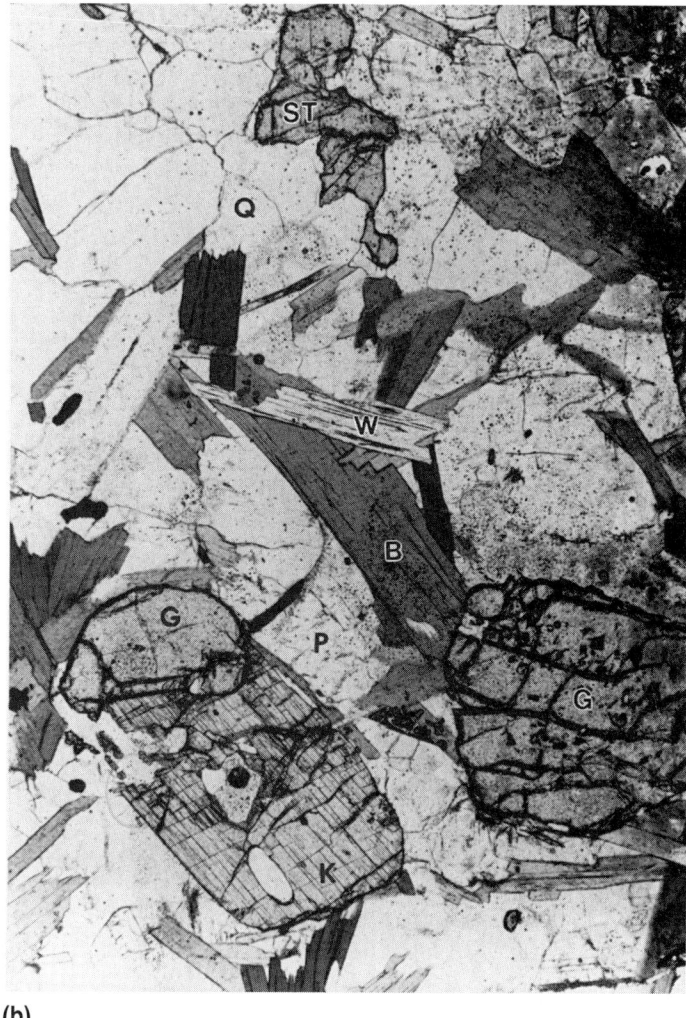

(b)

Figure 27.9 Photomicrographs of Barrovian Facies Series pelitic rocks. (a) Greenschist Facies metashale composed of white mica, chlorite, quartz, and other minerals, Grandfather Mountain Formation, North Carolina. (XN). (b) Amphibolite Facies pelitic schist, Ashe Metamorphic Suite, Ashe County, North Carolina. (PL). (c) Granulite Facies gneiss, Winding Stair Gap, North Carolina. (PL). B = biotite, CH = chlorite, G = garnet, K = kyanite, O = opaque minerals, P = plagioclase, Q = quartz, SI = sillimanite, ST = staurolite, W = white mica. Long dimension of photo (a) is 1.27 mm, of (b) is 3.25 mm and of (c) is 1.19 mm.

kyanite, ilmenite, and garnet. For rocks of the Snow Peak, Idaho, area, one such reaction is

$$\text{chlorite + garnet + muscovite + ilmenite} \Longleftrightarrow$$

$$\text{staurolite + biotite + plagioclase + quartz} + H_2O \quad (27.13)$$

(Lang and Rice, 1985b). Kyanite may arise by the conversion of pyrophyllite via the reaction

$$\text{pyrophyllite} \Longleftrightarrow \text{kyanite} + 3\,SiO_2 + H_2O \quad (27.14)$$

(Kerrick, 1968; Haas and Holdaway, 1973) or by more complex reactions involving other phases, such as

$$\text{staurolite + muscovite + quartz} \Longleftrightarrow \text{biotite + kyanite} + H_2O \quad (27.15)$$

565

(McLellan, 1985b). The particular reaction depends on the bulk composition and other factors. Sillimanite may form by the simple polymorphic transformation

$$\text{kyanite} <==> \text{sillimanite} \qquad (27.16)$$

as it appears to have done in western Labrador (Rivers, 1983) and in the Himalayan Orogen (Lal, Mukerji, and Ackermand, 1981), but reactions involving muscovite or biotite, such as

$$6 \text{ staurolite} + 4 \text{ muscovite} + 7 \text{ quartz} <==> 4 \text{ biotite} +$$
$$31 \text{ sillimanite} + 3 \text{ H}_2\text{O} \qquad (27.17)$$

(J. B. Thompson and Norton, 1968; McLellan, 1985a) may be more common.

The upper part of the Amphibolite Facies and the Granulite Facies are characterized locally by zones of *lit-par-lit* (literally, bed-by-bed) "injection" gneisses—banded rocks with alternating dark and light (granitoid) layers. Although such rocks may have begun as pelitic schists, partial melting has altered their character. They consist of a light-colored, quartz-feldspar-rich **leucosome**, produced by partial melting and/or the local injection and crystallization of a lower-temperature (eutectic) melt between layers or foliation planes, and a **melanosome**, a dark-colored, ferromagnesian mineral-rich, refractory rock. Leucosomes typically have granitoid mineral assemblages, whereas melanosomes are characterized by biotite- or amphibole-rich assemblages.[41]

The Granulite Facies is distinguished by the general absence of white mica, the presence of orthopyroxene in pelitic rocks, and by the occurrence of the assemblage cordierite + orthopyroxene.[42] Rocks of this facies are more common in Buchan Facies Series, but some occur in Barrovian Facies Series, for example, in Ontario (E. W. Reinhardt, 1968), the Adirondack Mountains of New York (A. E. J. Engel and Engel, 1958, 1960; Stoddard, 1980; Bohlen, Valley, and Essene, 1985; Metzger, 1987), the southern Appalachian Orogen (Force, 1976; Absher and McSween, 1985; Gulley, 1985), Antarctica (Sheraton, 1984/1985; Harley, 1986, 1987), and Calabria, Italy (Schenk, 1984). Granulite assemblages are also found in xenoliths in mafic volcanic rocks (Rudnick and Taylor, 1987). Pelitic assemblages (figure 27.9c) include

biotite–garnet–sillimanite–K-rich alkali feldspar–
andesine–quartz–magnetite–ilmeno-hematite,

biotite–K-rich alkali feldspar–
andesine–quartz–cordierite–garnet–iron oxides, and

garnet–andesine–quartz–rutile–ilmenite–orthopyroxene–
sillimanite–zircon.

The alkali feldspar is typically perthitic. Additional phases include apatite, spinel, and graphite.

These assemblages develop through reactions such as equation 27.8, which describes the breakdown of muscovite, and equation 27.9, which describes the breakdown of biotite to form orthopyroxene. Cordierite may arise in Granulite Facies rocks of the Barrovian Facies Series via reactions involving garnet or biotite, plus sillimanite and quartz (Holdaway and Lee, 1977; S. M. Lee and Holdaway, 1977).

Siliceous-Alkali-Calcic (SAC) Rocks

Metamorphosed sandstones, felsic to intermediate volcanic rocks, and granitoid rocks exhibit mineral assemblages that range from those identical to the assemblages of pelitic schists to those like assemblages in mafic rocks. The SAC rock modes, however, differ from those of pelitic rocks. Quartz and feldspar are the dominant phases, rather than the phyllosilicates, and calcium-bearing phases are common. Additional minerals that may occur in various assemblages include stilbite, epistilbite, calcite, pyrophyllite, stilpnomelane, garnet, Fe-Ti oxides, actinolite, and hornblende.

Typical mineral assemblages in SAC rocks for various facies of metamorphism are shown in table 27.2.[43] Compare the assemblages listed to those depicted in the AFM and ACF diagrams in figure 27.7. Notice that these rocks have an intermediate character, with compositions that are between those for which these diagrams are best suited. Figure 27.10 shows photomicrographs of SAC rocks metamorphosed under a variety of conditions.

Reactions involving the various phases present in SAC rocks include some of those described above for the aluminous rocks, as well as some applicable to the mafic rocks described below. Important reactions include equation 27.18, which describes the disappearance of laumontite and the upper limit of the Zeolite Facies, equation 27.21, which marks the appearance of actinolite and the Prehnite-Pumpellyite to Greenschist Facies transition, and equation 27.10 marking the appearance of stilpnomelane. At higher grades of metamorphism, reactions such as equations 27.4 and 27.13 mark the broad transition from the Greenschist Facies to the Amphibolite Facies, and reaction equation 27.8, defining the conversion of white mica to alkali feldspar, delimits the lower boundary of the Granulite Facies.

Assemblages and Reactions in Mafic Rocks

Mafic rocks of the Barrovian Facies Series have not been studied as extensively as have aluminous rocks, but their compositions are generally well known. The work of Liou, Maruyama, and Cho (1985, 1987) has been definitive in outlining the phase relations in metabasic rocks at low grade.[44] These rocks contain assemblages in the Zeolite Facies such as:

albite–chlorite/smectite–prehnite–pumpellyite–heulandite,
and laumontite–analcite–albite–pumpellyite–
epidote–chlorite–quartz–sphene.[45]

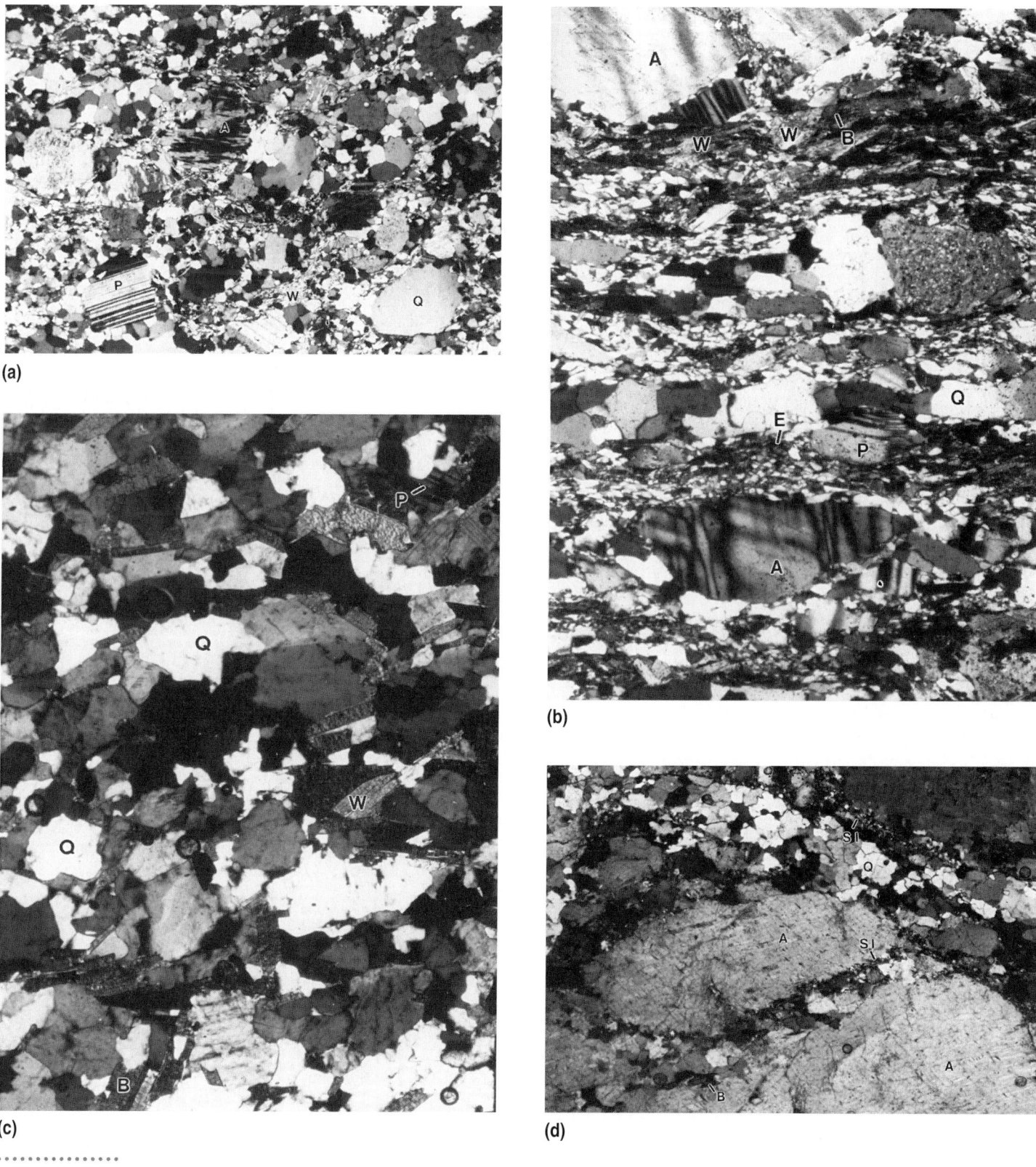

(a)

(b)

(c)

(d)

....

Figure 27.10 Photomicrographs of quartz-feldspar (SAC) rocks of the Barrovian Facies Series, metamorphosed under a variety of conditions. (a) Greenschist Facies metaquartz arenite, Grandfather Mountain Formation, Boone, North Carolina. (XN). (b) Greenschist Facies Gneiss, Caldwell County, North Carolina. (XN). (c) Amphibolite Facies metasandstone, Ashe Metamorphic Suite, near Little Switzerland, North Carolina. (XN). (d) Granulite Facies quartz-feldspar gneiss, Cloudland Gneiss, Roan Mountain, Tennessee. (XN). A = K-rich alkali feldspar, C = calcite, E = epidote. Other symbols as in figure 27.9. Long dimension of photos (a) – (c) is 3.25 mm and of (d) is 1.27 mm.

Notice that epidote, prehnite, and pumpellyite all may occur in the Zeolite Facies. The Zeolite Facies is distinguished, not by the presence or absence of these minerals, but by the presence of a zeolite in association with them. The Prehnite-Pumpellyite Facies is distinguished by associations containing these minerals but *lacking zeolites and actinolite*.

Important reactions that mark the transition from the Zeolite Facies to the Prehnite-Pumpellyite Facies include equation 26.15, for the conversion of analcite to albite, and the reactions

$$\text{laumontite + prehnite} <==> \text{clinozoisite + quartz} + H_2O$$
$$(27.18)$$

and

$$\text{laumontite + pumpellyite} <==> \text{clinozoisite + chlorite +}$$
$$\text{quartz} + H_2O \qquad (27.19)$$

which define the disappearance of laumontite (Liou, Maruyama, and Cho, 1985, 1987). These and similar reactions involving other zeolites occur at temperatures of about 200° C ± 75° C.

Phase assemblages of the Prehnite-Pumpellyite Facies are similar to those of the Zeolite Facies, but, as noted, lack the zeolites and analcite.[46] No new Ca-Al silicate minerals appear in the Prehnite-Pumpellyite Facies. Stilpnomelane does appear as a new phyllosilicate phase in some rocks of this facies, and sphene appears as a titanium-bearing phase.[47] Additional minerals that may occur include calcite, white mica, and alkali feldspar.

Phase assemblages of the Greenschist Facies are different. They are marked by the occurrence of actinolite with chlorite and either pumpellyite or epidote-clinozoisite (figure 27.11a). Pumpellyite-bearing Greenschist Facies rocks form at temperatures below about 350° C (except at high pressures) and thus characterize only the lowest-grade part of the facies. The pumpellyite-bearing zone is equivalent to the Pumpellyite-Actinolite Facies of Liou, Maruyama, and Cho (1985). Representative Greenschist Facies assemblages include

albite–actinolite–epidote–pumpellyite–calcite,

chlorite–epidote–quartz–calcite–garnet–prehnite–albite–actinolite,

actinolite–chlorite–albite–quartz–magnetite, and

epidote–albite–actinolite–calcite–chlorite–quartz–sphene.[48]

Additional phases reported include stilpnomelane, white mica, biotite, K-rich alkali feldspar, and pyrite.[49]

Reactions marking the transition from the Prehnite-Pumpellyite Facies to the Greenschist Facies, or the *lower boundary* of the Greenschist Facies, are those defining the appearance of actinolite. Such reactions as

$$\text{10 calcite + 3 chlorite + 21 quartz} <==> \text{2 epidote +}$$
$$\text{3 actinolite + 8 } H_2O + \text{10 } CO_2 \qquad (27.20)$$

and

$$\text{25 pumpellyite + 2 chlorite + 29 quartz} <==>$$
$$\text{43 epidote + 7 actinolite + 67 } H_2O \qquad (27.21)$$

are important (Nakajima, Banno, and Suzuki, 1977; Liou, Maruyama, and Cho, 1985, 1987; Cho and Liou, 1987).

The *upper* (high-*T*) *boundary* of the Greenschist Facies is marked by a transition zone in which (1) oligoclase appears, (2) chlorite disappears as a common phase, (3) hornblende appears, (4) garnet appears, (5) albite disappears, and (6) actinolite disappears.[50] The reactions involved are complex and continuous, involving changing compositions of amphibole, chlorite, garnet, and plagioclase (Laird, 1980; Spear, 1980, 1982; J. B. Thompson, Laird, and Thompson, 1982).[51] An example of such a reaction is

$$\text{amphibole}_1 + \text{chlorite}_1 + \text{epidote + albite} <==>$$
$$\text{amphibole}_2 + \text{chlorite}_2 + \text{oligoclase + quartz} + H_2O$$
$$(27.22)$$

(Laird, 1980). In this text, the disappearance of albite is considered to be a principal indicator of the upper limit of Greenschist Facies conditions in mafic rocks.

Because of these various reactions, rocks of the Amphibolite Facies are characterized by hornblende and plagioclase (oligoclase, andesine, bytownite, and/or anorthite) (figure 27.11b). Amphibolite Facies metabasites may also include garnet, quartz, epidote, cummingtonite, gedrite, anthophyllite, staurolite, clinopyroxene, biotite, paragonite, K-rich alkali feldspar, magnetite, ilmenite, rutile, calcite, dolomite, chlorite, and pyrite. Representative assemblages include

hornblende–epidote–andesine–biotite–quartz–sphene–ilmenite–apatite,

hornblende–garnet–plagioclase (bytownite)–quartz–ilmenite,

hornblende–garnet–cummingtonite–gedrite–anthophyllite–andesine–ilmenite–magnetite, and

hornblende–Ca-plagioclase–clinopyroxene–ilmenite–sphene.[52]

Table 27.2   Representative Mineral Assemblages in Metamorphosed Siliceous-Alkali-Calcic (SAC) Rocks of the Barrovian Facies Series

Zeolite Facies

quartz–kaolinite

quartz–analcite–heulandite–chlorite–illite–hematite

quartz–albite–laumontite–chlorite–illite–illite/chlorite–hematite

Prehnite-Pumpellyite Facies

quartz–albite–white mica–chlorite–stilpnomelane–prehnite–pumpellyite–hematite

quartz–albite–white mica–chlorite–chlorite/vermiculite–pumpellyite–stilpnomelane–sphene

quartz–albite–chlorite–epidote–prehnite–pumpellyite–calcite

Greenschist Facies

quartz–alkali feldspar–white mica–chlorite–magnetite

quartz–albite–epidote–white mica–chlorite–sphene

quartz–alkali feldspar–plagioclase–white mica–chlorite–biotite–magnetite

Amphibolite Facies

quartz–alkali feldspar–plagioclase–white mica–biotite–garnet–ilmenite

quartz–plagioclase–biotite–garnet–gedrite–ilmenite

quartz–plagioclase–biotite–garnet–sillimanite

Granulite Facies

plagioclase–quartz–alkali feldspar–biotite–garnet–sillimanite–ilmenite

quartz–alkali feldspar–plagioclase–orthopyroxene–ilmenite

quartz–alkali feldspar–plagioclase–orthopyroxene–clinopyroxene–ilmenite

Sources: See note 43.

Note both the absence of albite and the common (but not ubiquitous) occurrence of garnet.

Hornblende persists into the Granulite Facies, as does biotite. In metabasites, the Granulite Facies is characterized by the appearance of orthopyroxene, the coexistence of plagioclase with *both* orthopyroxene and clinopyroxene, and the occurrence of biotite *only* in magnesium-richer bulk compositions. In most metabasites, biotite and epidote are absent. Typical assemblages (figure 27.11c) include

hornblende–plagioclase–orthopyroxene–clinopyroxene–garnet–ilmenite and

orthopyroxene–labradorite–clinopyroxene–ilmenite–apatite–zircon.[53]

The appearance of orthopyroxene in these assemblages may arise from a reaction such as

$$\text{hornblende}_1 + \text{plagioclase}_1 <==> \text{hornblende}_2 + \text{plagioclase}_2 +$$
$$\text{clinopyroxene} + \text{orthopyroxene} + \text{ilmenite} + H_2O$$
$$(27.23)$$

(Spear, 1981b). Notice that this is a dehydration reaction. The generation of a fluid phase via such dehydration reactions may cause local anatexis and the formation of migmatites. (Recall that the rocks of the Granulite Facies develop above the minimum melting curve for wet granite.)

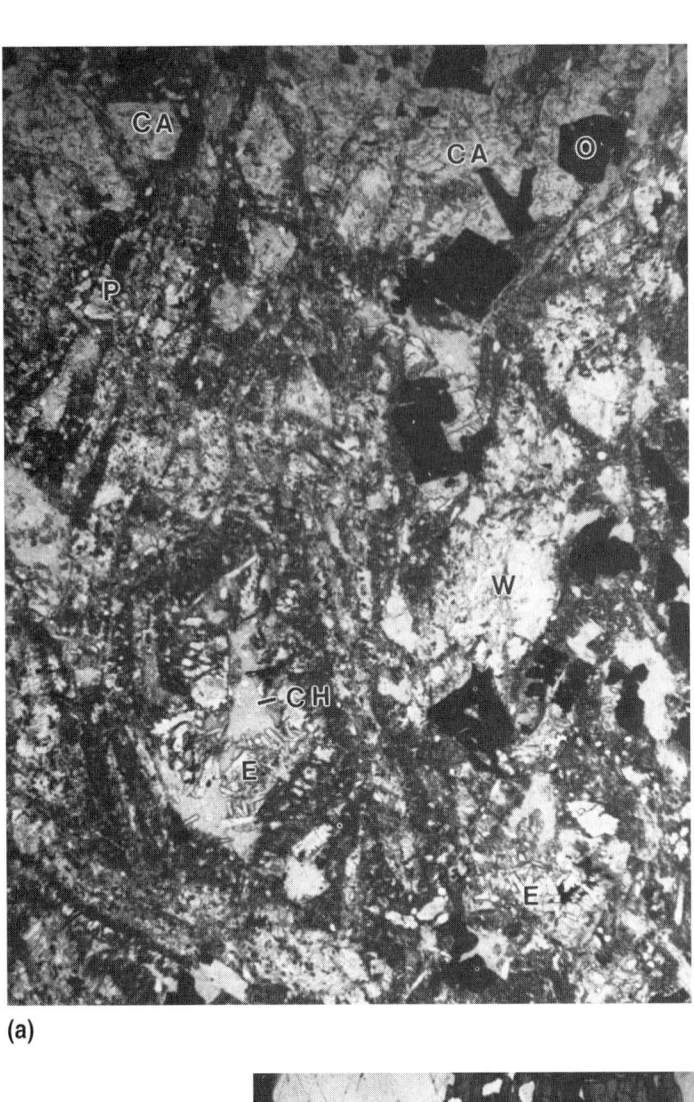

(a)

(b)

(c)

Figure 27.11 Photomicrographs of metabasites. (a) Metadiabase, Greenschist Facies, Linville Metadiabase, Linville, North Carolina. (PL). (b) Hornblende schist (amphibolite), Amphibolite Facies, Ashe Metamorphic Suite, Todd, North Carolina. (XN). (c) Hornblende-plagioclase granoblastite, Granulite Facies, Winding Stair Gap, southwestern North Carolina. (XN). CA = Ca-amphibole, CH = chlorite, E = epidote, G = garnet, H = hornblende, O = opaques (magnetite, ilmenite, etc.), OP = orthopyroxene, P = plagioclase, Q = quartz, S = sphene, W = white mica. Long dimension of photos (a) and (b) is 6.5 mm and of (c) is 4.9 mm..

Calcareous Rocks and Ultramafic Rocks

Regional metamorphism of limestones, dolostones, and calcareous shales yields marbles and calcareous schists that are mineralogically like contact-metamorphosed carbonate rocks.[54] Comparison of the phase assemblages in figures 27.7 and 26.13 reveals that similarity. At low grades of metamorphism, that is, within the Zeolite and Prehnite-Pumpellyite Facies, as well as within much of the Greenschist Facies, at moderate to high values of X_{CO_2}, the phase assemblage in marbles is essentially the same as the primary assemblage (calcite–dolomite–quartz). At low values of X_{CO_2}, Greenschist Facies assemblages may contain phases such as ankerite, talc, and chlorite (compare the Albite-Epidote Hornfels Facies, figure 26.13). At higher metamorphic grades, phases such as tremolite, phlogopite, diopside, idocrase, scapolite, garnet, hornblende, spinel, and forsterite will appear, depending on the pressure temperature, and the compositions of both the rock and the fluid phase.

In a similar way, the phase assemblages of ultramafic rocks parallel those of equivalent contact metamorphic assemblages.[55] Representative topologies for the various facies of metamorphism are shown in figure 27.7 and representative assemblages are listed in table 27.3. At low grades of metamorphism, serpentines (lizardite and chrysotile) typically are the dominant phases, with variable amounts of talc, carbonate minerals, and magnetite comprising the remainder of the mineralogy. Likewise, serpentine (antigorite) is commonly the principal phase within ultramafic rocks of the Greenschist Facies, but talc and chlorite may predominate in some schists. Talc, tremolite, chlorite, magnetite, brucite, and magnesite are important accessory minerals in Greenschist Facies ultramafic rocks (B. W. Evans, 1977).

Several reactions clearly mark the facies boundaries in terranes with ultramafic rocks. The reaction that marks the transition from the Zeolite to the Prehnite-Pumpellyite Facies is reaction equation 26.26:

$$17 \text{ chrysotile} <==> \text{antigorite} + 3 \text{ brucite}.$$

Serpentine may convert to forsterite in the upper part of the Greenschist Facies, near the beginning of the Amphibolite Facies, via reaction 26.27 (page 546). The upper limit of the Greenschist Facies is marked by a terminal reaction for antigorite:

$$\text{antigorite} <==> 18 \text{ forsterite} + 4 \text{ talc} + 27 \text{ H}_2\text{O} \quad (27.24)$$

(Johannes, 1975). Within the Amphibolite Facies, anthophyllite and enstatite appear. Anthophyllite forms via reaction equation 26.28 and enstatite forms through either the reaction

$$\text{anthophyllite} + \text{forsterite} <==> 9 \text{ enstatite} + \text{H}_2\text{O} \quad (27.25)$$

Table 27.3 Selected Metamorphic Mineral Assemblages in Barrovian Facies Series Ultramafic Rocks

Zeolite Facies

chrysotile–magnetite
chrysotile–talc–calcite–magnetite
chrysotile–lizardite–magnetite–brucite–dolomite
lizardite–chrysotile–chlorite–magnetite–magnesite–dolomite

Prehnite-Pumpellyite Facies

chrysotile–lizardite–brucite–chlorite–magnetite
antigorite–lizardite–tremolite–chlorite–magnetite

Greenschist Facies

antigorite–magnesite–magnetite
antigorite–chlorite–talc–tremolite–magnetite
antigorite–forsterite–diopside–chlorite–magnetite

Amphibolite Facies

forsterite–anthophyllite–talc–chlorite–magnetite
anthophyllite–tremolite–talc–chlorite–magnetite
forsterite–enstatite–chlorite–tremolite–chromite

Granulite Facies

forsterite–enstatite–diopside–spinel–chromite

Sources: See note 55.

at low pressures (< 6.5 kb) or

$$\text{talc} + \text{forsterite} <==> 5 \text{ enstatite} + \text{H}_2\text{O} \quad (27.26)$$

at high pressures (H. J. Greenwood, 1963).[56] In the Granulite Facies, anthophyllite and talc are absent, and assemblages containing coexisting enstatite, diopside, and forsterite characterize the rocks (B. W. Evans, 1977).

Example: Barrovian Metamorphism in the Southern Appalachian Orogen

The southern Appalachian Orogen extends from central Virginia to Alabama (figure 27.12a). It is a complex orogenic belt, parts of which have experienced regional metamorphism during each of four orogenic events. The ages of these events are Mesoproterozoic, Ordovician (the Taconic Orogeny), Devonian-Mississippian (the Acadian Orogeny), and Pennsylvanian-Permian (the Alleghanian/Appalachian

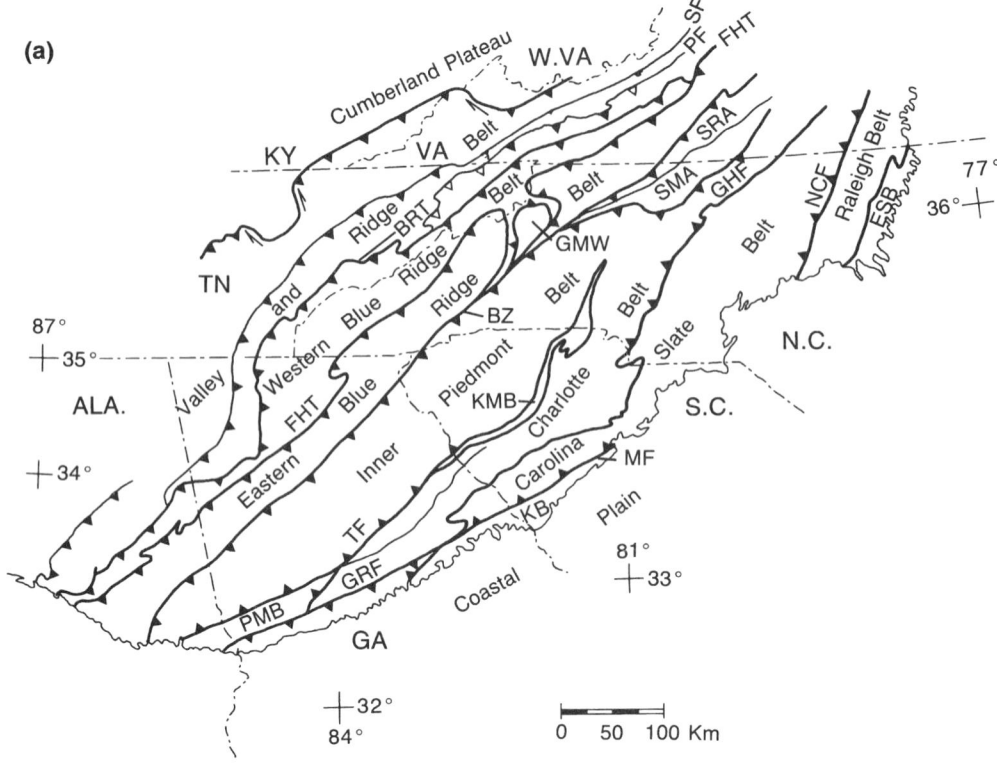

(a)

· · · · · · · · · · · · · · · · · ·

Figure 27.12 Maps of the Southern Appalachian Orogen.
(a) Structural map of the Southern Appalachian Orogen
showing major lithotectonic belts (based on many sources,
including Hatcher et al., 1979). From northwest to southeast,
abbreviations are as follows: SF = Saltville Fault, PF = Pulaski
Fault, BRT = various faults of the Blue Ridge Thrust Fault
system, FHT = Fries-Hayesville Fault, GMW = Grandfather
Mountain Window, BZ = Brevard Zone, SRA = Smith River
Allochthon, SMA = Sauratown Mountains Anticlinorium, TF
= Towaliga Fault, PMB = Pine Mountain Belt, KMB = Kings

Mountain Belt, GHF = Gold Hill Fault, GRF = Goat Rock Fault,
MF = Modoc Fault, KB = Kiokee Belt, NCF = Nutbush Creek
Fault Zone, and ESB = Eastern Slate Belt.
(b) Paleozoic metamorphic facies map for the Southern
Appalachian Orogen. The ages of the metamorphic events
that produced the various facies are not the same throughout
the orogen.
*(Sources include Abbott and Raymond (1984), Absher and McSween
(1985), Bartholomew (1983), Bearce (1973, 1982), Black and Fullagar
(1976), S. E. Boyer (1978), S. E. Boyer and Elliott (1982), B. Bryant*

Orogeny).[57] R. H. Carpenter (1970) and others[58] have delineated various zones of metamorphic rock and L. Glover et al. (1983), Hatcher (1987), and J. R. Butler (1991) have attempted to discriminate zones associated with the individual orogenies. A new map based on these and other works is shown in figure 27.12b.

While the Southern Appalachian Orogen is one of the major regions of Barrovian Facies Series rocks in North America, analysis of the metamorphism there has been confounded by several factors. First, the various tectonic belts (terranes) in the southern Appalachian Orogen have been juxtaposed by significant movements of various types along major faults—in several cases, *after* metamorphism had oc-

curred (Hatcher, 1978, 1987; Abbott and Raymond, 1984; Secor, Snoke, and Dallmeyer, 1986; Vauchez, 1987a, 1987b). This problem is particularly significant in the central and eastern parts of the Orogen. Second, the thermal significance of various metamorphic zones is open to question.

Within individual belts, various Paleozoic metamorphic events have been attributed either to a series of thermal maxima associated with the individual orogenic events[59] (figure 27.13a) or to cooling and uplift associated with a single event[60] (figure 27.13b). The first of these models requires that the metamorphic zones, as they now exist, are diachronous and polymetamorphic. In the west, peak metamorphism occurred during the Taconic Orogeny. Yet

(b)

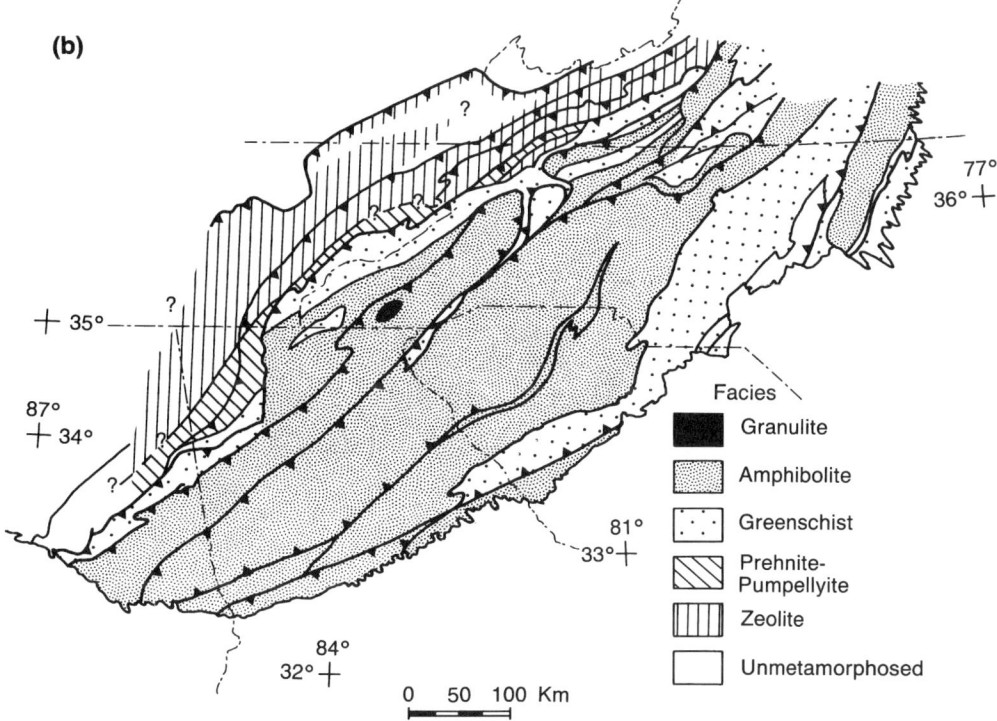

Facies

■ Granulite

▦ Amphibolite

⋰ Greenschist

▨ Prehnite-
Pumpellyite

▥ Zeolite

□ Unmetamorphosed

0   50   100 Km

Figure 27.12   Continued

and Reed (1970a, 1970b), J. R. Butler (1972, 1973; 1984 on Brown et al., 1985, 1991), R. H. Carpenter (1970), Conley (1987), Dallmeyer (1975a, 1975b, 1988), Dallmeyer et al. (1986), Dietrich, Fullagar, and Bottino (1969), Drake et al. (1989), Eckert et al. (1989), Espenshade et al. (1975), Farrar (1984), Force (1976), Fullagar and Dietrich (1976), Fullagar et al. (1980), Gillon (1989); Glover et al. (1983), Gulley (1985), Hadley and Nelson (1971), Hatcher (1976, 1987), Hatcher et al. (1979, 1980), Hatcher et al. (1989), Hopson, Hatcher, and Stieve (1989); Horton and Stearn (1983), V. J. Hurst (1970, 1973), Kish

(1990), L. E. Long, Kulp, and Eckelmann (1959), McConnell and Costello (1984), Mohr and Newton (1983), W. J. Morgan (1972), Mose and Nagel (1984), Nesbitt and Essene (1982), Noel, Spariosu, and Dallmeyer (1988), Osberg et al. (1989), Rankin, Espenshade, and Neuman (1972), Rankin, Espenshade, and Shaw (1973), G. S. Russell, Russell, and Farrar (1985), J. W. Sears and Cook (1984), Secor et al. (1986), Secor, Snoke, and Dallmeyer (1986), Sinha and Glover (1978), Snoke, Kish, and Secor (1980), Sundelius (1970), Tull (1980, 1982), C. E. Weaver et al. (1984), and unpublished observations of the author. See Morgan (1972), Hatcher (1987), and Drake et al.(1989) for additional relevant sources.)

most zones of Acadian- and Alleghenian-age metamorphism in the west and many in the central part of the orogen are of Greenschist or lower grade (L. Glover et al., 1983). This evidence is compatible with the hypothesis that post-Taconic metamorphism in the west-central orogen may have resulted from monotonic cooling. Such a cooling history, resulting from uplift punctuated by periods of deformation and local heating along fault zones, also suggests that the relatively regular zones of Barrovian Facies Series metamorphism present today are diachronous, but not polymetamorphic (in the sense that thermal maxima developed repeatedly). In the east and locally in the central and western parts of the orogen, thermal maxima

did yield Amphibolite Facies conditions during the Acadian deformation (Dallmeyer et al., 1986; Dallmeyer, 1988; Kish, 1989). Throughout the orogen, analysis of the metamorphic history is complicated where such younger metamorphic events have overprinted the mineral assemblages and textures generated by older events (J. R. Butler, 1973b, 1991; Abbott and Raymond, 1984).

Barrovian metamorphism in the *western* part of the orogen is the focus of this example.[61] In this region, strike-slip faulting is insignificant and the isograds are primarily cut by thrust faults that have telescoped, but not laterally offset, the metamorphic zones. A map of the orogen, showing the approximate positions of metamorphic facies of Paleozoic

(a)

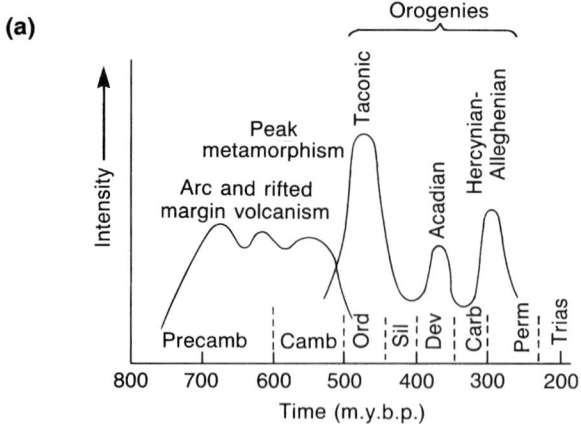

(b)

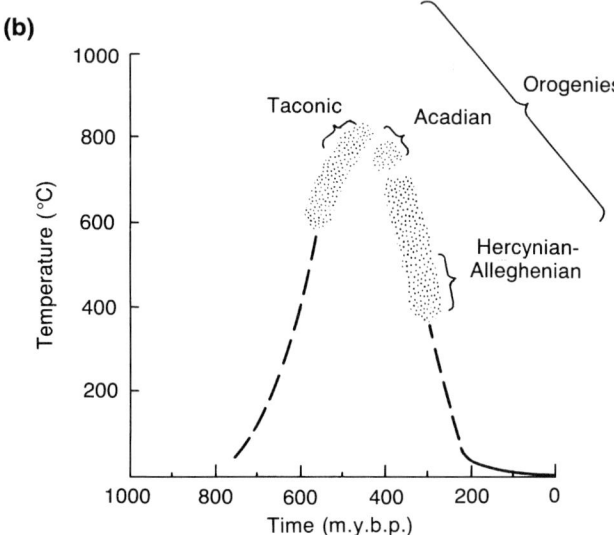

Figure 27.13 Graphs showing the timing of Paleozoic metamorphism and orogeny in the Southern Appalachian Orogen. (a) Time vs. intensity plot showing multiple peaks of metamorphism and orogenic activity (modified from Hatcher et al., 1979). (b) Time vs. temperature plot showing a single metamorphic peak and monotonic cooling of the orogen punctuated by orogenically induced recrystallization.
(Modified from Abbott and Raymond, 1984)

age, is presented in figure 27.12b. A broad range of rock types exists in the region, but carbonate rocks, especially impure carbonate rocks, are relatively rare in the higher-grade parts of the metamorphic belt, whereas mafic and ultramafic rocks are rare to nonexistent in the low-grade zones.

Rocks of the Zeolite and Prehnite-Pumpellyite Facies occur primarily in the Valley and Ridge Belt, but some occur along the northwestern edge of the Blue Ridge Belt. At these lowest grades of metamorphism, the pelites are characterized by clays and the carbonate rocks by calcite and/or dolomite ± quartz (figures 27.14a and 27.15). Rare basic rocks contain chlorite, calcite, and locally may contain rare pumpellyite. Greenschist Facies assemblages are distributed in the western Blue Ridge Belt and in structural windows in that belt. Rocks of this grade consist of younger (Neoproterozoic to Cambrian) sedimentary and igneous rocks and older (Mesoproterozoic) polymetamorphic rocks. Quartz-rich metaclastic rocks typically contain the assemblage quartz–white mica–chlorite–alkali feldspar–magnetite in this facies (figure 27.14b). Quartz-feldspar gneisses, in part probably products of retrograde metamorphism of Precambrian Amphibolite and Granulite facies rocks, contain similar assemblages, plus assemblages such as biotite–chlorite–quartz–albite–K-rich alkali feldspar–calcite–pyrite. Pelitic rocks, at the lower grades, typically are composed of the assemblage chlorite–white mica–quartz–albite–magnetite (figure 27.14c). In higher-grade assemblages, garnet is present. Metabasites contain assemblages such as chlorite–epidote–albite–quartz–hematite and actinolite–chlorite–albite–quartz–magnetite. Iron-rich carbonate rock and calcareous metasiltstones contain ferroan dolomite–calcite–quartz–plagioclase–biotite ± white mica (Bryant and Reed, 1970a; Raymond, Neton, and Cook, 1991).

Much of the eastern Blue Ridge Belt is composed of rocks of the Amphibolite Facies. Migmatites are common.[62] Pelitic mica schists consist of various assemblages containing staurolite, kyanite, and sillimanite (figures 27.14d and 27.15). Cordierite is present locally, in the central Blue Ridge,[63] but it is rare and its significance is unknown. Quartzo-feldpathic (SAC) rocks are composed predominantly of the assemblage plagioclase–quartz–biotite–white mica–garnet, and may contain magnetite, ilmenite, and K-rich alkali feldspar (figure 27.15). Some feldspar-rich rocks contain the assemblage plagioclase–gedrite–biotite–epidote–apatite–quartz–ilmenite ± garnet. Mafic rocks are typical amphibole schists and gneisses, with hornblende and plagioclase as the dominant phases. Garnet, epidote, biotite, quartz, ilmenite, magnetite, pyrite, pentlandite,[64] and chalcopyrite are accessory minerals in these rocks (figures 27.14f and 27.15). Diopside-calcite marble is rare. Granite to quartz diorite (trondhjemite) dikes are common throughout the Amphibolite Facies terrane, suggesting that local anatexis was a widespread event.[65] Geothermometry and geobarometry indicate that the Amphibolite Facies rocks of the central core of the Blue Ridge were metamorphosed at temperatures between 500° C and 850° C at pressures of 0.5–1.1 Gpa (5–11 kb).[66]

Paleozoic Granulite Facies rocks have been recognized at only a few localities (Force, 1976; Absher and McSween, 1985; Eckert, Hatcher, and Mohr, 1989). Absher and McSween (1985) give complete descriptions of a full range of

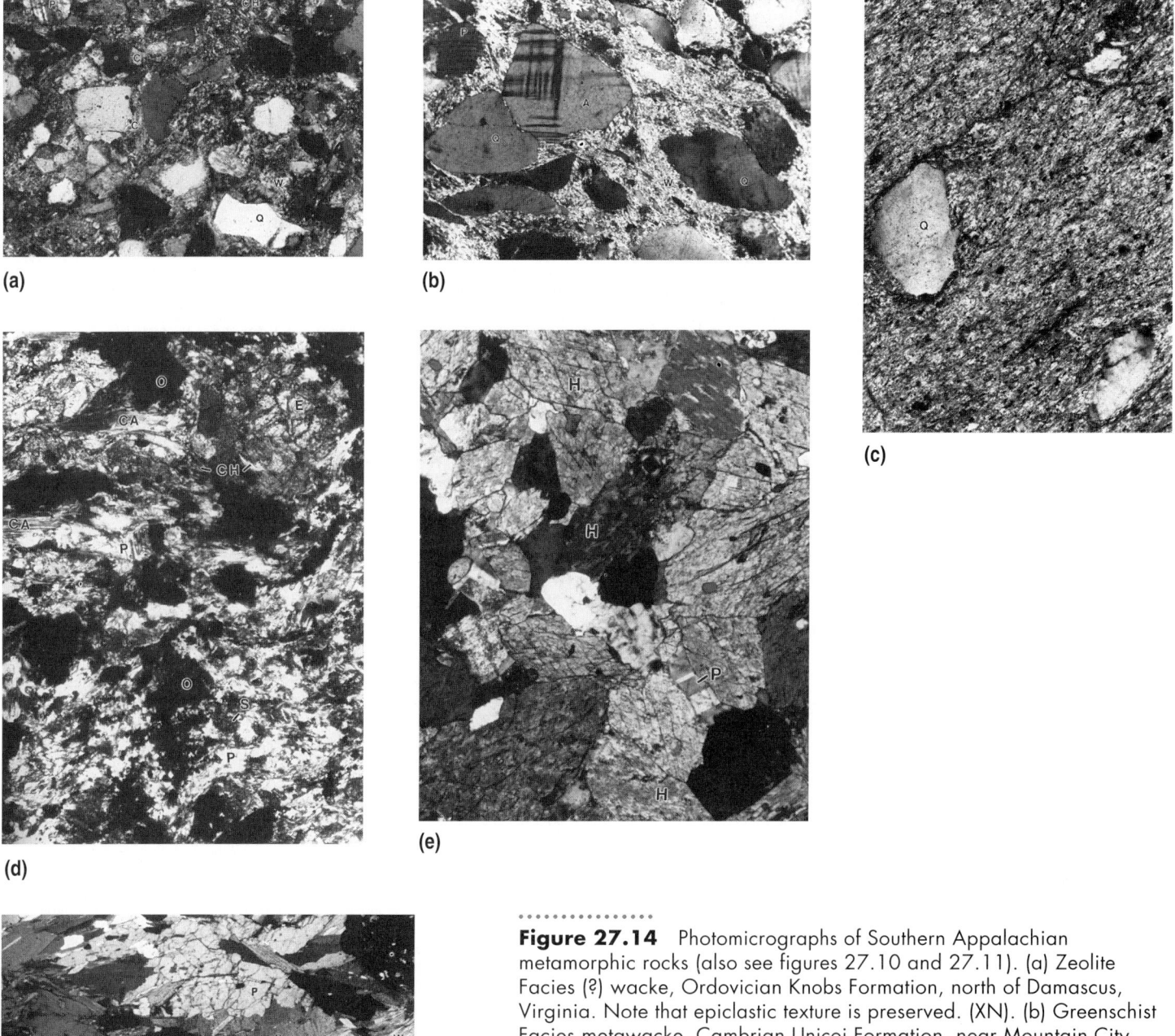

Figure 27.14 Photomicrographs of Southern Appalachian metamorphic rocks (also see figures 27.10 and 27.11). (a) Zeolite Facies (?) wacke, Ordovician Knobs Formation, north of Damascus, Virginia. Note that epiclastic texture is preserved. (XN). (b) Greenschist Facies metawacke, Cambrian Unicoi Formation, near Mountain City, Tennessee. (XN). (c) Greenschist Facies metasiltstone, Cambrian Unicoi Formation, near Mountain City, Tennessee. (XN). (d) Greenschist Facies metabasalt, Grandfather Mountain Formation, near Valle Crucis, North Carolina. (PL). (e) Amphibolite Facies metabasite (amphibolite), near Celo, North Carolina. (XN). (f) Amphibolite Facies pelitic schist, near Todd, Ashe County, North Carolina. (PL). C = calcite, K = kyanite, St = staurolite. Other abbreviations the same as those on earlier figures. Long dimension of photos is 0.33 mm in (a) and (c), 3.25 mm in (b), 2.7 mm in (e), and 1.27 mm in (d) and (f).

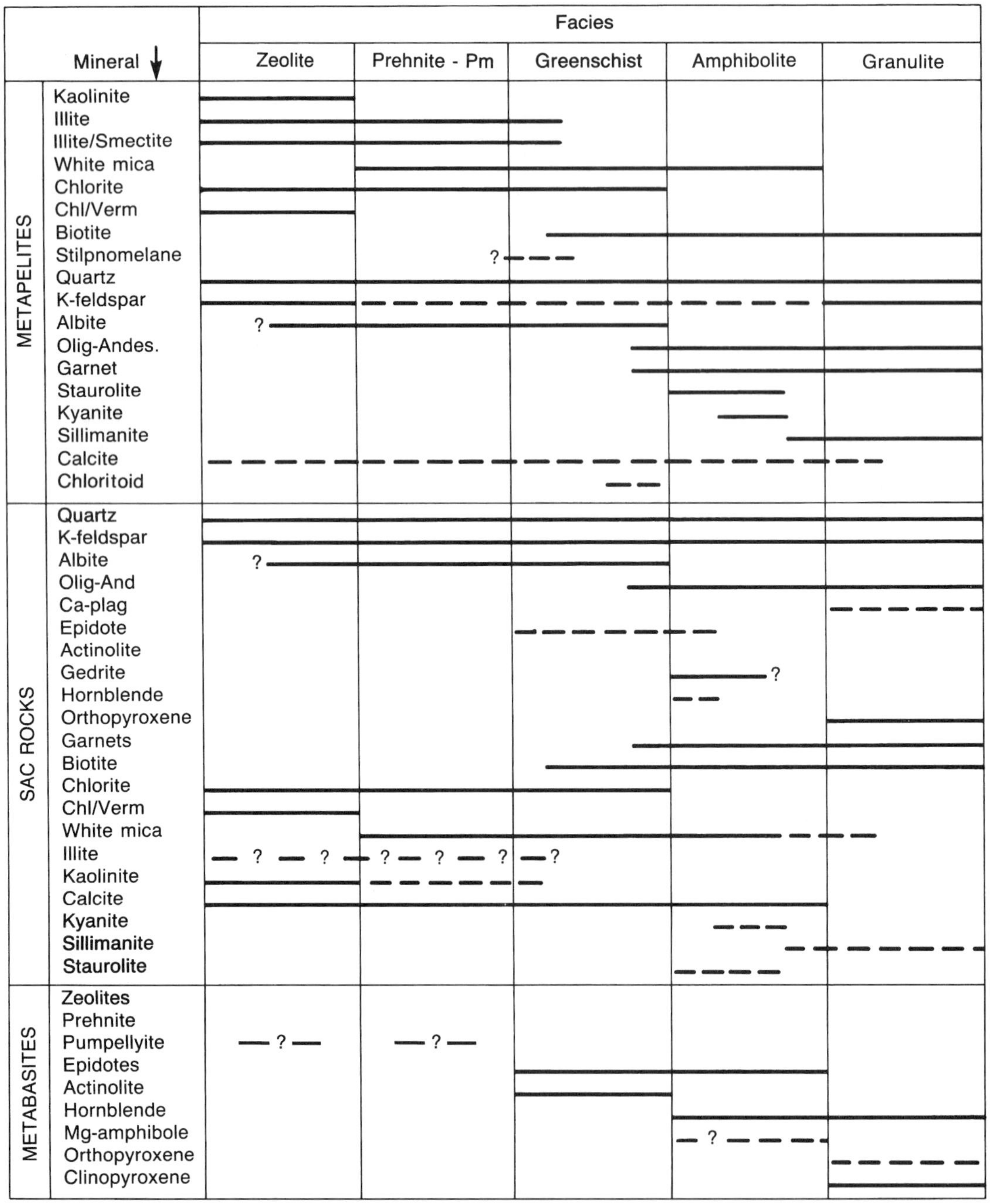

	Facies				
Mineral ↓	Zeolite	Prehnite - Pm	Greenschist	Amphibolite	Granulite

METAPELITES

Kaolinite, Illite, Illite/Smectite, White mica, Chlorite, Chl/Verm, Biotite, Stilpnomelane, Quartz, K-feldspar, Albite, Olig-Andes., Garnet, Staurolite, Kyanite, Sillimanite, Calcite, Chloritoid

SAC ROCKS

Quartz, K-feldspar, Albite, Olig-And, Ca-plag, Epidote, Actinolite, Gedrite, Hornblende, Orthopyroxene, Garnets, Biotite, Chlorite, Chl/Verm, White mica, Illite, Kaolinite, Calcite, Kyanite, Sillimanite, Staurolite

METABASITES

Zeolites, Prehnite, Pumpellyite, Epidotes, Actinolite, Hornblende, Mg-amphibole, Orthopyroxene, Clinopyroxene

Figure 27.15 Mineral-facies chart for Paleozoic metamorphic rocks of the western edge of the Southern Appalachian Orogen.

[Based on Abbott and Raymond (1984), Absher and McSween (1985), Bearce (1973), Brewer (1986), Bryant and Reed (1970a, 1970b), J. R. Butler (1972, 1973b), R. H. Carpenter (1970), Espenshade et al. (1975), Gillon (1989), Gulley (1985), Hadley and Nelson (1971), Hatcher (1976), Hatcher et al. (1979), Helms et al. (1987), V. J. Hurst (1973), McElhaney and McSween (1983), McSween and Hatcher (1985), Mohr and Newton (1983), Nesbitt and Essene (1982, 1983), Rankin, Espenshade, and Neuman (1972), Rankin et al. (1973), G. S. Russell, C. W. Russell, and Farrar (1985), S. E. Swanson (1980, 1981), Weaver et al. (1984), and unpublished observations of the author.]

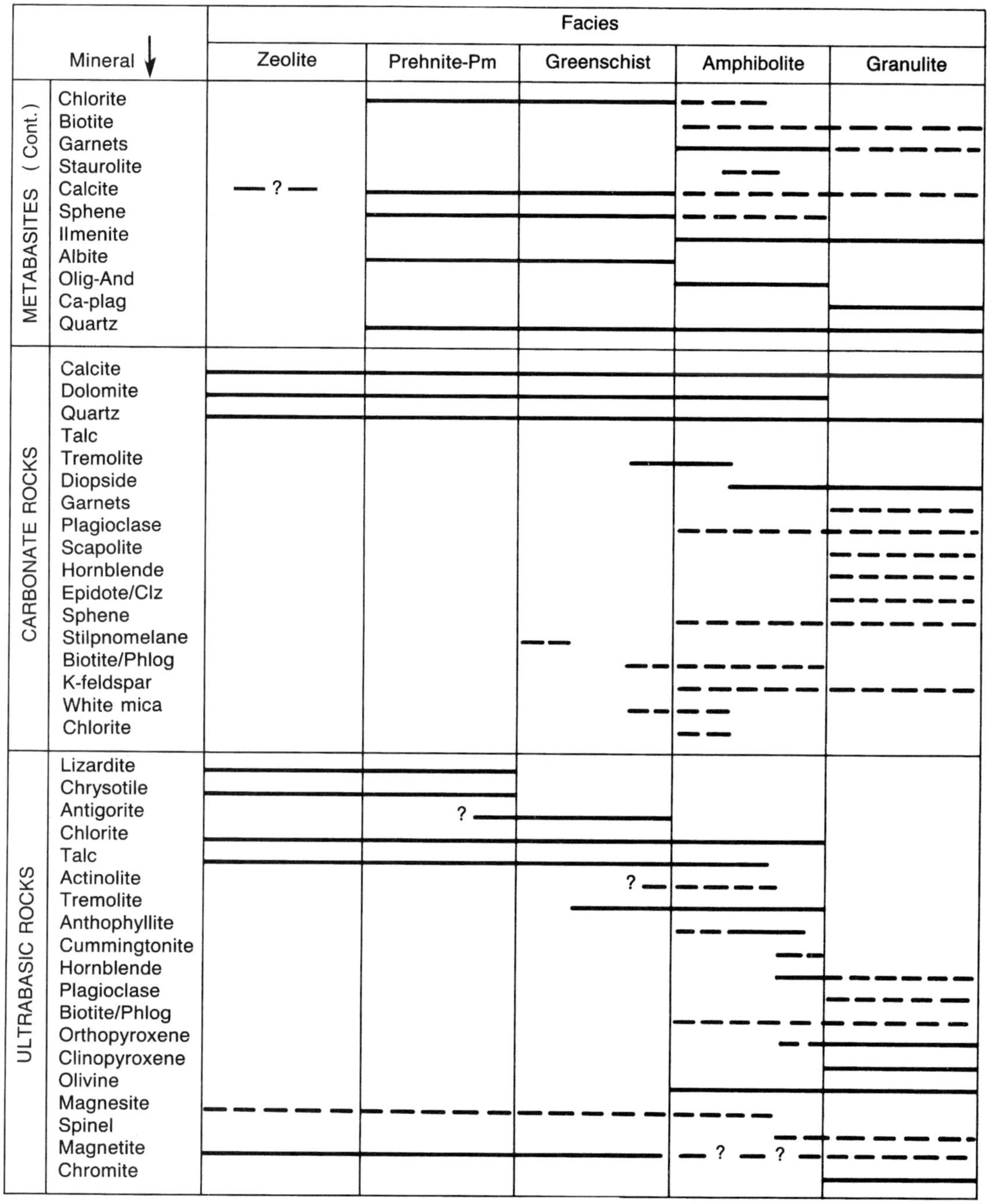

	Mineral ↓	Facies				
		Zeolite	Prehnite-Pm	Greenschist	Amphibolite	Granulite
METABASITES (Cont.)	Chlorite		▬▬▬▬▬	▬▬▬▬▬	– – –	
	Biotite				▬ – – –	▬ – – –
	Garnets				▬▬▬▬▬	– – – –
	Staurolite				▬ ▬	
	Calcite	— ? —	▬▬▬▬▬	▬▬▬▬▬	▬ – – –	▬ – – –
	Sphene		▬▬▬▬▬	▬▬▬▬▬	▬ – – –	▬ – – –
	Ilmenite				▬▬▬▬▬	▬▬▬▬▬
	Albite		▬▬▬▬▬	▬▬▬▬▬	▬▬▬	
	Olig-And					
	Ca-plag				▬▬▬▬▬	▬▬▬▬▬
	Quartz		▬▬▬▬▬	▬▬▬▬▬	▬▬▬▬▬	▬▬▬▬▬
CARBONATE ROCKS	Calcite	▬▬▬▬▬	▬▬▬▬▬	▬▬▬▬▬	▬▬▬▬▬	▬▬▬▬▬
	Dolomite	▬▬▬▬▬	▬▬▬▬▬	▬▬▬▬▬	▬▬▬▬▬	▬▬▬▬▬
	Quartz	▬▬▬▬▬	▬▬▬▬▬	▬▬▬▬▬	▬▬▬▬▬	▬▬▬▬▬
	Talc					
	Tremolite			▬▬▬	▬	
	Diopside				▬▬▬▬▬	▬▬▬▬▬
	Garnets					– – –
	Plagioclase				▬ – – –	▬ – – –
	Scapolite					– – –
	Hornblende				▬ – – –	▬ – – –
	Epidote/Clz					– – –
	Sphene				▬ – – –	▬ – – –
	Stilpnomelane			▬ ▬		
	Biotite/Phlog			– –	▬ – – –	▬ – – –
	K-feldspar				▬ – – –	▬ – – –
	White mica			– – ▬	▬	
	Chlorite				▬ ▬	
ULTRABASIC ROCKS	Lizardite	▬▬▬▬▬	▬▬▬▬▬	▬▬▬		
	Chrysotile	▬▬▬▬▬	▬▬▬▬▬			
	Antigorite		? ▬▬	▬▬▬▬		
	Chlorite	▬▬▬▬▬	▬▬▬▬▬	▬▬▬▬▬	▬▬▬▬▬	
	Talc	▬▬▬▬▬	▬▬▬▬▬	▬▬▬▬▬	▬▬▬▬▬	
	Actinolite			? ▬ – – –	– – –	
	Tremolite			▬▬▬		
	Anthophyllite				▬ – – –	
	Cummingtonite				▬	
	Hornblende				▬ – – –	– – –
	Plagioclase					▬ – – –
	Biotite/Phlog				▬ – – –	▬ – – –
	Orthopyroxene				▬ –	
	Clinopyroxene					▬▬▬▬▬
	Olivine					▬▬▬▬▬
	Magnesite	▬ – – –	▬ – – –		▬ – – –	
	Spinel					▬ ▬ – –
	Magnetite	▬▬▬▬▬	▬▬▬▬▬	▬▬▬▬▬	▬ ? — ? ▬	▬ – – –
	Chromite	▬▬▬▬▬	▬▬▬▬▬	▬▬▬▬▬	▬▬▬▬▬	▬▬▬▬▬

Figure 27.15 Continued

rock types present in what is arguably a melange metamorphosed to this grade. Aluminous schist consists of biotite–garnet–sillimanite–K-feldspar–andesine–quartz–magnetite–ilmeno-hematite (figure 27.15). SAC rocks contain assemblages such as andesine–quartz–K-feldspar–biotite–garnet–magnetite–ilmenite. A typical metabasite assemblage is hornblende–bytownite–biotite–orthopyroxene–quartz–magnetite–ilmenite. Calc-silicate gneisses consist of calcite–quartz–scapolite–bytownite, with accessory minerals that include hornblende, garnet, diopside, clinozoisite,

sphene, and apatite. Ultramafic rocks contain such assemblages as orthopyroxene–andesine–biotite–hornblende–cummingtonite–quartz–pyrrhotite–pyrite–chalcopyrite. It is likely that the quartz, andesine, biotite, hornblende, and cummingtonite are retrograde phases. Granitoid veins in many of the rocks suggest that, where fluids were available, anatexis occurred. Given that the estimated P-T conditions (T = 750–775° C, P = 0.65–0.7 Gpa, or 6.5–7 kb) do not differ significantly from those for Amphibolite Facies metamorphism in the area, the zones of Granulite Facies

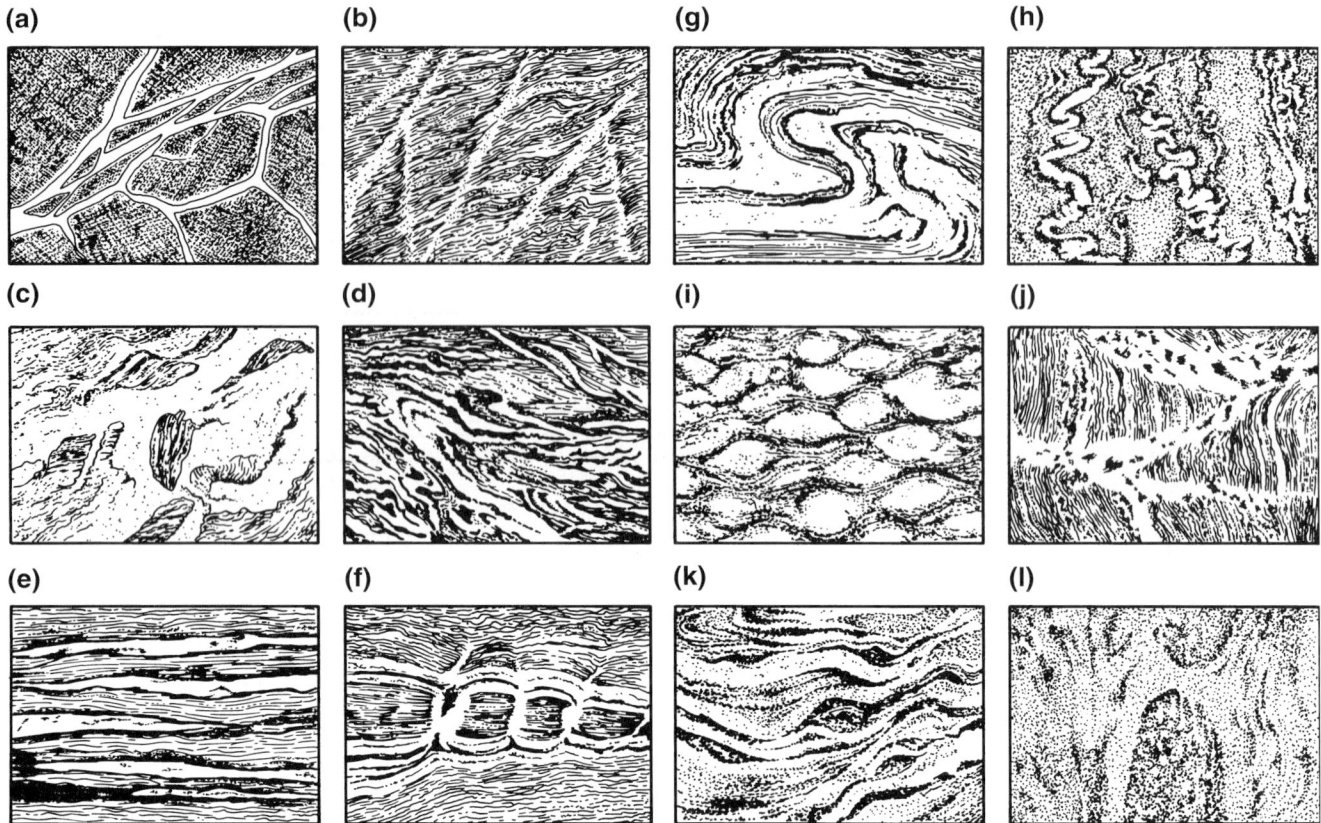

Figure 27.16 Sketches showing the various structural types of migmatites: (a) breccia (agmatic) structure, (b) diktyonitic structure, (c) raft (schollen) structure, (d) vein (phlebitic) structure, (e) layered (stromatic) structure, (f) boudinage (surreitic) structure, (g) folded structure, (h) ptygmatic structure, (i) augen (opthalmitic) structure, (j) stictolithic (fleck) structure, (k) Schlieren structure, (l) nebulitic structure.
(After Mehnert, 1968)

metamorphism probably represent local areas in which the rocks were dehydrated by previous or prograde metamorphic events.

Because the overall metamorphic pattern in the Southern Appalachian Orogen developed over a long period of time (i.e., it is diachronous), it is difficult, especially in both low-grade and thoroughly recrystallized metamorphic rocks, to discern the *complete* patterns of metamorphism *associated with each orogenic event*. In the western part of the orogen, that problem is increased where thrust faults have shortened the width of the orogen, concealing sections of the metamorphic belt. Nevertheless, the elongate metamorphic zones are typical of orogenic Barrovian Facies Series metamorphic belts.

MIGMATITES

Migmatites are quite common in the highest-grade terranes of Buchan and Barrovian Facies series. **Migmatites** are masses of crystalline, mixed rocks, consisting of various pro-

portions of dark, ferromagnesian mineral-rich rock and light quartz- or feldspar-rich rock, that occur in medium- to high-grade metamorphic terranes (Sederholm, 1967).[67] The light-shaded rock is referred to as the *leucosome*, whereas the dark-shaded rock is referred to as the *melanosome*. Together, the leucosome and melanosome comprise new rock formed by migmatization processes, rock referred to collectively as the **neosome.** Leucosomes commonly have a nonfoliated, igneous-like appearance, whereas melanosomes are usually foliated. The **mesosome** is rock typically of intermediate shade that has the appearance of ordinary metamorphic rock (Ashworth, 1985a). Some mesosomes may represent the metamorphic protolith of the migmatite. **Restite** is a term applied to the residual rock remaining after leucosome has been removed from a protolith.

Twelve types of migmatite structures are recognized (figure 27.16) (Mehnert, 1968, ch. 2). Among the more common of these are breccia (agmatic) structure, raft (schollen) structure, layered (stromatic) structure, vein (phlebitic) structure, folded structure, augen (opthalmitic) structure, and

Table 27.4 Migmatite-Forming Processes

Process	Open System Required	Open System Not Required
Magmatic Processes		
Magmatic injection	X	
Anatexis		X
Metamorphic Processes		
Metasomatism	X	
Metamorphic differentiation		X
Combination Processes		
Structural-metamorphic processes		
Melange formation + metamorphism	X	
Ductile deformation yielding boudinage, folding, or layering during metamorphism		X
Other combinations (many combinations are possible, including the following.)		
Melange formation + metamorphism + magmatic injection	X	
Ductile deformation followed by anatexis		X

Source: Based on Ashworth (1985a) and Raymond, Yurkovich, and McKinney (1989).

boudinage (surreitic) structure. In breccia structure, angular blocks of melanosome, mesosome, or protolith, the edges of which correspond, are separated by thin veins of leucosome. In contrast, in raft structure, commonly rounded and/or rotated blocks of darker material are enclosed in larger masses of leucosome. Augen, folded, boudinage, and layered structures have their usual meanings. In some cases, many of these structures may be found within a single outcrop.[68]

The textures of migmatites are typical of textures in other kinds of metamorphic and igneous rocks (Yardley, 1978; McLellan, 1983a; Ashworth and McLellan, 1985). Leucosomes are commonly medium- to very coarse-grained, are coarser-grained than adjacent nonleucosomes, and are typically hypidiomorphic-granular. Aplitic (allotriomorphic-granular) leucosomes also exist. Where leucosomes are deformed, polygonization of grain boundaries and other metamorphic textures overprint the igneous textures. Melanosomes and mesosomes are commonly fine- to medium-grained, lepidoblastic to nematoblastic foliated rocks, but granoblastic varieties occur in some terranes (Kenah and Hollister, 1983). Extraction of melt from some mesosomes apparently is accompanied by grain-size reduction in certain minerals, notably the micas (Dougan, 1983).

A number of processes have been suggested to account for the origin of various migmatites (table 27.4). These may be divided into single-stage and multistage processes. Both

categories include processes that require open systems and those that do not (Ashworth, 1985a).[69] Magmatic injection involves the emplacement and crystallization of small layers, lenses, dikes, or sills of leucosomal magma between masses of darker material. Where one phase of magmatic activity is responsible for the migmatite, the migmatization is a single-stage, open-system process. Where injection has occurred more than once, at significantly different times, the origin of the migmatite is diachronous and multistage.

Recall that metasomatic processes involve chemical change promoted by chemically active fluids derived either from nearby intrusions or from the country rock.[70] By definition, metasomatism is also an open-system process of migmatization.

Processes that do not require an open system (at the scale of ~ 1 m^3) include anatexis and metamorphic differentiation, both of which may be single-stage or multistage, "semiclosed" system processes.[71] Additional multistage processes include combinations of the above processes, plus combined structural-metamorphic processes that involve early histories of fragmentation by ductile or brittle deformation, followed by later events involving either regional metamorphism or one or more of the four migmatite-forming processes mentioned above (Raymond, Yurkovich, and McKinney, 1989).

In summary, examination of the various processes invoked to explain the origin of migmatites reveals that igneous, metamorphic, and structural processes are involved. Magmatic injection and anatexis are magmatic by definition, whereas metasomatism and metamorphic differentiation are metamorphic by definition. Structural processes include ductile deformation, which yields folded structure and boudinage, and brittle deformation, which produces breccia structure and melange protoliths for the migmatites.[72] Although ductile deformation is usually associated with regional metamorphism and brittle deformation is not, both may predate the regional metamorphic event that gives the final textural, structural, and mineralogical character to the migmatite.

Magmatic injection may arise either through emplacement of magmas formed elsewhere and intruded into a terrane (Sawyer, 1987) or from crystallization of locally derived anatectic melts (S. N. Olsen, 1982). In the former case, there is not a genetic relationship between the mesosome or melanosome and the intruded leucosome. The leucosome may have any composition.

In the case of leucosomes of local anatectic-magmatic origin, an unambiguous genetic relationship will exist *if* (1) crystal fractionation did not occur during melt migration and (2) postcrystallization metasomatism is absent. An anatectic-magmatic origin for leucosomes must be demonstrated, not only by a genetic link, but by chemical and/or experimental data that indicate both melt compositions and a melt origin for the leucosome. The experimental work of Tuttle and Bowen (1958) clearly delineated the composition of ternary minimum melts in the haplogranite system (SiO_2-$NaAlSi_3O_8$-$KAlSi_3O_8$-H_2O) (refer to figure 5.19, and Luth, Johns, and Tuttle, 1964).[73] The addition of calcic plagioclase to that system and increased degrees of partial melting may shift the compositions of anatectic melts away from the ternary minimum towards granodioritic or dioritic compositions (G. C. Brown and Fyfe, 1972). Similarly, melting experiments on shales yield melts of granitic to granodioritic composition (Wyllie and Tuttle, 1961). In any case, Johannes (1983b) found that the normative CIPW compositions of several leucosomes plot near either the eutectic or the quartz-orthoclase cotectic of the haplogranite system at 0.05 to 0.5 Gpa (0.5–5 kb) (figure 27.17a). This indicates near-minimum and cotectic melt compositions for the leucosomes and suggests that anatexis *may have been* a controlling process in the development of the migmatites studied.

Mass balance (the equivalence of *quantities* of elements) is also needed, in addition to genetic links between leucosome and mesosome or melanosome, to demonstrate a local anatectic origin. For example, a linear chemical relationship exists between leucosome, mesosome, and melanosome from migmatites of the Central Gneiss Complex of British Columbia (Kenah and Hollister, 1983), but simple anatectic formation of the leucosome is not indicated, because mass balance considerations preclude the derivation of the leucosome and melanosome solely and directly from the mesosome by simple separation of an anatectic melt. This is so, because the *amounts* of silica and other elements are not consistent with such a derivation. Instead, control of the composition of the migmatite layers either by the initial composition of the protolith layers or by postmelting fractionation processes is required to explain the differences in the bulk chemistry.

In some Colorado Front Range and Baltimore Gneiss Dome migmatites, mass balance *is* indicated (S. N. Olsen, 1977, 1982, 1985). The composition of the leucosome plus that of the melanosome selvage (bordering the leucosome) equals the composition of the mesosome, and their sum is equal to the predicted (estimated) value (figure 27.17b). That equality also suggests that the mesosome *was* the protolith for the migmatite and that the leucosome was derived directly from the mesosome. Yet it does not demonstrate an anatectic origin for the leucosome, because such chemical relationships are consistent with both anatectic and metamorphic differentiation origins for migmatites. Additional compositional and textural evidence is needed to demonstrate anatexis.

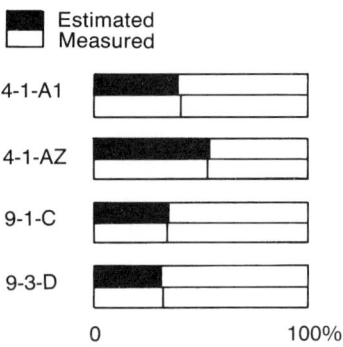

Figure 27.17 Evidence bearing on the origin of migmatites. Results of mass balance calculations on closed-system migmatites of the Front Range of Colorado (from S. N. Olsen, 1983). Rocks metamorphosed in closed systems will have estimated (calculated) volumes of leucosome plus marginal selvage material equal to the volume of the paleosome measured on slabs of the rock (as they do here). In open systems, the estimated and measured values will be significantly different, because some mass will have moved into or out of the system. See the works of S. N. Olsen (1977, 1982, 1983, 1985) for more detailed explanations of the procedures used to make such mass balance calculations.

Demonstration of a metamorphic origin for a leucosome requires the demonstration that magmas were not involved. In addition, one must show either that metasomatism has occurred or that metamorphic differentiation is responsible for compositional equivalence of leucosome + melanosome and protolith (or mesosome). The absence of a magma is suggested by the lack of igneous textures in the rocks and a composition different than minimum melt compositions (Misch, 1968). Metasomatism is indicated where mass balance calculations show a lack of equivalence between leucosome + melanosome and mesosome (S. N. Olsen, 1983, 1985).

The preponderance of evidence, including that provided by recent major and trace element analyses, seems to indicate an anatectic origin or a combined anatectic-metasomatic origin for most migmatites.[74] Nevertheless, some migmatites of apparent metamorphic origin do exist (McLellan, 1983a).

SUMMARY

Buchan and Barrovian Facies series are moderate to high T/P facies series of regional metamorphism. Both types of facies series develop in orogenic belts formed at convergent plate margins. Here, regional heat flow in the orogen is increased by upward movement of magmas and migration of metamorphic fluids. Thrust faulting thickens the crust, increasing pressure and telescoping previously formed metamorphic zones. Buchan Facies Series may also develop in association with divergent (rift) margins or major fault zones where crust or mantle are thinned.

The high temperatures at relatively low pressures, characteristic of Buchan metamorphism, yield mineral assemblages typified by a critical aluminum silicate sequence of kaolinite → pyrophyllite → andalusite → sillimanite. The association of andalusite with cordierite in the middle grades of metamorphism is particularly definitive of Buchan metamorphism. In contrast, Barrovian Facies Series are characterized by the critical sequence of aluminum silicates: pyrophyllite → kyanite → sillimanite. Calcium silicates of the middle to upper grades of metamorphism include garnet and plagioclase feldspar, whereas the phyllosilicates biotite and muscovite are potassium-bearing phases. Both Buchan and Barrovian Facies series consist of various assemblages composed of these and other minerals representing Zeolite to Granulite Facies conditions. Together, the combined subassemblages for various bulk compositions of rock indicate geothermal gradients in the range 20° C to 40° C/km for the Barrovian Facies Series and 40° C to 80° C/km for the Buchan Facies Series.

Most of the important reactions that occurred during the formation of Barrovian Facies Series are continuous reactions involving such phases as white micas, chlorites, biotites, garnets, and Ca-amphiboles. Similar reactions occurred as Buchan Facies Series formed. Some discontinuous reactions, as well as polymorphic changes, such as kyanite <==> sillimanite, are important locally.

Buchan rocks develop at higher pressure and lower maximum temperature than typical Contact Facies Series rocks. Yet there is considerable overlap between the conditions of formation of the two facies series. Mineralogically, the rocks of the Buchan Facies Series are quite similar to those of the Contact Facies Series, but the Buchan rocks are typically foliated and distributed over terranes of regional extent.

Migmatites are common in high-grade zones of all of the facies series of moderate to high temperature. Various types of migmatites are recognized on the basis of the structural character of the leucosome-melanosome mixtures. Where hydrous fluids are available and temperatures exceed the minimum melting temperature of wet granitoid rocks, the low melting (granitoid) fraction of the rocks will melt, coalesce, and migrate as a magma, crystallizing to form a light-colored, granitoid leucosome interlayered or intermixed with a dark-colored, refractory melanosome. The latter is composed of various combinations of minerals including amphiboles, biotite, calcic plagioclase, and quartz. Some leucosomes, particularly some that have unusual compositions and/or exhibit metamorphic, rather than igneous textures, may have had an origin by metasomatism or metamorphic differentiation. Other migmatites form via intense ductile flow and boudinage, in some cases coupled with folding. Still other migmatites may have had a premetamorphic history of fragmentation and mixing of felsic and mafic materials. Whatever their early history, migmatites generally owe their final character to magmatism and/or metamorphism, with attendant recrystallization and neocrystallization, under Amphibolite or Granulite Facies conditions.

EXPLANATORY NOTES

1. This is a generalization, as there is a definite overlap in the conditions of regional and contact metamorphism (Pattison and Tracy, 1991). Nevertheless, it is *generally* true that most contact metamorphism occurs at $P < 4$ kb, whereas most regional metamorphic terranes develop at pressures in the range of 3–12 kb.
2. Refer to chapter 1. Also, Wyllie (1971a, pp. 30–32) summarizes older work on the geothermal gradient of the Earth.

3. See note 2 and Ernst (1972b).
4. Also see Harker (1932, p. 230ff.), Chinner (1966),
   J. A. Winchester (1974), Harte and Hudson (1979, 1980),
   A. J. Baker and Droop (1983), A. J. Baker (1985), and
   N. F. C. Hudson (1985).
5. For example, see Lux et al. (1986), Osberg et al. (1989), and
   the description of the Buchan Facies Series of New England
   in this chapter.
6. See Miyashiro (1973a) for references. See Bard (1967a,
   1967b, 1969, 1970) and Bard and Moine (1979) for
   descriptions of a Buchan Facies Series in the Aracena, Spain
   area. Also see Vernon (1978) and Morand (1990) for
   Australian examples.
7. The exact P-T history of the Buchan rocks in Scotland is not
   altogether clear, as there seems to be evidence, in the form of
   andalusite replacing and being replaced by other aluminum
   silicates, that episodes of Barrovian metamorphism both
   predated and postdated the Buchan event. For the details of
   this problem, the interested reader may refer to Harte and
   Hudson (1979), Chinner (1980), A. J. Baker (1985), and
   Beddoe-Stephens (1990).
8. Harker (1932), Chinner (1966), Porteous (1973),
   J. A. Winchester (1974), Harte and Hudson (1979), N. F. C.
   Hudson (1980, 1985), A. J. Baker (1985), Leslie (1988).
9. The assemblages listed here are those described from terranes
   that form under identical conditions. For example, see
   Dunoyer de Segonzac (1970), M. Frey (1970, 1986, 1987b),
   C. E. Weaver, Beck, and Pollard (1971), Zen (1974b), Zen
   and Thompson (1974), Bathurst (1975), Hower et al. (1976),
   Kubler et al. (1979), Seki and Liou (1981), C. E. Weaver et
   al. (1984), and Kisch (1987), for descriptions of assemblages
   in this grade of rocks and for related discussions.
10. Hower et al. (1976), J. Hoffman and Hower (1979), Eslinger
    and Sellars (1981), E. C. Thornton and Seyfried (1985). Also
    see Grim (1968).
11. Zen (1961), J. Hoffman and Hower (1979), Juster (1987).
12. M. Frey (1970), C. E. Weaver, Beck, and Pollard (1971),
    Ghent (1979), C. E. Weaver et al. (1984, p. 191).
13. This reaction is generally consistent with the observations of
    Weaver, Beck, and Pollard (1971) and Hower et al. (1976).
    Also see experimental work by B. M. Sass, Rosenberg, and
    Kittrick (1987) and Aja, Rosenberg, and Kittrick (1991).
14. Also see Carmignani et al. (1982). McNamara (1966) and
    E. H. Brown (1975) discuss various biotite-forming reactions.
    See Guidotti and Sassi (1976) for a discussion of the
    variations in white mica chemistry as related to metamorphic
    grade.
15. See Osberg (1968, 1971, 1979), Osberg, Moench, and
    Warner (1968), Guidotti (1966, 1968, 1970a,b, 1973, 1978),
    Guidotti, Robinson, and Conatore (1975), Guidotti and

Cheney (1989) Guidotti et al. (1991), and Ferry (1976,
1980a, 1981, 1982, 1983a, 1984), and Ferry and Osberg
(1989). Also see Billings (1937), Albee (1968), J. B.
Thompson and Norton (1968), Boone (1970), Cheney and
Guidotti (1973, 1979), Guidotti et al. (1973), Bickel (1974),
Osberg (1974), C. T. Foster (1977), Guidotti, Robinson, and
Guggenheim (1977), Dallmeyer (1979), Laird (1980),
Holdaway et al. (1982), Lux et al. (1986), Holdaway,
Dutrow, and Hinton (1988), DeYoreo et al. (1989),
A. A. Drake et al. (1989), Osberg et al. (1989), and Hatcher
et al. (1989).

16. W. J. Morgan (1972) depicts some areas with the
    subassemblage sillimanite + K-feldspar, as Granulite Facies
    terranes. However, the presence of coexisting white mica in
    the assemblage sillimanite–K-feldspar–muscovite
    (B. W. Evans and Guidotti, 1966; Cheney and Guidotti,
    1979; Ferry, 1980b) indicates that these rocks actually
    represent the Amphibolite Facies.
17. Pavlides (1962, 1973), T. H. Clark and Eakins (1968),
    D. S. Coombs, Horodyski, and Naylor (1970), Richter and
    Roy (1974), J. R. Walker (1989).
18. Dallmeyer (1979), Holdaway et al. (1982), Lux et al. (1986),
    Hubacher and Lux (1987), Holdaway, Dutrow, and Hinton
    (1988), Spear and Harrison (1989).
19. Guidotti (1970b), Cheney and Guidotti (1979), Ferry
    (1980b, 1984), Holdaway et al. (1982), Lux et al. (1986),
    Holdaway, Dutrow, and Hinton (1988).
20. Also see the references in note 19.
21. Cheney and Guidotti (1979), Lux et al. (1986). DeYoreo et
    al. (1989), recognizing the thermal conditions, refer to the
    metamorphism as "deep-level contact metamorphism."
22. Migmatites are complexly layered metamorphic rocks with
    light-colored layers that represent partial melts. They are
    discussed at the end of this chapter.
23. J. B. Thompson and Norton (1968), Albee (1968), Hietanen
    (1956, 1968), Grambling and Williams (1985), Dusel-Bacon
    and Foster (1983).
24. Barrow (1893, 1912), Tilley (1925), Harker (1932),
    Wiseman (1934), Billings (1937), Chinner (1960, 1966,
    1978, 1980), Zen (1960), Albee (1968), J. B. Thompson and
    Norton (1968), Porteous (1973), Vidale (1974), J. A.
    Winchester (1974b), Zen (1974b), Thompson et al. (1977),
    Mason (1978), Rumble (1978), Abbott (1979b), Harte and
    Hudson (1979), P. R. A. Wells (1979), Laird (1980), Laird
    and Albee (1981), A. J. Baker and Droop (1983), C. M.
    Graham et al. (1983), A. J. Baker (1985), McLellan
    (1985a,b), Moles (1985), Spear and Rumble (1986),
    Hollocher (1987), Juster (1987), K. P. Watkins (1987), A. A.
    Drake et al. (1989), Hatcher et al. (1989), and Osberg et al.
    (1989). For the central and southern Appalachian Orogen,

see Stose and Stose (1957), Dietrich (1959), Bryant and Reed (1970a), R. H. Carpenter (1970), Fisher (1970), Sundelius (1970), Hadley and Nelson (1971), B. A. Morgan (1972), J. R. Butler (1972, 1973b), Rankin, Espenshade, and Neuman (1972), Rankin, Espenshade, and Shaw (1973), V. J. Hurst (1973), Dallmeyer (1975b), Espenshade et al. (1975), Hatcher et al. (1979), Hatcher et al. (1980), W. A. Thomas et al. (1980), Bearce (1982), Nesbitt and Essene (1982), L. Glover et al. (1983), Abbott and Raymond (1984), C. E. Weaver et al. (1984), Absher and McSween (1985), Farrar (1985), Gully (1985), Secor et al. (1986b), Hatcher (1987), A. A. Drake et al. (1989), Hatcher et al. (1989), and Osberg et al. (1989).

25. Miyashiro (1973a) reviews some occurrences of Barrovian Facies Series rocks, including a possible occurrence in the Hida Complex of Japan (p. 361). Asian metamorphic belts are reviewed in Sobolev, Lepezin, and Dobretsov (1965). References to particular occurrences include the following reports and additional sources may be found in the reference lists of these works: *Rocky Mountains (U.S.)*—Hietanen (1968), J. Hoffman and Hower (1979), Hyndman (1980), Lang and Rice (1985a,b); *British Columbia*—Crosby (1968), Pigage (1976), Simony et al. (1980); *Southeast Alaska*—Himmelberg, Drew, and Ford (1991); *Venezuela*—B. A. Morgan (1970); *Sistema Central, Spain*—Ruiz, Aparicio, and Cacho (1978); *Calabria, Italy*—Paglionico and Piccarretta (1978), Maccarrone et al. (1983), Schenk (1984); *Alps*—Hoernes (1973), Fry et al. (1974), C. Miller (1977), Ernst (1979), Zingg (1980); *Menderes Massif, Turkey*—Evirgen and Ashworth (1984); *The Himalayan Range*—Misch (1964), Dobretsov (1965), Frank et al. (1973), Kumar (1978, 1981), Lal, Mukerji, and Ackermand (1981), various papers in Saklani (1981), and Pognante and Lombardo (1989); *The Ural Mountains, Taimyr Fold Belt, and other Asian occurrences*—Lepezin (1965b).

26. For additional information, see the following sources and the works listed in their reference lists. *Adirondack Mountains of New York*—Buddington (1939), A. E. G. Engel and Engel (1953, 1958, 1960a,b, 1962), Stoddard (1980), Bohlen, Valley, and Essene (1985), R. L. Edwards and Essene (1988). *Black Hills*—J. A. Noble and Harder (1948), Redden (1963), Redden et al. (1982). *Rocky Mountains*—Eslinger and Sellars (1981). *Labrador*—C. Klein (1978), Callahan (1980), Rivers (1983). *Quebec*—Indares and Martignole (1984). *Ontario*—E. W. Reinhardt (1968).

27. *The former Soviet Union* (e.g., in the Mama-Bodaibo Synclinorium north of Lake Baikal and along the margin of the Siberian Platform)—Lepezin (1965c). *India and Sri Lanka*—Lepezin (1965a), Ghosh (1978), Sharma and MacRae (1981), Hansen et al. (1987). *China*—Lepezin (1965a).

28. In addition to Barrow's work, see Tilley (1925), Harker (1932), Wiseman (1934), Chinner (1960, 1966, 1978, 1980), Mather (1970), J. A. Winchester (1974), Harte and Hudson (1979), A. J. Baker and Droop (1983), C. M. Graham et al. (1983), A. J. Baker (1985), McLellan (1985a,b), Moles (1985), K. P. Watkins (1987), and the summary in Miyashiro (1973a). Mather (1970) discusses Greenschist Facies assemblages and the reaction producing biotite. K. P. Watkins (1987) discusses biotite and garnet-forming reactions.

29. See the depictions of a complete set of topologies for just one type of phase diagram, for partial sets of conditions, in J. B. Thompson and Thompson (1976) and Abbott (1982).

30. Many workers who study low-grade rocks have adopted terms that are different from the traditional facies terms used by metamorphic petrologists. This is because many of the changes that occur in low-grade rocks are gradational changes in the crystallinity of illite, the rank of coal, the vitrinite reflectance, and the proportions of clay minerals. M. Frey (1986), Kisch (1987, 1990), and Blenkinsop (1988) review and compare the available data on these various measures of low-grade metamorphism. The major zones recognized include various zones of diagenesis, the "anchizone" or "anchimetamorphic" zone, and the "epizone" or "epimetamorphic" zone (Kubler, 1967; M. Frey, 1978, 1986, 1987b; Kubler et al., 1979; C. E. Weaver et al., 1984; Kisch, 1987, 1990). The epizone corresponds to the Greenschist Facies. The Prehnite-Pumpellyite Facies generally encompasses the anchizone and the uppermost zone of diagenesis, the middle to upper zones of diagenesis designated by these workers correspond to the Zeolite Facies, as it is defined here.

31. Work on rocks of this grade is summarized in Dunoyer de Segonzac (1970), Zen (1974b), Zen and Thompson (1974), Kubler et al. (1979), Seki and Liou (1981), M. Frey (1987b), and Kisch (1987). Also see, M. Frey (1970, 1986), C. E. Weaver, Beck, and Pollard (1971), Bathurst (1975), Hower et al. (1976), and C. E. Weaver et al. (1984).

32. Hower et al. (1976), J. Hoffman and Hower (1979), Eslinger and Sellars (1981), E. C. Thornton and Seyfried (1985). Also see Keller (1967), Grim (1968), and Kisch (1990).

33. M. Frey (1970, 1978), C. E. Weaver, Beck, and Pollard (1971), Hower et al. (1976), Ghent (1979), J. Hoffman and Hower (1979), C. E. Weaver et al. (1984, p. 191).

34. Also see McNamara (1966).

35. C. E. Weaver (1960), Kubler (1967, 1968), J. Hoffman and Hower (1979). Also see the review in Kisch (1987).

36. For example, see the descriptions in Harker (1932), Brown (1958), Zen (1960), M. L. Crawford (1966), Albee (1968), Frank et al. (1973), Hoernes (1973), C. Klein (1978), Ruiz, Aparicio, and Cacho (1978), Lal, Mukerji, and Ackermand (1981), Rivers (1983), C. E. Weaver et al. (1984), and Ghent, Stout, and Ferri (1989).

37. Also see Carmignani et al. (1982) and note 14.

38. Because the assemblage albite–epidote–hornblende occurs in the *mafic* rocks of some areas (i.e., hornblende appears with albite and epidote) before the appearance of garnet, some workers have proposed the names Epidote Amphibolite Facies and Albite Epidote Amphibolite Facies to designate the rocks with that assemblage (F. J. Turner, 1948, p. 88; Turner, 1958; also see Liou, Kuniyoshi, and Ito, 1974). Garnet-bearing *pelitic* rocks may occur in association with such hornblende–albite–epidote rocks. Here, I acknowledge that a transition zone exists between facies and that assemblages in mafic rocks are marked more by continuous changes in mineral composition and modal abundance of minerals than by discontinuous reactions in which new phases appear (Laird, 1980; J. B. Thompson, Laird, and Thompson, 1982; Spear, 1982). Consequently, a separate facies for the mafic rocks containing the hornblende–albite–epidote assemblage is unnecessary. Here, the disappearance of albite is considered to mark the maximum upper limit of the Greenschist Facies (cf. Fyfe et al., 1958; deWaard, 1959).

39. For example, see Harker (1932), Brown (1958), Chinner (1960), Albee (1968), Frank et al. (1973), Hoernes (1973), Ruiz, Aparicio, and Cacho (1978), Abbott (1979b), Tracy and Robinson (1979), Lal, Mukerji, and Ackermand (1981), Rivers (1983), Abbott and Raymond (1984), E. Wenk and Wenk (1984), Lang and Rice (1985b), and McLellan (1985a).

40. For a sample of alternative reactions, see J. B. Thompson and Norton (1968), Thompson et al. (1977), Karabinos (1985), Lang and Rice (1985), and McLellan (1985).

41. These rocks are discussed in more detail below, under "Migmatites."

42. The confusion surrounding the terms granulite and Granulite Facies is reviewed by Winkler (1979, ch. 16). Charnokite is a term applied in India and elsewhere to rocks of granitoid to ultramafic composition that contain orthopyroxene (see comments in F. J. Turner, 1981, p. 401, and Winkler, 1979, p. 263). Because igneous and metamorphic charnokites may be impossible to distinguish, because the term charnokite encompasses rocks of a wide compositional range, and because the term signifies little more than the fact that the rocks to which it is assigned contain orthopyroxene, the term charnokite is not used for metamorphic rocks in this text.

43. M. Frey (1987a) summarizes some assemblages in SAC rocks, as does F. J. Turner (1981, ch. 9). The P-T conditions at the lower grades of Greenschist Facies metamorphism of the Buchan, Barrovian, and Sanbagawa Facies series are so similar that the assemblages are essentially the same. Many reports on Prehnite-Pumpellyite and Greenschist Facies rocks are based on the works of Coombs (1954, 1960) and his successors (e.g., E. H. Brown, 1967) on the rocks of South Island, New Zealand. Similar work on rocks of the Great Valley Group (Sequence) of California (Dickinson et al., 1969) is also of interest here. The latter rocks actually represent a facies series more akin to the Franciscan Facies Series. Assemblages reported here are based, in part, on the author's observations in the southern Appalachian Orogen and the California Coast Ranges, on the works cited above, and on additional relevant works including Schreyer and Chinner (1966), D. G. Bishop (1972), R. J. Stewart and Page (1974), Houghton (1982), Rivers (1983), Abbott and Raymond (1984), Absher and McSween (1985), Gulley (1985), Schenk (1984), and Vavra (1989).

44. Also see Liou (1971a, 1971b), Liou et al. (1974), S. E. Swanson and Schiffman (1979), Cho, Liou, and Maruyama (1986), and Cho and Liou (1987).

45. For additional phase assemblages and information, refer to Coombs, Horodyski, and Naylor (1970), Zen (1974b), W. Glassley (1975), Bevins (1978), Kirchner (1979), Offler et al. (1980), Bevins and Rowbotham (1983), Offler and Aguirre (1984), Liou, Maruyama, and Cho (1985, 1987), Cho, Liou, and Maruyama (1986), Cho and Liou (1987), Bettison and Schiffman (1988), Lucchetti, Cabella, and Cortesoguo (1990), Starkey and Frost (1990), and Ishizuka (1991).

46. Bevins and Rowbotham (1983), Liou, Maruyama, and Cho (1985, 1987).

47. Zen (1974b), Bevins (1978), Bevins and Rowbotham (1983).

48. E. H. Brown (1971), Coombs, Horodyski, and Naylor (1977), Offler et al. (1980), Laird (1980), Ernst, Liou, and Moore (1981), Liou (1981a), Houghton (1982), Offler and Aguirre (1984), Cho and Liou (1987), Bevins and Merriman (1988), Lucchetti, Cabella, and Cortesogno (1990). Additional assemblages are tabulated by F. J. Turner (1981, ch. 9).

49. Hornblende has also been reported from the Greenschist Facies (J. B. Lyons, 1955; E. H. Brown, 1971). Because the appearance of hornblende is generally considered to mark the transition to the Amphibolite Facies in mafic rocks (see note 17), the reported occurrences of hornblende in the Greenschist Facies raise some questions. For example: Is the amphibole actually hornblende? Is it a *stable* part of an equilibrium assemblage? J. B. Lyons (1955) reported such an occurrence, but amphibole compositions were not reported for the rocks he studied, the capacity to spot-analyze the compositions of amphiboles (thereby avoiding zones and impurities) did not exist at the time Lyons did his work, and the geologic history of the region in western New Hampshire in which Lyons worked is complex and polymetamorphic (F. S. Spear and Rumble, 1986). From an area near that in which Lyons worked, Brady (1974) reports coexisting hornblende and actinolite from rocks surrounding the staurolite isograd (the beginning of the Amphibolite Facies). Similarly, Hietanen (1968) reports coexisting actinolite and hornblende in calcareous rocks of the kyanite-staurolite zone from Idaho. These observations suggest that hornblende may appear, under certain conditions, either in the upper Greenschist Facies or in a transition zone between the Greenschist and Amphibolite Facies and that actinolite may occur in the Amphibolite Facies. (The problems associated with these gradational changes are discussed in detail by Laird, 1980, and J. B. Thompson, Laird, and Thompson, 1982.) All plagioclase reported by Brady is more calcic than albite, suggesting that the rocks are not of the Greenschist Facies, but rather those of a transition zone or the lower part of the Amphibolite Facies (cf. the lower-grade part of the Hornblende Hornfels Facies).

50. Refer again to notes 38 and 49. Note that some grossular-andradite garnets appear well within the Greenschist Facies, but the more common reactions yielding almandine-rich garnets occur near the top of the facies or at the Greenschist-Amphibolite Facies transition. Experimental analyses of this transition using real rock compositions have been carried out by Liou, Kuniyoshi, and Ito (1974), Spear (1981b), and Moody, Meyer, and Jenkins (1983). Also see Maruyama, Liou, and Suzuki (1982).

51. Also see A. F. Cooper (1972) and Maruyama, Liou, and Suzuki (1982).

52. For example, see Hietanen (1963, 1973a), J. P. Morgan (1970), Laird (1980), Spear (1982), Stephenson and Hensel (1982), Abbott and Raymond (1984), Sills and Tarney (1984), Gulley (1985), F. S. Spear and Rumble (1986), and Helms et al. (1987).

53. Granulite Facies assemblages are described in A. E. J. Engel and Engel (1958, 1960b), Paglionico and Piccarreta (1978), Schenk (1984), Absher and McSween (1985), Gulley (1985), and Rudnick and Taylor (1987). Hacker (1990) discusses problems of experimental analysis of this facies transition.

54. J. B. Lyons (1955), A. E. J. Engel and Engel (1958, 1960), Zen (1960), Misch (1964), Frank et al. (1973), Hoernes (1973), Ferry (1982; 1983), Absher and McSween (1985).

55. Chidester (1968), J. R. Carpenter and Phyfer (1969), Trommsdorff and Evans (1974), B. W. Evans (1977), Yurkovich (1977), Vance and Dungan (1977), Misra and Keller (1978), Raymond and Swanson (1981), Honeycutt and Heimlich (1980), S. E. Swanson (1980, 1981), Lan and Liou (1981), Sanford (1982), McElhaney and McSween (1983), Scotford and Williams (1983), Trommsdorff (1983), Abbott and Raymond (1984), Absher and McSween (1985), Raymond and Abbott (1985), Raymond (1987). See chapter 31 for additional information.

56. Greenwood (1971), Chernosky (1976), Hemley et al. (1977), Trommsdorff (1983).

57. Hadley (1964), Butler (1972, 1973b, 1991), Fullagar and Odom (1973), Dallmeyer (1975a,b, 1978), Fullagar and Dietrich (1976), Sinha and Glover (1978), R. D. Hatcher et al. (1979), R. D. Hatcher et al. (1980), W. A. Thomas et al. (1980), Tull (1980), Fullagar and Bartholemew (1983), L. Glover et al. (1983), Abbott and Raymond (1984), C. E. Weaver et al. (1984), Gulley (1985), A. S. Russell, Farrar, and Russell (1985), Secor et al. (1986a), Secor, Snoke, and Dallmeyer (1986b), Dallmeyer et al. (1986), R. D. Hatcher (1987), Dallmeyer (1988), A. A. Drake et al. (1989), Osberg et al. (1989), R. D. Hatcher et al. (1989), Kish (1990).

58. B. A. Morgan (1972), L. Glover et al. (1983), J. R. Butler (1984, 1991), C. E. Weaver et al. (1984), R. D. Hatcher (1987).

59. J. R. Butler (1973b), R. D. Hatcher et. al. (1979), Swanson et al. (1985), Butler (1991).

60. Hadley (1964), Abbott and Raymond (1984), and Absher and McSween (1985).

61. See Kulp and Poldervaart (1956), Stose and Stose (1957), P. B. King and Ferguson (1960), Brobst (1962), Hadley and Goldsmith (1963), Bryant and Reed (1970a,b), R. H. Carpenter (1970), Hadley and Nelson (1971), J. R. Butler (1972, 1973), Rankin, Espenshade, and Neuman (1972), Rankin, Espenshade, and Shaw (1973), V. J. Hurst (1973), Dallmeyer (1975b, 1978), Espenshade et al. (1975), Force (1976), Fullagar and Dietrich (1976), R. D. Hatcher (1978, 1987), R. D. Hatcher et al. (1979), R. D. Hatcher et al. (1980), Raymond and Abbott (1980), W. A. Thomas et al. (1980), Tull (1980), Raymond and Swanson (1981), Nesbitt and Essene (1982, 1983), Bartholomew et al. (1983), L. Glover et al. (1983), Abbott and Raymond (1984), Bartholomew and Lewis (1984), C. E. Weaver et al. (1984), Mohr and Newton (1983), Absher and McSween (1985), Gulley (1985), Raymond and Abbott (1985), Brewer (1986), Raymond (1987), Dallmeyer (1988), Eckert, Hatcher, and Mohr (1989), Gillon (1989), J. L. Hopson, Hatcher, and Stieve (1989), McSween, Abbott, and Raymond (1989), Raymond, Yurkovich, and McKinney (1989), Merschat and Wiener (1990), J. W. Miller (1990), J. R. Butler (1991), Raymond et al. (1991), and the references therein.

62. Hadley and Nelson (1971), Merschat (1977), Brewer and Woodward (1988), Raymond, Yurkovich, and McKinney (1989), Merschat and Wiener (1990). For a discussion of migmatites, see the last section of this chapter.

63. L. A. Raymond (unpublished data from the Spruce Pine District, North Carolina).

64. J. W. Miller (1990).

65. Yurkovich and Butkovich (1982), Z. A. Brown et al. (1985), J. L. Hopson, Hatcher, and Stieve (1989), Kish (1989), McSween, Spear, and Fullagar (1991).

66. Geothermometry and geobarometry on Blue Ridge rocks is reported by Nesbitt and Essene (1982), Mohr and Newton (1983), Absher and McSween (1985), Helms et al. (1987), Eckert, Hatcher, and Mohr (1989), and McSween, Abbott, and Raymond (1989).

67. Key sources that form the basis for this discussion and definitions include Sederholm (1967), Mehnert (1968), Atherton and Gribble (1983), and Ashworth (1985b). See the collected works in the latter two sources and the references therein for additional information. In particular, see Tuttle and Bowen (1958), Grant (1968, 1985b), Misch (1968), C. G. Brown and Fyfe (1972), S. N. Olsen (1977, 1983, 1985), Johannes and Gupta (1982), Dougan (1983), Johannes (1983a,b), Kenah and Hollister (1983), McLellan (1983a,b), Ashworth (1985a), and Ashworth and McLellan (1985). Also see Hedge (1972), Kays (1976), Sighinolfi and Gorgoni (1978), Touret and Dietvorst (1983), Abbott (1985), Barr (1985), Touret and Olsen (1985), Tracy (1985), C. Weber et al. (1985), P. Robinson et al. (1986), van Gaans et al. (1987), and Sawyer (1987). *Note:* The term *paleosome* is used by some workers in the same ways that *protolith* and *mesosome* are used here.

68. For example, see Raymond, Yurkovich, and McKinney (1989).

69. Here the terms open system and closed system are not used in a strict thermodynamic sense.

70. See the brief reviews in Ashworth (1985a) and S. N. Olsen (1985) describing the protracted debate on the origin of metasomatic fluids. Also see the works of Sederhom, in Sederhom (1967), and the works of H. Ramberg (1952, pp. 182–188) and Wickham (1987).

71. The term semiclosed system is used, following similar uses of closed system by S. N. Olsen (1983) and Ashworth (1985a), to refer to systems closed to all components other than the volatiles.

72. Raymond, Yurkovich, and McKinney (1989) propose that some migmatites are diachronous bodies of rock that have had an early history of fragmentation and mixing to yield a melange, followed by a later history involving high-grade metamorphism.

73. Also see Winkler, Boese, and Marcopoulos (1975) and Winkler and Breitbart (1978).

74. See the more recent references in note 67.

PROBLEMS

27.1. (a) Draw the necessary ACF topologies for a step-by-step series of reactions between the Zeolite and Greenschist Facies diagrams shown in figure 27.7. (b) Write and balance a series of reactions representing those topologic changes.

27.2. Using the CFM diagram of Abbott (1982, 1984), explain why epidote is not a typical constituent of Granulite Facies metabasites.

27.3. (a) What kinds of evidence would make a convincing case for the polymetamorphic model for Southern Appalachian Blue Ridge metamorphic history (figure 27.13a)? (b) What kinds of evidence would make a convincing case for the monotonic cooling model for Southern Appalachian Blue Ridge metamorphic history (figure 27.13b)?

28

High P/T Metamorphism: Franciscan and Sanbagawa Facies Series and the Origin of Blueschists

INTRODUCTION

Glaucophane in abundance imparts an attractive blue hue to rocks. This feature undoubtedly accounts for the considerable interest given to the relatively uncommon glaucophane schists (the "blueschists") of the California Coast Ranges, the Alpine-Himalayan orogenic belt, and some western Pacific Islands.[1] The blue color also serves as the basis for the name Blueschist Facies, even though this facies contains large volumes of rock that are neither blue nor schistose. It is also true that all rocks containing blue amphibole do not belong to the Blueschist Facies.[2]

The Blueschist Facies is one of six facies that develop in terranes in which the geothermal gradient is low (< 20° C/km) or the overall P/T is moderate to high (Ernst, 1971a, 1973, 1977, 1988).[3] Two types of facies series are recognized in such terranes—the Sanbagawa Facies Series and the Franciscan Facies Series. In the Sanbagawa Facies Series, in which the maximum temperatures are somewhat higher than in the Franciscan Facies Series, the facies sequence is Zeolite → Prehnite-Pumpellyite → Blueschist → Greenschist → Amphibolite.[4] In the Franciscan Facies Series, the facies sequence is Zeolite → Prehnite-Pumpellyite → Blueschist → Eclogite. As is the case with other facies series, not all metamorphic belts with a particular facies series exhibit all of the facies of that series.

The P-T conditions and histories represented by Sanbagawa and Franciscan facies series have been deduced from experimentally determined phase stabilities, geothermobarometry, vitrinite reflectance, apatite fission track analyses, and isotope studies.[5] These, together with microstructural and paleogeographic analyses, indicate that the rocks were metamorphosed where temperatures were maintained at low levels while pressures were elevated, and, in many cases, where deviatoric stresses were low to nonexistent. Rapid burial through sedimentation represents one process that will produce these static-type conditions and for that reason some petrologists have referred to such metamorphism as "burial metamorphism."[6] Yet Franciscan and Sanbagawa Facies series include rocks metamorphosed under both static and regional conditions; thus a name that implies that only static metamorphism is responsible for their petrogenesis is inappropriate. In this chapter, after assessing the occurrences, mineral assemblages, and textures of the rocks of these facies series, we examine a diversity of proposed origins, including metasomatic and tectonic burial hypotheses.

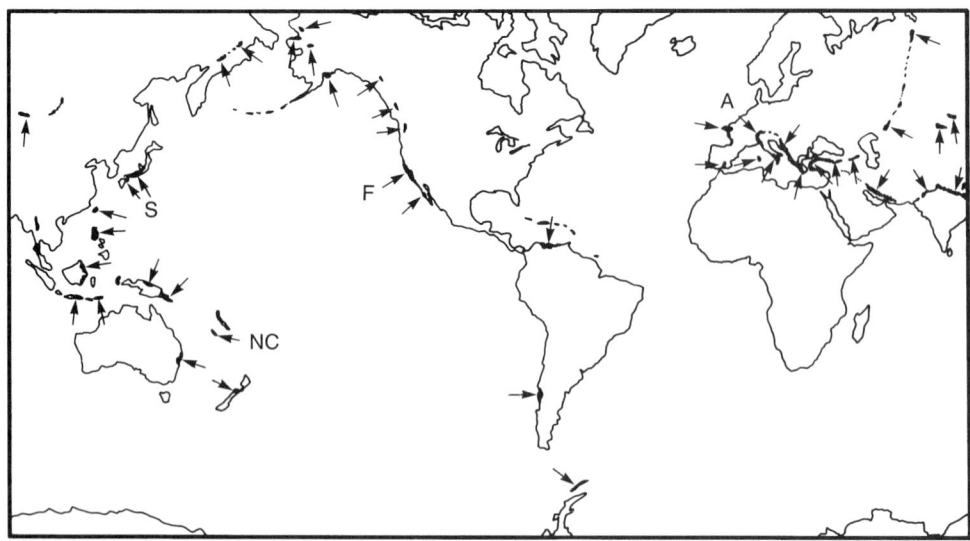

Figure 28.1   World map showing the location of high PT metamorphic belts, including those mentioned prominently in this chapter. A = Alps of southern Europe, F = Franciscan Complex of California, NC = New Caledonia, S = Sanbagawa Belt, Japan.
(Modified from R. G. Coleman (1972).)

OCCURRENCES

Franciscan and Sanbagawa facies series are widely distributed on the Earth (figure 28.1).[7] They occur in North, Central, and South America, in the Caribbean region, in Europe, especially in the Alps and along the northern margin of the Mediterranean Sea, in the Middle East, in Asia, and in the circum-Pacific region. Typically, these facies series form on the outer (trench) side of a paired metamorphic belt associated with a subduction zone (Miyashiro, 1961, 1973b). In some cases, high P/T (low-temperature) rocks form where subduction-induced collision between a continent and island arc, microcontinental block, or another continent is inferred.[8]

Young mountain belts contain the majority of Sanbagawa and Franciscan facies series rocks, but early Paleozoic and rare Precambrian Blueschist Facies rocks are known.[9] In many cases, Blueschist Facies mineral assemblages are overprinted by younger, higher-temperature or lower-pressure assemblages.[10] In part, such overprinting is made possible by the fact that, in general, low-temperature rocks contain hydrous phase assemblages that yield a fluid phase which facilitates recrystallization. In addition, during orogenesis, there is a tendency for cooler rocks to be heated, which also promotes recrystallization. In some cases, rocks formed under lower-pressure or higher-temperature conditions are remetamorphosed under Blueschist Facies conditions.[11]

The two facies series of high P/T metamorphism take their names from well-studied examples on opposite sides of the Pacific Ocean. The Franciscan Facies Series is named for the Franciscan Complex of western California and southern Oregon (E. H. Bailey, Irwin, and Jones, 1964; Berkland et al., 1972). Similar rocks are known from Baja California and Washington.[12] The Sanbagawa Facies Series takes its name from rocks exposed in southeastern Japan (Seki, 1958; Miyashiro, 1961; Ernst et al., 1970).[13]

MINERAL ASSEMBLAGES, FACIES, AND TEXTURES

Mineral assemblages, facies, and textures set the high P/T facies series apart from those of lower P/T. Minerals such as lawsonite occur only at high T and low P. In the lower grades of metamorphism, and locally in the Blueschist Facies, rock textures differ from the textures exhibited by rocks of other facies series.

Textures in Franciscan and Sanbagawa Facies Series Rocks

The textures of Franciscan and Sanbagawa Facies Series rocks are highly varied. In many localities, sedimentary or structural burial provides a P_{load} that is essentially hydrostatic. In addition, high pore-fluid pressures are common in rocks and sediments that descend from the ocean floor, where fluids are abundant, into a subduction zone. In the absence of high deviatoric stresses, these pressures promote recrystallization and neocrystallization without accompanying development of pronounced foliations. Under these circumstances, rocks commonly retain *relict textures* and structures, such as epiclastic texture, porphyritic texture, diabasic texture, amydaloidal structure, and pillow structure. The outlines of radiolaria may

Table 28.1 Textures in Sanbagawa and Franciscan Facies Series Rocks

Texture Type	Cataclastic and Related Textures (C types) (in SF-tectonites)	Foliated-Crystalline Textures (S types) (in S- and L-tectonites)
Texture I	*Relict textures (IC).* Includes epiclastic, porphyritic, intergranular, ophitic, diabasic, and other textures, with or without slight breaking of grains.	*Relict texture (IS).* Includes epiclastic, porphyritic, intergranular, ophitic, diabasic, and other textures, with some neocrystallized or recrystallized grains.
Texture II	*Breccia texture (IIC).* Texture dominated by broken grains. Rocks lack foliation but may contain bands of microbreccia.	*Semischistose and semislaty textures (IIS).* Weakly foliated, phaneritic, and aphanitic textures, respectively, produced by aligned, neocrystallied phyllosilicates, plus elongation and flattening of rock fragments.
Texture III	*Foliated cataclastic texture (IIIC).* Foliation produced by microfolia composed of phyllosilicates that separate microlithons (m) and clasts (c)(m + c = 50–99% of rock).	*Schistose texture (IIIS).* Well-developed foliation, with or without lineation of inosilicates and phyllosilicates. Incipient segregation of grains into foliated and granoblastic layers.
Texture IV	*Protomylonitic texture (IVC).* Texture with well-developed microfolia with shear surfaces; syntectonically recrystallized bands of mica, quartz, or other minerals; with or without recovery (granoblastic) texture. (m + c > 50% of rock).	*Gneissose texture (IVS).* Foliated texture in which segregation banding predominates. Dark bands typically dominated by ferromagnesian inosilicates and phyllosilicates. Light bands typically dominated by mosaic-textured quartz and feldspar.

Source: Modified from Raymond (1973a).

even be preserved in cherts, in spite of the fact that the opal-A has thoroughly recrystallized to quartz. Where neocrystallization and recrystallization are pronounced and acicular to platy minerals crystallize, *diablastic textures* develop. In contrast, *foliated textures* form where deviatoric stresses are important.

Foliated textures belong to two main categories and range from weak to pronounced (table 28.1). Because of the tectonic setting in which high P/T rocks form, many of these rocks, are SF-tectonites that have *cataclastic-type textures*, including brecciated, foliated-cataclastic, or protomylonitic textures (figure 28.2a)(Raymond, 1975, 1984a; S. E. Lucas and Moore, 1986; J. C. Moore et al., 1986). The latter two textures are foliated types. Foliated-cataclastic textures are particularly characteristic of melanges and related rock bodies, which are common in subduction zones. In such rock bodies, the rocks may consist of sheared and broken clastic grains and microlithons separated by microfoliated zones that are characterized by aligned phyllosilicate grains (Cowan, 1982a; J. C. Moore et al., 1986). Shear-induced dislocation has occurred along many of these microfoliated zones. In contrast, in S- and L-tectonites, where recrystallization dominates, *crystalline-foliated textures* occur, ranging from aphanitic

semislaty and fine-grained semischistose types to coarse-grained gneissose types. A variety of textural classifications of foliated rocks have been proposed and used for mapping different textural zones in high P/T terranes (F. J. Turner, 1948, p. 38; 1968, pp. 31–32; Blake, Irwin, and Coleman, 1967, 1969).[14] Three types of foliated-crystalline textures are recognized here (table 28.1). These are semischistose/semislaty texture, schistose texture, and gneissose texture (figure 28.2b and d).

Characteristic Minerals, Mineral Assemblages, and Facies

Because the rocks in outer metamorphic belts are metamorphosed pieces of ocean crust and overlying sediments, metabasites, metawackes, and metapelites are the dominant rocks of the high P/T facies series. Of these, the metabasites and the metawackes are the most widely studied. The Sanbagawa Belt of Japan is exceptional, however, in that, rather than metawackes, there are abundant pelitic rocks that have been studied in detail. In the Sanbagawa Belt and elsewhere, the less abundant metacherts, metacarbonate rocks, and ultramafic rocks have been studied less thoroughly.

589

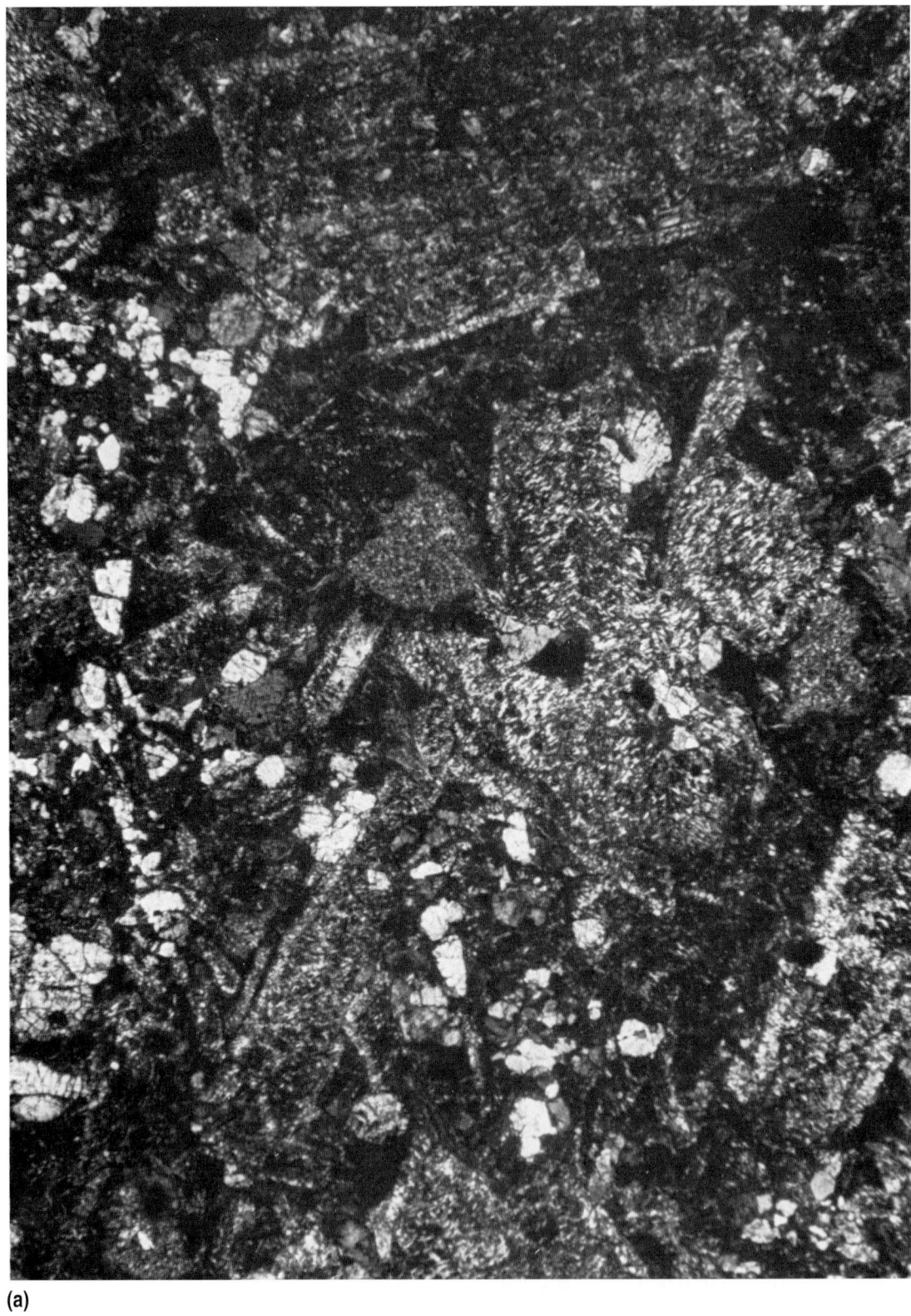

(a)

Figure 28.2 Photomicrographs of textures in high P/T rocks. a) Blastoporphyritic (relict porphyritic) texture (type IS) in metabasite, northeastern Diablo Range, California, (XN). b) Semischistose texture (type IIS) in metawacke, northeastern Diablo Range, California, (XN). c) Foliated cataclastic texture (type IIIC) in melange matrix, melange of Blue Rock, northeastern Diablo Range, California, (PL). d) Gneissose texture (type IVS), South Fork Mountain Schist, northern Coast Ranges, California, (XN). Long dimension of all photos is 3.25 mm.

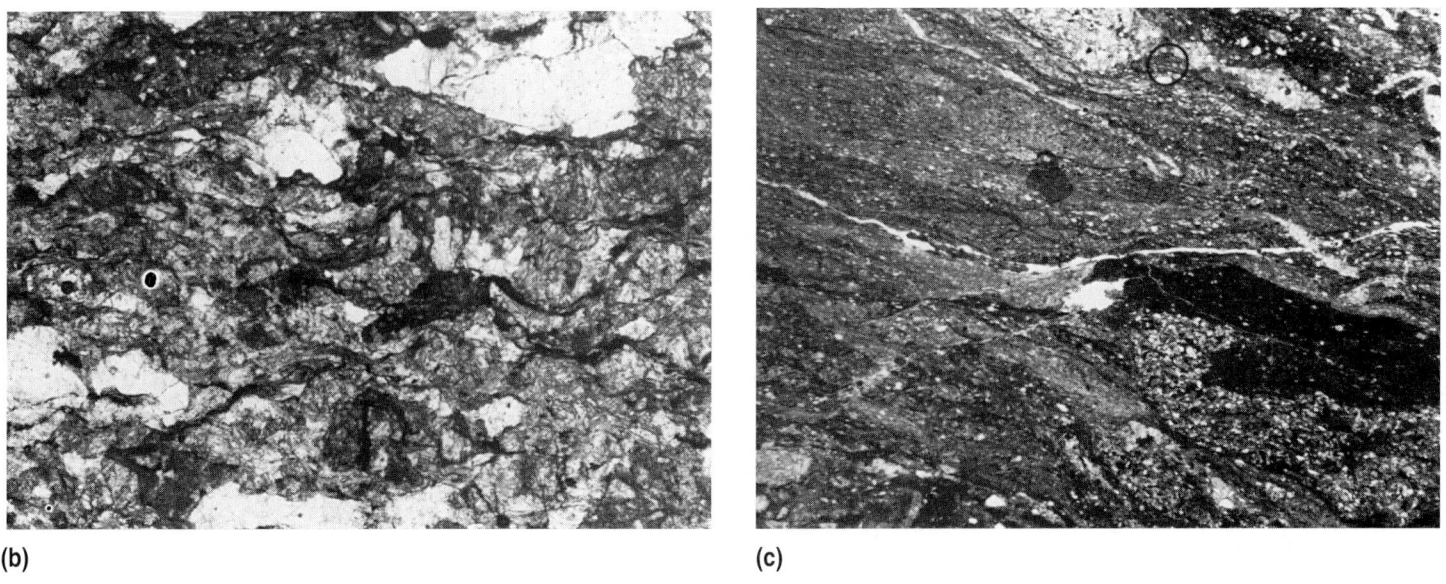

(b)

(c)

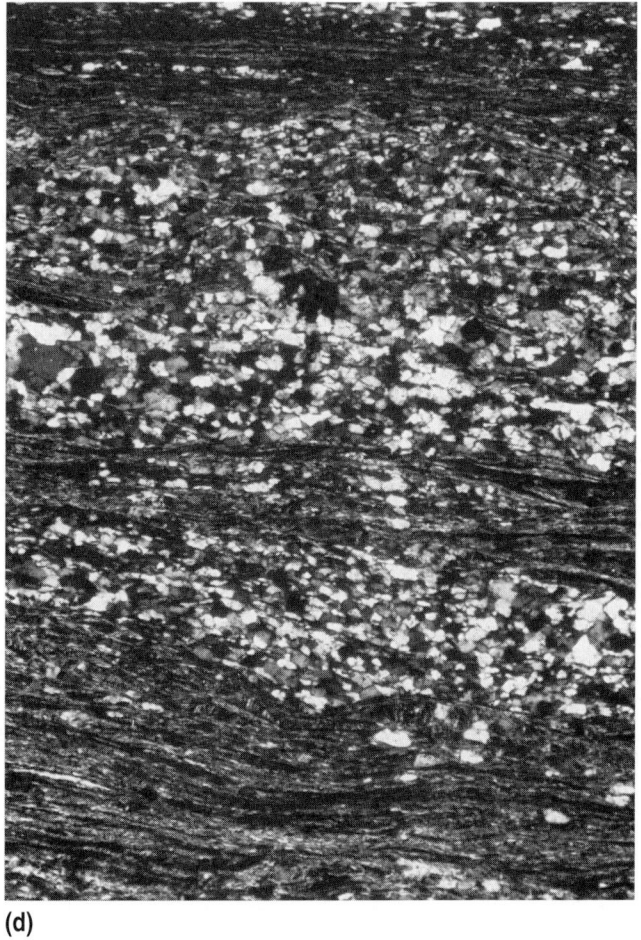

Figure 28.2 Continued

(d)

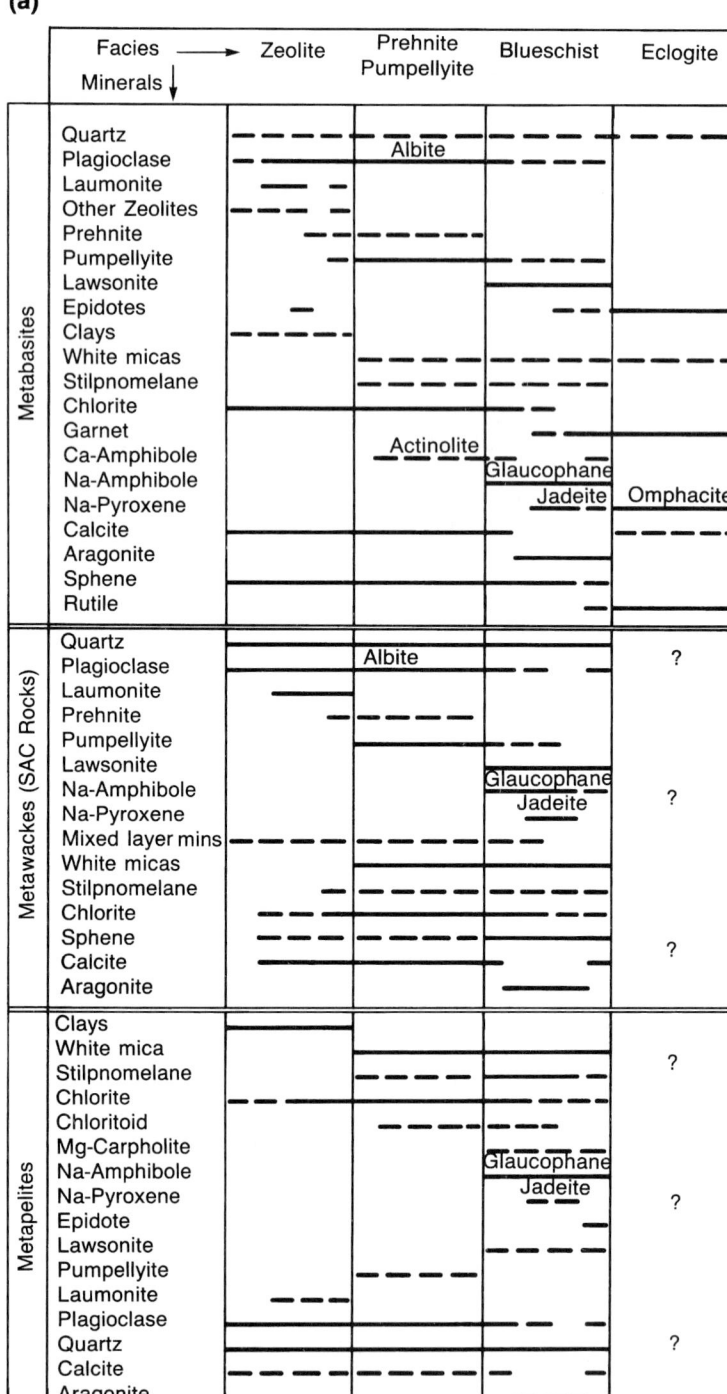

Figure 28.3 Idealized mineral-facies charts for Franciscan (a) and Sanbagawa (b) Facies series.

[Based in part on D. S. Coombs (1960), Seki et al. (1969), Ernst, Onuki, and Gilbert (1970), Ernst (1977a; 1984), Enami (1983), Blake et al. (1988), Higashino (1990), Otsuki and Banno (1990), Oh, Liou, and Maruyama (1991), observations of the author, and additional sources in notes 18 and 25–28]

Representative mineral-facies charts for the Sanbagawa and Franciscan Facies series are presented in figure 28.3. For the Franciscan example, note that all of the facies of the series—Zeolite, Prehnite-Pumpellyite, Blueschist, and Eclogite—are represented. Rocks of the Eclogite Facies, however, do not comprise large-scale units or regional terranes.

The most common of the critical minerals that appear include laumontite, pumpellyite, glaucophane/crossite, lawsonite, aragonite, jadeitic pyroxene, and omphacite (E. H. Bailey, Irwin, and Jones, 1964; Blake, Irwin, and Coleman, 1967; Ernst, 1971c, 1977a). As we shall see, jadeitic pyroxene occurs only in the highest-grade Blueschist Facies zones. In

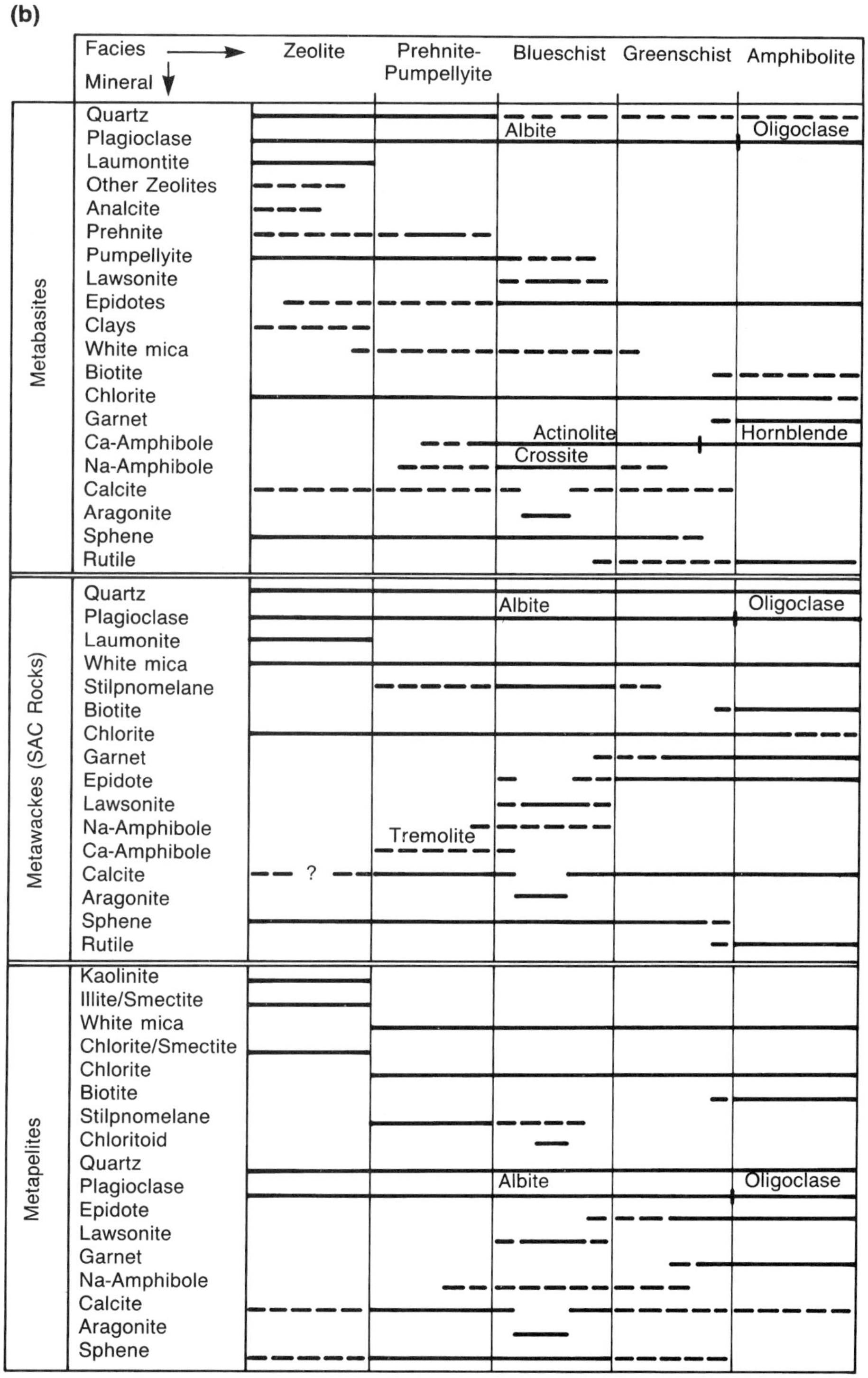

(b)

Facies → Mineral ↓	Zeolite	Prehnite- Pumpellyite	Blueschist	Greenschist	Amphibolite

Metabasites

Mineral	Zeolite	Prehnite-Pumpellyite	Blueschist	Greenschist	Amphibolite
Quartz	———	———	———	———	———
Plagioclase			Albite		Oligoclase
Laumontite	———				
Other Zeolites	– – –				
Analcite	– – –				
Prehnite	– – –	– – –			
Pumpellyite	———	———	– – –		
Lawsonite			———		
Epidotes	– – –	———	———	———	———
Clays	– – –				
White mica		– – –	– – –	– – –	
Biotite					– – –
Chlorite	———	———	———	———	
Garnet					———
Ca-Amphibole		– – –	Actinolite ———	———	Hornblende
Na-Amphibole		– – –	Crossite ———	– –	
Calcite	– – –	– – –	———		
Aragonite			———		
Sphene	———	———	———	———	———
Rutile			– – –	– – –	———

Metawackes (SAC Rocks)

Mineral	Zeolite	Prehnite-Pumpellyite	Blueschist	Greenschist	Amphibolite
Quartz	———	———	———	———	———
Plagioclase			Albite		Oligoclase
Laumonite	———				
White mica	———	———	———	———	———
Stilpnomelane		– – –	– – –	– –	
Biotite					– – –
Chlorite	———	———	———	———	———
Garnet				– – –	———
Epidote			———	———	———
Lawsonite			———		
Na-Amphibole			———		
Ca-Amphibole		Tremolite – – –	– – –		
Calcite	– – ? – –	———	———	———	———
Aragonite			———		
Sphene	———	———	———	———	———
Rutile					———

Metapelites

Mineral	Zeolite	Prehnite-Pumpellyite	Blueschist	Greenschist	Amphibolite
Kaolinite	———				
Illite/Smectite	———				
White mica	———	———	———	———	———
Chlorite/Smectite	———				
Chlorite	———	———	———	———	———
Biotite					–
Stilpnomelane		– – –	– – –		
Chloritoid			———		
Quartz	———	———	———	———	———
Plagioclase			Albite		Oligoclase
Epidote				– – –	———
Lawsonite			———		
Garnet				———	———
Na-Amphibole		– – –	———		
Calcite	– – –	– – –	———	———	———
Aragonite			———		
Sphene	– – –	———	———	– – –	———

Figure 28.3 Continued

central California, the Blueschist Facies is extensive and is subdivided into several zones marked by the appearance (a) or disappearance (d) of several phases, including lawsonite (a), aragonite (a), albite (d), jadeitic pyroxene (a), and pumpellyite (d).[15]

Sanbagawa Facies Series rocks are assigned to the Zeolite, Prehnite-Pumpellyite, Blueschist, Greenschist, or Amphibolite Facies. In the Sanbagawa Belt of Japan, Zeolite Facies rocks apparently do not occur (Ernst, 1977a; Liou, Maruyama, and Cho, 1987; Otsuki and Banno, 1990), though they were reported earlier when the definition of the

Table 28.2   Representative Mineral Assemblages of Sanbagawa Facies Series Rocks

Zeolite Facies

quartz–white mica–chlorite–albite–hematite	(S)
quartz–albite–laumontite–white mica–chlorite–sphene	(SAC)
illite/smectite–kaolinite–chlorite/smectite–quartz–hematite	(P)
laumontite–analcite–pumpellyite–epidote–quartz–chlorite–white mica–sphene	(MB)
lizardite–magnetite–magnesite–dolomite	(U)
calcite–dolomite–quartz	(C)

Prehnite-Pumpellyite Facies

quartz–stilpnomelane–white mica–chlorite–hematite	(S)
quartz–albite–white mica–chlorite–calcite–graphite–sphene	(SAC)
white mica–chlorite–stilpnomelane–quartz–albite–Na-amphibole–sphene	(P)
Prehnite-pumpellyite–epidote–calcite–albite–quartz–chlorite–sphene	(MB)
antigorite–brucite–magnetite–dolomite	(U)
calcite–prehnite–chlorite–quartz	(C)

Blueschist Facies

quartz–crossite–aegirine-augite–stilpnomelane–garnet	(S)
quartz–albite–lawsonite–chlorite–white mica–calcite–glaucophane	(SAC)
white mica–chlorite–chloritoid–quartz–Na-amphibole–sphene–hematite	(P)
epidote–Na-amphibole–actinolite–chlorite–albite–quartz–sphene	(MB)
antigorite–magnetite–magnesite–dolomite	(U)
aragonite–chlorite–hematite	(C)

Greenschist Facies

quartz–albite–white mica–stilpnomelane–Na-amphibole–garnet	(S)
quartz–albite–epidote–white mica–chlorite–graphite	(SAC)
white mica–chlorite–quartz–albite–garnet–graphite	(P)
crossite–epidote–chlorite–albite–garnet–quartz–white mica–rutile	(MB)
antigorite–magnesite–brucite	(U)
calcite–tremolite–quartz	(C)

Amphibolite Facies

quartz–white mica–biotite–garnet	(S)
quartz–oligoclase–white mica–epidote–biotite–garnet–calcite–rutile	(SAC)
white mica–biotite–chlorite–quartz–garnet–oligoclase–epidote–graphite	(P)
hornblende–epidote–garnet–oligoclase–quartz–hematite	(MB)
olivine–diopside–tremolite–antigorite[1]	(U)
calcite–diopside–quartz–garnet[1]	(C)

Sources: See note 18.

[1] Assemblages are based on chemographic analyses. C = carbonate rock, MB = metabasite, P = pelitic rock, S = siliceous rock, SAC = siliceous-alkali-calcic (quartzo-feldspathic) rock, U = ultramafic rock.

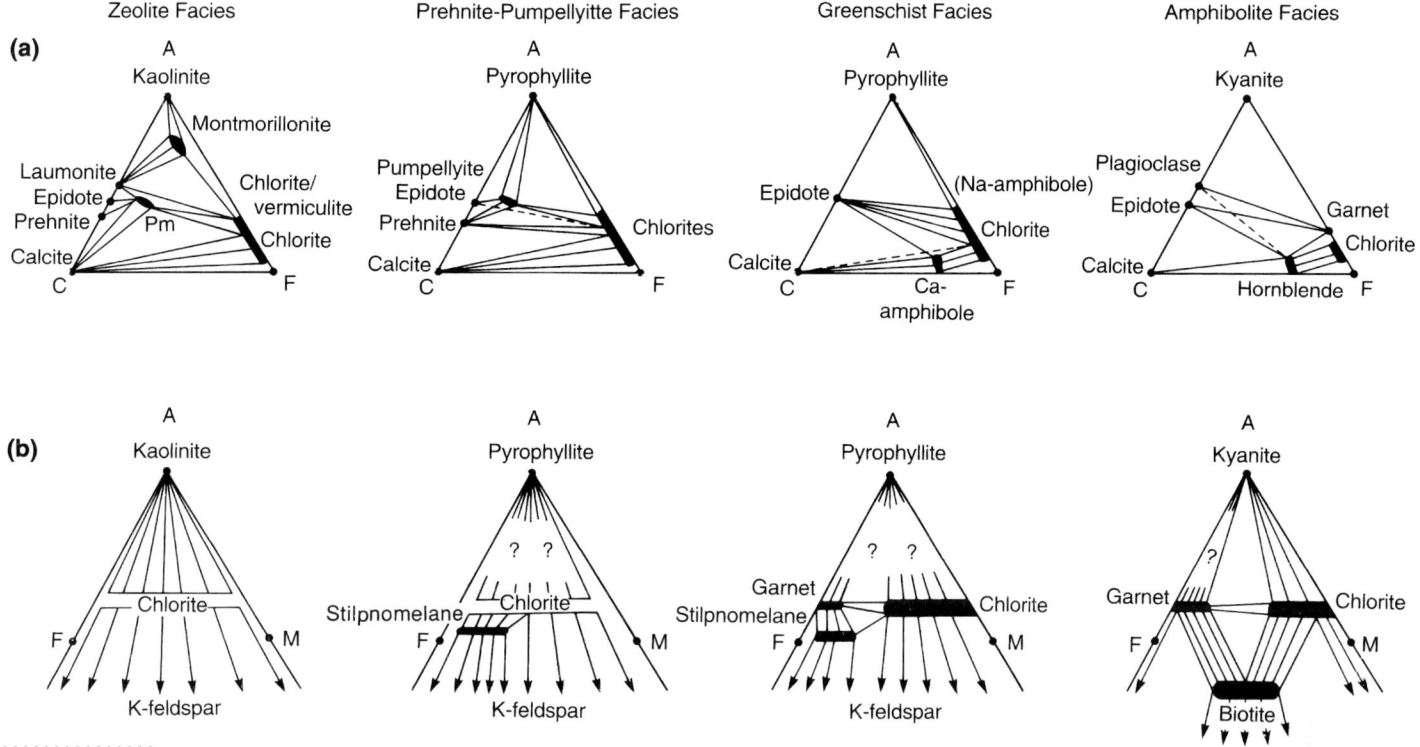

Figure 28.4 Selected schematic phase diagrams for the Sanbagawa Facies Series. (a) ACF diagrams for SAC and metabasic rocks. (b) AFM diagrams for pelitic rocks.

Sanbagawa Belt was broader.[16] In some Sanbagawa-like terranes (e.g., in the Alps), Eclogite Facies rocks occur in association with other rocks of the Sanbagawa Facies Series (Oberhansli, 1986). Jadeitic pyroxene is absent in the Sanbagawa Belt (but not in all rocks of the Sanbagawa Facies Series). Lawsonite is rare, occurring only locally in pelitic schists and gabbros.[17] Critical minerals in basic schists include pumpellyite with actinolite, epidote with actinolite, pumpellyite with actinolite and magnesioriebeckite (sodic amphibole), lawsonite and actinolite, winchite (sodic amphibole), crossite (sodic amphibole), barroisite (sodic amphibole), albite with hornblende, and oligoclase with hornblende (Banno, 1986).

Examination of common assemblages composed of the critical minerals and associated phases allows the construction of phase diagrams for each facies series. Typical phase assemblages for the Sanbagawa Facies Series are listed in table 28.2 and selected phase diagrams are shown in figure 28.4.[18] The rocks of the Zeolite, Prehnite-Pumpellyite, and Greenschist Facies are mineralogically like those of lower-pressure facies series. Prehnite-Pumpellyite and Greenschist Facies assemblages are virtually identical. Wairakite, however, is generally absent from Zeolite Facies rocks, whereas laumontite and heulandite are common.

In the Sanbagawa Belt of Japan (see figure 24.10a), four zones are recognized in pelitic rocks and seven or eight zones are defined in metabasites (Toriumi, 1975; Banno, 1986).[19] Characteristic assemblages of the pelitic zones are as follows:

Chlorite Zone

chlorite–white mica–quartz–albite–stilpnomelane–calcite–sphene–graphite

chlorite–white mica–quartz–albite–epidote–calcite–sphene–pyrite–graphite

Garnet Zone

chlorite–epidote–white mica–garnet–quartz–albite–sphene–graphite

Albite-Biotite Zone

white mica–chlorite–biotite–garnet–quartz–albite–epidote–graphite

Oligoclase-Biotite Zone

white mica–biotite–chlorite–garnet–oligoclase–quartz–rutile–graphite.[20]

The lowest part of the Chlorite Zone corresponds to the Prehnite-Pumpellyite Facies and the lower part of the

Blueschist Facies, whereas the middle and upper parts represent the lower Greenschist Facies.[21] The Garnet and Albite-Biotite zones correspond generally to the middle and upper Greenschist Facies, respectively, and the Oligoclase-Biotite Zone is equivalent to the lower Amphibolite Facies.

Critical reactions marking the isograds for the appearance of garnet, biotite, and oligoclase are probably continuous reactions[22] of the form

$$\text{chlorite}_1 + \text{epidote} \rightarrow \text{chlorite}_2 + \text{garnet} + \text{fluid} \quad (28.1)$$

in which the composition of chlorite becomes more magnesian,

$$\text{chlorite} + \text{white mica}_1 \rightarrow \text{biotite} + \text{white mica}_2 \quad (28.2)$$

in which the white mica composition changes (Brown, 1971), and

$$\text{albite} + \text{epidote} + \text{garnet}_1 + \text{biotite}_1 \rightarrow \text{oligoclase} + \text{garnet}_2 + \text{biotite}_2. \quad (28.3)$$

Alternative reactions for the appearance of garnet include

$$\text{chlorite} + \text{quartz} \rightarrow \text{garnet} + H_2O \quad (28.4)$$

(Banno, Sakai, and Higashino, 1986) and

$$\text{chlorite} + \text{quartz} + \text{magnetite} \rightarrow \text{garnet} + H_2O \quad (28.5)$$

(Hsu, 1968). The specific reactions for phase assemblage changes are not entirely resolved.

Two metabasite sequences are recognized in the Sanbagawa Belt, one with hematite-bearing rocks and one with rocks lacking hematite (table 28.3) (Banno, 1986; Otsuki and Banno, 1990). Assemblages in metabasites include the following:

Prehnite-Pumpellyite (Zone) Facies

prehnite–pumpellyite–chlorite–albite–calcite–quartz

Hematite-Pumpellyite-Actinolite Zone

chlorite–epidote–albite–quartz–Na-amphibole–sphene–hematite

pumpellyite–albite–chlorite–actinolite–epidote–stilpnomelane–quartz–sphene

chlorite–Na-amphibole–albite–quartz–calcite–white mica

chlorite–actinolite–lawsonite–epidote–albite–quartz

Hematite-Epidote-Actinolite Zone

chlorite–epidote–actinolite–albite–quartz–sphene

chlorite–epidote–crossite–albite–quartz–sphene–hematite

Winchite Zone

chlorite–epidote–actinolite–albite–quartz–sphene

chlorite–epidote–winchite–albite–quartz–sphene–hematite

Crossite Zone

chlorite–epidote–actinolite–albite–quartz–rutile

chlorite–epidote–crossite–albite–quartz–sphene–hematite–magnetite

Barroisite Zone

chlorite–barroisite–epidote–biotite–albite–quartz–sphene

chlorite–crossite–barroisite–epidote–albite–quartz–ilmenite–hematite

barroisite–chlorite–epidote–albite–quartz–rutile

Albite-Hornblende Zone

hornblende–chlorite–epidote–albite–quartz–hematite–ilmenite

hornblende–chlorite–epidote–albite–quartz–ilmenite

Oligoclase-Hornblende Zone

hornblende–epidote–oligoclase–quartz–magnetite–hematite

hornblende–epidote–chlorite–garnet–oligoclase–quartz–ilmenite[23]

Sodic amphibole and minor lawsonite show clearly that some Sanbagawa Belt rocks were metamorphosed under high-pressure, Blueschist Facies conditions (tables 28.2 and 28.3).

Reactions marking isograds between metabasite zones of the Sanbagawa Belt involve terminal and nonterminal, discontinuous and continuous reactions for the production of actinolite, epidote–actinolite–pumpellyite assemblages, winchite, barroisite, hornblende, and oligoclase. For example, the change from a Prehnite-Pumpellyite Facies assemblage to an actinolite-bearing, Hematite-Pumpellyite-Actinolite Zone assemblage may involve a reaction such as

$$\text{prehnite} + \text{chlorite} + \text{quartz} \rightarrow \text{pumpellyite} + \text{actinolite} + H_2O \quad (28.6)$$

(cf. Liou, Maruyama, and Cho, 1985). Lawsonite may be produced in this zone by a reaction such as

$$\text{prehnite} + \text{chlorite} + \text{quartz} + H_2O \rightarrow \text{lawsonite} + \text{actinolite} \quad (28.7)$$

(cf. Liou, Maruyama, and Cho, 1987). The transition to the Pumpellyite-Epidote-Actinolite Zone (in non-hematitic rocks) may be represented by the reaction

$$\text{pumpellyite} + \text{chlorite} + \text{quartz} \rightarrow \text{epidote} + \text{actinolite} + H_2O \quad (28.8)$$

(Toriumi, 1975; Brown, 1977b; Liou, Maruyama, and Cho, 1985). In successively higher-grade zones, the compositions of the amphiboles change via continuous reactions such as

$$\text{magnesioriebeckite} + \text{chlorite} + \text{quartz} + \text{albite} \rightarrow \text{glaucophane} + \text{hematite} + H_2O \quad (28.9)$$

(Hosotani and Banno, 1986),

$$\text{crossite} + \text{epidote} + \text{chlorite} + \text{albite} + \text{quartz} \rightarrow \text{barroisite} + \text{hematite} + H_2O \quad (28.10)$$

(Otsuki and Banno, 1990), or

$$\text{glaucophane} + \text{clinozoisite} + \text{quartz} + H_2O \rightarrow \text{chlorite} + \text{tremolite} + \text{albite} \quad (28.11)$$

596

Table 28.3 Mineral Zones in the Sanbagawa Belt of Japan

Pelitic Schists	Hematitic Metabasites	Metabasites Lacking Hematite	Facies
Oligoclase-Biotite Zone	Oligoclase-Hornblende Zone	Oligoclase-Hornblende Zone	Amphibolite Facies (lower)
Albite-Biotite Zone	Albite-Hornblende Zone	Albite-Hornblende Zone	Upper Greenschist Facies (high P, higher T)
	Barroisite Zone	Barroisite Zone	
Garnet Zone			Middle Greenschist
	Crossite Zone	Epidote-Actinolite Zone	Facies (high P, moderate T)
Chlorite Zone	Winchite Zone	Pumpellyite-Epidote-Actinolite Zone	Lower Greenschist Facies (high P, lower T)
	Hematite-Epidote-Actinolite Zone		
	Hematite-Pumpellyite-Actinolite Zone	Pumpellyite-Stilpnomelane Zone	Blueschist Facies
			Prehnite-Pumpellyite Facies

Source: Modified from Banno (1986).

(Liou, Maruyama, and Cho, 1985). Hornblende is similarly produced by a complex reaction involving barroisite, epidote, chlorite, albite, and quartz (Otsuki and Banno, 1990). Albite and hornblende probably yield oligoclase through a continuous reaction like equation 26.18.

The Blueschist-Greenschist Facies transition exhibited in the Sanbagawa Facies Series has been of considerable interest to petrologists. E. H. Brown (1974, 1977b) and Laird (1980), in revealing mineralogical and topological changes that occur over this transition zone, laid the groundwork for many studies that followed.[24] Although at least 120 ternary topologies are theoretically possible, observed phase assemblages are more limited. A few topologies, based on observed assemblages, are shown in figures 28.4 and 28.5 and are implied by the petrogenetic grid for low- to moderate-temperature rocks that encompasses the transition (see appendix C). The upper thermal limit of the Blueschist Facies is marked by reactions in which epidote group minerals appear *at the expense of pumpellyite and lawsonite.*

The Franciscan Facies Series, which includes the Blueschist Facies but not the Greenschist Facies, exhibits more extensive development of the Blueschist Facies rocks than does the Sanbagawa Facies Series. In particular,

assemblages containing jadeitic pyroxene, lawsonite, and aragonite are common. Franciscan Facies Series SAC rocks, typically metawackes, are more thoroughly studied than are other compositional types. The phase assemblages for these are represented on some ACF diagrams in figure 28.6, which presents selected ACF and other phase diagrams for the Zeolite, Prehnite-Pumpellyite, and Blueschist Facies of this facies series.

In the Zeolite Facies, common phase assemblages in SAC rocks are

heulandite–quartz–analcite–chlorite/vermiculite–white mica–sphene,
laumontite–quartz–albite–chlorite/vermiculite–white mica–calcite, and
laumontite–heulandite–quartz–albite–white mica–chlorite/vermiculite.[25]

These are replaced in the Prehnite-Pumpellyite Facies by assemblages such as

quartz–albite–pumpellyite–white mica–chlorite–stilpnomelane–calcite, and
quartz–albite–prehnite–chlorite–white mica–hematite.[26]

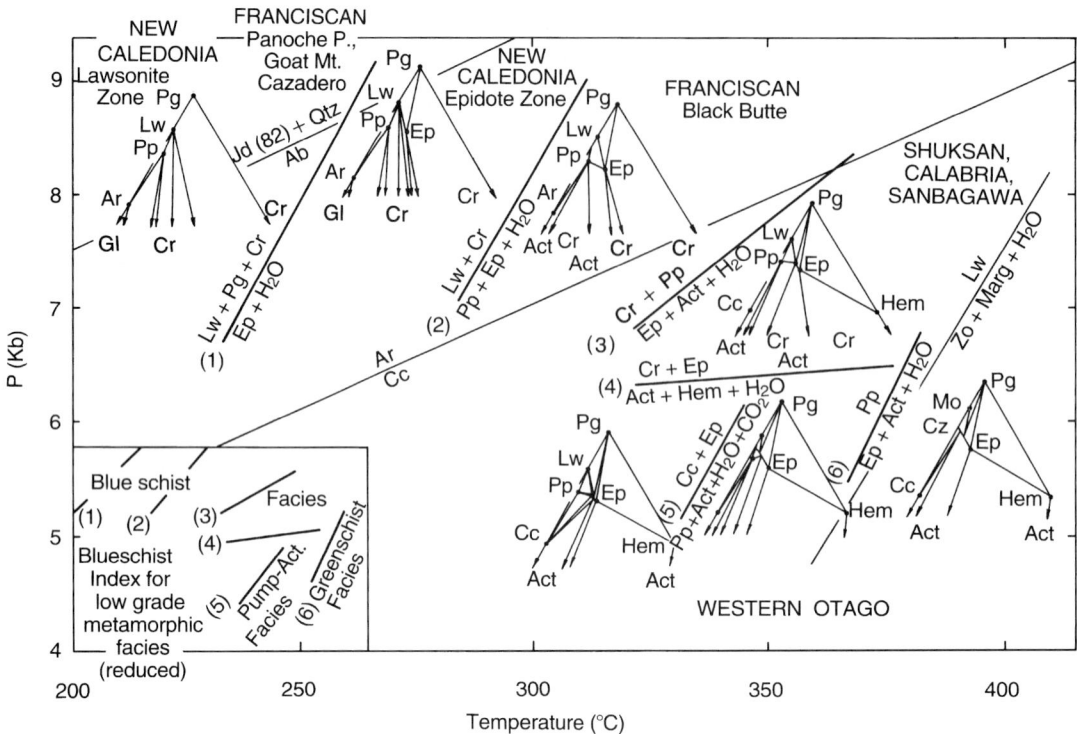

Figure 28.5   Petrogenetic grid with Al-Ca-Fe^3 phase diagrams for the Blueschist-Greenschist Facies transition. *(From E. H. Brown, 1977)*

Blueschist Facies SAC assemblages include

quartz–albite–lawsonite–pumpellyite–chlorite–white
mica–calcite, and

quartz–jadeitic pyroxene–lawsonite–white mica–
glaucophane–aragonite.[27]

Few Eclogite Facies rocks have SAC chemistries, but the assemblage

quartz–white mica–omphacite–glaucophane–garnet–
epidote

is a representative phase assemblage under lower-grade Eclogite Facies conditions (Yokoyama , Brothers, and Black, 1986). More aluminous Eclogite Facies rocks may contain kyanite. Representative assemblages of the Franciscan Facies Series for other bulk compositions are presented in table 28.4.[28]

Some reactions that may yield critical phases in SAC rocks of the Franciscan Facies Series include the following:

$$\text{plagioclase (detrital)} + \text{quartz} + H_2O \rightarrow$$
$$\text{laumontite} + \text{albite,} \qquad (28.12)$$

$$\text{laumontite} + \text{calcite} \rightarrow \text{prehnite} + \text{quartz} + H_2O + CO_2 \qquad (28.13)$$

(A. B. Thompson, 1971),

$$\text{laumontite} \rightarrow \text{lawsonite} + \text{quartz} + H_2O \qquad (28.14)$$

(Liou, 1971b),

$$\text{laumontite} + \text{prehnite} + \text{chlorite} \rightarrow \text{pumpellyite} + \text{quartz} + H_2O \qquad (28.15)$$

(Cho, Liou, and Maruyama, 1986),

$$\text{plagioclase (detrital)} + \text{quartz} + \text{calcite} + \text{hematite} + H_2O \\ \rightarrow \text{pumpellyite} + \text{albite} + CO_2, \qquad (28.16)$$

$$\text{plagioclase (detrital)} + H_2O \rightarrow \text{lawsonite} + \text{albite,} \qquad (28.17)$$

$$\text{albite} \rightarrow \text{jadeite} + \text{quartz} \qquad (28.18)$$

(Newton and Smith, 1967; Maruyama, Liou, and Sasakura, 1985), and

$$\text{albite} + \text{chlorite} + \text{pumpellyite} + \text{hematite} + H_2O \rightarrow \\ \text{jadeitic pyroxene} + \text{glaucophane} + \text{lawsonite} + \text{quartz} \qquad (28.19)$$

(Patrick and Day, 1989). Note that some of these reactions involve the direct reaction of detrital plagioclase feldspar to form a phase present in a facies of higher grade than the Zeolite Facies. Petrographic evidence suggests that such reactions occur in Franciscan Facies Series rocks of the Franciscan Complex of California (Raymond, 1973c). These may occur because of very slow reaction rates in the rocks at low temperatures and rapid burial rates in convergent margin settings, a combination that allows sediments or rocks to reach higher-grade conditions before undergoing metamorphic changes.

Reactions in other bulk compositions, such as the carbonate rocks and metabasites, range from simple to complex.

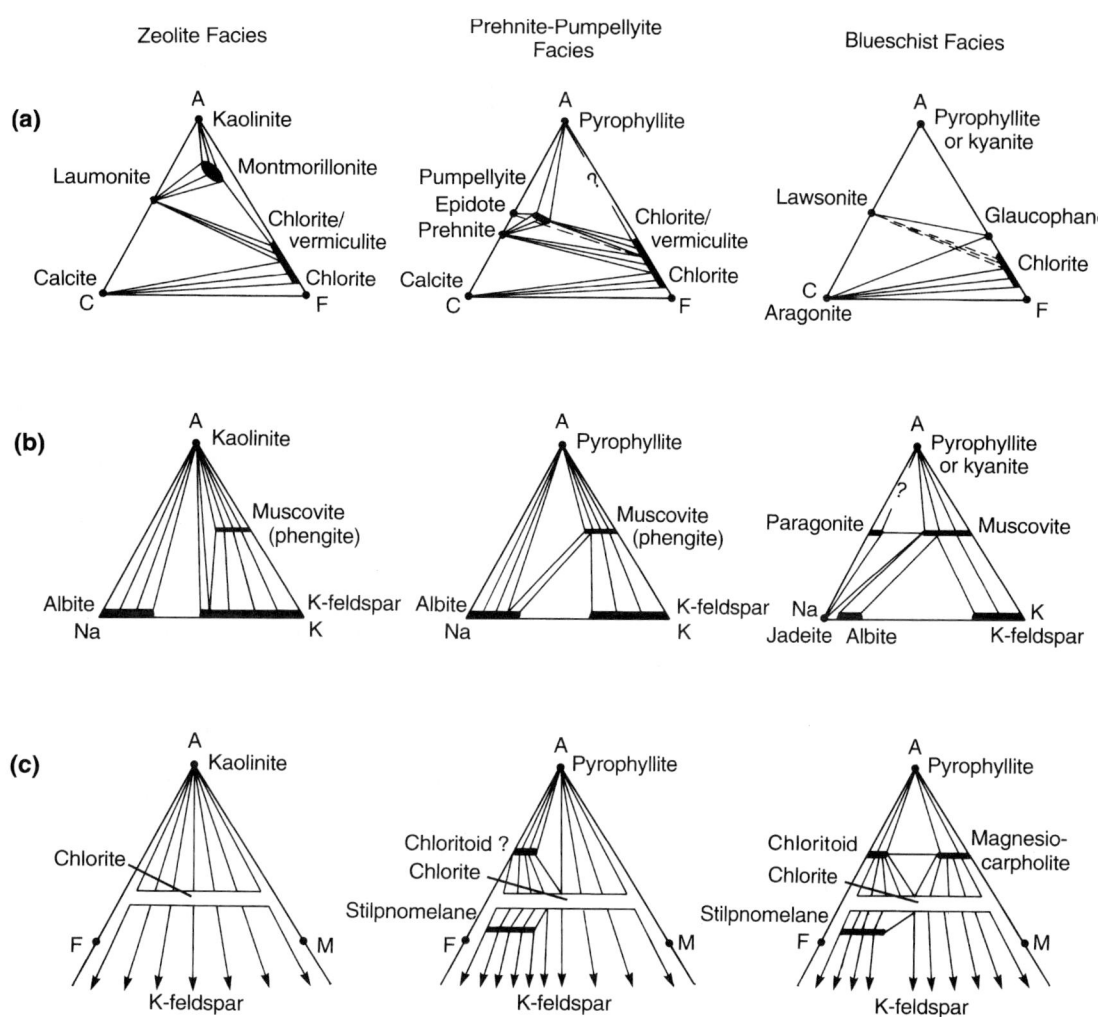

Figure 28.6 Selected phase diagrams for the Franciscan Facies Series. (a) ACF diagrams for SAC and metabasic rocks. (b) AKN diagrams for pelitic rocks. (c) AFM diagrams for pelitic rocks. Compare and contrast these with the diagrams in figure 28.4.

In the carbonate rocks, the most important reaction is the simple polymorphic phase transformation

$$\text{calcite} \rightarrow \text{aragonite.} \qquad (28.20)$$

Via this reaction, limestones are converted to aragonite marbles. In contrast, exchange of Fe and Mg among phases is indicated by complex reactions in metabasites, such as the discontinuous reactions

$$\text{laumontite} + \text{pumpellyite} + \text{quartz} \rightarrow \text{prehnite} + \text{epidote}$$
$$+ \text{chlorite} + H_2O \qquad (28.21)$$

(Cho, Liou, and Saskura, 1986), which marks the Zeolite/Prehnite-Pumpellyite Facies boundary;

$$\text{pumpellyite} + \text{epidote} + \text{chlorite} + \text{albite} + \text{quartz} + H_2O$$
$$\rightarrow \text{lawsonite} + \text{crossite} \qquad (28.22)$$

(Brown, 1977b), which marks the Prehnite-Pumpellyite Blueschist Facies boundary;

$$\text{lawsonite} + \text{Na-amphibole} \rightarrow \text{clinopyroxene} +$$
$$\text{pumpellyite} + H_2O \qquad (28.23)$$

(Maruyama and Liou, 1987);

$$\text{lawsonite} + \text{Na-amphibole} \rightarrow \text{epidote} + \text{chlorite} + \text{albite}$$
$$+ \text{quartz} + H_2O \qquad (28.24)$$

(E. H. Brown and Ghent, 1983); and the continuous reaction equation 28.10.

Table 28.4   Representative Phase Assemblages for the Franciscan Facies Series

Zeolite Facies

quartz–albite–calcite–hematite	(S)
illite/smectite–kaolinite–chlorite/smectite–quartz–hematite	(P)
laumontite–albite–pumpellyite–quartz–chlorite–sphene	(MB)

Prehnite-Pumpellyite Facies

quartz–albite–pumpellyite–chlorite	(S)
white mica–chlorite/smectite–stilpnomelane–quartz–albite - sphene	(P)
prehnite–pumpellyite–epidote–albite–quartz–chlorite–sphene	(MB)

Blueschist Facies

quartz–crossite–aegirine–stilpnomelane–garnet–hematite	(S)
white mica–quartz–lawsonite–Na-amphibole–garnet–hematite	(P)
glaucophane–lawsonite–jadeitic pyroxene–aragonite–magnetite–white mica–sphene	(MB)

Eclogite Facies

quartz–omphacite–white mica	(S)
quartz–sodic pyroxene–garnet–white mica–kyanite[1]	(SAC)
omphacite–garnet–kyanite–rutile	(MB)

Sources: See notes 25–28.
[1] From chemographic analysis.
MB = metabasites, P = pelitic rocks, S = siliceous rocks.

Experimental Investigations and Mineral Stabilities

Although the tectonic setting in which Blueschist Facies rocks occur suggests high pressures of metamorphism, alternative low-pressure models of formation have been proposed for these rocks. A high-pressure origin is supported by the available experimental data. The stability relations of several minerals important to both the Franciscan and Sanbagawa Facies series have been investigated in a number of experiments over the years. Representative curves are shown in figure 28.5 and in appendix C.

Petrogenetic grids such as figure 28.5, if based only on experimental data, are somewhat misleading, because (1) experiments have not been done on all of the important reactions, (2) the addition of components to these systems may alter the stability fields of the phases, and (3) the composition of the fluid phase can have a profound effect on the stabilities of phases. For example, the presence of various minerals including prehnite, pumpellyite, and other Ca-Al silicates is a function of X_{CO_2}.[29] In a rock of a certain bulk composition, the Prehnite-Pumpellyite Facies assemblage epidote–prehnite–pumpellyite, stable at lower values of X_{CO_2}, is

replaced under conditions of high X_{CO_2} by the Greenschist Facies assemblage epidote–chlorite–calcite (W. Glassley, 1974; Ivanov and Gurevich, 1975; Liou, Maruyama, and Cho, 1987). The validity and scope of such grids can be enhanced via the addition of calculated reaction curves (Guiraud, Holland, and Powell, 1990; B. W. Evans, 1990).

The available data from the reactions shown in figure 28.5 and appendix C reveal that critical minerals of the Franciscan and Sanbagawa Facies series are stable under generally high P/T conditions.[30] The Zeolite Facies represents conditions of $P < 0.4$ Gpa (4 kb) and $T < 250°$ C. The Prehnite-Pumpellyite Facies occurs at slightly higher temperatures and similar pressures under conditions of about $P = 2$–3.5 kb and $T = 200$–350° C. At the low to moderate temperatures of $T = 100$–450° C and pressures of 0.35–1.0 Gpa (3.5–10 kb) the Blueschist Facies has an even higher P to T ratio. Greenschist Facies assemblages are stable between about 300° C and 550° C at pressures of about 0.2 to 0.8 Gpa (2 to 8 kb). The Eclogite Facies has the highest pressures, at $P > 0.8$ Gpa, but temperatures vary from about 200° C to more than 600° C.

Relating the experimental studies to natural occurrences is sometimes difficult. In addition to the limitations of the experimental studies noted above, natural phases are seldom pure (i.e., they do not have stoichiometric compositions), and many cases of metastable crystallization, incomplete recrystallization, and polymetamorphism are known.[31] For example, pyroxenes in some metabasites from the Franciscan Complex are zoned, reflecting incomplete neocrystallization under differing metamorphic conditions associated with progressive metamorphic events (Maruyama and Liou, 1987). Metastable persistence of lower-grade phases in higher-grade rocks is relatively common in the Sanbagawa and Franciscan Facies series. In part, this is the result of reactions being sluggish at low temperatures, but variations in the amount and character of a fluid phase may also control the neocrystallization process.[32] In addition, pseudomorphs and incomplete replacement of phases reflect polymetamorphic histories.[33] All of these factors make it difficult to relate mapped isograds in high P/T terranes to experimentally determined stability fields.

PETROGENETIC MODELS

Increased knowledge about Blueschist Facies rocks obtained through field and laboratory studies during the 1950s and 1960s spurred interest in their petrogenesis. That interest led to a controversy that was outlined and discussed by van der Kaaden (1969) and Ernst (1971a). Four hypotheses for the origin of these rocks were proposed and advocated by various geologists. Two of them—the tectonic overpressure and the burial hypotheses—accepted the high pressures suggested by the experimental data. Two others—the metasomatic and the metastable recrystallization hypotheses—appealed to processes other than neocrystallization under high-pressure conditions.[34]

Metasomatic and Metastable Recrystallization Hypotheses

The metastable recrystallization and metasomatic hypotheses assign little importance to the experimental studies on the P-T stability ranges of Blueschist Facies minerals. Taliaferro (1943) recognized the common field association of glaucophane schists and serpentinites before the stabilities of minerals such as jadeitic pyroxene and glaucophane were known. He and other advocates of metasomatism[35] were followed by Gresens (1969, 1970), who argued that blueschists result from low-pressure metasomatism induced by highly concentrated, reducing, saline pore fluids created during serpentinization. Gresens cited several arguments and lines of evidence, including the following, in support of his thesis.

1. Glaucophane-rich rocks and serpentinites are commonly associated in field settings.

2. Minerals such as aragonite, glaucophane, pumpellyite, and lawsonite commonly occur in veins that presumably formed at low pressures.

3. The predominance of ferrous iron over ferric iron in Blueschist Facies minerals, especially in minerals lacking sodium, indicates the presence of a reducing ore fluid.

4. Sodium metasomatism is known, notably in the Green River Formation, where sodic pyroxene and blue amphibole formed in a concentrated, sodium-rich solution.[36]

5. Reactions that involve saline serpentinizing fluids, presumed to carry the appropriate chemical species, may be written to show the production of Blueschist Facies minerals.

Ernst (1971a) and others present convincing evidence and arguments against the metasomatic hypothesis.[37] First, although serpentinite and glaucophane schists are commonly associated in the field, large tracts of Blueschist Facies rocks in Washington, Italy, southwestern Japan, and California contain little or no ultramafic rock.[38] At some locales where serpentinite *is* present, serpentinization is known to have occurred *after* ultramafic rocks were emplaced; yet the contacts with surrounding rocks are not marked by an intervening layer of glaucophane schist.[39] Second, there is no reason to assume that veins can only form under low-pressure conditions. Third, ferrous iron does not predominate in all Blueschist Facies minerals. Rather, ferric iron dominates in some non- and low-sodium minerals, such as garnet, epidote, muscovite, and pumpellyite, that occur in the Blueschist Facies.[40] Fourth, although local metasomatism may have occurred in some blueschist-bearing terranes, the important questions, as noted by Ernst (1971a), are (1) is metasomatism required to produce Blueschist Facies mineralogies? and (2) did regional metasomatism take place? Available evidence, including the lack of appropriate metamorphic gradients away from serpentinite bodies, does not support an affirmative response to either question.[41] Finally, the fact that reactions may be written does not mean that those reactions took place in the rocks, nor is there reason to assume that the chemical species needed for the reactions will be present in serpentinizing or metasomatizing fluids. In short, because the metasomatic hypothesis is based on assumptions and conclusions not consistent with available data, it may be ruled out as a viable, general hypothesis for the petrogenesis of Blueschist Facies rocks.

Metastable recrystallization arguments have not been vigorously promoted. Ernst (1971a), however, summarized the two possibilities:

Metastable Recrystallization Model 1. Blueschist Facies mineral assemblages form from precursor mineral suites of higher energy state under conditions of low pressure.

Metastable Recrystallization Model 2. Blueschist Facies minerals, normally unstable under low-pressure conditions, are stabilized under these conditions by active pore fluids. The latter hypothesis is a version of the metasomatic hypothesis rejected above.

Model 1 was proposed to explain the occurrence of single mineral species, such as jadeitic pyroxene and aragonite, rather than an entire mineral assemblage. For example, simply stated, high albite has a higher energy state than low albite, and therefore the reaction

$$\text{high albite} = \text{jadeite} + \text{quartz} \qquad (28.25)$$

will occur (under low-temperature conditions) at lower pressures than will the reaction

$$\text{low albite} = \text{jadeite} + \text{quartz} \qquad (28.26)$$

Stated another way, at low pressures, the energy difference between the left and right sides of equation 28.25 is greater than that between the two sides of equation 28.26. This favors reaction equation 28.25. Therefore, high albite, which is metastable in relation to low albite, will react to form jadeite via the metastable reaction equation 28.25 *if* metastable high albite is present in the rocks.

In the Franciscan Complex of California, Ernst (1971a) found no evidence that (highly strained or disordered) high albite was an important premetamorphic component of jadeitic-pyroxene-bearing rocks. The absence of high albite and the fact that metastable recrystallization (according to model 1) only applies to single minerals, rather than to complete Blueschist Facies assemblages, argue against the general application of this hypothesis as an explanation for petrogenesis of Blueschist Facies rocks. The hypothesis also fails to explain regional distributions of facies and mineral zones.

The Tectonic Overpressure Model

Hypotheses that require relatively high pressures of formation for Blueschist Facies mineral assemblages seem more reasonable, considering the experimental data that indicate that relatively high pressures are required to stabilize jadeitic pyroxene, glaucophane, aragonite, and lawsonite in metamorphic rocks. These hypotheses become even more compelling in the absence of data supporting low-pressure hypotheses. Coleman, Blake, and others[42] argued that **tectonic overpressures,** i.e., pressures in excess of P_{load}, cause Franciscan Facies Series metamorphism (Blake, Irwin, and Coleman, 1967, 1969). They recognized relationships in Northern California and New

Caledonia in which metamorphic grade decreases away (and down) from major regional thrust faults (i.e., the metamorphism is "upside down," with higher grade rocks at shallower depths).[43] These relationships are consistent with the tectonic overpressure hypothesis. For the California case, Blake, Irwin, and Coleman (1967) argued that tectonic overpressures developed below a regional thrust fault that is capped by serpentinite. The impervious cap trapped water, creating the overpressures (figure 28.7).

The sum of the available evidence does not strongly support the tectonic overpressure hypothesis. Although the regional metamorphic patterns recognized in northern California *seem* to support the idea, lower-grade terranes, in many cases, are separated from adjoining higher-grade terranes by faults, and the isograds are cut by the faults (B. L. Wood, 1971; Jayko, Blake, and Brothers, 1986).[44] This suggests that the upside-down pattern may be a structural rather than a metamorphic phenomenon. In New Caledonia, in the South Pacific, the regional metamorphic patterns also apparently resulted from tectonic rather than strictly metamorphic processes.[45] Additional negative evidence is provided by regional and detailed studies in central California, which reveal that no relationship exists between the fault and the highest-grade (jadeitic) rocks or the associated isograds (Ernst, 1971c; Raymond, 1973a,c; Patrick and Day, 1989).[46] Furthermore, laboratory analyses of rock strengths suggest that the Franciscan rocks of California, under the appropriate metamorphic conditions, may not be able to maintain overpressures in excess of about 1 kb, particularly if the rocks contained a pore fluid (Brace, Ernst, and Kallberg, 1970; E. C. Robertson, 1972). Thus, field and laboratory evidence, taken together, suggest that tectonic overpressures are not generally responsible for regional Franciscan Facies Series metamorphism.

The Burial Metamorphism Hypothesis

Deep burial may result from either sedimentation or tectonic thickening of the crust via faulting. In either case, P_{load} will be high. Ernst (1965, 1971a, 1973a, 1975, 1984, 1988) and several other petrologists advocate some form of deep burial as a method of generating high pressures of metamorphism in Franciscan and Sanbagawa Facies series rocks (figure 28.8).[47] Experimental data, *which require high pressures* of metamorphism for Blueschist Facies mineral assemblages, are the primary evidence favoring the deep burial hypothesis. Thermal modelling also supports it (Oxburgh and Turcotte, 1971; C. Y. Wang and Shi, 1984; Dumitru, 1991). In addition, the paleogeographic and tectonic setting of high P/T metamorphic belts is consistent with the hypothesis. In particular, the presence of Sanbagawa and

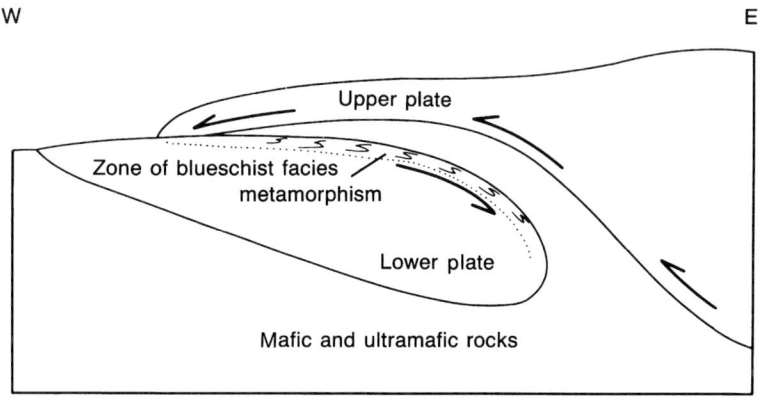

Figure 28.7 The tectonic overpressure model of Blueschist Facies petrogenesis. *(Redrawn from Blake, Irwin, and Coleman, 1967)*

Franciscan Facies series rocks in paired metamorphic belts (which form in response to subduction) suggests that subduction and associated accretion of subducted sediments and rocks (i.e., tectonic burial), are generally responsible for the petrogenesis of Blueschist Facies and related rocks (figure 28. 8)(Ernst, 1984). Subduction-induced tectonic burial provides high pressures and low temperatures of metamorphism.[48] It also explains the common age progression in high P/T metamorphic belts, in which the age of the rocks decreases away from the associated high-temperature belt of the pair. Finally, the subduction burial hypothesis explains inverted sequences of Amphibolite, Greenschist, and Blueschist Facies rocks that apparently form on the hanging walls of subduction zones (Platt, 1975; Peacock, 1987a, 1988; Sorenson, Bebout, and Barton, 1991).

Although both sedimentation and tectonic burial can yield high pressures and low temperatures, the best model for any particular belt will be that which best explains regional patterns and other phenomena related to the metamorphic event. The regional patterns within both Franciscan and Sanbagawa Facies series belts typically are linear, coast-parallel patterns, with higher-grade rocks landward towards the core of the orogen (e.g., see figure 24.10). Tilting of a sedimentary sequence will produce such a pattern. Likewise, the imbricate faulting and accretion associated with subduction parallel to a coastline will yield a pattern of parallel metamorphic belts. In the former case, tilting and folding may yield a coast-parallel set of metamorphic zones without major, intervening faults—a condition that apparently exists in the San-bagawa Belt of Japan.[49] Alternatively, individually subducted imbricate rock slabs and melange masses, separated by faults, may each develop with a distinctive metamorphic zone or facies, as is the case, in part, in northern California (Blake et al., 1988).[50]

One of the unresolved problems of Blueschist Facies metamorphism is the problem of preservation and uplift of Blueschist Facies rocks. Either prolonged burial or slow subduction may result in progressively higher temperatures, over time, in a buried mass of rock.[51] Progressive heating will result in conversion of Blueschist and Prehnite-Pumpellyite Facies rocks to Greenschist or Amphibolite Facies rocks. It follows that if blueschists are preserved at the surface, they must have been protected from heating, either by rapid uplift following metamorphism or by some form of "refrigeration" (i.e., continued cooling while at depth) (Peacock, 1987a; Ernst, 1988; Dumitru, 1991). If they were refrigerated, they must have remained cool *while they moved towards the surface*.

Two general types of subduction tectonics hypotheses that allow for the origin, preservation, and uplift of Blueschist Facies rocks have been proposed. One type of hypothesis advocates coeval subduction and uplift, that is synsubduction uplift, uplift that occurs while subduction continues. The second type assumes metmorphism during subduction, but relies on postsubduction isostatic rise for uplift of the blueschists. Both types of hypothesis accept the premise that, during subduction, masses of rock are successively underplated beneath the overriding plate. The fact that blueschists occur in two different structural settings, as regionally extensive slabs and as relatively small tectonic blocks and slabs in a matrix of serpentinite or metashale tectonite (i.e., in melanges), may mean that two kinds of uplift processes were operative.

Synsubduction models may be divided into five subtypes. The subduction zone is the avenue of uplift in the first, the S-type model. In S-type models, subduction zones are considered to be "two-way streets" along which materials are being subducted (carried down) *at the same time* that previously subducted materials are moved up (Suppe, 1972; Cloos, 1982; Ernst, 1984). Both Suppe (1972) and Ernst

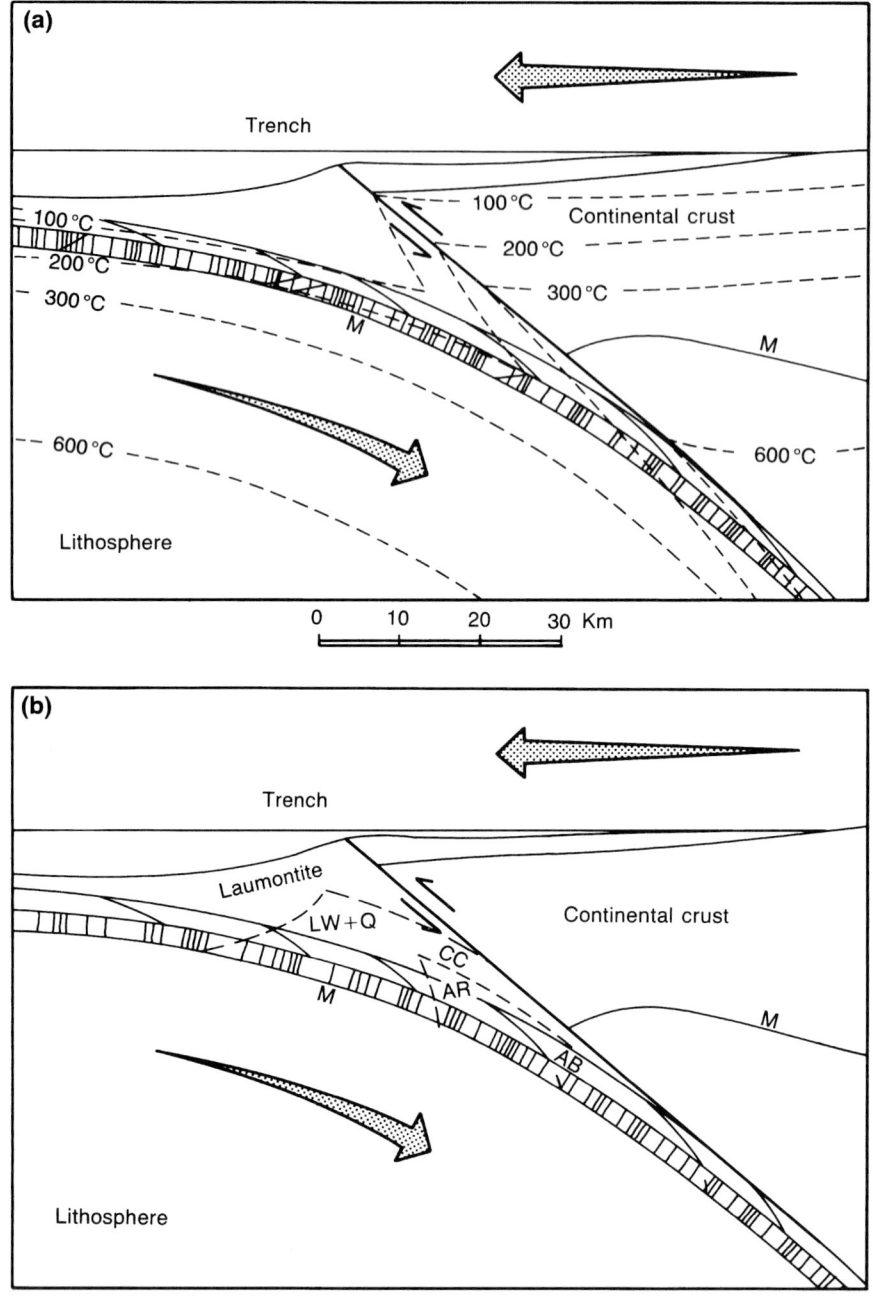

Figure 28.8 Tectonic burial (subduction zone) model of Blueschist Facies petrogenesis. (a) Thermal structure of an active subduction zone. M = Mohorovicic discontinuity. (b) Subduction zone showing location of selected isograds.

LW = lawsonite, Q = quartz, CC = calcite, AR = aragonite, AB = albite, JD = jadeite.
(Redrawn from Ernst, 1973a)

(1971a, 1977, 1988) adopt an imbricate thrust model of sub-duction accretion[52] with the proviso that large coherent *slabs* move up within the subduction zone to be emplaced at shallow levels of the crust, where they no longer would be subject to heating. Aragonite, which converts to calcite rather quickly upon heating (W. D. Carlson and Rosenfeld, 1981), is present in many blueschists, indicating that the

blueschists were never heated on their way to the surface. Because the subduction zone itself remains cool as long as the subduction rate does not fall below a certain limiting value or the heat content of the subducting plate does not exceed a specific level (C. Y. Wang and Shi, 1984; Peacock, 1987; Dumitru, 1991), materials moving *within* the subduction zone will remain cool. Ernst (1971a, 1988)

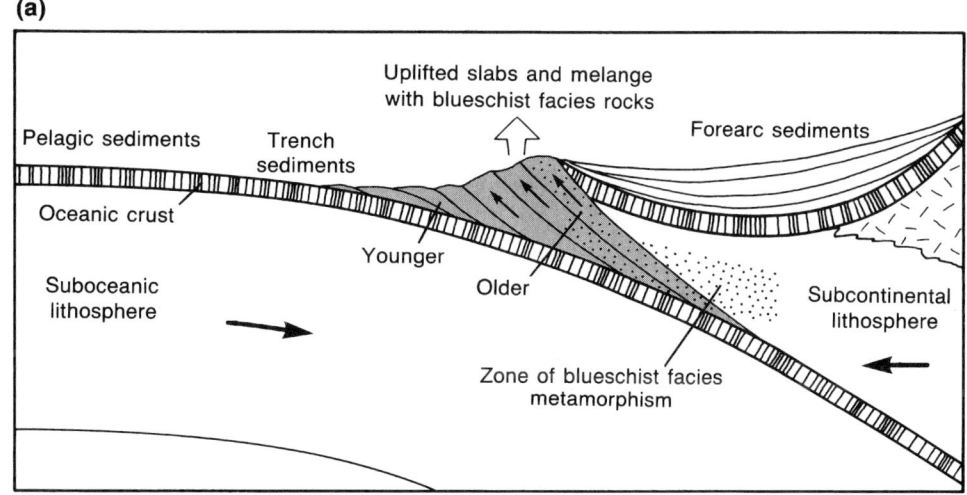

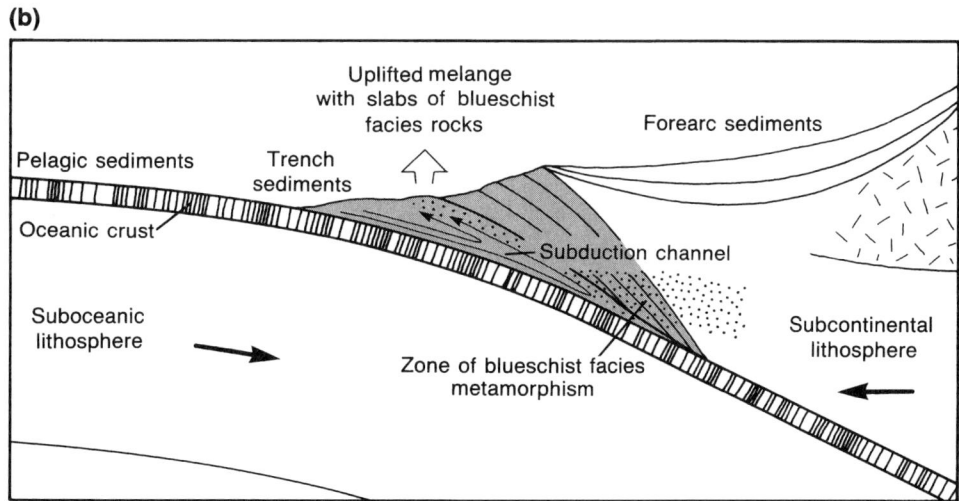

Figure 28.9 Various models for uplift of Blueschist Facies rocks (based on selected hypotheses). (a) Imbricate slab (S-type) model (cf. Ernst, 1977a, 1977b). (b) Subduction channel (S-type) model (cf. Cloos, 1982, 1984; Cloos and Shreve, 1988a,b). (c) Forearc extension (E-type) model (cf. Platt, 1986). (d) Postsubduction, diapiric uplift model.

suggests that, if the rate of subduction decreases, previously subducted slabs will move up due to buoyant forces (subducted rocks are considered to be of lower density than the rocks of the overlying plate). Clearly, such a decrease cannot fall below the limiting value, if refrigeration is to be maintained. Ernst (1971a, 1975) depicted the upward-moving slabs as fault-bounded, imbricate masses (figure 28.9a). These would appear at the surface as major (coherent) fault blocks.

The emplacement of relatively small tectonic blocks of eclogite and blueschist poses a different problem. In general, these are incorporated in shaly or serpentinous SF-tectonites. Adopting the view that the Central Belt of the Franciscan Complex (described below) is a single large melange and

drawing on the experimental work of D. S. Cowan and Silling (1978), Cloos (1982, 1984)[53] detailed an S-type synsubduction model, the subduction channel model, which requires that upward movement of blocks and slabs be a function of circulating flow in a shale-based tectonic melange that dominates the active part of the subduction zone (figure 28.9b). In this model, blueschists and eclogites metamorphosed deep within the subduction zone are fragmented and transported up the subduction zone, where they may mix with both descending Zeolite and Prehnite-Pumpellyite Facies rocks and ascending Amphibolite and Greenschist Facies rocks metamorphosed along the hotter hanging wall of the zone. Via subduction channel flow, some relatively small blocks, metamorphosed in the cool depths of the subduction zone, are

(c)

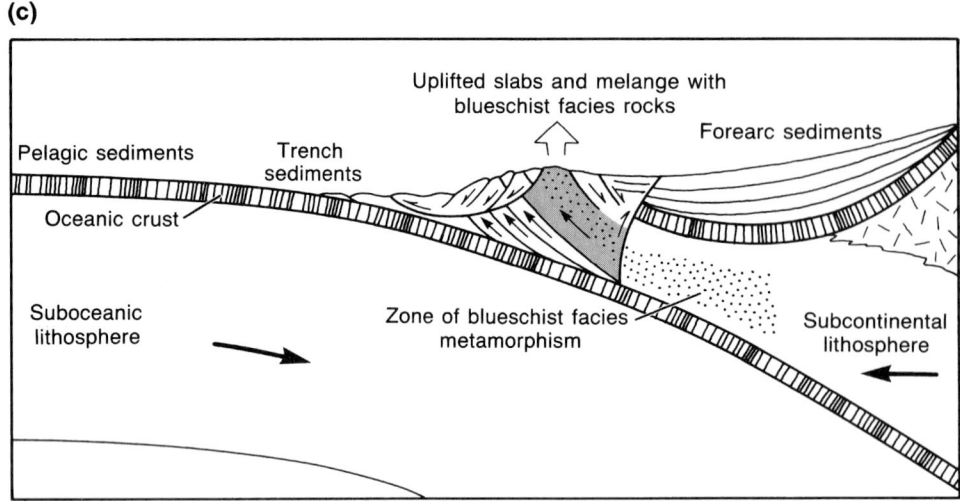

(d)

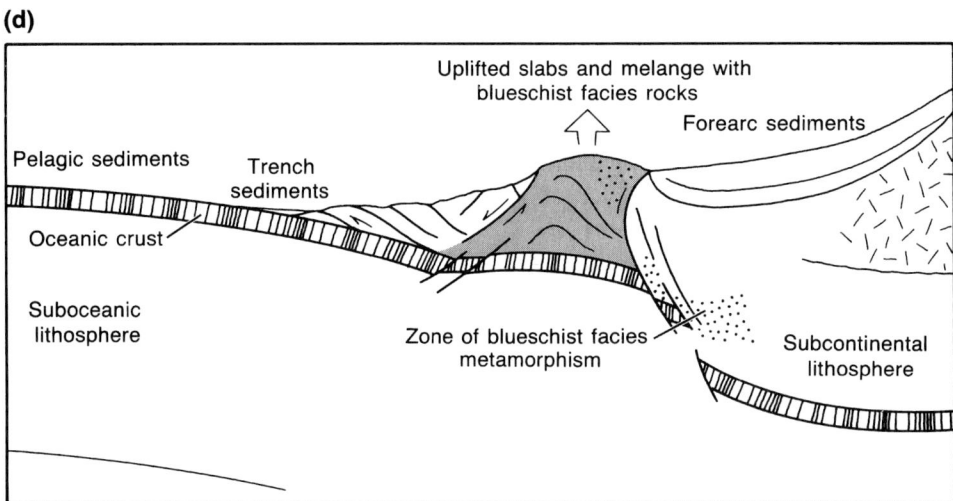

....
Figure 28.9   Continued

transported relatively quickly back to the surface. This rapid transport within the subduction zone prevents the reheating that yields retrograde or polymetamorphic effects in some blocks, while allowing those effects to develop in other blocks before they are transported up. The resulting melange contains a mix of low- to high-grade rocks, ranging from eclogites and high-grade blueschists metamorphosed at great depth to barely metamorphosed zeolitic metawackes that were altered at shallow depths.

In these S-type models, subduction must be fast and continuous enough to preserve the low geothermal gradients required for high P/T metmamorphism (C. Y. Wang and Shi, 1984; Dumitru, 1990, 1991). Rates of uplift, however, need

not be rapid. In some cases, uplift was relatively slow or prolonged, as indicated (1) by the absence of overprint assemblages, (2) by fission track data, and (3) by regional structural relationships coupled with a disparity between sedimentological and metamorphic uplift ages of the rocks (Raymond, 1981; S. L. Baldwin and Harrison, 1989; Dumitru, 1989).

A second type of synsubduction model, the E-type model, requires that accretion and underplating in the subduction zone generate an instability in the overlying accretionary wedge that causes *extension* and (listric) normal faulting (figure 28.9c) (Platt, 1986; Krueger and Jones, 1989; Harms, Jayko, and Blake, 1992).[54] The faulting removes the overburden from the buried Blueschist Facies rocks. Continued underplating (accretion) via imbricate thrusting of slabs below the extensionally faulted zone provides the driving force for uplift. Eventually, blueschist masses are driven to the surface and exposed by faulting and by erosion.

Three additional types of synsubduction, Blueschist Facies uplift models have been proposed, none of which is yet widely accepted. The T-type model, which involves *transpression* and strike-slip faulting, was proposed by Karig (1980).[55] According to this model, blueschists are exposed to erosion where an imbricate thrust margin is cut by strike-slip faults. Vertical movement, as well as lateral movement along the faults, is envisioned to transport deeply buried blueschists along strike to sites where less erosion is required to expose them at the surface. Such a model requires that blueschist terranes be bounded, at least in part, by high-angle (strike-slip) faults. Another model involving *oblique* plate convergence, the O-type model, was proposed by Avé Lallement and Guth (1990). In this model—which admittedly requires special circumstances—uplift of blueschists occurs when a decrease in subduction angle is caused by a change in the angle of plate convergence of an obliquely converging subducting plate. A final synsubduction model, the J-type model, like the T-type, involves lateral faulting, in this case associated with a *triple junction*. K. F. Fox (1976, 1983) proposed that, at a triple junction where obliquely converging plates are replacing a transform fault boundary along a continental margin, rocks of the overthrust continental edge would be deformed and uplifted.[56] Through these processes, tracts of Blueschist Facies rocks would be uplifted and exposed to gravity sliding that would yield blueschist-bearing melanges. Thus, both large terranes and isolated blocks of blueschist would be uplifted.

Post-subduction models of blueschist uplift rely on one of three processes. Uplift results from (1) bouyant rise and diapiric uplift of the subducted accretionary slabs and melanges caused by the disparity in density between subducted rocks and those of the overlying plate, (2) erosion induced isostatic uplift and diapirism, or (3) orogenic uplift (figure 28.9d)(Ernst, 1965, 1988; Dobretsov, 1991).[57] In the first and second processes, buoyancy would have to drive the uplifted material rapidly toward the surface to prevent a Greenschist Facies overprint from developing on the blueschists. Either entire deformed slabs or block-bearing melanges could be moved to the surface via this mechanism, but whether regional terranes of Blueschist Facies rocks could reach the surface *before* being overprinted by higher-grade assemblages is debatable (Draper and Bone, 1981). In the case of the Franciscan Complex of California, fission track data indicate that postsubduction uplift was, in fact, minor (Dumitru, 1989).

EXAMPLE: REGIONAL HIGH P/T METAMORPHISM OF THE FRANCISCAN COMPLEX, CALIFORNIA

The Franciscan Complex forms the structurally complicated, locally chaotic basement of much of the California and southwestern Oregon Coast Ranges (figure 28.10).[58] It is composed of a wide variety of rock types, not all of which are metamorphosed. As a group, however, metamorphic rocks dominate. Wacke and metawacke and associated shale and metashale are the most abundant rock types.[59] Red, green, and multicolored radiolarian chert; massive to pillowed basaltic rocks; gray and pink limestones; conglomerates; ultramafic rocks; and the metamorphic equivalents of all of these also occur at numerous localities. Well known among the metamorphic rocks are eclogites, glaucophane schists and gneisses, and actinolite and hornblende schist and gneiss that occur in isolated blocks and sheets.[60]

Both regional terranes and the isolated masses of metamorphic rock are present in the Franciscan Complex. The isolated masses—including glaucophane schists, eclogites, and related rocks, the first discovered and thoroughly studied of the high-pressure rocks (A. C. Lawson, 1895; Switzer, 1945, 1951; Borg, 1956)—most commonly occur in melanges. In addition, Eclogite, Blueschist, Amphibolite, and rare Greenschist Facies rocks form slabs and tectonic blocks along faults. The blue amphibole-bearing rocks, including eclogites, are typically divided into higher-temperature "high-grade" blocks and lower-temperature "low-grade" blocks. Both types may contain sodic pyroxene, but the high-grade blocks are generally coarser-grained, texture IIIS and IVS rocks, whereas the low-grade blocks are finer-grained, texture IIS to IIIS rocks. In contrast, texture I to texture III rocks are typical of the regional terranes. M. E. Maddock (1955) discovered and Bloxam (1956) and McKee (1962) reported that jadeitic pyroxene is widely distributed in rocks previously believed to be unmetamorphosed. Subsequent studies revealed that Blueschist, Prehnite-Pumpellyite, and Zeolite Facies assemblages characterize the rocks within large areas.[61]

In the northern Coast Ranges, rocks of the six metamorphic facies are distributed across three major, fault bounded belts—the Eastern Belt, the Central Belt, and the Coastal Belt—that are successively younger from east to west (figure 28.10) (Blake, Irwin, and Coleman, 1969; Berkland et al., 1972).[62] High-grade schists and gneisses, in tectonic blocks and slabs, form a fourth unit that locally caps the Franciscan Complex along its eastern edge. Each belt is

607

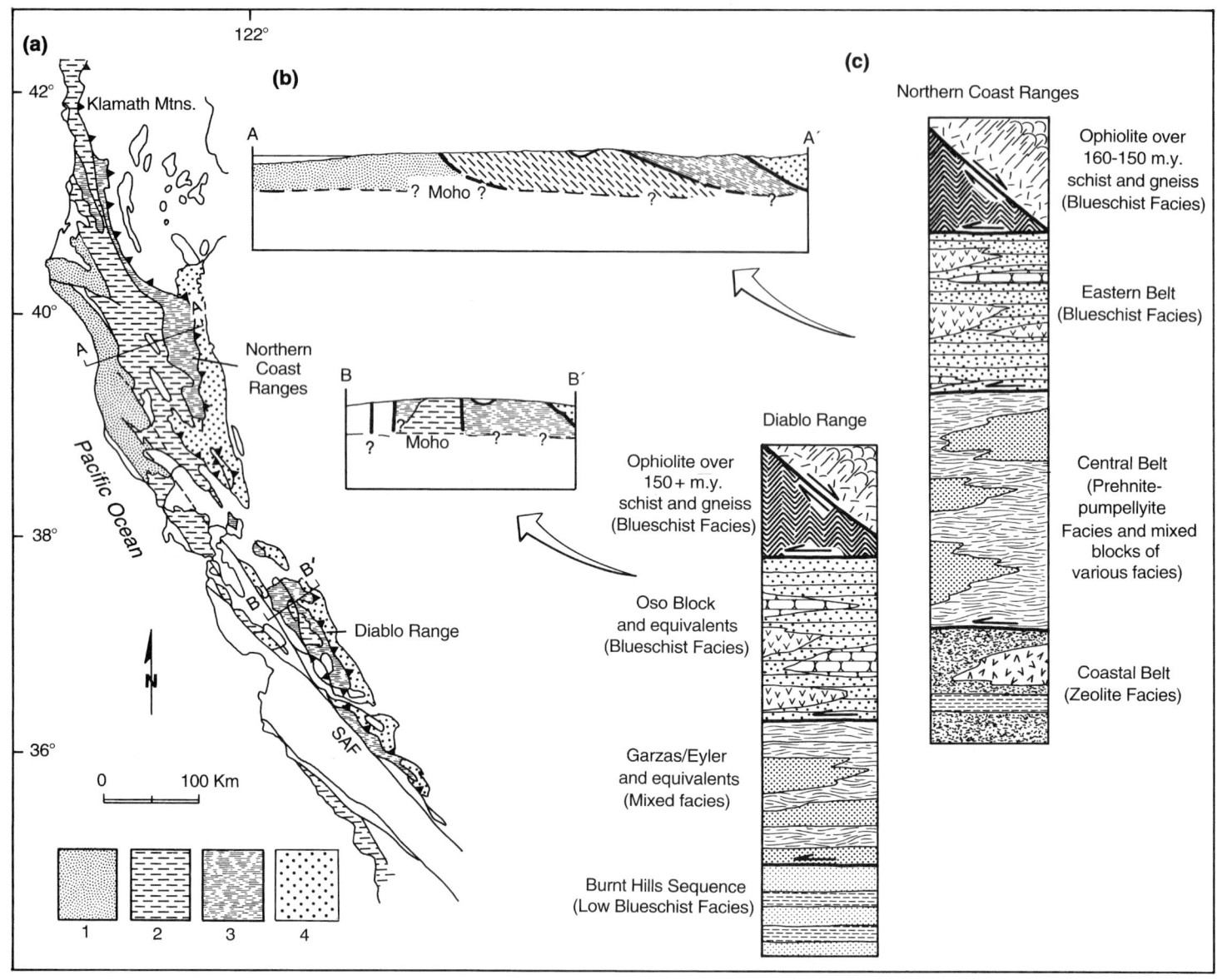

Figure 28.10 The Franciscan Complex of California.
a) Generalized map of western California showing the three structural-metamorphic belts of the Franciscan Complex.
1 = Coastal Belt, 2 = Central Belt, 3 = Eastern Belt, 4 = Great Valley Group (modified from Blake, Irwin, and Coleman, 1969; R. G. Coleman and Lanphere, 1971; Raymond and Swanson, 1980; Ernst, 1980). (b) Simplified, cross-sectional sketches across the northern coast ranges and the Diablo Range. (c) Diagrammatic, columnar structural sections showing the main structural units of the complex.

subdivided into several thrust sheets or fault blocks (commonly designated as terranes) that include various formations, broken formations, dismembered formations, and melanges.[63] The Central Belt is largely melange. In contrast, the adjoining Eastern and Coastal belts, though locally containing melange, consist predominantly of rock bodies with greater internal coherence. Similar structural units exist in the Diablo Range of central California, but relationships there are more complex and the Coastal Belt is missing.[64]

Some Franciscan rocks are also exposed along the south-central coast.[65] In the area at the southern end of the Northern Coast Ranges, in the San Francisco Bay area and to the north for several tens of kilometers, the structural and metamorphic patterns are highly disrupted by Cenozoic faulting.[66]

The metamorphic patterns of the northern Coast Ranges are more regular than the patterns in the south. In the north, the westernmost belt, the Coastal Belt, is a metawacke- and metashale-dominated, *Zeolite Facies* metamorphic belt

(a)

Figure 28.11 Photomicrographs of Franciscan Facies Series rocks from the Franciscan Complex of California. (a) Zeolite Facies metawacke from the Coastal Belt, near Willits, California, showing laumontite (Lm) replacing plagioclase (P), (XN). (b) Blueschist Facies metawacke from the northeastern Diablo Range containing lawsonite (Lw) and jadeitic pyroxene (J), (XN). (c) Blueschist Facies metachert containing crossite and aegirine needles (CR); from a

(figure 28.10).[67] The metawackes range from incipiently metamorphosed, texture IS rocks that retain their primary epiclastic textures and minerals to significantly neocrystallized texture IIS types. Locally, especially at the western edge of the area, broken formations and melanges with texture IIC to IVC rocks are present (R. J. McLaughlin et al., 1982). The metawackes in the north (and along the southern coast) contain laumontite, prehnite, or pumpellyite as incipient neocrystallized grains in feldspars, as complete replacements of feldspars, and as vein minerals. Thorough petrologic analyses of these weakly and incompletely metamorphosed rocks have not been published, but representative phase assemblages (figure 28.11a) appear to include

quartz–albite–laumontite–chlorite–white mica–hematite,
quartz–albite–laumontite–chlorite/smectite–white mica–calcite–hematite,
quartz–albite–prehnite–chlorite–white mica–calcite, and
quartz–albite–alkali feldspar–chlorite–white mica–stilpnomelane–calcite.[68]

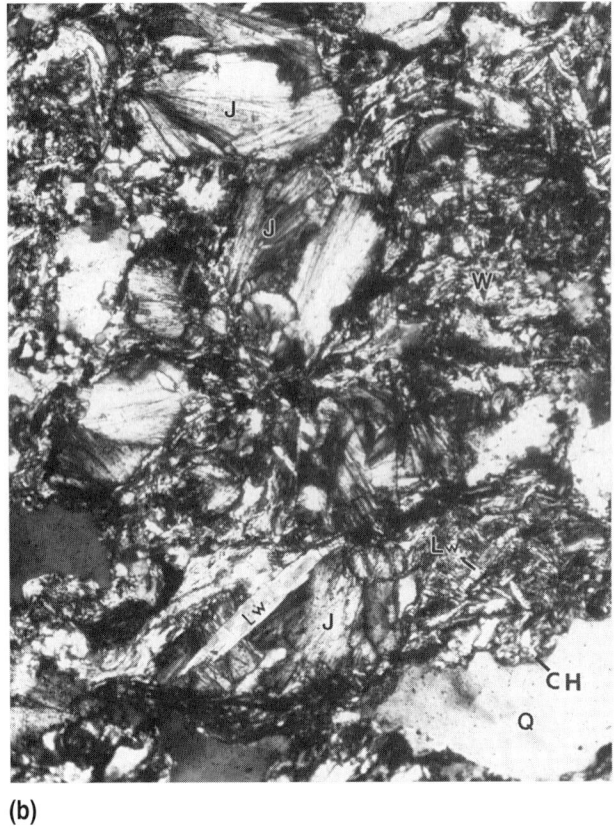

(b)

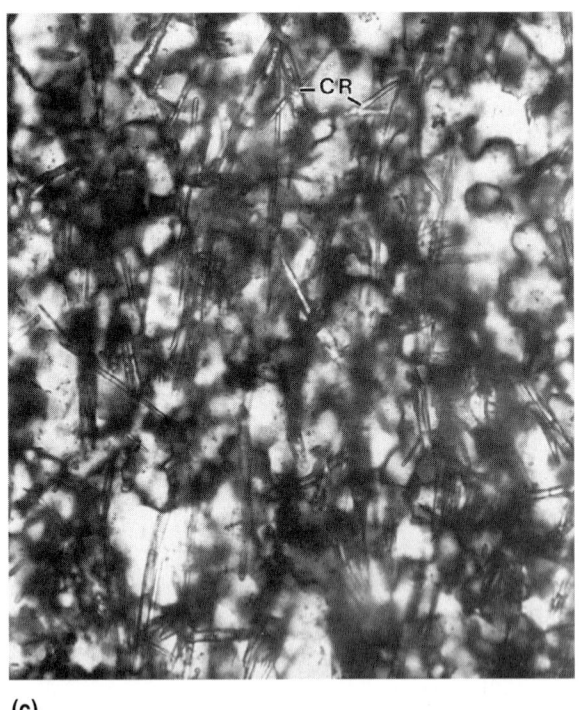

(c)

(d)

Figure 28.11 Continued
melange in the northeastern Diablo Range, California, (PL). (d) Glaucophane-bearing eclogite, Jenner, Sonoma County, California, (XN). CH = chlorite, G = garnet, GL = glaucophane, OM = omphacite, Q = quartz, W = white mica. Long dimension of photos (a) and (b) is 1.27 mm and of photos (c) and (d) is 1.06 mm.

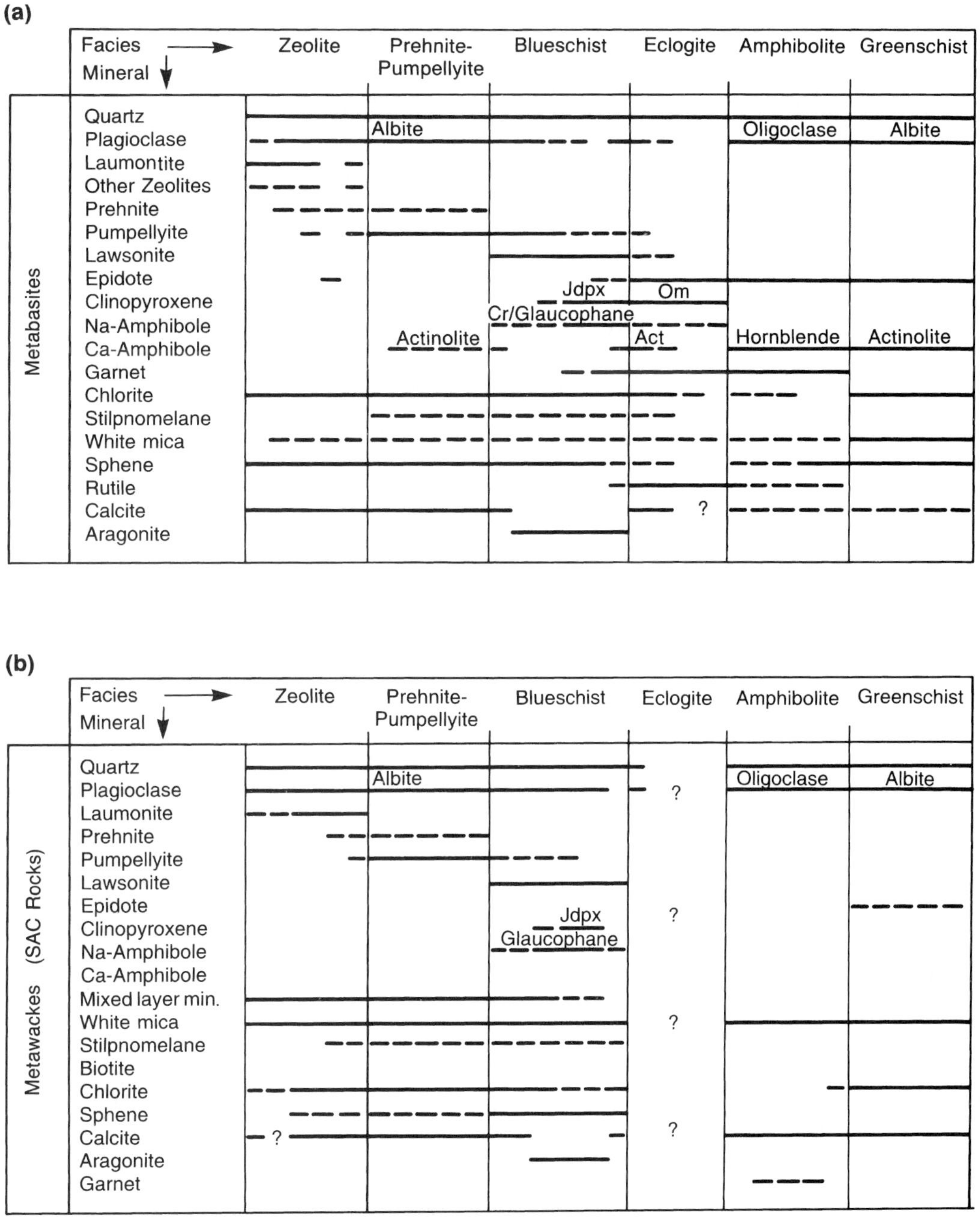

(a)

Facies → / Mineral ↓	Zeolite	Prehnite-Pumpellyite	Blueschist	Eclogite	Amphibolite	Greenschist
Quartz						
Plagioclase		Albite			Oligoclase	Albite
Laumontite						
Other Zeolites						
Prehnite						
Pumpellyite						
Lawsonite						
Epidote						
Clinopyroxene			Jdpx	Om		
Na-Amphibole			Cr/Glaucophane			
Ca-Amphibole		Actinolite		Act	Hornblende	Actinolite
Garnet						
Chlorite						
Stilpnomelane						
White mica						
Sphene						
Rutile						
Calcite				?		
Aragonite						

(Row group label: Metabasites)

(b)

Facies → / Mineral ↓	Zeolite	Prehnite-Pumpellyite	Blueschist	Eclogite	Amphibolite	Greenschist
Quartz						
Plagioclase		Albite		?	Oligoclase	Albite
Laumonite						
Prehnite						
Pumpellyite						
Lawsonite						
Epidote				?		
Clinopyroxene			Jdpx			
Na-Amphibole			Glaucophane			
Ca-Amphibole						
Mixed layer min.						
White mica				?		
Stilpnomelane						
Biotite						
Chlorite						
Sphene						
Calcite	?			?		
Aragonite						
Garnet						

(Row group label: Metawackes (SAC Rocks))

Figure 28.12 Mineral-facies charts for the Franciscan Complex: (a) metabasites, (b) metawackes, (c) metapelites, (d) metacherts.

(Sources include those in notes 25–28 and observations of the author)

Zeolite Facies metabasites include laumontite-, prehnite-, epidote-, and pumpellyite-bearing assemblages (figure 28.12). Reactions such as equation 28.12 probably produced the laumontite in metawackes.

The Central Belt melanges structurally overlie the Coastal Belt rocks. Most rocks of the Belt are considered to belong to the *Prehnite-Pumpellyite Facies*. As is the case for the Coastal Belt rocks, however, the petrogenesis of Central Belt rocks has not been adequately investigated. Furthermore, because the Central Belt consists primarily of an assemblage of melanges containing a variety of blocks and slabs of rock (J. C. Maxwell, 1974; Gucwa, 1975; Blake et al., 1988),[69] the metamorphic grade within the belt shows no regular pattern.

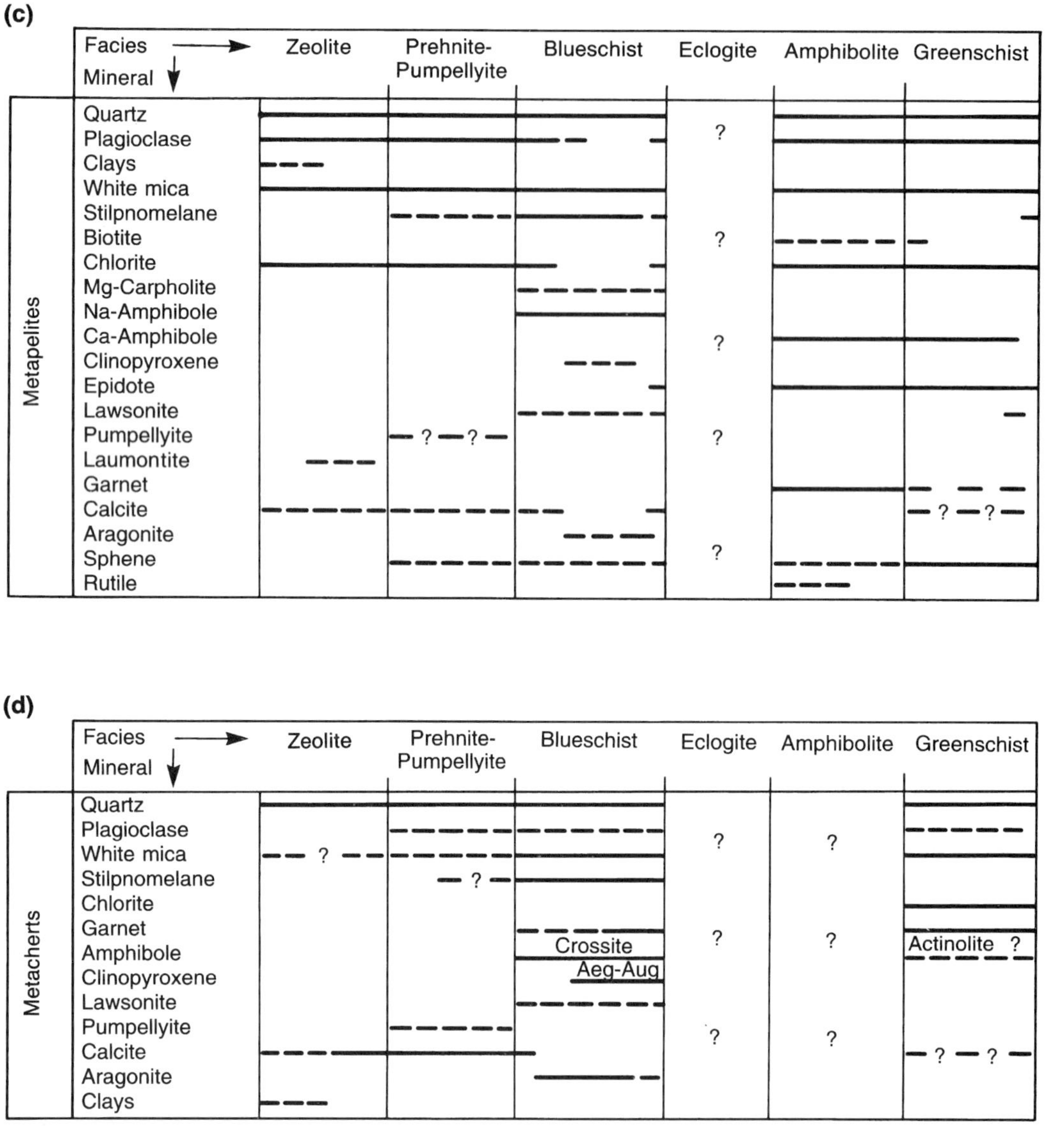

(c)

Mineral ↓ / Facies →	Zeolite	Prehnite-Pumpellyite	Blueschist	Eclogite	Amphibolite	Greenschist
Quartz	—	—	—	—	—	—
Plagioclase	—	—	—	?	—	—
Clays	- - -					
White mica	—	—	—	—	—	—
Stilpnomelane		- - -	- -			—
Biotite				?	- - - -	—
Chlorite	—	—	—			
Mg-Carpholite			- - -			
Na-Amphibole			—			
Ca-Amphibole				?	—	—
Clinopyroxene			- - -			
Epidote			—			
Lawsonite			- - -			—
Pumpellyite		— ? — ? —		?		
Laumontite	- - -					
Garnet					- -	- -
Calcite	- - -	- - -	- - -			— ? — ? —
Aragonite			- - -	?		
Sphene		- - -	- - -	?	- - -	
Rutile					- - -	

(Metapelites)

(d)

Mineral ↓ / Facies →	Zeolite	Prehnite-Pumpellyite	Blueschist	Eclogite	Amphibolite	Greenschist
Quartz	—	—	—	—	—	—
Plagioclase		- - -	- - -	?	?	- - -
White mica	- - ? - -	—	—	—	—	—
Stilpnomelane		— ? —				
Chlorite						—
Garnet			—	?	?	
Amphibole			Crossite			Actinolite ?
Clinopyroxene			Aeg-Aug			
Lawsonite			—			
Pumpellyite		- - - -	—	?	?	
Calcite	- - -					— ? — ? —
Aragonite			—			
Clays	- -					

(Metacherts)

Figure 28.12 Continued

Blocks in the melange range from very low-grade Zeolite Facies rocks to Eclogite and Amphibolite Facies metabasites. Most rocks retain some detrital or primary minerals, that is, they are only partly recrystallized and they exhibit a variety of phase assemblages. Texturally, metawackes range from IS and IC types to IIIS and IIIC types. Some rocks appear nearly unmetamorphosed, others contain assemblages such as

quartz–albite–alkali feldspar–chlorite–white mica (with veins of laumontite and prehnite),

quartz–albite–pumpellyite–chlorite–white mica–calcite, and

quartz–albite–prehnite–pumpellyite–chlorite–white mica–hematite,

whereas others, notably in the east, contain the Blueschist Facies assemblage

quartz–albite–white mica–chlorite–lawsonite–pumpellyite

(Jayko, Blake, and Brothers, 1986; Blake et al., 1988).[70] Prehnite is uncommon and jadeitic pyroxene is present in only a few metawacke blocks.[71] Metapelites that form the melange matrix contain the assemblages

white mica–chlorite–quartz–albite
white mica–chlorite–quartz–albite–pumpellyite, and
white mica–chlorite–quartz–albite–lawsonite

(Cloos, 1983). Like the metawackes, the metabasites show a range of mineral and metamorphic grades. The characteristic assemblage is

chlorite–pumpellyite–albite–quartz–calcite,

but actinolite-, lawsonite-, and glaucophane-bearing assemblages are present locally (see figure 28.12)(Blake et al., 1988).

To the east and structurally overlying the Central Belt is a faulted Blueschist Facies belt dominated by metasedimentary rocks and containing a variety of pumpellyite-, lawsonite-, and jadeitic-pyroxene-bearing assemblages (Blake, 1965; Suppe, 1973; Jayko, Blake, and Brothers, 1986).[72] Metawackes, metashales, and metacherts predominate. Typical metawacke assemblages include

quartz–albite–white mica–chlorite–lawsonite–aragonite,
quartz–albite–white mica–chlorite–lawsonite–jadeitic
    pyroxene–aragonite, and
quartz–jadeitic pyroxene–white mica–chlorite–lawsonite–
    aragonite–stilpnomelane–glaucophane.

In the Diablo Range of the central Coast Ranges (see figure 28.10), somewhat similar rocks are present. Blueschist Facies metawackes there include, in addition to the above assemblages, the assemblages

quartz–albite–white mica–chlorite–chlorite/vermiculite–
    lawsonite–calcite,
quartz–albite–white mica–chlorite–chlorite/vermiculite–
    lawsonite–glaucophane–pumpellyite, and
quartz–albite–white mica–chlorite–lawsonite–jadeitic
    pyroxene–pumpellyite–aragonite–sphene–hematite

(see figure 28.11b) (Ernst, 1965; 1971c; Raymond, 1973a, 1973c; Patrick and Day, 1989).[73] In this region, jadeitic-pyroxene-bearing metawackes are widespread and well developed. Jadeitic pyroxenes have problably formed via reactions such as

$$\text{albite + chlorite/vermiculite + calcite + hematite} \rightarrow$$
$$\text{jadeitic pyroxene + chlorite + white mica + quartz + H}_2\text{O +}$$
$$\text{CO}_2, \qquad (28.27)$$

$$\text{albite + chlorite}_1 \text{ + lawsonite + hematite} \rightarrow \text{jadeitic}$$
$$\text{pyroxene + chlorite}_2 \text{ + quartz + H}_2\text{O} \qquad (28.28)$$

(cf. Kerrick and Cotton, 1971), equation 28.18 (Maruyama, Liou, and Sasakura, 1985), and equation 28.19 (Patrick and Day, 1989).

Other rocks include metapelites, marbles, and metacherts. Metapelites contain the assemblage

white mica–chlorite–quartz–albite–lawsonite

(Cloos, 1983). Metacherts contain such assemblages as

quartz–crossite–aegirine–hematite and
quartz–stilpnomelane–garnet

(see figures 28.11c and 28.12). Where a carbonate mineral is present in the cherts, the stable carbonate is typically aragonite. Aragonite marbles may also occur, but locally aragonite has reverted to calcite.

Blueschist Facies metabasites in the Eastern Belt and its equivalent in the Diablo Range contain such assemblages as

glaucophane–lawsonite–albite– sphene,
glaucophane–lawsonite–stilpnomelane–chlorite–albite–
    quartz,
glaucophane–albite–quartz–garnet–white mica,
blue amphibole–lawsonite–pumpellyite–chlorite–albite–
    garnet, and
chlorite–lawsonite–jadeitic
    pyroxene–glaucophane–quartz–sphene.[74]

The assemblages reflect reactions in which plagioclase is converted to lawsonite (e.g., equation 28.17) and pumpellyite and pyroxenes incorporate Na to become sodic amphiboles. These rocks have textures that range from diablastic to schistose, but locally, relict diabasic and porphyritic textures are preserved.

In the Northern Coast Ranges, at the eastern edge of the Eastern Belt (figure 28.10) within a unit composed of metasedimentary and metavolcanic rocks, there is a small zone of *epidote-bearing*, textural zone IIIS–IVS rocks (Suppe, 1973; E. H. Brown and Ghent, 1983; Jayko, Blake, and Brothers, 1986). These rocks occur adjacent to the Coast Range Fault, which bounds the Franciscan Complex on the east. A lawsonite-out isograd separates the epidote-bearing rocks from other Blueschist Facies rocks to the west (Jayko, Blake, and Brothers, 1986). Typical assemblages in metabasites (figure 28.12) include

blue amphibole–albite–chlorite–pumpellyite–epidote,
blue amphibole–chlorite–pumpellyite–epidote–quartz–
    sphene,
blue amphibole–albite–chlorite–epidote–quartz–white
    mica–sphene,
blue amphibole–actinolite–albite–stilpnomelane, and
actinolite–albite–chlorite–pumpellyite–aragonite.

The reaction marking the isograd may be equation 28.24, in which lawsonite and Na-amphibole react to form epidote and other phases (E. H. Brown and Ghent, 1983). In this belt, typical metawacke phase assemblages include

quartz–albite–muscovite–paragonite–chlorite–lawsonite,
quartz–albite–white mica–chlorite–lawsonite–aragonite,
    and
quartz–albite–white mica–chlorite–lawsonite–stilpnome-
    lane.

Similar rocks occur in Oregon, to the north (B. L. Wood, 1971; R. G. Coleman, 1972; Blake and Jayko, 1986).

The high-grade blueschists and the Eclogite, Amphibolite, and Greenschist Facies metabasites that occur as blocks and slabs in Franciscan melanges have generally experienced polymetamorphism and a more complex history than the rocks in more coherent terranes (S. S. Sorenson, 1986;

613

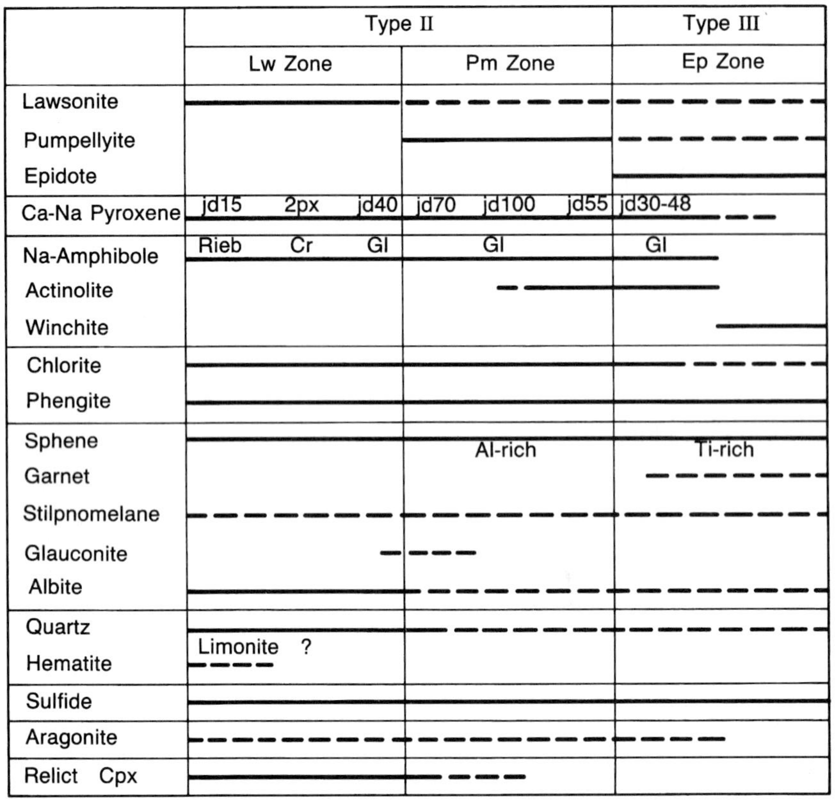

	Type II		Type III
	Lw Zone	Pm Zone	Ep Zone
Lawsonite	——————	– – – –	– – – –
Pumpellyite		——————	– – – –
Epidote			——————
Ca-Na Pyroxene	jd15 2px jd40	jd70 jd100 jd55	jd30-48 – –
Na-Amphibole	Rieb Cr Gl	Gl	Gl
Actinolite		— —	
Winchite			————
Chlorite	——————————		– – – –
Phengite	——————————		
Sphene	——————	Al-rich	Ti-rich
Garnet			– – – –
Stilpnomelane	– – – –	– – – –	
Glauconite		— – – –	
Albite	——————	– – – –	– – – –
Quartz	——————	– – – –	– – – –
Hematite	Limonite ?		
	– – –		
Sulfide	——————————————		
Aragonite	– – – –	– – – –	– – – –
Relict Cpx		— – –	

Figure 28.13

Mineral-facies chart for the metabasites of Ward Creek, California.

(From S. Maruyama and J. G. Liou, "Petrology of Franciscan metabasites along the jadeite-glaucophane type facies series, Cazadero, California" in Journal of Petrology, *29:1-37, 1988. Copyright © 1988 Oxford University Press, Oxford England. Reprinted by permission.)*

D. E. Moore and Blake, 1989; Wakabayashi, 1990). *In situ* terranes for all but the eclogites have been recognized in southern California (Platt, 1975; Bebout and Barton, 1989; S. S. Sorenson, Bebout, and Barton, 1991). Typical assemblages in the various facies are

glaucophane–lawsonite–chlorite–white
  mica–quartz–sphene (Blueschist Facies)
actinolite–epidote–albite–chlorite–quartz–white
  mica–sphene (Greenschist Facies)
hornblende–epidote–garnet–clinopyroxene–quartz–rutile
  (Amphibolite Facies)
omphacite–garnet–epidote–glaucophane–actinolite–
  phengite–sphene (Eclogite Facies)

(see figure 28.11d).[75] In many cases, early-crystallized assemblages formed in one facies are overprinted by assemblages formed later under conditions of another facies.

The most thoroughly studied of these high-grade blocks are those of the Ward Creek/Cazadero area of California (R. G. Coleman and Lee, 1963; Maruyama and Liou, 1988; Oh, Liou, and Maruyama, 1991).[76] At this locality, a large metabasite mass and some tectonic blocks are engulfed by Central Belt melange. R. G. Coleman and Lee (1963) divided metabasites here into four types, primarily on the basis of texture, but also on degree of metamorphism. These types include Type I (unmetamorphosed), Type II (metamorphosed, fine-grained, nonfoliated), Type III (metamorphosed, medium-grained, foliated), and Type IV (metamorphosed, coarse-grained, foliated; the high-grade tectonic blocks of blueschist and eclogite). The large metabasite mass at Ward Creek includes Type II and Type III blueschists and contains a series of three zones—a lawsonite zone, a pumpellyite zone, and an epidote zone—representing three metamorphic grades of increasing temperature (figure 28.13) (R. G. Coleman and Lee, 1963; Maruyama and Liou, 1988; Oh, Liou, and Maruyama, 1991). Reactions marking the zone boundaries are discontinuous and include a pumpellyite-in reaction, an actinolite-in reaction, and an epidote-in reaction.

Analyses of the metamorphic conditions that produced the metamorphic rocks in the Franciscan Complex are based on experimental phase equilibria, isotopic analyses, virtrinite

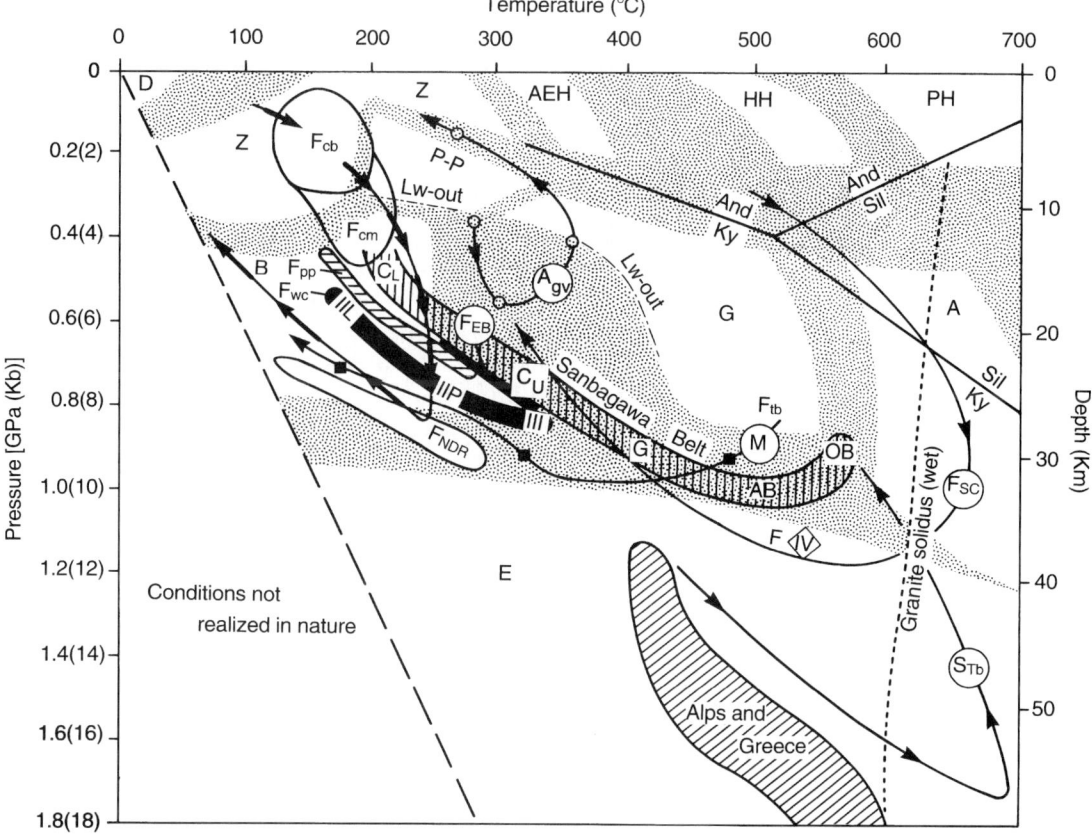

Figure 28.14 P-T-t paths and estimated P-T conditions under which Franciscan and Sanbagawa Facies series rocks were metamorphosed. Elliptical areas represent generalized conditions of metamorphism for various terranes or masses of rock.

(And = andalusite, Ky = kyanite, Sil = sillimanite. A_{gv} = P-T path for carpholite-bearing blueschists from the western Alps (Goffe and Velde, 1984). F_{Cb} = Franciscan Complex, Coastal Belt (Underwood, 1985; Underwood, Blake, and Howell, 1987; Blake et al., 1988; Dumitru, 1991). F_{CM} = Franciscan Complex, Central Belt melange (Cloos, 1983; Underwood, Blake, and Howell, 1987; Blake et al., 1988; Dumitru, 1991). F_{EB} (higher -T, dark band, left of center) = Franciscan Complex, Eastern Belt of the Northern Coast Ranges (Blake et al., 1988). F_{NDR} = Franciscan Complex rocks of the northern Diablo Range (Bostick, 1974; Patrick and Day, 1989). F_{pp} = Franciscan metawackes of Pacheco Pass, central Diablo Range, California. F_{WC} = Franciscan metabasites of Ward Creek. IIL = Lawsonite zone; IIP = Pumpellyite

Zone; III = Epidote Zone (Maruyama and Liou, 1988; Oh, Liou, and Maruyama, 1991). F-IV = Franciscan tectonic blocks of Type IV (Oh, Liou, and Maruyama, 1991). F_{SC} = Generalized path for Franciscan amphibolitic tectonic blocks (Moore and Blake, 1989; Wakabayashi, 1991) and Santa Catalina schists from subduction-zone hanging wall (Sorenson, Bebout, and Barton 1991). F_{tb} = Franciscan tectonic block retrograde metamorphic path (from D. E. Moore, 1984). In the Sanbagawa Belt, C_L = lower Chlorite Zone; C_U = upper Chlorite Zone; G = Garnet Zone; AB = Albite-biotite Zone; and OB = Oligoclase-biotite Zone for pelitic rocks (Takasu, 1989). S_{Tb} = generalized Sanbagawa tectonic block metamorphic path (Takasu, 1989). Field designated Alps and Greece represents P-T conditions indicated by some eclogitic blueschists from these regions (Schleistedt and Matthews, 1987; Brocker, 1990; Pognante, 1989; 1991). Facies fields are the same as those shown in figure 24.6. Facies designations as follows: A = Amphibolite, AEH = Albite-Epidote Hornfels, B = Blueschist, Ec = Eclogite, G = Greenschist, HH = Hornblende Hornfels, PH = Pyroxene Hornfels, P-P = Prehnite-pumpellyite, Z = Zeolite.)

reflectance, and fission tracks.[77] Estimated P-T conditions are shown in figure 28.14. In summary, Eclogite Facies rocks were metamorphosed under conditions of $T = 290–540°$ C and $P = 0.6–1.4$ Gpa (6–14kb). Metamorphism of Eastern Belt rocks occurred at $P = 0.4–1.0$ Gpa and $T = 125–350°$ C, whereas Central Belt melange metamorphism resulted from pressures of 0.2–0.6 Gpa and temperatures of 125–300° C. Zeolite Facies metamorphism of Coastal Belt rocks occurred at about

$P = 0.1–0.3$ Gpa and $T = 100–200°$ C. Projected P-T-t paths are shown in figure 28.14. Note that the rocks of the eastern Diablo Range remained at low T conditions, whereas the metamorphic conditions for the Eastern Belt in the northern Coast Ranges reaches somewhat higher temperature conditions. Note also that there is not a single geothermal gradient for the entire complex, but rather, different path histories for different blocks or belts.[78]

615

The origin of the Franciscan Facies Series in California served as a focal point for the debates about the origin of blueschists that occurred during the 1960s and 1970s. It is now generally agreed that subduction carried the rocks of each of the belts (or terranes within the belts) to the depth appropriate for metamorphism of that belt. Among the petrologic problems that remain are the problems of preservation and uplift of these rocks and problems associated with the fact that in most of the rocks recrystallization is incomplete.

As noted above, evidence of disequilibrium, incomplete neocrystallization, and polymetamorphism abounds in Franciscan rocks.[79] Some of the evidence of disequilibrium gave rise to the suggestion that the jadeitic pyroxene and other minerals in the jadeitized metawackes of the Diablo Range are detrital rather than metamorphic (Brothers and Grapes, 1989). The disequilibrium and lack of homogeneity in jadeitic pyroxenes is well known,[80] as are detrital Blueschist Facies materials (O'Day and Kramer, 1972; D. S. Cowan and Page, 1975; D. E. Moore and Liou, 1980). In the case of the jadeitic pyroxenes of the Diablo Range, euhedral crystals that have grown across grain boundaries, sympathetic changes in entire phase assemblages coincident with inferred metamorphic grade, and jadeitic pyroxene aspect ratios clearly demonstrate that most (if not all) jadeitic pyroxene in these metawackes is metamorphic in origin (Raymond, 1991; Raymond and Nall, 1991).

How the Franciscan Blueschist Facies rocks reached the surface without being recrystallized under higher-T conditions is a matter of some debate. Models currently favored include both S- and E-type models of synsubduction uplift. Cloos (1982, 1984) has argued, using the S-type subduction channel model, that high-grade blocks and eclogites in the Central Belt have been uplifted via flow in a subduction channel, as well as by local diapiric intrusion of melange into layered sequences of sediment. Krueger and Jones (1989) and T. A. Harms, Jayko, and Blake (1992) favor shallowing of the subduction angle, resulting in E-type exhumation. Ernst (1988) accepts both of these models for given situations, but argues that buoyant rise of accreted wedges may also take place.

SUMMARY

Sanbagawa and Franciscan Facies series are high P/T facies series distinguished by the presence of blue, sodic amphiboles. The Sanbagawa Facies Series displays the facies sequence Zeolite → Prehnite-Pumpellyite → Blueschist → Greenschist → Amphibolite, corresponding to a geothermal gradient of about 10–20° C (figure 28.14). In the Franciscan Facies Series, the facies sequence Zeolite → Prehnite-Pumpellyite → Blueschist → Eclogite reflects lower temperatures at low to high pressures. These facies series occur worldwide in orogenic belts, but are almost entirely confined to orogens of Phanerozoic age.

The key facies for recognizing high P/T facies series is the Blueschist Facies. Critical minerals indicative of this facies are lawsonite, aragonite, and jadeitic pyroxene. In addition, blue, sodic amphiboles, abundant in blueschists and present in other rocks, reflect high-pressure conditions. A variety of continuous and discontinuous, terminal and nonterminal reactions involving phases such as plagioclase, laumontite, prehnite, pumpellyite, epidote, actinolite, glaucophane, crossite, jadeitic pyroxene, omphacite, chlorites, white micas, and garnet yield these and other index phases. Aragonite forms by a simple polymorphic transformation. Although Zeolite, Prehnite-Pumpellyite, Greenschist, and Amphibolite Facies rocks occur in the high P/T facies series, their phase assemblages are generally like those found in the same facies of other (lower P/T) facies series. Eclogites may occur in both Franciscan and Sanbagawa Facies series, but they are characteristic of Franciscan Facies Series. The metamorphic conditions under which the rocks of these facies series generally form are $P = 0.2–1.4$ Gpa and $T = 100–550°$ C.

High P/T facies series form in subduction zone settings, where they characterize the rocks of the low-temperature outer belt of a paired metamorphic belt. How Blueschist Facies rocks are preserved over time and during uplift, when they would normally be heated and converted to Greenschist or other higher-temperature facies, is a problem. Postsubduction buoyant uplift, if quite rapid, can provide a mechanism for quick exposure of the rocks, but it is unlikely that this process is generally operative in the history of Franciscan Facies Series blueschists. The S-type imbricate thrust model of subduction, accretion, and uplift, in which subduction zones act as two-way streets, provides an explanation for ascent of slabs of rock up a cold subduction zone. As long as a relatively rapid subduction rate involving a cool slab is maintained, the rocks remain "refrigerated" during subduction. The S-type subduction channel model explains the similar upward movement of blueschist and eclogite blocks in tectonic melanges formed within the subduction zone. Alternatively, E-type exhumation of blueschists may occur where subduction and accretion create uplift and a resulting instability in the accretionary wedge, causing listric normal faulting at the top of the wedge and exposure of rocks formerly deeply buried in the subduction zone. In rare cases, other processes involving oblique subduction and strike-slip or transform faulting may also result in exposure of the blueschists along a plate margin.

EXPLANATORY NOTES

1. The term blueschist is used in this text in an informal way to refer to blue amphibole-bearing schists.
2. The use of the name Blueschist Facies in this text follows the original definition of E. H. Bailey (1962; E. H. Bailey, Irwin, and Jones, 1964). The critical mineral in SAC and metabasic rocks is lawsonite and assemblages such as lawsonite + albite + glaucophane and lawsonite + jadeitic pyroxene + glaucophane are characteristic. Brothers and Blake (1987)

noted and lamented the tendency of some petrologists to stray from this definition. In contrast, Taylor and Coleman (1968), B. W. Evans and Brown (1987), and B. W. Evans (1990) proposed broadening the definition of Blueschist Facies to include assemblages of glaucophane/crossite + epidote. Although B. W. Evans and Brown (1987) and B. W. Evans (1990) make a good case for distinguishing between high- and low-pressure Greenschist Facies rocks and perhaps for adding a new facies name for rocks formed at conditions of about $T = 300–550°$ C and $P = 0.6–1.1$ Gpa (6–11 kb), the use of the name "epidote blueschist" and the extension of the name Blueschist Facies to encompass all glaucophane (or crossite)-bearing rocks is not a good idea and is bound to create confusion, because it redefines an established and widely used term. If a new facies term is needed, a name such as High-Pressure Greenschist Facies, Blue-Green Facies, or Epidote–Blue Amphibole Facies would distinguish such a facies without creating unnecessary confusion and controversy.

Other names that have been applied to facies defined similarly to the Blueschist Facies include Glaucophane Schist Facies (F. J. Turner and Verhoogen, 1960), Lawsonite-Glaucophane-Jadeite Facies, and Lawsonite-Albite Facies (Winkler, 1965, 1967). Winkler (1974, 1976, 1979) later abandoned the latter terms.

3. Also see Seki (1960), Essene, Fyfe, and Turner (1965), Ernst and Seki (1967), Landis and Coombs (1967), H. P. Taylor and Coleman (1968), Liou (1971b), Landis and Bishop (1972), Black (1974b), Bostick (1974), Platt (1975), Caron, Kienast, and Triboulet (1981), M. A. Carpenter (1981), Seki and Liou (1981), Brothers and Yokoyama (1982), Schleistedt and Matthews (1987), Schreyer (1988), Patrick and Evans (1989), Wakabayashi (1990), and Wallis and Banno (1990).

4. In some cases, Eclogite Facies rocks are found in Sanbagawa Facies Series, for example, in the Kargi Massif of Turkey (Okay, 1986) and the southern Alps (Barnicoat and Fry, 1989).

5. Refer to the references in note 3 and appendix C for phase stability studies and papers on isotopic analyses. Vitinite reflectance studies are studies on the light-reflecting quality of carbonaceous (coal) fragments. The reflectance increases with temperature (For applications in the Franciscan Facies Series, see Bostick, 1974; Underwood, Blake, and Howell, 1987; Underwood, O'Leary, and Strong, 1988). Fission track studies are studies of the microscopic trails made in minerals by radiation produced by radioactive decay of included elements (see Dumitru, 1988, for an example). P-T conditions are discussed by Seki (1960), Ernst (1965, 1971a), Ernst and Seki (1967), Landis and Coombs (1967), Newton and Smith (1967), H. P. Taylor and Coleman (1968), D. G. Bishop (1972), Landis and Bishop (1972), Black (1974b), Caron, Kienast, and Triboulet (1981), M. A. Carpenter (1981), Seki and Liou (1981), Brothers and Yokoyama (1982), Kienast and Ranguin (1982), Underwood (1985, 1989), Newton (1986), Yokoyama, Brothers, and Black (1986), Matthews and Schleistedt (1987), Pognante and Kienast (1987), Underwood and Howell (1984), Maruyama and Liou (1988), Sedlock (1988), Dumitru (1989; 1991), Patrick and Day (1989), Wakabayashi (1990), Oh, Liou, and Maruyama (1991), Pognante (1991), and others.

6. For example, see Winkler (1965, 1967), Dickinson et al. (1969), Miyashiro (1973a), and Yardley (1989). Offler et al. (1980) use the term "burial metamorphism," to refer to low-grade rocks metamorphosed under a high geothermal gradient.

7. The distribution of Franciscan and Sanbagawa Facies series rocks or the specifics of the petrology of these facies series at various localities are discussed in Seki (1958), Miyashiro (1961), E. H. Bailey, Irwin, and Jones (1964), Ernst (1965, 1973a, 1977a), van der Kaaden (1969), Brothers (1970, 1974), Guitard and Saliot (1971), Makanjuola and Howie (1972), Black (1973, 1974a,b, 1975, 1977), Hotz (1973), Miyashiro (1973a,b), Herve et al. (1974), Brothers (1974), Nagle (1974), Triboulet (1974), Roy (1977a), Wood (1979c), Okay (1980, 1986), Shams, Jones, and Kempe (1980), E. H. Brown et al. (1981), Caron, Kienast, and Triboulet (1981), Higashino et al. (1981), Liou (1981a,b), Brothers and Yokoyama (1982), Kienast and Rangin (1982), Lippard (1983), Forbes, Evans, and Thurston (1984), Goffe and Velde (1984), Banno, Sakai, and Higashino (1986), B. W. Evans and Brown (1986), Cotkin (1987), Pognante and Kienast (1987), Sedlock (1988), Maruyama and Liou (1988), Banno and Sakai (1989), Barnicoat and Fry (1989), D. E. Moore and Blake (1989), Patrick and Evans (1989), Pognante (1989a, 1991), Shibakusa (1989), Smelik and Veblen (1989), Takasu (1989), Brocker (1990), Higashino (1990), Lucchetti, Cabella, and Cortesogno (1990), Shenbao (1990), Wakabayashi (1990), Wallis and Banno (1990), El-Shazly and Liou (1991), and Oh, Liou, and Maruyama (1991). Also see the additional references in these sources.

8. For example, Zhang and Liou (1987), Patrick and Evans (1989), and Schermer (1990).

9. See Ernst (1972b, 1988) and Dobretsov et al. (1982) for reviews. Also see I. A. Paterson and Harakal (1974) and Armstrong et al. (1986).

10. For example, see Ernst (1973b, 1977a), Laird and Albee (1981), Goffe and Velde (1984), Dal Piaz and Lombardo (1986), Schliestedt and Matthews (1987), Patrick and Lieberman (1988), Takasu (1989), and Brocker (1990).

11. E. H. Brown et al. (1981), Ross and Sharp (1988), D. E. Moore and Blake (1989), Wakabayashi (1990).

12. W. Glassley (1974), E. H. Brown et al. (1981), Kienast and Ranguin (1982), E. H. Brown (1986), T. E. Moore (1986), Sedlock (1988).

13. Sanbagawa is sometimes spelled Sambagawa. The Sanbagawa Facies Series, as used here, differs from the Sanbagawa "jadeite-glaucophane type" originally defined by Miyashiro (1961), as it does not include the very low temperature rocks, which are assigned to the Franciscan Facies Series. Many lawsonite-glaucophane schists and the jadeite-bearing rocks, once assigned to the Sanbagawa belt in Japan, are now recognized to be blocks in a tectonic melange along the Kurosegawa zone (Maruyama et al., 1984; Banno, 1986). In fact, the Sanbagawa rocks more closely match Miyashiro's description of "high pressure intermediate type" facies series, in that the Sanbagawa belt, as now defined, lacks jadeite-quartz assemblages and contains sporadic occurrences of sodic amphibole. The blue amphibole, commonly crossite, occurs with epidote and actinolite (Ernst et al., 1970; Higashino, 1990; Otsuki and Banno, 1990). The Sanbagawa Facies

Series, as defined here, reflects the petrology of the Sanbagawa belt as it is now known. For reviews of the Sanbagawa rocks, see Banno (1986) and Banno and Sakai (1989), as well as Wallis and Banno (1990) and articles in the *Journal of Metamorphic Petrology*, v. 8, p. 393ff (1990).

14. Also see R. G. Coleman and Lee (1963), Raymond (1973a), and Jayko, Blake, and Brothers (1986).

15. McKee (1962), Ernst (1971c), Raymond (1973c), D. S. Cowan (1974), Maruyama, Liou, Sasakura (1985), Patrick and Day (1989).

16. Seki et al. (1971, in Liou, Maruyama, and Cho, 1987).

17. Ernst et al. (1970), Watanabe and Kobayashi (1984), Banno (1986), Liou, Maruyama, and Cho (1987), but see Higashino (1990) and Otsuki and Banno (1990). Toriumi (1975) reports the sodic pyroxene aegirine-augite in some basic rocks.

18. Phase assemblages in Sanbagawa-like Facies Series rocks are listed or discussed by Seki (1958), D. S. Coombs (1960), Ernst and Seki (1967), Seki, Ernst, and Onuki (1969), Brothers (1970, 1974), Ernst et al. (1970), Black (1973, 1974a,b, 1975, 1977), E. H. Brown (1974), Toriumi (1975), D. S. Coombs et al. (1977), Ernst (1977a, 1982, 1984), Roy (1977a), R. M. Briggs (1978), Liou (1979), Shams, Jones, and Kempe (1980), Higashino et al. (1981, 1984a,b), Brothers and Yokoyama (1982), Enami (1983), Lippard (1983), Goffé and Velde (1984), Watanabe and Kobayashi (1984), Sakai et al. (1985), Cho, Liou, and Maruyama (1986), Jayko, Blake, and Brothers (1986), Kunugiza, Takasu, and Banno (1986), Oberhansli (1986), Cotkin (1987), Cho and Liou (1987), Liou, Maruyama, and Cho (1987), Banno and Sakai (1989), Barnicoat and Fry (1989), Shibakusa (1989), Brocker (1990), Higashino (1990), Lucchetti, Cabella, and Cortesogno (1990), Otsuki and Banno (1990), Shenbao (1990), and El-Shazly and Liou (1991).

19. Prehnite-Pumpellyite Facies rocks, recognized in the Sanbagawa Belt of the Kanto Mountains of Central Japan are not specifically included in zones in the south (compare Toriumi, 1975 with Otsuki and Banno, 1990 and Wallis and Banno, 1990).

20. For example, see Seki (1958), Ernst et al. (1970), Toriumi (1975), Enami (1983), Watanabe and Kobayashi (1984), Sakai et al. (1985), and Higashino (1990). Zones are based on Banno (1986).

21. This lower Greenschist Facies is equivalent to the Pumpellyite-Actinolite Facies of Liou, Maruyama, and Cho (1987).

22. Ernst (1964) discusses exchange reactions involving some of these phases.

23. Ernst et al. (1970), Toriumi (1975), Banno (1964, in Brown, 1977b), Nakajima, Banno, and Suzuki (1977), Liou, Maruyama, and Cho (1987), Otsuki and Banno (1990).

24. Also see Nakajima, Banno, and Suzuki (1977), E. H. Brown et al. (1981), J. B. Thompson, Laird, and Thompson (1982), E. H. Brown (1986), Maruyama, Cho, and Liou (1986), Schliestedt and Matthews (1987), and B. W. Evans (1990). In addition see Owen (1989).

25. Zeolite Facies minerals and assemblages of Franciscan-type Facies series are described by D. S. Coombs (1960), R. J. Stewart and Page (1974), Ernst (1977a; 1984), R. J. McLaughlin et al. (1982), Liou et al. (1987), and Blake et al. (1988). Data also from L. A. Raymond (unpublished data, Franciscan Complex, figure 28.3 and table 28.4). Also see Cho, Liou, and Maruyama (1986).

26. Prehnite-Pumpellyite Facies minerals and assemblages from Franciscan-type Facies Series are discussed by Ernst (1971b, 1980), D. S. Cowan (1974), W. E. Glassley (1975), Raymond (unpublished data, Franciscan Complex, this text), Nakajima, Banno, and Suzuki (1977), E. H. Brown et al. (1981), J. B. Thompson, Laird, and Thompson (1982), and Blake et al. (1988). Also see Cho, Liou, and Maruyama (1986) and Liou et al. (1987).

27. Blueschist Facies minerals and assemblages of the Franciscan Facies Series, are discussed, for example, by M. E. Maddock (1955), E. H. Bailey, Irwin, and Jones (1964), Ernst (1965, 1971c), Seki, Ernst, and Onuki (1969), Guitard and Saliot (1971), Black (1973), Hotz (1973), Raymond (1973a,c), Suppe (1973), Caron, Kienast, and Triboulet (1981), Maruyama, Liou, and Sasakura (1985), Jayko, Blake, and Brothers (1986), S. S. Sorenson (1986), Cotkin (1987), Schliestedt and Matthews (1987), Liou and Maruyama (1987), Blake et al. (1988), Maruyama and Liou (1988), Sedlock (1988), Bebout and Barton (1989), Pognante (1991), and S. S. Sorenson, Bebout, and Barton (1991).

28. Additional discussion of Franciscan Facies Series phases or phase assemblages may be found in McKee (1962), Onuki and Ernst (1969), Black (1973, 1974a,b, 1975), C. Hoffman and Keller (1979), D. E. Moore and Liou (1979), Mevel and Kienast (1980), Wood (1980), Brothers and Yokoyama (1982), D. E. Moore (1984), Oh et al. (1991) Shau et al. (1991), and in other sources listed in notes 68, 70, 73, 74, and 75.

29. Ernst (1972a), W. Glassley (1974), Ivanov and Gurevich (1975), Liou (1981b), Liou, Maruyama, and Cho (1987). Also see Schliestedt and Matthews (1987), Bebout and Barton (1989), and S. S. Sorenson, Bebout, and Barton (1991) for discussions of the influence of a fluid phase on particular assemblages.

30. Refer to the references in appendix C and to sources such as Liou (1971a,b), Maresch (1977), Luckscheiter and Monteani (1980), M. A. Carpenter (1981), F. J. Turner (1981), Maruyama, Cho, and Liou (1986), Newton (1986), Heinrich and Althaus (1988), and Oh, Liou, and Maruyama (1991) for the data on which the P-T limits are based. Also see figure 28.13 and note 77 below.

31. For example, R. G. Coleman and Lee (1963), Ernst et al. (1970), R. G. Coleman and Lanphere (1971), Kerrick and Cotton (1971), Ernst (1973b), Raymond (1973a,b,c), S. E. Swanson and Schiffman (1979), Wood (1979b); D. E. Moore (1984), Maruyama, Liou, and Sasakura (1985), Maruyama and Liou (1988), Barnicoat and Fry (1989), D. E. Moore and Blake (1989), Pognante (1989a), Takasu (1989), Wakabayashi (1990).

32. For example, see Kerrick and Cotton (1971), Ernst (1972a), and Raymond (1973a).

618

33. Ernst (1971b), Raymond (1973a,c), D. E. Moore (1984), Maruyama and Liou (1988), D. E. Moore and Blake (1989).

34. For example, see Taliaferro (1943), Gresens (1969, 1970), Ernst (1971a), and also see Hlabse and Kleppa (1968).

35. Brothers (1954), Bloxam (1959, 1966), Essene, Fyfe, and Turner (1965).

36. See Milton and Eugster (1959).

37. See Suppe (1970, 1973), Raymond (1973a, 1974b).

38. Misch (1959, 1966), E. H. Bailey, Irwin, and Jones (1964), Ernst (1971a), E. H. Brown et al. (1981).

39. For example, see Raymond (1973a,c, 1974b), S. E. Swanson (1981), Abbott and Raymond (1984).

40. Seki (1958), Ernst (1964).

41. Experimental studies plus evidence that Blueschist Facies metamorphism was isochemical (R. G. Coleman and Lee, 1963; Ernst, 1963a; Ghent, 1965) indicate a no answer to question 1. There is simply no evidence available that supports a yes answer to question 2.

42. R. G. Coleman and Lee (1962), Blake (1965), Blake, Irwin, and Coleman (1967, 1969), Blake and Cotton (1969), Brothers (1970), R. G. Coleman (1972), van Bemmelen (1974). Also see de Roever (1967 in van der Kaaden, 1969).

43. Although thrust faulting apparently occurred in some places along the eastern boundary of the Franciscan Complex (Raymond, 1973a,c, Suppe and Foland, 1978; Roure, 1981; Roure and Blanchet, 1983; Jayko, Blake, and Harms, 1987), several studies suggest that the "regional thrust fault" has a complex history and may be a normal or strike-slip fault in many places (Raymond, 1969, 1970, 1973a,c, Worrall, 1981; C. M. Wentworth et al., 1984; Jayko, Blake, and Harms 1987; T. A. Harms and Dunlap, 1991; Unruh et al., 1991).

44. J. C. Maxwell (1974), Bachman (1978), Underwood (1983), Blake and Jayko (1986), Blake et al. (1988).

45. Brothers and Blake (1973).

46. Also see D. S. Cowan (1974), Crawford (1975), and Maruyama, Liou, and Sasakura (1985) whose work reveals no regular metamorphic patterns or upside-down metamorphism *related to the fault.*

47. For example, E. H. Bailey, Irwin, and Jones (1964), Miyashiro (1967), Suppe (1969), van der Kaaden (1969), Brothers and Blake (1973), and van Bemmelen (1974, 1976).

48. Ernst (1970, 1971a, 1984), R. G. Coleman (1972), C. Y. Wang and Shi (1984). Also see references in notes 30 and 77.

49. For example, see Ernst et al. (1970) and Wallis and Banno (1990).

50. Also see the description of the Franciscan Complex in the example at the end of this chapter.

51. Wood (1979a), Draper and Bone (1981), C. Y. Wang and Shi (1984), Cloos and Shreve (1988a), Ernst (1988).

52. Seely, Vail, and Walton (1974), Karig and Sharman (1975).

53. Also see Cloos and Shreve (1988a,b).

54. Also see Schermer (1990) and Schermer, Lux, and Burchfiel (1990).

55. Ernst (1971a) also suggested that strike-slip faulting may be important in exposing blueschists.

56. There are some structural problems with this model relating to sense of shear, thrust geometry, regional structure, and age of melange formation, but a discussion of these is beyond the scope of this text.

57. Van Bemmelen's mantle diapirism (undation) model (1974, 1976) also provides such an explanation.

58. This description of the Franciscan Complex is based on the works of Taliaferro (1943), M. E. Maddock (1955, 1964), R. G. Coleman and Lee (1963), E. H. Bailey, Irwin, and Jones (1964), Ernst (1965, 1970, 1971b,c, 1980), B. M. Page (1966, 1981), Blake, Irwin, and Coleman (1967, 1969), Ernst et al. (1970), Kerrick and Cotton (1971), Berkland et al. (1972), W. G. Gilbert (1973, 1974), Raymond (1973a,c, 1974a,b, 1977), Suppe (1973), D. S. Cowan (1974), J. C. Maxwell (1974), K. E. Crawford (1975, 1976), Suppe and Foland (1978), R. J. McLaughlin et al. (1982), Blake, Howell, and Jayko (1984), R. J. McLaughlin and Ohlin (1984), D. E. Moore (1984), D. E. Moore and Blake (1989), Underwood (1984, 1989), Maruyama, Liou, and Sasakura (1985), Blake and Jayko (1986), Cloos (1986), Jayko, Blake, and Brothers (1986), Liou and Maruyama (1987), Blake et al. (1988), Dumitru (1988, 1991), Maruyama and Liou (1988), Patrick and Day (1989), Wakabayashi (1990), Oh, Liou, and Maruyama (1991), and the many additional works referenced in the notes below.

59. Metawackes are usually referred to as metagraywackes in the literature.

60. These high-pressure rocks drew much attention during the 1950s and 1960s from petrologists and mineralogists at the University of California, Berkeley, and the U.S. Geological Survey (Brothers, 1954; Borg, 1956; Bloxam, 1959, 1960; G. A. Davis and Pabst, 1960; R. G. Coleman and Lee, 1963; Essene et al.,1965; R. G. Coleman and Papike, 1968). Their focus on glaucophane and related schists led many geologists who had not been to California to the mistaken view that the Coast Ranges were dominated by glaucophane schists. In fact, these "high grade" rocks comprise less than 1% of the rocks of the Coast Ranges.

61. E. H. Bailey, Irwin, and Jones (1964), Ernst (1965, 1971a,b,c, 1980), Blake, Irwin, and Coleman (1967, 1969), Blake et al. (1988).

62. E. H. Bailey, Irwin, and Jones (1964), Ernst (1971b), Suppe (1972), O'Day and Kramer (1972), Blake and Jones (1974), Evitt and Pierce (1975), Bishop (1977), R. J. McLaughlin et al. (1982), Underwood (1985), Blake et al. (1988). Whether or not all of the contacts are faults has been a matter of controversy. See Maxwell (1974) and J. C. Maxwell et al. (1981) for alternative interpretations.

63. For studies showing subdivision of the Franciscan Complex, see Hsu (1969), Raymond and Christensen (1971), Raymond (1973a, 1974a), Suppe (1973), D. S. Cowan (1974), Maxwell (1974) and the references therein, K. E. Crawford (1975, 1976), Gucwa (1975), M. A. Jordan (1978), B. M. Page (1981), R. J. McLaughlin et al. (1982), Blake, Howell, and Jayko (1984), R. J. McLaughlin and Ohlin (1984), Underwood (1984), Blake, Jayko, and McLaughlin (1985), Jayko, Blake, and Brothers (1986), and Wakabayashi (1990). Also see Wagner, Bortugno, and McJunkin (1990).

64. Ernst (1965, 1971c), Cotton (1972), Raymond (1973a,c), D. S. Cowan (1974), Crawford (1975). Also see summaries in B. M. Page (1981) and Wagner, Bortugno, and McJunkin (1990).

65. E. H. Bailey, Irwin, and Jones (1964), Hsu (1969), W. G. Gilbert (1973, 1974), Ernst (1980), Page (1981).

66. For example, see Blake, Howell, and Jayko (1984) and R. J. McLaughlin and Ohlin (1984).

67. E. H. Bailey, Irwin, and Jones (1964), R. J. McLaughlin et al. (1982), Underwood (1989).

68. Coastal Belt metawacke assemblages are based on observations by the author and the modes or generalized reports of E. H. Bailey, Irwin, and Jones (1964), O'Day and Kramer (1972), Ernst (1980), R. J. McLaughlin et al. (1982), Blake, Howell, and Jayko (1984), Cloos (1986), Underwood (1989), and Blake et al. (1988). Pumpellyite is reported from this belt by R. J. McLaughlin et al. (1982) and prehnite and pumpellyite are reported in metabasites by Blake et al. (1988), but complete phase assemblages are not reported, making it impossible to assess whether the assemblages are Zeolite Facies or Prehnite-Pumpellyite Facies assemblages (both can contain these phases).

69. Also see Raymond and Christensen (1971), F. W. McDowell et al. (1984), and Blake et al. (1988).

70. Central Belt assemblages are based on Raymond and Christensen (1971), Jayko, Blake, and Brothers (1986), Blake et al. (1988), Underwood (1989), L. A. Raymond (unpublished data).

71. For example, see Cloos (1986) and Blake et al. (1988).

72. Also see Ghent (1965), Blake, Irwin, and Coleman (1967, 1969), B. L. Wood (1971), J. C. Maxwell (1974), Suppe and Foland (1978), and Blake et al. (1988).

73. Also see Maddock (1964), Kerrick and Cotton (1971), D. S. Cowan (1974), Crawford (1975), and D. E. Moore and Liou (1979).

74. Eastern Belt metabasite phase assemblages are listed or discussed in Ernst (1965), Ghent (1965), Seki, Ernst, and Onuki (1969), B. L. Wood (1971), Suppe (1973), E. H. Brown and Bradshaw (1979), Cloos (1982), E. H. Brown and Ghent (1983), Jayko, Blake, and Brothers (1986), and Blake et al. (1988).

75. For descriptions of eclogites, hornblende schists, Greenschist Facies metabasites, and high-grade blueschists, see works such as Switzer (1945, 1951), Gealey (1951), Brothers (1954), Borg (1956), R. G. Coleman and Lee (1963), R. G. Coleman et al. (1965), E. H. Brown and Bradshaw (1979), D. E. Moore (1984), S. S. Sorenson (1986), Liou and Maruyama (1987), Maruyama and Liou (1987, 1988), Bebout and Barton (1989), D. E. Moore and Blake (1989), Wakabayashi (1990), and S. S. Sorenson, Bebout, and Barton (1991).

76. Also see R. G. Coleman and Lee (1962), R. G. Coleman and Clark (1968), R. G. Coleman and Papike (1968), H. P. Taylor and Coleman (1968), E. H. Brown and Bradshaw (1979), Liou and Maruyama (1987), and Maruyama and Liou (1987).

77. Newton and Smith (1967), Ernst (1963, 1971a, 1988), Liou (1971a,b), Nitsch (1972), Bostick (1974), Maresch (1977), Cloos (1983), D. E. Moore (1984), Liou, Maruyama, and Cho (1985), Maruyama, Liou, and Sasakura (1985), Underwood (1985), Liou et al. (1987), Underwood, Blake, and Howell (1987), Blake (1988), Underwood, O'Leary, and Strong (1988), Dumitru (1988, 1991), Maruyama and Liou (1988), D. E. Moore and Blake (1989), Patrick and Day (1989), Wakabayashi (1990), S. S. Sorenson, Bebout, and Barton (1991), Oh, Liou, and Maruyama (1991).

78. M. M. Earle (1980) discusses this problem. Geothermal gradients in orogenic belts are often composites rather than true path histories (P-T-t paths). The practice of referring to them as geothermal gradients should be abandoned, except in cases where the terrane involved is coherent.

79. R. G. Coleman and Lee (1963), Ernst et al. (1970), R. G. Coleman and Lanphere (1971), Kerrick and Cotton (1971), Ernst (1973b), Raymond (1973a,b,c), S. E. Swanson and Schiffman (1979), D. E. Moore (1984), Maruyama, Liou, and Sasakura (1985), Maruyama and Liou (1988), D. E. Moore and Blake (1989).

80. For example, see Kerrick and Cotton (1971), Raymond (1973a,b), D. E. Moore (1984), Maruyama, Liou, and Sasakura (1985), and D. E. Moore and Blake (1989).

PROBLEMS

28.1 In the low-pressure part of the Zeolite Facies, prehnite may appear with laumontite. Using the ACF diagram for the Zeolite Facies in figure 28.6 as a starting point, write a balanced reaction, draw the new topology, and determine the type of reaction (terminal or nonterminal) for the addition of this new phase.

28.2. Using a copy of figure 28.14 as a base, (a) plot the P-T-t path for the New Caledonian blueschist belt using the data in Yokoyama, Brothers, and Black (1986). (b) Compare and contrast this path and the phase assemblages of the rocks with the P-T conditions and phase assemblages of the Sanbagawa Belt (from this chapter). Suggest reasons for the differences.

29

Eclogites

INTRODUCTION

Eclogites are eye-catching rocks composed predominantly of green, omphacitic pyroxene [(Na,Ca,Fe,Mg,Al)Si$_2$O$_6$] and red or red-brown garnets (figure 29.1). Chemically, they are basic rocks (table 29.1). The distinctive character of these rocks—which in spite of their basic chemistry, lack plagioclase—probably led Eskola (1920, 1939) to assign them to a separate facies.

Eskola interpreted the eclogites to be high-pressure rocks. Yet, while experimental work and field relations generally support that interpretation, the question of whether or not the eclogites should be assigned to a separate facies of crustal metamorphism remains open. The opposing views hinge in part on assessment of the significance of the varying mineral assemblages of various eclogites and on the meanings of the relationships that exist between the eclogites and the rocks associated with them. This chapter reviews the nature of eclogites, the conditions of eclogite petrogenesis, and the question of the existence of an Eclogite Facies.

OCCURRENCES AND MINERALOGY

The term eclogite, as used here, refers to plagioclase-free rocks that are *dominated* by Na-bearing, Ca-Mg clinopyroxene (omphacite) and Mg-Fe (low-Ca) garnet. The two essential minerals together should comprise more than 67% of the primary phase assemblage and garnet alone typically makes up more than 30% of the rock . Varieties of eclogite may be distinguished on the basis of chemistry, textures, or the presence of accessory minerals (e.g., kyanite eclogite, quartz eclogite), but these accessory minerals do not dominate the mode (cf. R. G. Coleman et al., 1965; Banno, 1970; I. D. MacGregor and Carter, 1970). Note that *not all* rocks containing abundant clinopyroxene and garnet are eclogites, especially those rich in carbonate minerals or quartz, which are more appropriately called marble, skarn, or tactite.[1]

Eclogites are found in the crust in four different associations. These include occurrences as

1. Xenoliths in basalts and kimberlites,
2. Exotic blocks in serpentinite- and shale-matrix melanges and fault breccias,
3. Interlayers and lenses in high P/T blueschist-bearing terranes,
4. Interlayers and lenses in medium to high T/P terranes characterized by Granulite Facies, migmatitic, gneissic, and ultramafic rocks.

In the first two associations, the eclogites are exotic or xenolithic; that is, they are transported and petrogenetically distinct from the matrix in which they occur. In the latter two associations, the eclogites are

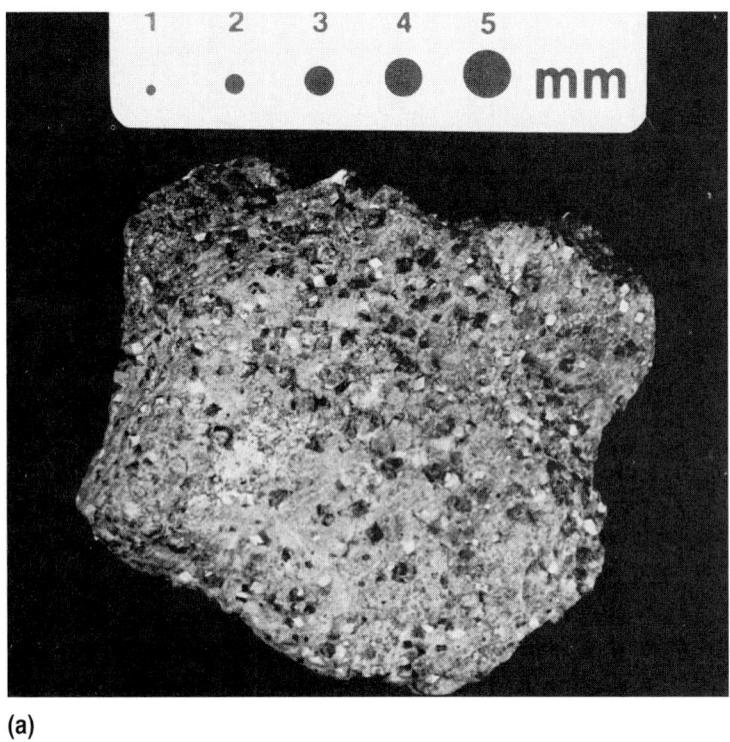

(a)

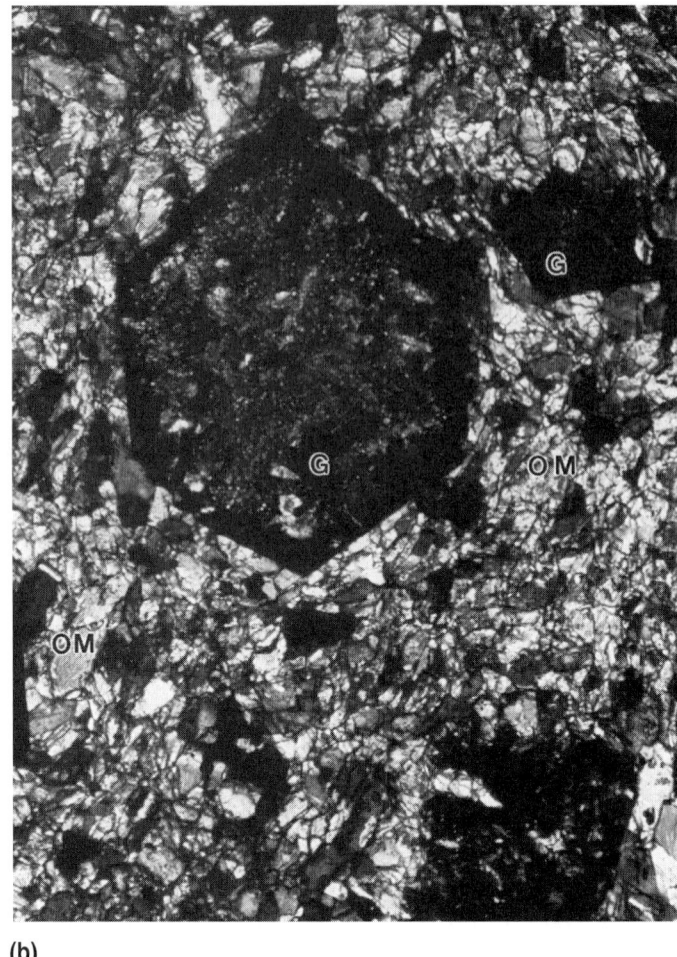

(b)

Figure 29.1   Photographs of eclogite. (a) Handspecimen, Franciscan Complex, Sonoma County, California. Note euhedral garnets in matrix of pyroxene. (b) Photomicrograph of specimen shown in (a).(XN). G = garnet, OM = omphacite. Long dimension of photo is 1.27 mm.

closely associated with, and in most cases petrogenetically related to, the enclosing rocks. Because some petrologists consider eclogite to be a major rock type in the upper mantle,[2] eclogites of occurrence 1, as well as some of those of occurrences 2, 3, and 4, may represent mantle rocks.

The four types of eclogite occurrence have been assigned by R. G. Coleman et al. (1965) to three major groups (Groups A, B, and C). Group A eclogites include the xenolithic types, whereas Group B eclogites occur in medium to high T/P terranes.[3] Associations 2 and 3 are combined as Group C eclogites. R. G. Coleman et al. (1965) use garnet chemistries in conjunction with *mode of occurrence* to divide eclogites into the three groups. These chemistries apparently form a continuum from low to high pyrope (Mg) content and do not exclusively distinguish rocks of the various groups. Nevertheless, these group names are used in some contemporary literature.

Xenolithic eclogites are rare rocks. Examples are reported from Hawaii,[5] the Colorado Plateau, the Colorado-Wyoming Front Ranges, Michigan, Japan, India, Australia, the former Soviet Union, South Africa, Botswana, and Angola.[6] These rocks typically occur in kimberlites, but are also reported from alkali olivine basalts. Mineralogically, xenolithic eclogites are composed of omphacitic to diopsidic pyroxene and Mg-rich garnet.[7] Additional minerals are listed in table 29.2. The garnets typically contain >55% pyrope component, 10–20% Ca-garnet component (grossularite-andradite), and 25–35% Fe-Mn-Al–garnet component (almandine-spessartite). In a few cases, the pyrope contents are lower (Hills and Haggerty, 1989). Omphacites contain from 1–37% jadeite component, a compositional range duplicated by pyroxenes from eclogites of high T/P terranes.

Table 29.1 Chemical Analyses and Modes of Selected Eclogites

	1	2	3	4	5
SiO_2	44.57[a]	44.6	45.21	47.2	51.51
TiO_2	1.76	1.8	1.43	2.1	1.20
Al_2O_3	13.61	16.3	14.90	15.9	13.92
Fe_2O_3	4.17	5.9	4.78	5.2	9.75[b]
FeO	8.49	8.8	10.15	4.5	—
MnO	0.21	0.26	0.19	0.15	0.15
MgO	13.34	5.9	10.91	6.1	8.36
CaO	11.42	11.7	7.98	7.3	11.23
Na_2O	1.69	3.1	2.83	3.5	3.26
K_2O	0.02	0.22	0.02	1.6	0.09
P_2O_5	0.02	0.12	0.24	0.43	0.23
Other	0.57	1.77	—	6.38	0.27
Total	99.87	100.35	98.63	100.36	99.98

Modes

	1	2	3	4	5
Cpx	61.2	42.6	57.0[d]	34.4[c]	30[c]
Gt	36.5	28.9	39.0	28.9	25
Ol	0.4	—	—	—	—
Qtz	—	—	—	8.2	trace
Ky	—	—	—	—	1
Sp	0.5	—	—	—	—
Rt	—	1.0	0.6	2.0	4
Sph	—	3.2	—	1.5	—
Ep	—	1.4	—	3.9	3
Lw	—	6.6	—	—	—
Il	0.3	—	?1.3	—	—
Mt	0.6	—	—	—	—
P-B	trace	—	0.5	—	1
Wm	—	2.2	—	0.7	—
Chl	—	14.1	—	0.5	—
CaA	—	—	?0.5	—	1
NaA	—	—	—	19.9	—
Pl	—	—	X	—	10
Sul	0.5	—	—	—	—
Other	—	—	—	—	25[e]

Sources:

1. Garnet pyroxenite (eclogite xenolith), sample 68SAL-11, Salt Lake Crater, Hawaii (Beeson and Jackson, 1970).
2. Eclogite 100-RGC-58, Tiburon, California (from melange) (R. G. Coleman et al., 1965).
3. Eclogite associated with gneisses, sample 82-42, Sunnmore, Norway (Jamtveit, 1987).
4. Eclogite from glaucophane-bearing schist terrane, New Caledonia, sample 36-NC-62 (R. G. Coleman et al., 1965).
5. Eclogite from garnet lherzolite, sample F-53, Alpe Arami, Switzerland (Ernst, 1977b).

[a] Values in weight percent

[b] Total iron as Fe_2O_3.

[c] Modal pyroxene and garnet do not always total 67+% because of secondary alteration.

[d] Pyroxene and amphibole include some plagioclase in symplectic intergrowths.

[e] Pyroxene-garnet symplectite.

X = plagioclase as symplectite with pyroxene.

Cpx = clinopyroxene, Gt = garnet, Ol = olivine, Qtz = quartz, Ky = kyanite, Sp = spinel, Rt = rutile, Sph = sphene, Ep = epidote-clinozoisite, Lw = lawsonite, Il = ilmenite, Mt = magnetite, P-B = phlogopite-biotite, Wm = white mica, Chl = chlorite, CaA and NaA = Ca-and Na-amphiboles, Pl = plagioclase, Su = sulfides.

Table 29.2 Minerals of Eclogites

Xenolithic Eclogites	Exotic Block Eclogites	Blueschist-related Eclogites	Granulite-related Eclogites
Primary			
omphacite	omphacite	omphacite	omphacite
garnet ($Py_{>30}$)	garnet ($Py_{<50}$)	garnet ($Py_{<30}$)	garnet (Py_{5-60})
kyanite	glaucophane	quartz	quartz
corundum	barroisite	kyanite	coesite
rutile	lawsonite	talc	kyanite
diamond	epidote	white mica	talc
graphite	quartz	phlogopite	white mica
amphibole	white mica	epidote	phlogopite/biotite
phlogopite	rutile	lawsonite	Ca-amphibole
lawsonite	sphene	barroisite	orthopyroxene
	apatite	zoisite	zoisite
	pyrite	rutile	rutile
			sphene
Secondary			
serpentine	glaucophane	glaucophane	hornblende
phlogopite	actinolite	actinolite	actinolite
amphiboles	hornblende	hornblende	augite
ilmenite	lawsonite	epidote	plagioclase
plagioclase	epidote	lawsonite	biotite
spinel	jadeitic pyroxene	jadeitic pyroxene	sphene
analcite	plagioclase	plagioclase	white mica
calcite	pumpellyite	chlorite	epidote
chlorite	chlorite	quartz	diopside
goethite	quartz	calcite	ilmenite
pyrite	aragonite	aragonite	
pyrrhotite	calcite	rutile	
chalcopyrite	rutile	sphene	
	sphene		
	white mica		
	montmorillonite		

Sources: See notes 6, 8, 9, 12, and 14.

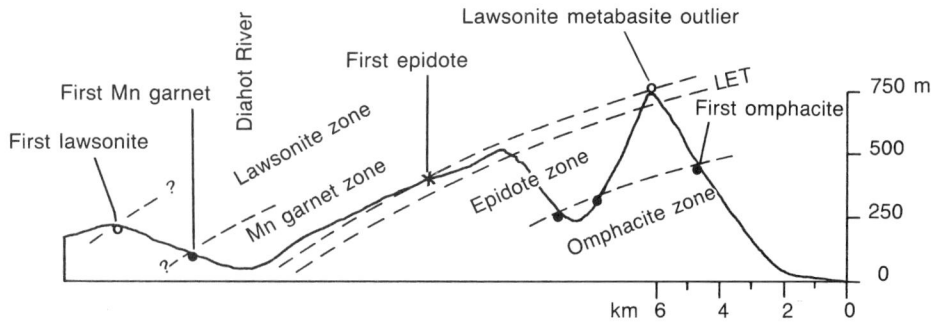

Figure 29.2 Cross section across part of northern New Caledonia showing metamorphic zones, including Omphacite Zone (Eclogite Facies) located structurally well below the Lawsonite Zone of the (high-pressure) Blueschist Facies. LET = lawsonite-bearing part of the Epidote Zone. *(From Yokoyama, Brothers, and Black, 1986)*

Serpentinite- and pelitic-matrix melanges and breccias exposed along fault zones in high P/T belts, locally contain exotic, tectonic blocks[8] of eclogite engulfed by matrix materials. R. G. Coleman et al. (1965) assigned eclogite blocks of this type to Group C. As required by the definition of eclogite, these rocks are composed of omphacitic pyroxene and garnet.[9] Garnets are high Fe-Al varieties (almandine) and are low in Mg (pyrope-component). The pyroxenes typically are relatively sodium-rich (Jd_{30-40}). Additional phases are listed in table 29.2.

Perhaps the best-known examples of exotic-tectonic eclogites occur in the Franciscan Complex of western California and southwestern Oregon.[10] Here, in association with blocks of many other rock types—including glaucophane schist, metabasalt ("greenstones"), metachert, metaconglomerate and metabreccia, limestone, ophiolite fragments, serpentinites, and voluminous metasandstones—the eclogites form mesoscopic-scale masses in pelitic- and serpentinite-matrix melanges. The blocks of eclogite are typically one to a few meters in diameter. Commonly they are veined and the primary mineralogy is partially replaced by more hydrous assemblages of the Greenschist and Blueschist facies. Similar eclogites have been reported from Japan and the southern Ural Mountains of Russia.[11]

Eclogites interlayered and closely associated with blueschists, ophiolites, and related rocks are reported from Greece, western France, the Alps, Asia, Venezuela, and New Caledonia (figure 29.2).[12] In size, the eclogite bodies range from thin lenses to macroscopic sheets. These rocks were also assigned to Group C by R. G. Coleman et al. (1965). Mineralogically, they contain primary omphacite and garnet ($Py_{<30}$), plus quartz, kyanite, rutile, zoisite, white mica, talc, and/or lawsonite.[13]

Eclogite interlayers and lenses also occur in medium to high T/P metamorphic terranes. Like the lenses and layers in glaucophane schists and related rocks, these occurrences range in scale from small mesoscopic masses to macroscopic sheets of a kilometer or more in length (W. L. Griffin, 1987; Jamtveit, 1987). Associated rocks include granitic gneisses, migmatites, marbles, metaquartz arenites, metamorphosed ultramafic rocks, anorthosites, and amphibole schists. Notable occurrences have been described in northern Europe from Sweden, Scotland, Poland, Germany, the former Soviet Union, and especially Norway, as well as in France and the Swiss and Austrian Alps and in Tasmania, Africa, Japan, and Newfoundland (Canada).[14]

The high T/P eclogites were assigned by R. G. Coleman et al. (1965) to Group B. Yet the garnets from some of these occurrences show a wide range of chemistry and plot in both Group A and Group B (Mysen and Heier, 1972). Pyrope contents range from about Py_5 to Py_{60}. The primary modes of the eclogites reveal, in addition to omphacite and garnet, such minerals as kyanite, talc, quartz, and orthopyroxene (see table 29.2).[15]

Because eclogites are composed primarily of *anhydrous* phases stable at medium to high temperatures and high pressures (see page 627), in cases where they are transported to higher levels of the crust and encounter hydrous, lower-grade environments, they tend to alter readily to *hydrous* phase assemblages. Hydrous phases commonly produced by such reactions include white micas (muscovite and paragonite), brown to black micas (phlogopite and biotite), sodic amphiboles (e.g., glaucophane and crossite), calcic amphiboles (hornblendes, including barroisite, and tremolite and actinolite), epidote-clinozoisite-zoisite, lawsonite, pumpellyite, and chlorite. Nonhydrous phases produced by these reactions may include aragonite, calcite, plagioclase, calcic pyroxenes, sphene, rutile, ilmenite, and magnetite. The particular phase assemblage produced by any of the retrograde or prograde reactions that yield these phases depends on the bulk chemistry of the eclogite, the nature of the fluid phase, and the P-T conditions present at the time of the reaction.

DO ECLOGITES REPRESENT A SEPARATE FACIES?

The Eclogite Facies is one of the facies originally defined by Eskola (1920). Yet, the existence of the Eclogite Facies as a *crustal* metamorphic facies (rathar than a mantle facies) is debatable. Conditions of metamorphism in the mantle are generally beyond the scope of this text and the bulk chemistries of mantle rocks are largely restricted to basic and ultrabasic compositions.

Arguments Against the Existence of a Crustal Eclogite Facies

R. G. Coleman et al. (1965) suggested that the designation of a separate Eclogite Facies be abandoned. They argued that:

1. The interlayering of eclogites with glaucophane schists, Amphibolite Facies rocks, and Granulite Facies rocks indicates that eclogites form over a range of P-T conditions that overlap the conditions associated with other facies (eclogites, in part, may simply represent anhydrous equivalents of hydrous metabasic rocks, such as hornblende- or glaucophane-schist);[16]
2. Partitioning of Ca, Mg, and Fe between pyroxene and garnets varies significantly between eclogites of different associations, indicating various P-T conditions of formation; and
3. The eclogite facies cannot be mapped as a facies in regionally metamorphosed terranes, as can other facies.

An additional argument that may be raised is that:

4. The eclogite facies, as defined and recognized traditionally, involves *only one bulk composition*, basic rocks, and therefore does not represent the set of all mineral assemblages, representing all bulk compositions, indicative of a particular and restricted set of P-T conditions. Thus, its character is not consistent with the definition of a metamorphic facies.

The suggestion of R. G. Coleman et al. (1965) to abandon the use of a separate Eclogite Facies has not been universally accepted. Nevertheless, some petrologists have supported it. For example, Winkler (1979) implies that the use of an Eclogite Facies is unjustified and he simply describes eclogitic rocks and their origins without assigning them to a facies.

Arguments in Favor of the Existence of a Crustal Eclogite Facies

A number of petrologists have argued in favor of retaining the Eclogite Facies. Fyfe et al. (1958) argued strongly that the eclogites should be recognized as a facies because of the distinctive mineral assemblages of these rocks. Hyndman (1985, p. 537) and Yardley (1989) continue to use the Eclogite Facies, as does F. J. Turner (1981, p. 410), who explicitly rejects the suggestion of R. G. Coleman et al. (1965) to abandon it. Miyashiro (1973a, p. 310ff.) adopts the Eclogite Facies, but specifies that quartz- and kyanite-eclogites are petrographic indicators of the facies; that is, he restricts the definition of the facies. E. H. Brown and Bradshaw (1979) argue that eclogites represent distinct phase topologies and metamorphic conditions and that the concept of the Eclogite Facies is a valid one.

Arguments in favor of retaining an Eclogite Facies include the following.

1. The phase assemblages of eclogites are so distinctive and different from those of other metabasites that they must represent a unique facies (Fyfe et al., 1958, p. 235; F. J. Turner, 1981, p. 410).
2. Although rare, rocks representing *a range of bulk chemistries* and exhibiting some unique mineral assemblages that develop at conditions of $P \geq 1.0$ Gpa(10 kb) and $T > 200°$ C do exist and should be used to define an Eclogite Facies.
3. Rare regional terranes composed of or including eclogites do exist (e.g., Yokoyama, Brothers, and Black, 1986; X. Wang and Liou, 1991).

To evaluate the direct conflicts indicated by these pro and con arguments, it is necessary to address two questions. (1) Are there distinctive phase assemblages present in crustal rocks, which can be represented by phase topologies that *uniquely* define a high-pressure facies? (2) Are all eclogites representative of similar conditions of metamorphism?

Natural High-Pressure Phase Assemblages and Associated Phase Topologies

Natural occurrences, chemical data, and experimental data can be focussed on the solution to the first of these two questions. First, eclogites associated with both glaucophane schists and rocks of medium to high T/P terranes appear as concordant masses. As such, they are commonly assumed to form under similar P-T conditions, generally in excess of 0.8 Gpa (8 kb) and 200° C (refer to the petrogenetic grids in chapter 28 and appendix C). Although these P-T limits do not precisely constrain the petrogenetic conditions for eclogite formation, they do eliminate all low-pressure conditions. The presence of kyanite as the stable aluminum silicate phase in eclogites provides a similar constraint. Rare coesite in eclogites (X. Wang, Liou, and Mao, 1989) is an unequivocal indicator of high pressures of formation. From another perspective, an indication that pressures of formation are substantially higher than 0.3–0.4 Gpa (3–4 kb) is provided by occurrences such as that in New Caledonia, where eclogites and associated rocks form a metamorphic ("omphacite") zone that lies structurally *below* zones exhibiting typical, high-pressure phase assemblages of the Blueschist Facies (figure 29.2) (Yokoyama, Brothers, and Black, 1986; Ghent et al., 1987). Similarly, in eastern China, coesite-bearing eclogites are apparently part of a regional metamorphic terrane that is flanked by lower-grade rocks (X. Wang and Liou, 1991).

Specific experimental studies of systems relevant to the formation of eclogites provide better constraints on the P-T conditions under which eclogites form. The absence of plagioclase in rocks of basalt-gabbro composition provides an important constraint. The disappearance of plagioclase is governed by reactions[17] such as

$$forsterite + anorthite = garnet (pyrope), \quad (29.1)$$

$$enstatite + anorthite = garnet (pyrope) + diopside + quartz, \quad (29.2)$$

$$anorthite = garnet (grossularite) + kyanite + quartz, and \quad (29.3)$$

$$albite = jadeite + quartz. \quad (29.4)$$

Experimental determinations of reactions 29.3 and 29.4 suggest limits for the conditions of formation of eclogite (at realistic temperatures) to pressures above about 0.5 or 0.6 Gpa (figure 29.3a).

A number of workers have evaluated the stability relations of eclogites and plagioclase-bearing basic rocks.[18] Experimental studies on basaltic compositions are synthesized in figure 29.3b. Under wet conditions in the crust and upper mantle, amphibole-free eclogite is stable only above about 1.8–2.7 Gpa, depending on the temperature and composition of the basalt. Amphibole-bearing eclogite is stable above 0.5–1.8 Gpa. Under dry conditions (above the wet basalt solidus), eclogite is only stable above about 650° C at pressures of 1.0–1.8 Gpa. The lower limit of amphibole-eclogite stability is well within the limits of crustal metamorphism, but the amphibole-free eclogite stability field occurs at pressures realized in the mantle, at mantle depths in subduction zones, or in the basal regions of thick crust like that beneath high mountain ranges.

Geothermometry and geobarometry document the high-pressure, low- to high-temperature conditions of formation of eclogites (Banno, 1970; Raheim and Green, 1974; Newton, 1986).[19] Pressure and temperature estimates for petrogenesis of Group B and C eclogites have been summarized by Newton (1986). Metamorphic temperatures ranged from 450° C to 750° C, whereas metamorphic pressures were between 1.2 and 2.0 Gpa. These values encompass those recently determined for specific localities (Schliestedt, 1986; Austrheim, 1987; Ghent et al., 1987), are a bit higher than older analyses suggested, and are quite compatible with the experimental results cited above.

ACF phase topologies for rocks metamorphosed under the wet and dry conditions necessary to yield eclogites are shown in figure 29.4. Recall that the bulk rock chemistry of basalts plots in the lower right third of the diagram. By comparing these topologies to ACF topologies for the Blueschist, Amphibolite, and Granulite facies, it is clear that a unique topology does exist for "dry" eclogites, one in which there is no tie line between Mg-garnet and plagioclase, epidote group minerals, or lawsonite. The diagram for "wet" eclogites does have such a tie line, making some of the phase assemblages of this topology like those of the Greenschist and Amphibolite facies. The latter, however, are characterized by plagioclase—albite in the Greenschist Facies and more calcic plagioclases in the Amphibolite Facies. The CFM diagram cannot be used to depict assemblages for eclogites, as the saturating phase plagioclase is absent from all eclogite assemblages, and the remaining saturating phases quartz, H_2O, and magnetite are commonly absent as well.

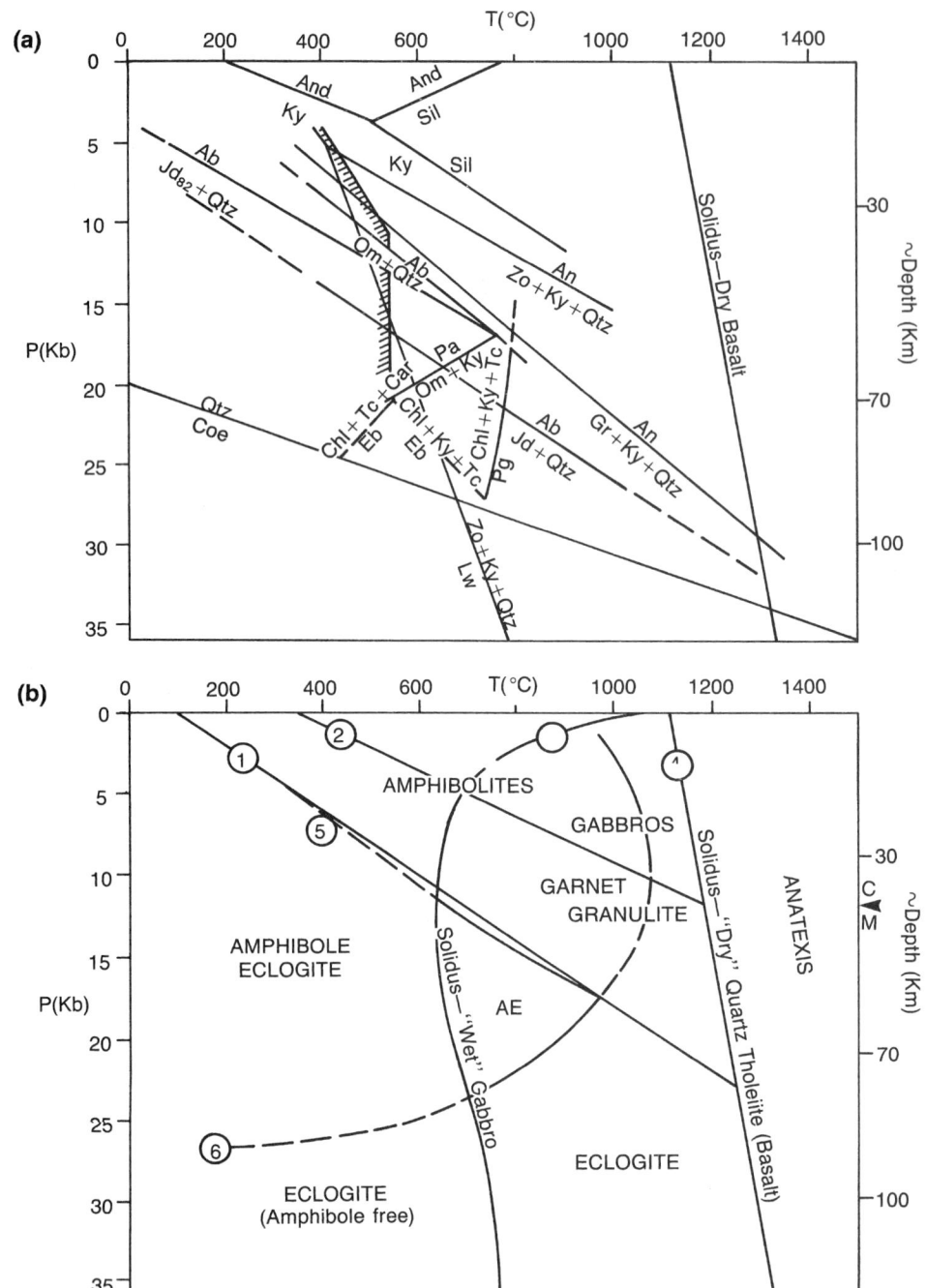

Figure 29.3 Eclogite stability. (a) Petrogenetic grid showing some reaction curves important to the stability of eclogites and related rocks.

And (andalusite), Ky (Kyanite), and Sil (sillimanite) curves from Holdaway (1971). An = Zo + Ky + Q+z (quartz) and Zo + Ky + Qtz = Lw (lawsonite) curves from Newton and Kennedy (1963). An = Gr + Ky + Qtz curve from Hariya and Kennedy (1968). Ab (albite) = Om (omphacite) + Qtz and Pa (paragonite) = Om + Ky curves (calculated) from Newton (1986). Ab = Jd$_{82}$ (jadeitic pyroxene, with 82% jadeite) + Qtz curve from Newton and Smith (1967). Ab = Jd + Qtz curve from Birch and LeCompte (1960) and Holland (1980).

Qtz = Coe (coesite) curve after Boyd and England (1960) and Ostrovsky (1966, 1967). Chl + Tc (talc) + Car (Mg-carpholite) = Eb (ellenbergite), Chl + Tc + Ky = Eb, and Chl + Ky + Tc = Py (pyrope)

curves are from Chopin (1986). The solidus for dry basalt is from Green and Ringwood (1967a).

(b) P-T grid showing experimentally determined melting and stability relations for rocks of basaltic composition.

Curves 1 and 2, bounding the upper and lower limits of garnet granulite stability are from Green and Ringwood (1972). Curve 3 is from I. B. Lambert and Wyllie (1972). Curve 4 is from D. H. Green and Ringwood (1967a) and Ito and Kennedy (1971). Curve 5 is from Essene, Henson, and Green (1970) and represents the lower limit of stability of amphibole-bearing eclogite (AE). Curve 6, which is a compound curve, represents the upper stability limit of amphibole in eclogite at pressures above 17 kb (from Essene, Henson, and Green 1970) and the upper limit of stability of hornblende in rocks of basaltic composition at pressures below 10 kb (from Holloway and Burnham, 1972).

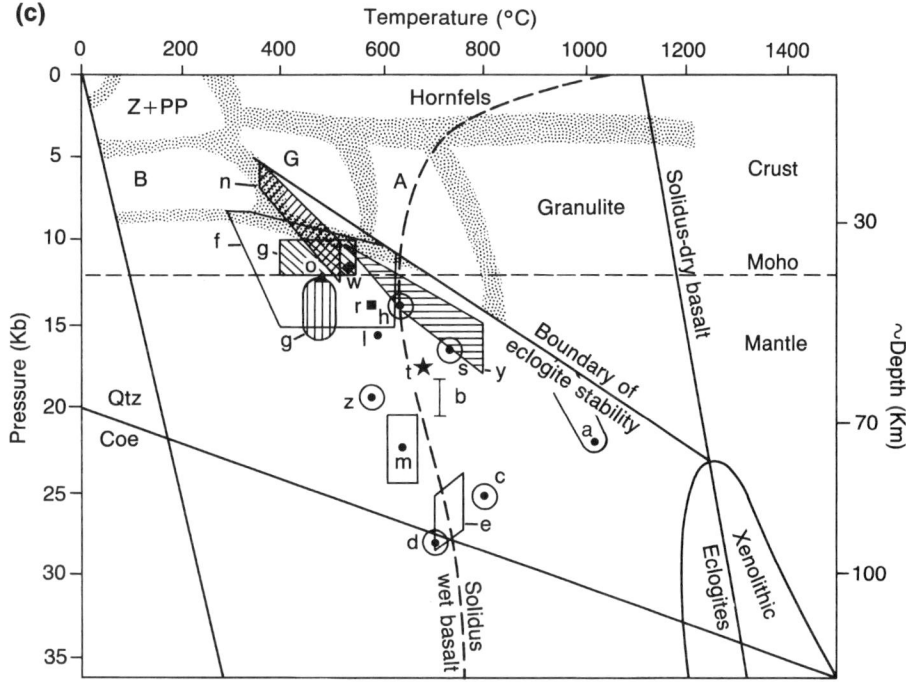

Figure 29.3 Continued

(c) P-T grid showing the crystallization fields of various eclogitic rocks superimposed on selected curves from figures 29.3a and b and various metamorphic facies fields.

Z + Pp = Zeolite + Prehnite-Pumpellyite Facies. G = Greenschist Facies. B = Blueschist Facies A = Amphibolite Facies. Stability fields are as follows: a = eastern Australian margin (Pearson et al., 1991); b = Bergen Arcs, Norway (Jamtveit et al., 1990); c = Cima di Gagnone, Ticino, Switzerland (Evans et al., 1979); d = Dora-Maira Nappe, Alps (Pognante, 1991); e = coesite and ellenbergite rocks of the western Alps (several sources, including Chopin, 1986); f = Franciscan Complex (see text and Oh et al., 1991); g = Sifnos

Island, Greece (Matthews and Schleistedt, 1984; also see Schleistedt, 1986); h = Hareidland, Norway (Mysen and Heier, 1972); l = Alpine Sesia-Lanzo Zone (Castelli, 1991); m = Munchberg, Germany (Klemd, 1989); n = Newfoundland (Jamieson, 1990, and others); o = Alpine ophiolite eclogites (Pognante and Kienast, 1987); q = New Caledonia (Yokoyama et al., 1986); r = Rhodope Zone, northern Greece (Liati and Mposkos, 1990); s = Glenelg, Scotland (Sanders, 1989); t(star) = Troms, Norway (Krogh et al., 1990); w = Ward Creek Type IV (Oh et al., 1991); y = Norwegian eclogites (Griffin, 1985, 1987; Jamtveit et al., 1990); z = Zermatt-Saas Zone, Alps (Pognante, 1991). Refer to text for additional references.

Do other bulk compositions metamorphosed under conditions like those that yield eclogites occur in the crust and, if so, do they reflect unique phase topologies? The answer to these questions, which in the past seemed to receive a resounding "no," seems now to be a qualified "yes." A number of interesting occurrences of distinctive rocks are worth noting here. These include "whiteschists," coesite-bearing rocks, tactites, gneisses, garnet lherzolites, and sodic pyroxene + paragonite-bearing rocks. Some of these occurrences reveal unique phase assemblages. Others exhibit assemblages that are mineralogically indistinguishable from those of normal Blueschist, Amphibolite, or Granulite facies rocks, but geothermometry and geobarometry yield metamorphic P-T conditions equivalent to those of the mantle.

Whiteschists have been described by Schreyer (1973, 1977) and Munz (1990). These rocks, characterized by the assemblage kyanite + talc, are rare, but are found in Brazil, Tasmania, central Africa, Europe, and Afganistan. The bulk chemistry of the whiteschists is unusual and may represent that of evaporitic mudstones. The significance of these rocks to this discussion is that the assemblage kyanite + talc is stable between 600° C and 850° C at pressures above 1.0 Gpa. At lower and higher temperatures, the pressures of stability are higher. The presence of a hydrous fluid phase may lower the limit of stability somewhat to about 0.7–0.8 Gpa. Thus, whiteschists represent phase assemblages of nonbasic rocks that reflect P-T conditions like those under which many eclogites are formed.

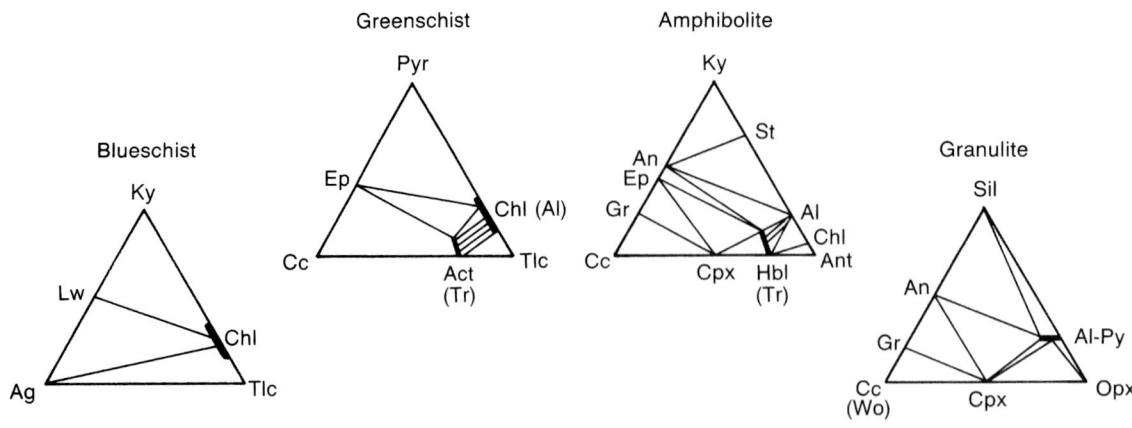

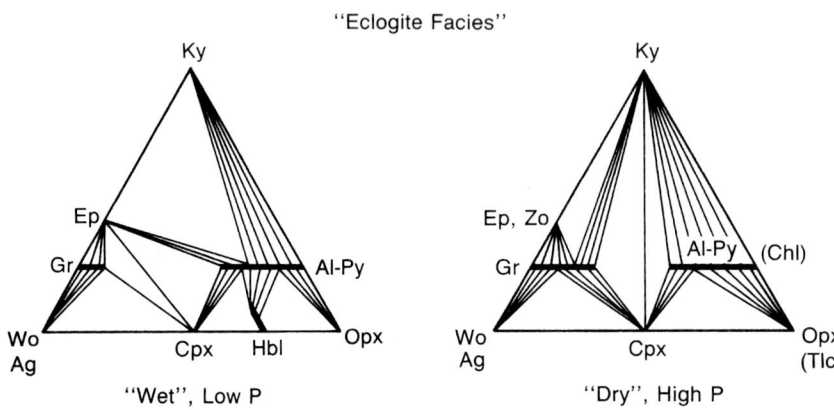

Figure 29.4 ACF topologies for "wet" and "dry" eclogites compared to typical topologies for Blueschist, Greenschist, Amphibolite, and Granulite facies rocks.
Act = actinolite, Ag = aragonite, Al = almandine, An = anorthite, ant = antigorite, Cc = calcite, Chl = chlorite, Cpx = clinopyroxene, Ep = epidote, Gr = grossularite, Hble = hornblende, Ky = kyanite, Lw = lawsonite, Opx = orthopyroxene, Py = pyrope, Pyr = pyrophyllite, Sil = sillmanite, St = staurolite, Tlc = talc, Tr = tremolite, Wo = wollastonite, Zo = zoisite

A pyrope-quartzite exposed in the western Alps near Parigi, Italy, contains both coesite (the high-pressure silica polymorph) and the recently discovered Mg-Al-Ti–silicate, ellenbergite (also stable only at high pressures)(Chopin, 1984, 1986). The assemblages pyrope (Py_{90-98}) + coesite + talc and kyanite + talc + chlorite + ellenbergite + rutile + zircon indicate conditions of metmorphism of about 2.5 Gpa and 700° C.

In the Lepontine Alps of Switzerland, lenses of eclogite occur with garnet-bearing lherzolite (Ernst, 1977b; Evans et al., 1979). Garnet lherzolite (an ultramafic rock) is stable under conditions like those of eclogite stability (>1.0 Gpa at $T \geq 500°$ C) (see chapter 31). The phase assemblage for the ultramafic rock is distinctive and reflects a unique topology. P-T conditions indicated for the eclogites are reported to be 800–1000° C and 2.5–5.0 Gpa (Ernst, 1977; Evans et al., 1979).

Carbonate rocks, SAC rocks, and pelitic schists were metamorphosed with eclogites in the Sesia-Lanzo Zone of the western Alps (Koons et al., 1987; Pognante, 1989, 1991; Castelli, 1991). Here, pelitic schists contain phases such as quartz, white mica, lawsonite, jadeitic pyroxene, and glaucophane, and SAC rocks are characterized by the high-pressure assemblage jadeitic pyroxene–zoisite–quartz–garnet–white mica. The carbonate rocks contain phases such as aragonite, dolomite, quartz, garnet, white mica, clinopyroxene, zoisite, and sphene. P-T conditions for the eclogite-forming event are estimated to have been 575° C and 1.5 Gpa (Castelli, 1991).

Calcite + garnet + clinopyroxene assemblages have been described by Swainbank and Forbes (1975) and E. H. Brown and Forbes (1986) from rocks of the Fairbanks, Alaska area. These rocks are calcite-bearing and/or quartz-rich and differ chemically from normal eclogites. They are more appropriately referred to as tactites or skarns. Associated with the tactites are garnet-glaucophane-barroisite-plagioclase schists and granoblastites (garnet-amphibolites), calcite granoblastites and schists, garnet-quartz-mica schists, and metaquartz arenites. None of the phase assemblages in these rocks seems to differ from those found in normal medium- to high-temperature crustal metamorphic rocks. Geothermometry and geobarometry on the various rocks, however, indicate P-T conditions of about 600° C and 1.5 Gpa (E. H. Brown and Forbes, 1986).

Eclogite of Hareidland, Norway, is associated with kyanite-bearing, dioritic gneiss and quartz-orthopyroxene-amphibole-plagioclase gneiss (Mysen and Heier, 1972). Mineralogically, these rocks are normal Amphibolite to Granulite Facies rocks, but thermobarometry reveals metamorphic P-T conditions of 625° C and 1.4 Gpa.

High-pressure, meta-SAC and metabasic rocks occur together in New Caledonia (Brothers and Blake, 1973; Yokoyama, Brothers, and Black, 1986; Ghent et al., 1987). Eclogites are associated with the more siliceous rocks, which exhibit phase assemblages such as quartz + omphacite + almandine + paragonite + muscovite + glaucophane + clinozoisite + rutile + sphene and quartz + omphacite + orthoclase + muscovite + ferroglaucophane + stilpnomelane. Except for the occurrence of pyroxenes more calcic than those of the Blueschist Facies and the presence of garnet in the assemblage, these rocks are mineralogically like those of the Blueschist Facies. These new phases, however, change the topologies. P-T conditions appear to have been about 1.0–1.2 Gpa and 400–550° C (Yokoyama, Brothers, and Black, 1986; Maruyama, Cho, and Liou, 1986; Ghent et al., 1987).

Like New Caledonia, Sifnos Island, Greece, contains eclogitic rocks associated with rocks having a range of bulk compositions (Matthews and Schliestedt, 1984; Schliestedt, 1986). These rocks include quartz-epidote-glaucophane schists, omphacite-actinolite-glaucophane schists, garnet-paragonite-glaucophane-quartz-jadeite gneisses, dolomite-calcite marble granoblastite, chloritoid metaquartz arenite, epidote-muscovite (phengite)-quartz-glaucophane-garnet schist, and a number of other rock types. The P-T conditions indicated by the associations are 470° C and 1.5 Gpa.

Coesite-bearing eclogites occur in eastern China (X. Wang, Liou, and Mao, 1989; X. Wang and Liou, 1991). The eclogites are present as interlayers in white mica schists, biotite gneiss, and marble, all of which also have quartz pseudomorphs after coesite. Marbles are composed of the high-pressure assemblage calcite–garnet–epidote–clinopyroxene–white mica–quartz (coesite)–rutile. Biotite gneiss contains assemblages such as biotite–muscovite–garnet–epidote–K-feldspar–plagioclase–quartz (coesite)–rutile–sphene. Mica and pyroxene compositions from these rocks yield P-T conditions compatible with the 600° C and >2.0 Gpa estimated for the eclogites.

These descriptions indicate that *locally* in the crust there are rocks of a range of bulk compositions that were metamorphosed under high-grade conditions different from those of the normal crust (see figure 29.3c). The phase assemblages of some of these rocks suggest unique topologies that represent these higher-grade conditions. Based on these observations and on theoretical projections, phase diagrams for a proposed Eclogite Facies may be constructed. Figure 29.5 presents schematic ACF, AKN, and AFM diagrams for such a facies. Reaction curves shown in the petrogenetic grid (see appendix C) constrain the probable P-T conditions for such a facies to conditions of $P \geq 1.0$ Gpa and $T = 300–1200°$ C.

The formation of eclogites is not confined to the conditions of such an Eclogite Facies. Apparently many eclogites do form under these conditions in both anhydrous and H_2O-bearing environments (E. H. Brown and Bradshaw, 1979; Newton, 1986), but others reportedly occur in close association with Amphibolite, Greenschist, Blueschist, and Granulite Facies rocks. Eclogites found in the latter associations apparently owe their existence to different bulk compositions and/or to very low partial pressures of water (D. H. Green and Ringwood, 1967a; Fry and Fyfe, 1969; B. A. Morgan, 1970; De Wit and Strong, 1975). The latter eclogites represent the facies with which they are associated, rather than the Eclogite Facies.

In summary, there are some unique phase assemblages, representing *some* bulk compositions of crustal rocks, that formed under higher-grade conditions than did the phase assemblages of the typical crust. The existence of these rocks argues for the existence of an Eclogite Facies. Such a facies represents conditions indicative of mantle conditions. In crustal rocks, those "mantle conditions" are only developed in the deepest levels of thickened crust or in subduction zones, where sialic crust is transported to depths *in excess of* 35–40 km.

In spite of the above points, it may be argued that, given that the conditions necessary for the formation of most eclogites and similarly metamorphosed rocks *exceed* those typical of normal crust, a *separate* Eclogite Facies of *crustal* metamorphism does not exist. Clearly, in some rocks associated with eclogites, the phase assemblages do not differ from those of the Greenschist or Amphibolite facies (see figure 29.3c). In addition, rocks of bulk compositions other than metabasite compositions, metamorphosed under the conditions of the proposed Eclogite Facies, are extremely rare and may not form an adequate foundation for a separate facies.

The arguments for and against the designation of an Eclogite Facies remain unresolved and adoption of the Eclogite Facies as a facies of crustal metamorphism is still controversial. Here, the Eclogite Facies is recognized as a crustal metamorphic facies.

631

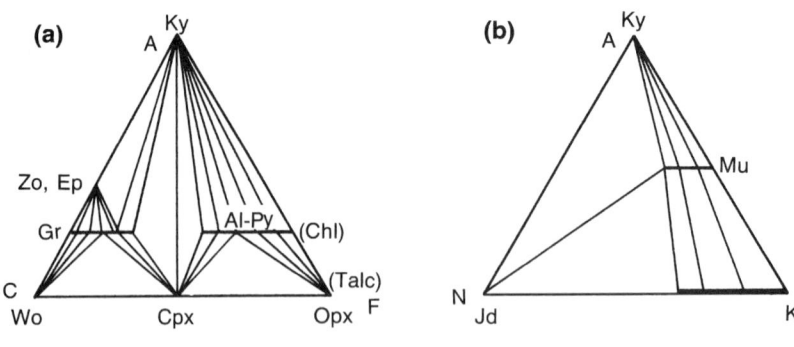

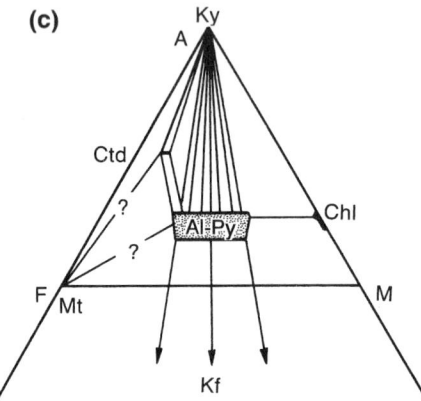

Figure 29.5 Schematic diagrams for the proposed Eclogite Facies: (a) ACF, (b) AKN, (c) AFM. Mu = muscovite, Kf = K-feldspar, Jd = jadeitic pyroxene, Ctd = chloritoid, Mt = magnetite. Other symbols as in figure 29.4.

EXAMPLES OF ECLOGITE OCCURRENCES

Eclogite of the East Pond Metamorphic Suite, Newfoundland

In northwest Newfoundland, a thick sequence of metamorphosed Eocambrian- to Ordovician-aged rocks, the Fleur de Lys Supergroup, overlies a Neoproterozoic to Cambrian metasedimentary sequence, that, in turn, overlies a Proterozoic basement composed of gneiss (figure 29.6).[20] Both the basement and the overlying metasediments contain metabasites and both are assigned to the East Pond Metamorphic suite. The metabasites apparently represent metamorphosed tholeiitic basalt dikes. Where they occur in the metasediments, they are now metamorphosed to amphibole schists, amphibole diablastites (?), and similar rocks, collectively referred to as amphibolites. Where they occur in the basement gneisses, they typically form boudins of amphibolite-eclogite rock that nowhere exceed 10 m in

diameter. Eclogite is not known to be in contact with the gneisses, being confined instead to the cores of the boudins, with a layer of amphibolite separating eclogite and gneiss.

The eclogites are of typical composition, with euhedral red to pink garnets in a granoblastic, fine-grained matrix of omphacite. Although this occurrence is of the medium to high T/P type, the garnet is 18–19% pyrope, and therefore plots in the Group C field of R. G. Coleman et al. (1965). The pyroxene has about 35% jadeite component. Additional phases include hornblende, quartz, white mica, zoisite, Mg-calcite, and rutile. Secondary minerals include diopside, hornblende, biotite, plagioclase, epidote, chlorite, calcite, apatite, and sphene. Hornblende, plagioclase, and epidote locally form wormy intergrowths (symplectite). Kelphytic rims of plagioclase, hornblende, and other minerals rim the garnets in altered eclogites.

The surrounding amphibolites are hornblende-plagioclase rocks with minor amounts of such minerals as epidote, quartz, biotite, sphene, and garnet. Some are schistose, whereas others are nonfoliated. DeWit and Strong (1975) report that phase

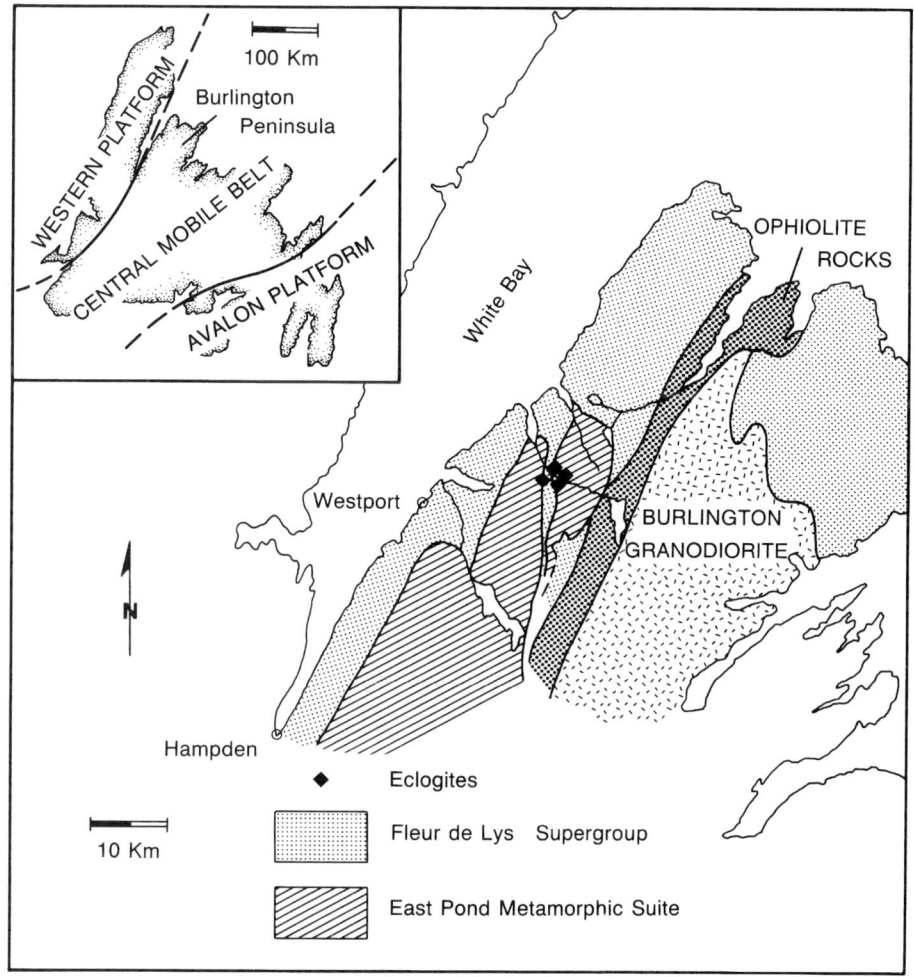

Figure 29.6 Regional geologic map showing terrane in which eclogite-bearing amphibolites occur in East Pond Metamorphic Suite of Newfoundland. Diamonds show locations of eclogites.
(Simplified from Jamieson, 1990)

relations indicate a temperature and pressure of formation of about 350° C and 0.5–0.7 Gpa, conditions that lie within the upper Greenschist Facies. Yet the phase assemblages of the amphibolites suggest amphibolite facies conditions. Reanalysis of the data using the plagioclase-hornblende geothermobarometer of Plyusnina (1982), in fact, does indicate lower Amphibolite Facies P-T conditions of about 450–500° C and 0.8–0.9 Gpa (figure 29.3c). For the eclogite, similar temperatures (T = 350–450° C), but higher pressures (P = 0.8–1.2 Gpa) were inferred. Jamieson (1990), using several geothermometers, obtained a temperature in the range of T = 450–500° C for the eclogites, but amphibolite overprint temperatures of 600–750° C. Her geobarometric analysis produced a minimum pressure of 1.0–1.2 Gpa for the eclogite and 0.7–0.9 Gpa for the overprint in amphibole-rich rocks.

The origin of the Newfoundland eclogites is thought by DeWit and Strong (1975) to have occurred through metamorphism of basalts under "dry" conditions. The amphibolites are considered to be chemically equivalent, hydrous rocks, an assertion supported by major element chemistry and all of the minor element abundances, except Sr, for which analyses are available. Sr tends to be slightly lower in eclogites than in amphibolites. These chemical data and the petrographic data support the hypothesis that the eclogites and amphibolites were metamorphosed under similar conditions, except for the partial pressure of water, which was lower in those rocks that became eclogites. Jamieson (1990), however, points out that the eclogites occur in "wet" country rocks composed of metapelites and metasandstones, and further, that the eclogites contain hydrous phases such as zoisite and white mica. Consequently, she argues that while water was present and facilitated reaction progress, it did not serve as a thermodynamic agent in the metamorphic process.

Eclogites of the Franciscan Complex, California and Oregon

Franciscan eclogites occur as widely distributed blocks in melanges, fault zones, and serpentinites (figure 29.7).[21] They occur in association with blocks of glaucophane schist; glaucophane-metabasalt and -metagabbro; zeolitic-, pumpellyitic- lawsonitic-, and jadeitic-metabasalt ("greenstones"); chert and metachert; conglomerate, metaconglomerate, breccia and metabreccia; siliceous metavolcanic rocks; limestone and marble; ophiolite fragments; serpentinites; various schists; and voluminous metasandstones—all of which also form tectonic blocks. Commonly, the matrix materials originally surrounding the eclogites (and other rocks) have been eroded away and the eclogite masses rest on soil or other surficial deposits. The ages of the eclogite blocks and associated high-grade glaucophane and amphibole schists range from about 135 to 165 m.y.b.p. (late Jurassic), an age that is older than that of the enclosing Franciscan rocks.[22] No known source terrane exists for the eclogites, rendering their petrogenesis particularly enigmatic.

The eclogites appear to be of two types, those associated or interlayered with glaucophane schist and gneiss and those associated or interlayered with barroisitic amphibole schist and gneiss (Oh, Liou, and Maruyama, 1987, 1991). The former are more abundant and are characterized by primary assemblages such as

omphacite + garnet + rutile,
omphacite + garnet + epidote + rutile,
omphacite + garnet + white mica + rutile, and
omphacite + garnet + glaucophane + rutile

(R. G. Coleman et al., 1965).[23] The latter are characterized by assemblages such as

omphacite + garnet + epidote, and
omphacite + garnet + barroisitic hornblende

(Oh, Liou, and Maruyama, 1987, 1991). Geothermometry and geobarometry indicate that assemblages of the first type formed at temperatures between 290° C and 350° C and pressures between 0.8 Gpa and 0.9 Gpa (Oh, Liou, and Maruyama, 1991). P-T conditions under which the second type (Type IV tectonic eclogites)[24] formed were 400° C<T<715° C and P>0.7–1.3 Gpa (see figure 29.3c)

(D. E. Moore and Blake, 1989; Wakabayashi, 1990; Oh, Liou, and Maruyama, 1991). As is typical of eclogites, veins and rinds of retrograde minerals, as well as bands and patches of these minerals, have rendered many of the eclogites heterogeneous and their histories complex (figure 29.8) (cf. D. E. Moore, 1984; D. E. Moore and Blake, 1989; Wakabayashi, 1990).

Currently accepted tectonic models suggest that oceanic crust was subducted in western California during the late Jurassic Period.[25] Metamorphism of basaltic oceanic crust, under the conditions of high pressure and moderate temperature expected in a subduction zone environment, yields eclogite and amphibole eclogite (note that the stability field of Franciscan eclogites in figure 29.3c straddles the upper stability limit of glaucophane defined by Maresch (1977) and would thus include glaucophane-bearing assemblages at low T and hornblende-bearing assemblages at higher T). The Franciscan eclogites probably formed in such an environment and were then transported to the surface in thrust slices of serpentinite, in serpentinite diapirs, and in obducted, diapirically emplaced, subduction channel, or uplifted melanges. Either metamorphism during transport or resubduction with accompanying metamorphism may account for the various retrograde and prograde assemblages present in these rock masses.

PETROGENESIS OF ECLOGITES

Although the two examples described above provide some perspective on the origins of eclogites, a number of processes have been proposed to explain the origin of specific eclogite occurrences. These include (1) crustal or subcrustal metamorphism of basalt or gabbro under drier conditions than those that normally yield amphibole-rich rocks such as glaucophane or hornblende schist; (2) high-pressure metamorphism of a basaltic or gabbroic rock crystallized from the fusion-generated magma derived from a parental upper-mantle peridotite; (3) contact metasomatism of basalt at moderate P-T conditions; (4) direct (primary) crystallization of eclogite from alkaline magmas in the mantle, with or without subsequent recrystallization under lower grade conditions; (5) crystallization of pyroxenites from alkaline magmas at mantle depths, followed by exsolution of garnet at lower P-T conditions; and (6) fractional crystallization of alkaline magmas in the mantle, at the base of the crust or within the crust, to produce eclogitic crystal cumulates.[26] Some eclogites are polymetamorphic rocks with early histories that are at least partially obscure, making the nature of the protolith

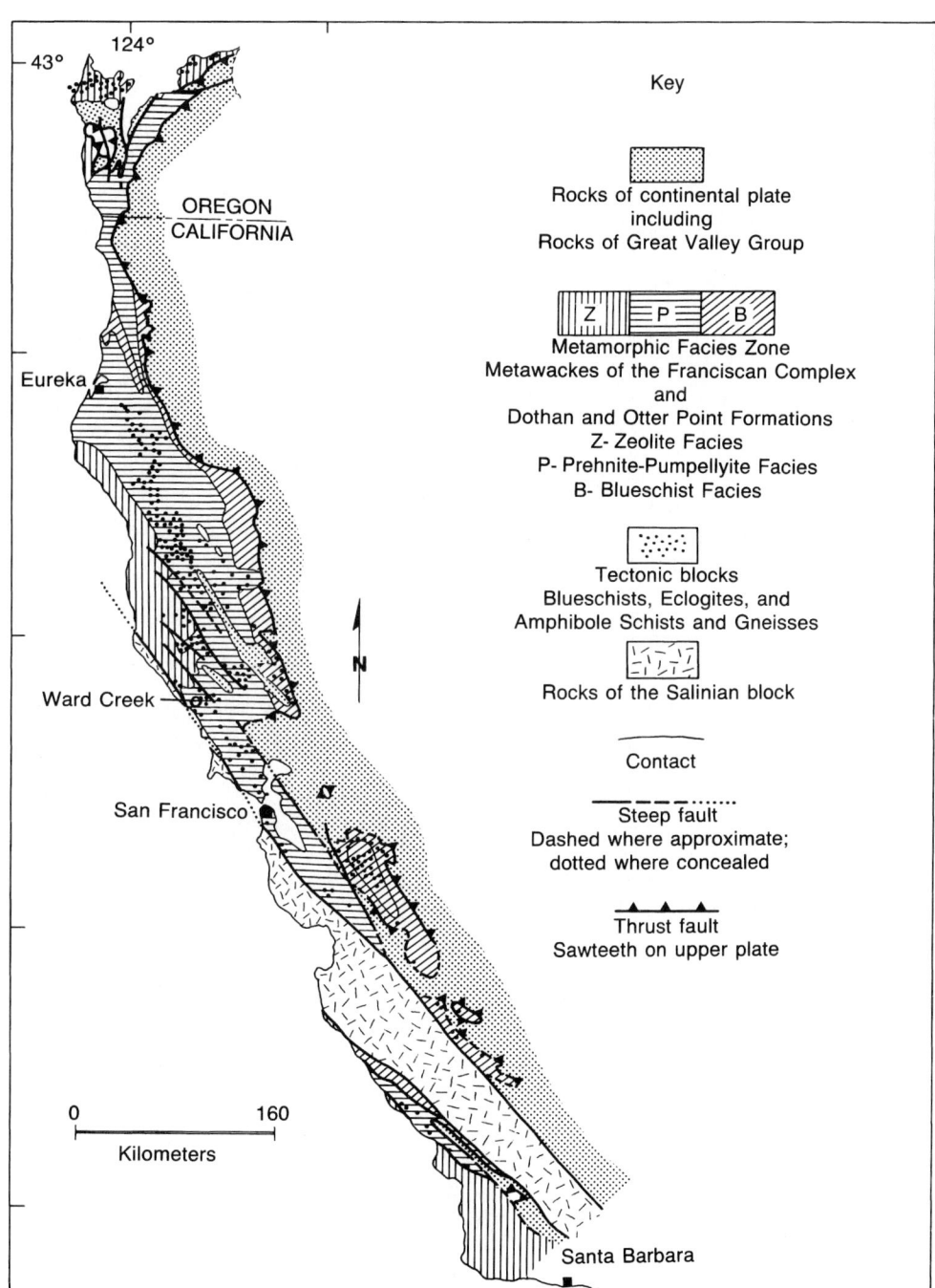

Figure 29.7 Generalized map of western California and southwestern Oregon showing the locations of the tectonic blocks of eclogite and associated glaucophane and amphibole schist and gneiss with respect to the regional belts of Blueschist Facies (B), Prehnite-pumpellyite Facies (P), and Zeolite Facies (Z) rocks.

[Modified from R. G. Coleman and Lanphere, 1971; based on data from E. H. Bailey, Irwin and Jones (1964), Blake, Irwin, and Coleman (1969), Raymond (1973c), Ernst (1980), and others]

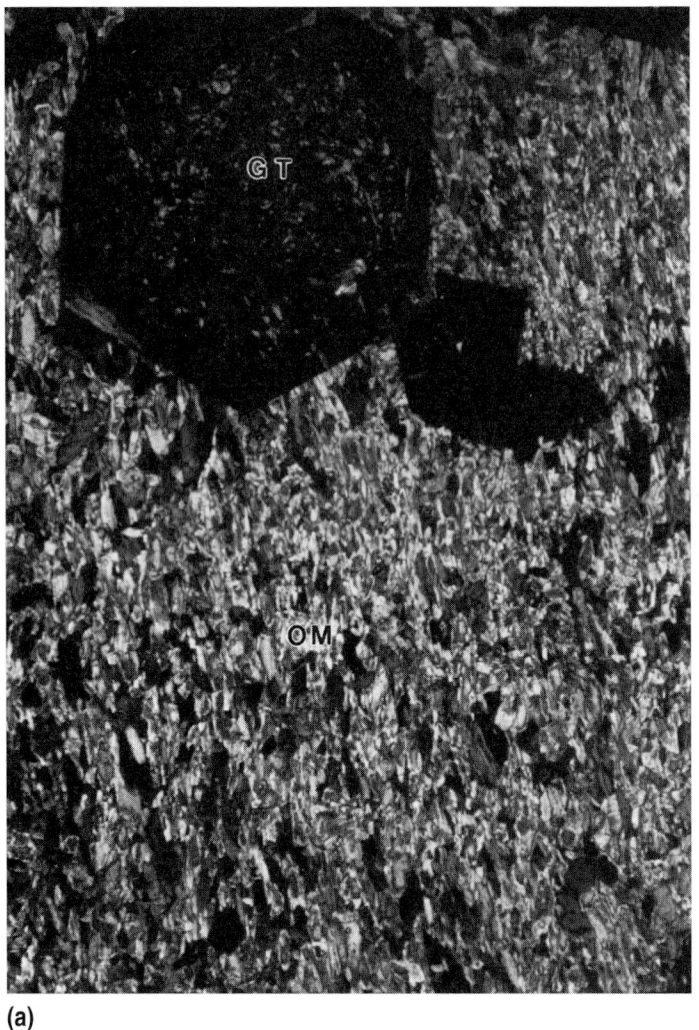

(a)

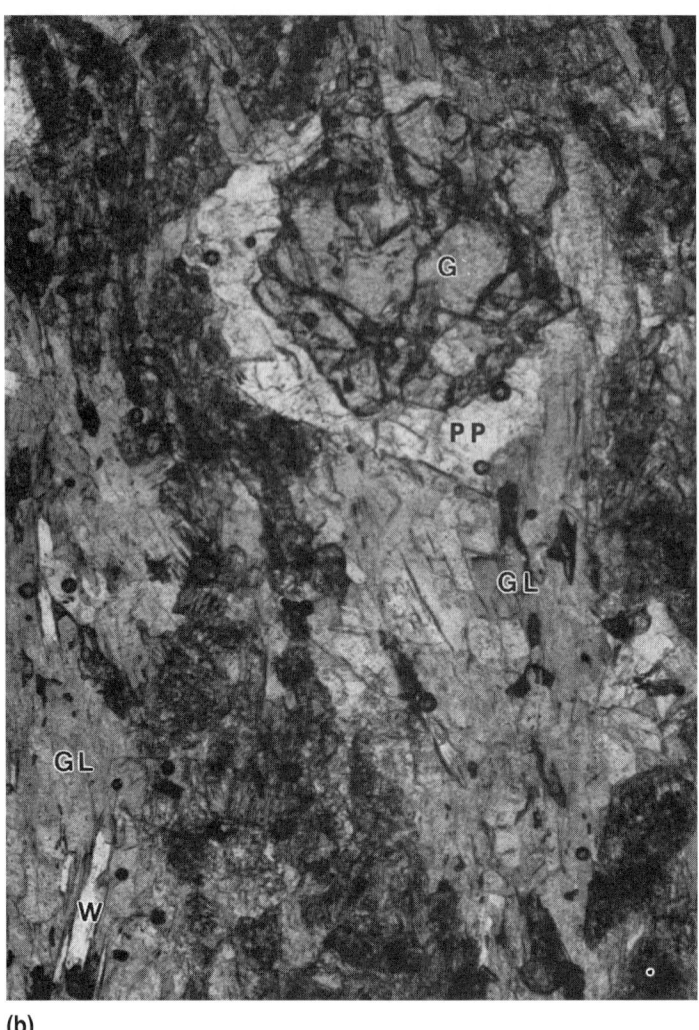

(b)

··················

Figure 29.8   Photomicrographs of eclogites from the Franciscan Complex. (a) Relatively unaltered eclogite, Tiburon, California.(XN). (b) Glaucophane-bearing eclogite, Tiburon, California.(PL) (c) Veined and altered eclogite with pumpellyite (radiating fibrous mineral) in calcite-prehnite-pumpellyite veins, Narrows melange, near Hospital Creek, northeastern Diablo Range, California. (PL)

(C = calcite, CH=chlorite, E=epidote, G=garnet, GL=glaucophane, J=jadeitic pyroxene, L=lawsonite, OM=omphacite, PP=pumpellyite, PR=prehnite, W=white mica. Long dimension of all photos is 1.27 mm.)

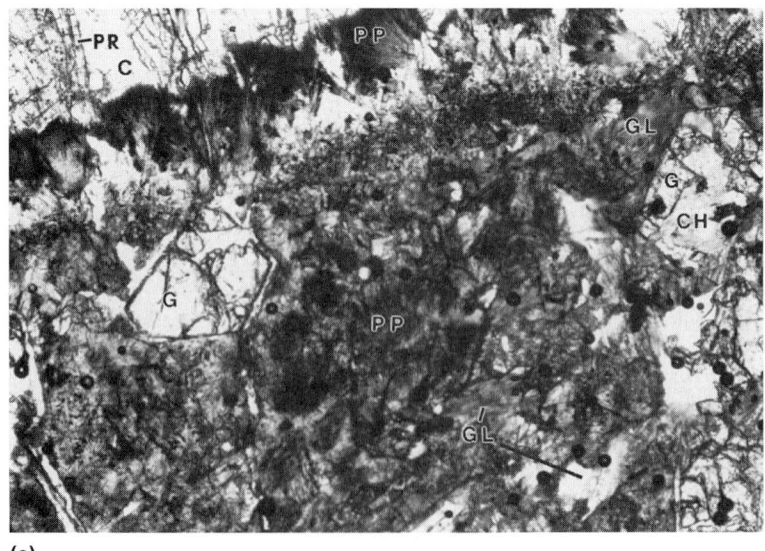

(c)

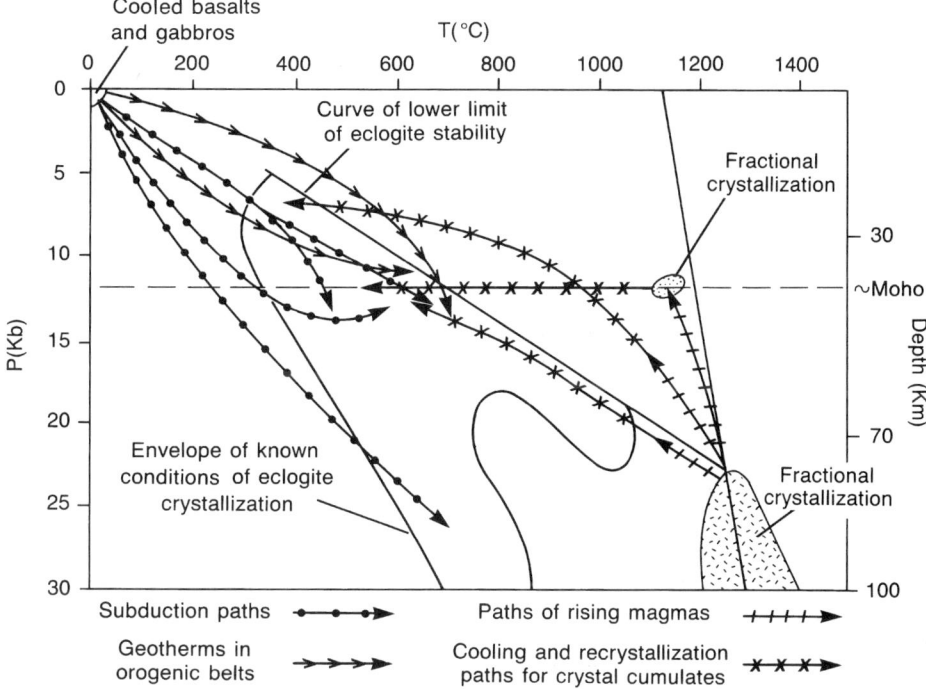

Figure 29.9
P-T grid showing the multiple paths by which rocks may reach eclogite compositions.

problematical (Jamtveit, Bucher-Nurminen, and Austrheim, 1990; Wakabyashi, 1990). Model 3 has been rejected, because mineral stability studies do not support the hypothesis. In models 1 and 2, which are very similar, the eclogites are derived from basaltic or gabbroic rocks that represent crystallized partial fusion products (the liquid separate) derived from mantle ultramafic rocks. In hypotheses 4, 5, and 6, the eclogites are derived from the products of mantle crystallization (the crystal separate) of all or fractions of alkaline magmas. Except for hypotheses 1, 3, and 6, as well as in some versions of those models, emplacement of the eclogites into the crust involves some kind of transport, either via magmatic transport (for xenoliths) or tectonic transport in thrust plates, diapirs, or tectonic melanges.

Considering the range of P-T conditions under which eclogites have formed, it is likely that different eclogites formed via different processes under different conditions. Which of the proposed petrogenetic hypotheses are valid is the question. Examination of figure 29.3c suggests that eclogites form along two general, curved geotherms. One extends down at 8 to 13° C/km through the New Caledonian, Sifnos, and Franciscan eclogite fields to the site of Alpine eclogite and coesite-pyrope generation (f, a, g to c, d, e of figure 29.3c). Such a geotherm is typical of subduction zones. The other geotherm has a slope of about 13–17° C/km and ex-

tends through the Newfoundland and Norwegian eclogite fields and on into the field of xenolithic eclogites. Such geotherms are representative of normal crust, some orogenic belts, and perhaps "hot" subduction zones. These observations, together with chemical and field data reported in the various sources cited in the notes, suggest the following.

Eclogite protoliths are of two kinds—those derived from fusion products and those derived from crystal cumulates. Eclogites are produced from these protoliths by *either* postcooling subduction, which yields high pressures, or postcrystallization cooling (± decompression), which yields lower temperatures (and pressures) than the crystallization temperatures (figure 29.9). Major and trace element chemistry and thermobarometric analyses (Menzies, 1983; F. A. Frey, 1980; Herzberg, 1978b) are consistent with a model of *xenolithic eclogite petrogenesis* in which the eclogites represent a crystal residuum derived from crystal fractionation of alkaline magmas (hypothesis 4 above). Fractionation occurs at mantle depths ($P > 2.0$ Gpa) and at temperatures in the range of 1200–1450° C. Although some re-equilibration and exsolution may have occurred at lower P-T conditions, those events likely occurred within the field of eclogite stability along the high-temperature geotherm (figure 29.9).

Some *layered to lenticular eclogites*, interlayered with ultramafic rocks, amphibole gneisses and schists, and anorthosites, may have originated as crystal cumulates in magma chambers (Griffin, 1987) in the mantle, the crust, or at or near the base of the crust. Ponding of basalt at or near the base of the crust is commonly suggested in models of rhyolite genesis and ocean crust formation. Fractional crystallization of the ponded mafic magmas (at T = 1100–1200° C, P = 0.1–1.2 Gpa), followed by metamorphism in the eclogite field (at T = 400–800° C, P = 1.0–1.5 Gpa, $P_{H_2O} < P_{total}$) will yield eclogites. Magmas crystallized at the base of the crust simply cool isobarically until they cross the eclogite stability boundary. Those crystallized at shallow depths (e.g., in the oceanic crust) are transported to depth via subduction.

Eclogites and related rocks formed from *crystallized fusion products*, that is, from typical basalts and from some gabbros, may occur in three associations. Some may form from protoliths crystallized from melts trapped in their parental, mantle ultramafic rocks (Beeson and Jackson, 1970; Dickey, Obata, and Suen, 1977), whereas others form from mafic dikes intruded into crustal country rock (DeWit and Strong, 1975; Griffin, 1987). Many, however, may form from magmas erupted and crystallized at the surface and later transported along the low-temperature geotherm by subduction, until they attain the pressures and temperatures necessary for the formation of eclogite.

Eclogite formation may have been isochemical in some cases, but the basaltic compositions revealed by analyses of eclogites commonly indicate losses of K, Sr, or other mobile elements, or other changes in chemistry.[27] In route along any of the postcrystallization paths, mafic crystal cumulates or crystallized fusion products may experience alteration; dynamothermal, contact, or other isochemical metamorphic changes; or metasomatism, leading to bulk compositions that differ from those of typical basalts. These changes in chemistry do not, however, appear to be necessary for the formation of eclogite.

eclogite as a result of increased pressure realized along subduction zones characterized by a low-temperature geotherm. Other eclogites form where mafic rocks, crystallized at depth, re-equilibrate to conditions between 300° C at 0.5 Gpa and 800° C at 1.8 Gpa, at $P_{H_2O} < P_{total}$. These conditions are, in part, typical of the crustal Greenschist and Amphibolite facies and, in part, represent mantle-like conditions that occur in both upper mantle rocks and in thickened or subducted crust.

The conditions under which eclogites form are clearly variable. This fact plus the observations that eclogites (1) represent a single kind of protolith (mafic igneous rock), (2) occur in some cases in close association with rocks of the Blueschist, Greenschist, and Amphibolite facies, (3) form over a wide range of conditions, and (4) cannot be mapped as regional terranes in most cases (as can other facies) have led some petrologists to suggest that the Eclogite Facies is not a legitimate facies of crustal metamorphism. Some eclogites seem to have formed under low fluid-pressure conditions at pressures and temperatures essentially identical to those that characterize the Greenschist and Amphibolite facies.

A contrasting view is founded on other data. In New Caledonia, Norway, and China, regional eclogite-bearing terranes are recognized. Geothermometry and geobarometry indicate that these and other eclogites crystallized under conditions typical of the mantle ($P > 1.2$ Gpa and $T > 400°$ C), but also characteristic of thickened crust. Rare, very high pressure metamorphic rocks of nonbasic character are known and are locally associated with eclogites. Together, the eclogites and these rocks may represent a range of protoliths formed under the conditions of an Eclogite Facies developed in thickened crust or in subduction zones. This evidence argues for the need to recognize the Eclogite Facies as a legitimate facies of high-pressure metamorphism. Because the legitimacy of the Eclogite Facies remains controversial, there is a clear need for further study of these interesting and unusual rocks.

SUMMARY

Eclogites (*sensu stricto*) are rocks composed of essential Mg-Fe-rich garnet and omphacitic pyroxene. Although they have basic protoliths, eclogites lack primary plagioclase feldspar. The absence of plagioclase, the presence of accessory minerals such as kyanite and coesite, and an association with rocks such as talc-kyanite schist, garnet lherzolite, and barroisitic hornblende-bearing gneiss indicate that eclogites are high-pressure rocks.

Eclogites apparently form from two types of mafic protolith—crystal cumulates derived from alkaline magmas and crystallized fusion products generated by partial melting of ultramafic rocks within the mantle. Some rocks develop into

EXPLANATORY NOTES

1. Fyfe et al. (1958, p. 236) note correctly that the presence of red garnet and green pyroxene in a rock does not make it an eclogite. They point out that skarns, ultramafic rocks, and granoblastites of the Granulite Facies may also contain that particular association of minerals. Nevertheless, some workers have called calcite-garnet-pyroxene rocks and other garnet-pyroxene-bearing rocks eclogites. For example, Swainbank and Forbes (1975) designate as eclogite some calcite-garnet-pyroxene and garnet-pyroxene-quartz rocks that are interlayered with marble and amphibole schist near Fairbanks, Alaska. None of the rocks for which modes are presented contains more than 62% garnet + pyroxene and all contain significant amounts of calcite or abundant quartz.

Chemically, as Swainbank and Forbes (1975) point out, the rocks differ from normal eclogites. Thus, regardless of the P-T conditions under which such rocks formed (E. H. Brown and Forbes, 1986), they are mineralogically and chemically different from eclogites and should not be called eclogites. A. J. R. White (1964) used pyroxene chemistry to distinguish between eclogites and basic "granulites."

2. For example, see Kuno (1969), Wyllie (1971a, ch. 6), and Kennedy and Ito (1972). I. D. MacGregor (1970) and Dawson (1981) provide brief reviews and Wyllie (1971, ch. 3, 5, 6) provides an extensive review that discusses eclogite as a mantle rock.

3. R. G. Coleman et al. (1965). See note 6 for references to xenolithic eclogites and note 14 for references to eclogite occurrences in high T/P terranes.

4. Mysen and Heier (1972) found that garnets from a high T/P Norwegian eclogite plot over a chemical range that spans the three groups. DeWit and Strong (1975) found eclogites associated with amphibolites and gneisses in a medium to high T/P terrane to contain garnets that plot in Group C, rather than Group B. J. Ferguson and Sheraton (1979) discovered that some xenolithic garnets plot in Group B rather than in Group A.

5. Hawaiian xenoliths composed primarily of clinopyroxene and garnet were called eclogite by Kuno (1969) and Shimizu (1975), but garnet pyroxenite by Green (1966), Beeson and Jackson (1970), and F. A. Frey (1980). D. H. Green (1966) suggested that the garnet in these rocks was exsolved from the pyroxene during cooling and that the exsolution occurred under conditions outside of the stability field of eclogite. Consequently, he suggested that the rocks are not eclogites *sensu stricto*. These rocks are here called eclogites for three reasons. First, the phase assemblage and chemistry are like those of other eclogites. Because the rocks have an eclogitic composition, in my view, they should be called eclogites and presumed genesis should not be used to disqualify them as eclogites. Second, at least some of these rocks may, in fact, contain primary garnet (see the data of Beeson and Jackson, 1970, and F. A. Frey, 1980). Third, the conditions of initial crystallization of clinopyroxenes that exsolved garnet may have been of higher P than supposed by D. H. Green (1966; see Herzberg, 1978) and subsequent exsolution may well have occurred *within* the stability field of eclogite. This case emphasizes the need to name rocks based on their textures and mineralogy or chemistry, leaving presumed genesis out of the naming process. See F. A. Frey (1980) for additional references to works on the much-studied Hawaiian xenoliths.

6. Xenolithic eclogites are discussed by Yoder and Tilley (1962), R. G. Coleman et al. (1965), M. J. O'Hara and Mercy (1966), Kuno (1969), Lovering and White (1969), I. D. MacGregor and Carter (1970), Sobolev (1970), Helmstaedt et al. (1972), Harte and Gurney (1975), Helmstaedt and Doig (1975), Lappin and Dawson (1975), M. E. McCallum, Eggler, and Burns (1975), M. J. O'Hara, Saunders, and Mercy (1975), Akella et al. (1979), J. Ferguson and Sheraton (1979), Boyd and Danchin (1980), Dawson (1980), McGee and Hearn (1984), D. N. Robinson, Gurney, and Shee (1984), Smyth,

McCormick, and Caporuscio (1984), Smyth and Caporuscio (1984), Schultze (1987), Hills and Haggerty (1989), Smyth, Caporuscio, and McCormick (1989), Pearson, O'Reilly, and Griffin (1991), and others. Also see I. D. MacGregor (1968, 1970) and Dawson (1981).

7. The minerals and chemistry cited in this paragraph are based on a compilation from the works cited in note 6. See particularly R. G. Coleman et al. (1965).

8. Berkland et al. (1972) define exotic and tectonic blocks. A *tectonic block* is a mass of rock that has been transported with respect to adjacent masses of rock through the operation of tectonic processes. An exotic block is a mass of rock occurring in a lithologic association foreign to that in which the mass formed. Also see Raymond (1984a). For descriptions of such blocks, see Brothers (1954), Borg (1956), Bloxam (1959), R. G. Coleman and Lee (1963), E. H. Bailey, Irwin, and Jones (1964), R. G. Coleman et al. (1965), Ernst et al. (1970), R. G. Coleman and Lanphere (1971), Raymond (1973a), D. E. Moore (1984), Cloos (1986), Dal Piaz and Lombardo (1986), N. V. Sobolev et al. (1986), D. E. Moore and Blake (1989), and Wakabyashi (1990).

9. The mineralogy cited here is based on descriptions in the references cited in note 8, on D. E. Moore (1984), and on the author's observations.

10. R. G. Coleman and Lanphere (1971) and Cloos (1986) provide reviews. See the examples described below for a more detailed discussion of the Franciscan-related occurrences.

11. R. G. Coleman et al. (1965), N. V. Sobolev et al. (1986), Dobretsov (1991).

12. Matthews and Schliestedt (1984), Schliestedt (1986), C. Miller (1977), N. V. Sobolev et al. (1986), Ohta, Hirajima, and Hiroi (1986), Ernst et al. (1970), B. A. Morgan (1970), Ridley (1984), Yokoyama, Brothers, and Black (1986), Ghent et al. (1987), Pognante and Kienast (1989), Bouchardon et al. (1989), Pognante, (1989, 1991), Pognante and Sandrone (1989), Liati and Mposkos (1990). Philippot and Selverstone (1991) discuss fluid compositions and veins in eclogites, and Castelli (1991) describes Eclogite Facies metamorphism in carbonate rocks.

13. Minerals are based on the references cited in note 12.

14. Eskola (1921, in R. G. Coleman et al., 1965), Yoder and Tilley (1962), R. G. Coleman et al. (1965), Mysen and Heier (1972), Mori and Banno (1973), DeWit and Strong (1975), Ernst (1977b), C. Miller (1977), N. V. Sobolev et al. (1986), Austrheim (1987, 1990), Griffin (1987), Jamtveit (1987), Bouchardon et al. (1989), Klemd (1989), I. S. Sander (1989), X. Wang, Liou, and Mao (1989), Krogh et al. (1990), Jamtveit, Bucher-Nurminen, and Austrheim (1990), X. Wang and Liou (1991). Also see Liati and Mposkos (1990) for a polymetamorphic example from northern Greece.

15. Minerals are based on the reports cited in note 14.

16. Contrast D. H. Green and Ringwood (1967a, 1972), Fry and Fyfe (1969), DeWit and Strong (1975), and Austrheim (1987) with B. A. Morgan (1970) and E. H. Brown and Bradshaw (1979).

17. See Dawson (1981) and Miyashiro (1973a, p. 313). Also see Newton and Smith (1967), Boettcher (1970), and Ghent et al. (1987).

18. For example, Yoder and Tilley (1962), D. H. Green and Ringwood (1967a, 1972), Essene, Hensen, and Green (1970), K. Ito and Kennedy (1970, 1971), Howells et al. (1975), and Saxena and Eriksson (1985).

19. Also see Ryburn, Raheim, and Green (1976) and Jamieson (1990).

20. This description is based essentially on the reports of DeWit and Strong (1975) and Jamieson (1990). DeWit and Strong assigned some eclogites to the Fleur de Lys Supergroup, but Jamieson assigns all eclogites to the underlying East Pond Metamorphic Suite. See also Church (1969) and Neale and Kennedy (1967).

21. E. H. Bailey, Irwin, and Jones (1964) and R. G. Coleman and Lanphere (1971) provide overviews. R. G. Coleman et al. (1965) give some detailed descriptions, as do Switzer (1945, 1951), Borg (1956), and Bloxam (1959). Also see Gealey (1951), Essene, Fyfe, and Turner (1965), Ernst et al. (1970), Raymond (1973a, 1977), D. E. Moore (1984), Cloos (1986), D. E. Moore and Blake (1989), Wakabyashi (1990), and Oh, Liou, and Maruyama (1987, 1991).

22. R. G. Coleman and Lanphere (1971), Suppe and Armstrong (1972), J. C. Maxwell (1974), Lanphere, Blake, and Irwin (1978), and Mattinson (1986, 1988).

23. Also see the references cited in note 21.

24. See the discussion of R. G. Coleman and Lee's (1963) subdivision of metabasites into Types I, II, III, and IV in chapter 28.

25. Ernst (1970, 1984), Schweickert and Cowan (1975), Evarts (1978), Dickinson (1981), Cloos (1982, 1984).

26. Most of these hypotheses are summarized by J. F. G. Wilkinson (1976), Ernst (1977b), or F. A. Frey (1980), all of whom supply additional references. See also B. A. Morgan (1970), DeWit and Strong (1975), and Griffin (1987) for discussions of hypothesis 1. See Kornprobst (1969), Beeson and Jackson (1970), Dickey, Obata, and Suen (1977), Cabanis and Godard (1987), Paquette, Menot, and Peucat (1989), and Pearson, O'Reilly, and Griffin (1991) for discussions of hypothesis 2. Switzer (1945) discusses hypothesis 3. Hypothesis 4 is discussed by Yoder and Tilley (1962) and Kuno (1969). See D. H. Green (1966) and Smyth, Caporuscio, and McCormick (1989) for expositions of hypothesis 5. Also see J. F. G. Wilkinson (1976). Hypothesis 6 is discussed or supported by Shimizu (1975), Dickey, Obata, and Suen (1977), and F. A. Frey (1980). Menzies (1983) provides a general discussion of xenolith-forming processes.

27. B. A. Morgan (1970), Mysen and Heier (1972), Evans, Trommsdorf, and Richter (1979), Evans, Trommsdorf, and Goles (1981), Jamtveit (1987), Jamtveit, Bucher-Nurminen and Austrheim (1990), Philippot and Selverstone (1991).

PROBLEMS

29.1. (a) Using ACF plots of eclogites, plot the position represented by the chemistry of sample 2, table 29.1, on a copy of the ACF diagram for "wet" eclogites in figure 29.4. (b) What minerals should occur in this rock? (c) Compare this predicted mineralogy to the mode reported in R. G. Coleman et al. (1965, p. 487).

29.2. (a) Write the discontinuous reactions necessary to explain the conversion of rocks of the Granulite Facies to rocks of the "dry" Eclogite Facies (see figure 29.4), assuming kyanite appears before zoisite. (b) Write reactions to explain the conversion of the Blueschist Facies assemblages shown in figure 29.4 to the "wet" Eclogite Facies assemblages shown in the same diagram.

29.3. Using a copy of figure 29.9 and the data of Wakabyashi (1990) for sample TEC2, (a) trace the P-T-t path for the sample. (b) Which of the six models of eclogite formation best explains the history of this rock?

30

Dynamic Metamorphism

INTRODUCTION

Dynamic metamorphism is metamorphism of rock masses caused primarily by deviatoric stresses that yield relatively high strain rates. Thus, it is metamorphism resulting from deformation. The deformation may be dominantly **brittle,** in which case rock and mineral grains are broken, or it may be dominantly **ductile,** in which case plastic behavior and flow occur via structural changes within and between grains.[1] Temperatures during dynamic metamorphism are typically elevated and may be created by the deformation process. Fluids commonly contribute to the metamorphic process, both by altering phase chemistry and by facilitating recrystallization.

Both local and regional dynamic metamorphism are recognized. At the local scale, in narrow zones from less than 1 mm to several meters wide, brittle or ductile deformation along localized zones of deformation (e.g., along faults and fold limbs) causes rock to break, to recrystallize, and even to melt locally. Similarly, both brittle and ductile deformation, as well as melting, occur during impacts of extraterrestrial bodies. Brittle and ductile deformation processes also operate at the regional scale, but at this scale, mappable rock masses of regional extent are affected.

The rocks produced at all scales by dynamic metamorphism are rocks composed of *clasts*, or fragments of preexisting material (porphyroclasts and microlithons), surrounded by a deformed matrix, the texture or mineral composition of which was produced by metamorphic processes. Such rocks, which fit into the broad category of clastic rocks, are here referred to as **dynamoblastic rocks.**[2]

OCCURRENCES OF DYNAMOBLASTIC ROCKS

Faults pervade the crust of the Earth. Inasmuch as faults are deformation zones, dynamoblastic rocks associated with faults are a ubiquitous feature of the crust. In addition, folds and related deformation zones are relatively common in the root zones of orogenic belts. Even in zones in which newly formed rocks are only partially lithified, for example, in soft sediments on the seafloor or in crustal rocks on lava lakes, deformation may yield dynamically metamorphosed rocks.[3] Particularly noteworthy among the local- to regional-scale zones of dynamoblastic rock are the mylonite zones associated with metamorphic core complexes (G. A. Davis, 1980, 1988; J. L. Anderson, 1988)[4] and the melanges of outer metamorphic belts.[5] Melanges are, in fact, mappable masses of dynamoblastic rock of local to regional dimensions. Impact structures with dynamoblastic rocks include Meteor Crater in Arizona; the Ries Basin of Germany, the Manicouagan impact structure of Quebec, and the Boltysh impact crater of Asia.[6]

Regional zones of dynamoblastic rocks occur at plate boundaries. Along spreading ridges, regional stress may be widespread enough, particularly at mantle depths beneath the spreading axis, to yield dynamically metamorphosed zones of rock. Perhaps more commonly, ductile deformation is concentrated in *narrow zones* within a regional terrane of semischistose ultramafic tectonite (Girardeau and Mercier, 1988). Most local and regional zones of this type are probably subducted and are not preserved. Nevertheless, evidence of their existence is preserved locally in mantle slabs and the basal tectonites of ophiolites.[7] More commonly, oceanic crustal rocks are deformed along transform faults, yielding regional zones (large scale ductile shear zones), localized zones, or bodies (melanges) of dynamoblastic rock.[8] Subaerial examples of rocks deformed in this way are exposed in the Sierra Nevada of California (Saleeby, 1982, 1984), in northern Italy (Gianelli, 1977), and on the island of Cyprus (Spray and Roddick, 1981; Murton, 1986). Exposures of transform faults that transect the continents also reveal brittly and ductilely deformed rocks, such as those along faults of the San Andreas Fault System in California (J. L. Anderson, Osborne, and Palmer, 1983; Chester, Friedman, and Logan, 1985)[9] and faults in the British Isles (Flinn, 1977).

The most extensive development of regional, dynamically metamorphosed rocks occurs in the orogenic belts. Rocks of the transform fault zones may be accreted here, but most commonly, the regional zones of dynamoblastic rock are produced by deformation associated with the plate (and continent) collisions that yield the orogen. At the shallower and cooler levels of Phanerozoic orogens, *melanges*, formed by brittle deformation, ductile deformation, or both, are widespread (Raymond and Terranova, 1984). These rock bodies are less commonly recognized in Precambrian terranes and the internal zones of orogens, where they have been metamorphosed.[10] Well-known examples include the melanges of the Franciscan Complex of California, the Dunnage Melange of Newfoundland, the Gwna Melange in Wales, and the melanges of the Apennine Mountains of Italy.[11]

Ductile deformation zones of regional extent are common in the internal, high-temperature zones of the orogenic belts. Here, discrete fault lines are replaced by extensive zones of recrystallization and flow, attenuated limbs of folds may exhibit mylonitic fabrics, and mylonitic rocks may pervade entire terranes. Examples of such ductile deformation zones include some of the more regionally extensive mylonitic zones associated with metamorphic core complexes in the Rocky Mountain region, thrust faults of the Peninsular and Transverse ranges of California, the Brevard Zone of the Southern Appalachian Orogen, faults in the Grenville Front Tectonic Zone in Ontario, the Moine Thrust of the Scottish Highlands, thrust faults in southern Sweden and Norway, Hercynian deformation of the Pyrenees Mountains along the French-Spanish border, shear and nappe zones in the Armorican region of France, massifs of the Alps, Precambrian shear zones in the African craton, and the Woodroffe Thrust of central Australia.[12]

ROCK TYPES, TEXTURES, AND STRUCTURES

Dynamoblastic rocks exhibit either foliated or nonfoliated textures. Brittle deformation produces *cataclastic textures*, the generally nonfoliated metamorphic textures characterized by fragmented rock or mineral grains (figure 30.1a; also see figure 23.10d).[13] A matrix of fine-grained fragmental material may surround the clasts, frictionally produced glass may form a matrix, or there may be almost no matrix, the clasts being cemented by precipitated minerals. In some breccias and most gouges there are no cements and the rocks are incoherent and unlithified. *Foliated cataclastic textures*, a transitional type of texture between cataclastic textures *sensu stricto* and mylonitic textures, are characterized by sets of shear fractures, clasts and microlithons in preferred orientations, grains showing deformation bands, and compositional layering consisting of mineral and microlithon segregations (Raymond, 1975, 1985; Chester, Friedman, and Logan, 1985; S. M. Agar, Prior, and Behrmann, 1989).[14] Some gouges and the quasimylonites of many melanges and fault zones display these textures. *Mylonitic textures*, the foliated textures produced by ductile deformation, typically contain either numerous porphyroclasts of protolith minerals or microlithons of the protolith (figure 30.1c). Highly elongated "ribbon" quartz grains, grains with deformation bands, grains with mosaic-textured recrystallization zones and margins, and porphyroclasts occur in a fine mosaic of recrystallized minerals (figure 30.1c)(Higgins, 1971, Dell'Angelo and Tullis, 1989).[15] In both handspecimen and thin section, mylonites usually are characterized by intersecting planar fabric elements enhanced by phyllosilicates. These may include a schistocity (S), "cisaillement" or shear planes (C), and shear bands (SB) (figure 30.2)(Berthe, Choukroune, and Jegouzo, 1979; C. Simpson, 1986; Dell'Angelo and Tullis, 1989). These intersecting elements yield lenticular or phacoidal domains that differentially weather in many rocks to yield a "fish-scale" or "button-like" character.

Because dynamoblastic rocks are associated with fault zones and other zones of deformation, they exhibit structures characteristic of these zones. The mylonitic types commonly are banded and may contain crenulation cleavages, boudin, or flow folds (figure 30.3). Foliation and banding occur at all scales. The relationships between thin-section, handspecimen, outcrop, and regional textures and structures in mylonites are

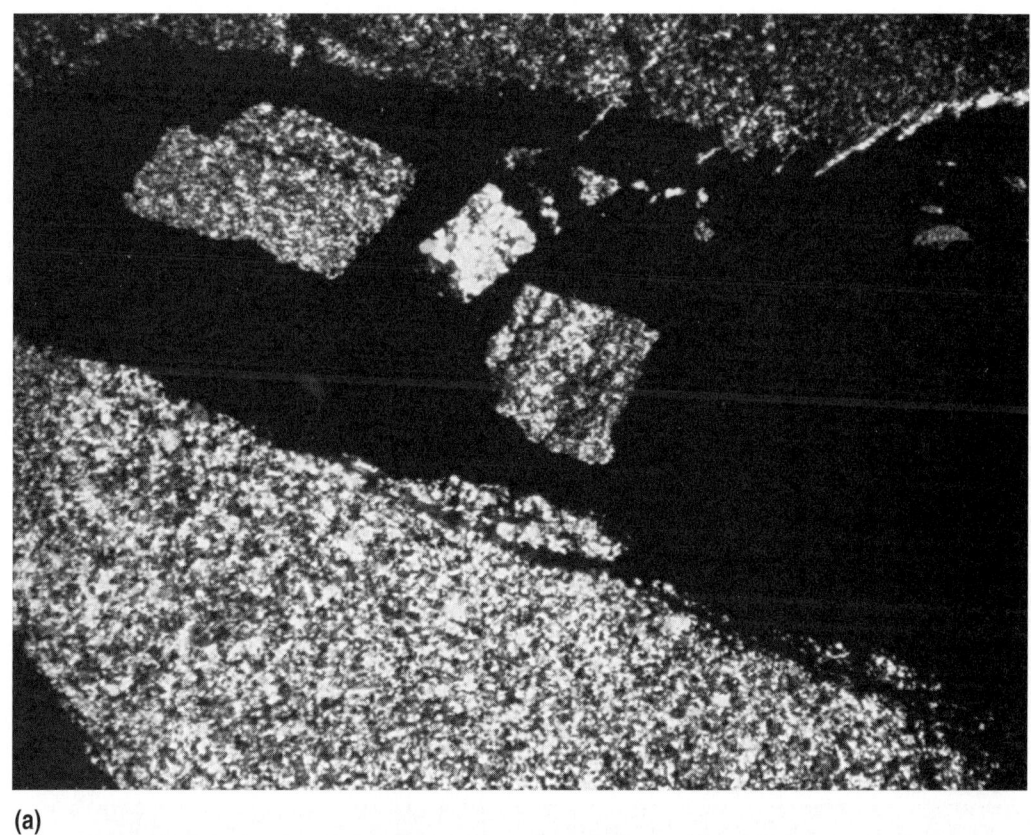

(a)

(b)

Figure 30.1 Photomicrographs of textures in dynamoblastic rocks. (a) Manganese-oxide (black) cemented, cataclastic chert breccia, Rome Formation, Mountain City, Tennessee. Long dimension of photo is 6.5 mm. (XN).
(b) Metafeldspathic arenite protomylonite, Stone Mountain Fault Zone, Highway 421, 4 miles north of Trade, Tennessee. (XN). (c) Orthomylonite, southwest flank of the Santa Catalina Mountains, Arizona. (XN). (d) Ultramylonite, Rosman Fault, Brevard Zone, North Carolina. Long dimension of photos in (b)–(d) is 1.27 mm.
((d) Stop 2 in Horton and Butler, 1986.)

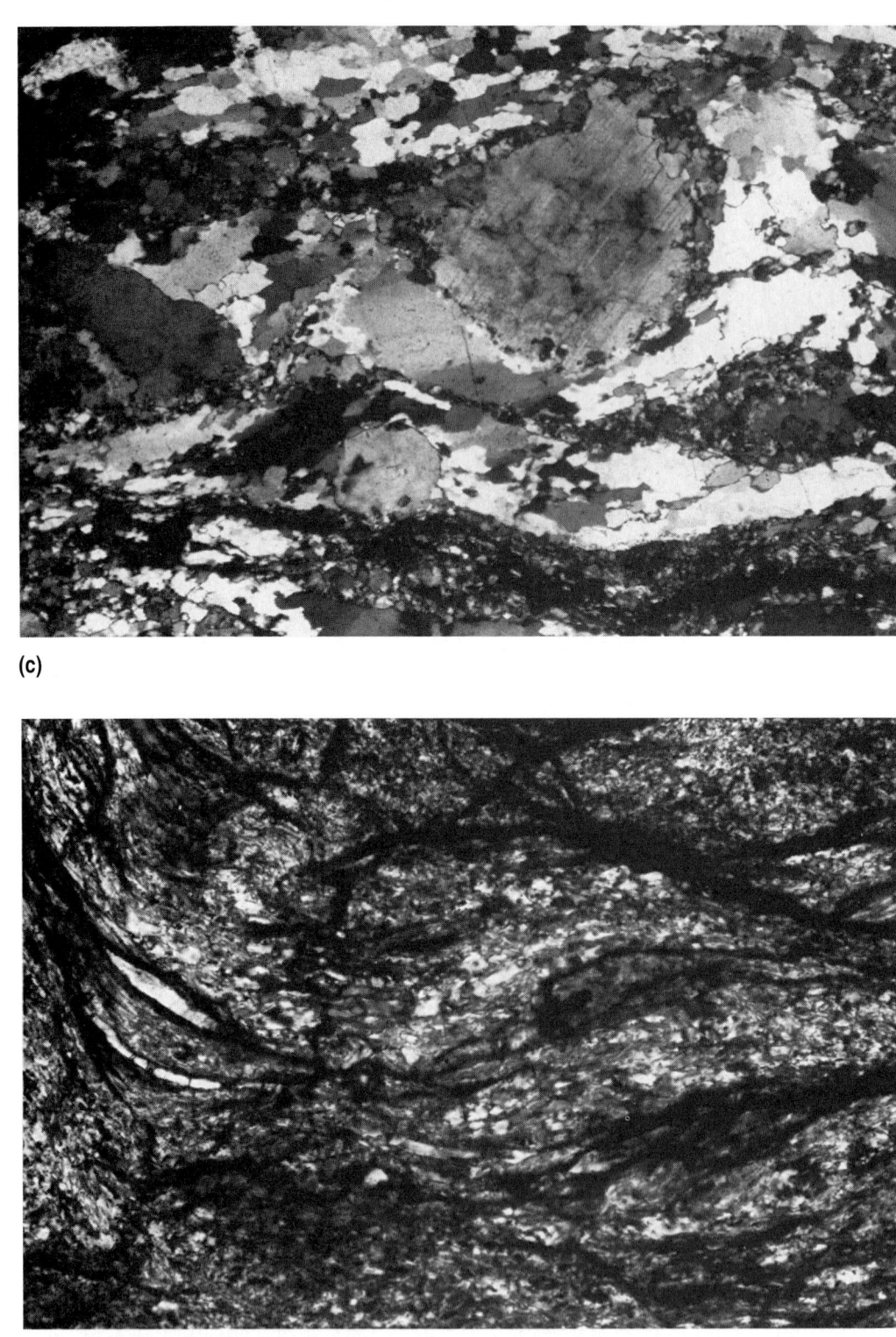

(c)

(d)

Figure 30.1   Continued

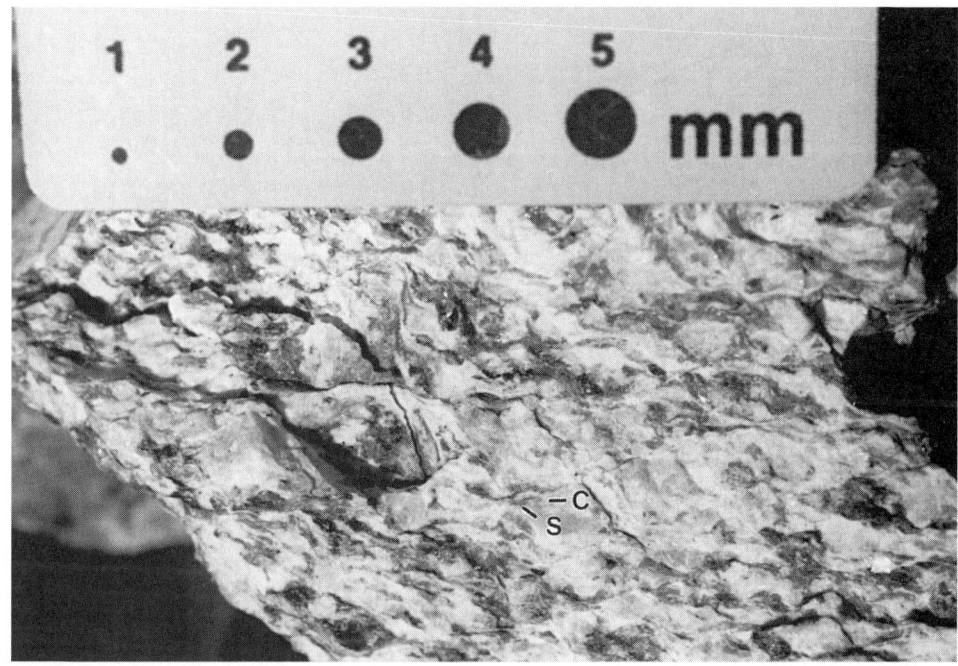

Figure 30.2 Mylonite showing S-C fabric. Southwest flank of Santa Catalina Mountains, Arizona. S = schistocity, C = spaced cleavage, shear planes.

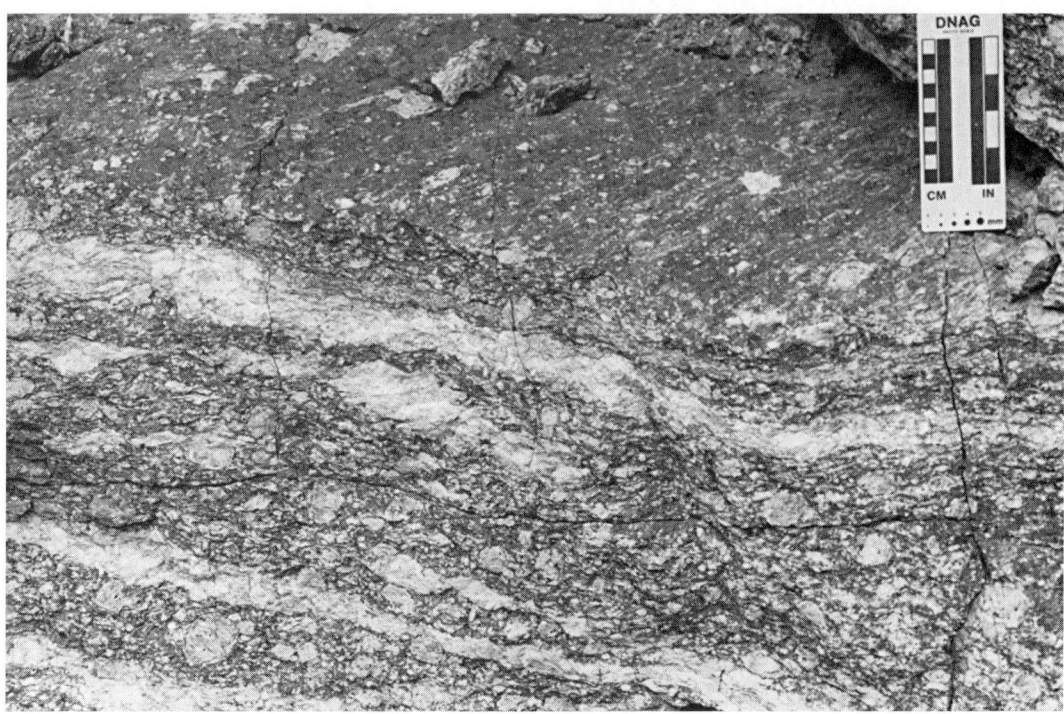

Figure 30.3 Structures in outcrop of mylonite on the southwest flank of the Santa Catalina Mountains, Arizona. Note lineation on the top surface and foliation, boudinage in veins, and augen on the front surface of the exposure.

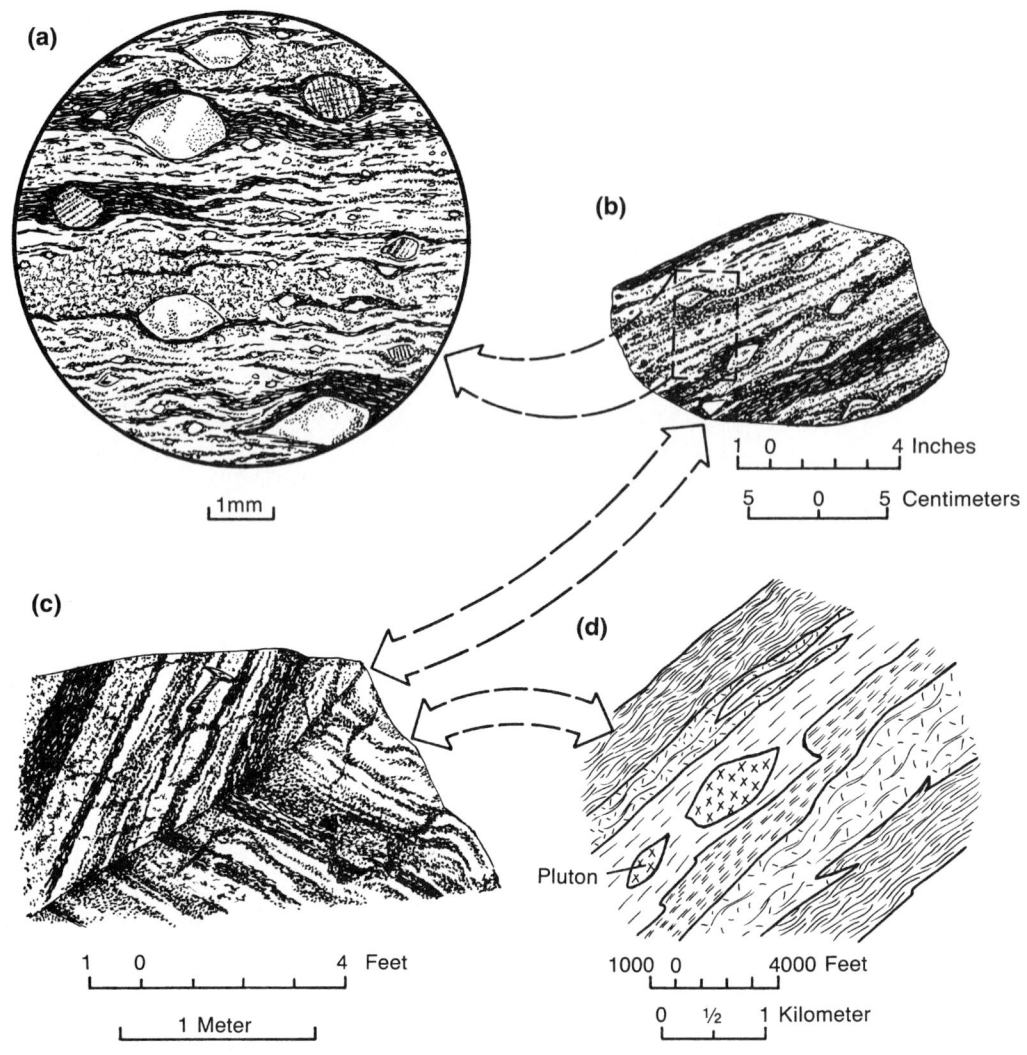

Figure 30.4   Relationship between textures and structures in mylonites at different scales: (a) thin section, (b) handspecimen, (c) outcrop, and (d) regional. Note that foliation and banding pervade all scales.
(Source: M. W. Higgins, "Cataclastic Rocks" in U S Geological Survey PP. 687, 1971)

shown in figure 30.4 and the site of mylonite formation in deep (hotter) fault zones is depicted in figure 30.5. Cataclastic rock types may appear more massive than mylonites, but commonly form tabular zones that contain associated slickensides, minor faults, and buckle folds. In fault zones, cataclasites form near the surface and exhibit increased milling of rock fragments and grains from the margins to the center of the zone (figure 30.5, top three circles). Transitional rocks with weak to moderate, foliated-cataclastic and protomylonitic foliations (quasimylonites and the protomylonites) form at moderate depths (J. L. Anderson, Osborne, and Palmer, 1983; Babaie et al., 1991).[16]

Names for rocks formed in zones of high strain were developed long before the processes involved in their origin were understood (Lapworth, 1885). Undoubtedly the most serious error made by early workers was that they thought that fine-grained rocks found in ductile shear zones owe their small crystal size to physical breakdown via brittle failure—cushing, grinding, and breaking—of the minerals of the protolith (Lapworth, 1885; Waters and Campbell, 1935). That such processes occur during near-surface faulting is evidenced by breccia and gouge that lack any coherence between grains. This is not the case at depth.

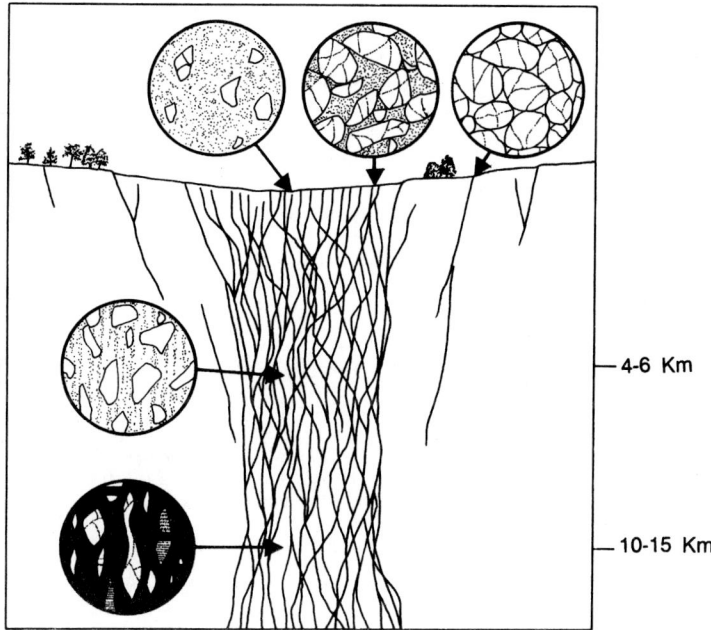

Figure 30.5 Relationship between depth of formation, position in shear zone, and dynamoblastic textures. Thin-section views are shown in circles. Upper right circle shows incipiently fractured sandstone. Upper center circle shows cataclastic breccia. Upper left circle shows gouge with a few larger clasts remaining. Weakly foliated quasimylonite occurs at intermediate depths. Well-developed mylonite (lowest circle) containing local ultramylonite bands (not shown) occurs at greatest depth.
(Based on the ideas of Sibson, 1977, and J. L. Anderson, Osborne, and Palmer, 1983)

In recent years, experimental deformation of rocks, plus the associated textural studies, have shown conclusively that the fine-grained textures of the mylonites result from syntectonic recrystallization associated with flow and intragranular deformation and from post-tectonic recovery and recrystallization, rather than from brittle failure of the constituent grains of the rock (N. L. Carter, Christie, and Griggs, 1964; T. H. Bell and Etheridge, 1973; Dell'Angelo and Tullis, 1989). Matrix textures are crystalline, not clastic. Thus, the term mylonite, based on the Greek root *mule* for mill (to grind down), is a misnomer, but it is retained because it is entrenched in the literature and is used widely for rocks of ductile deformation zones.

The main types of dynamoblastic rock have been defined and redefined by several workers, as knowledge of rock origins increased over the years.[17] Three classifications, one proposed by Higgins (1971), another proposed by Wise et al. (1985), and a third created by the author, are shown in figure 30.6. Higgins's (1971) classification, admirably, used observable criteria (cohesion, grain size, percent clasts, and foliation or fluxion structure) to define various types of dynamoblastic rocks. Because it was developed prior to our modern understanding of the roles of intragranular deformation and syntectonic recrystallization, however, it defines all dynamoblastic rocks as cataclastic (produced by fracture rather than ductile deformation processes) and therefore conveys meanings at odds with contemporary understanding of the rock-forming processes. Nevertheless, Higgins's (1971) subdivisions based on porphyroclast percentage are useful and have been adopted by D. U. Wise et al. (1984) and Raymond (1993). The classification of D. U. Wise et al. (1984) (1) focusses on fault-related rocks, (2) uses coherence, porphyroclast percentages, the presence or absence of foliation, and the presence or absence of evidence of "syntectonic crystal-plastic processes" as criteria for distinguishing various rock types, and (3) emphasizes relative rates of strain versus recovery in rock genesis. Yet the classification has drawbacks. It implies that dynamically metamorphosed rocks are not metamorphic, it does not clearly distinguish between recovery and recrystallization, it fails to recognize foliated quasimylonites, and it specifies that all mylonites are coherent.[18] The simplified classification proposed

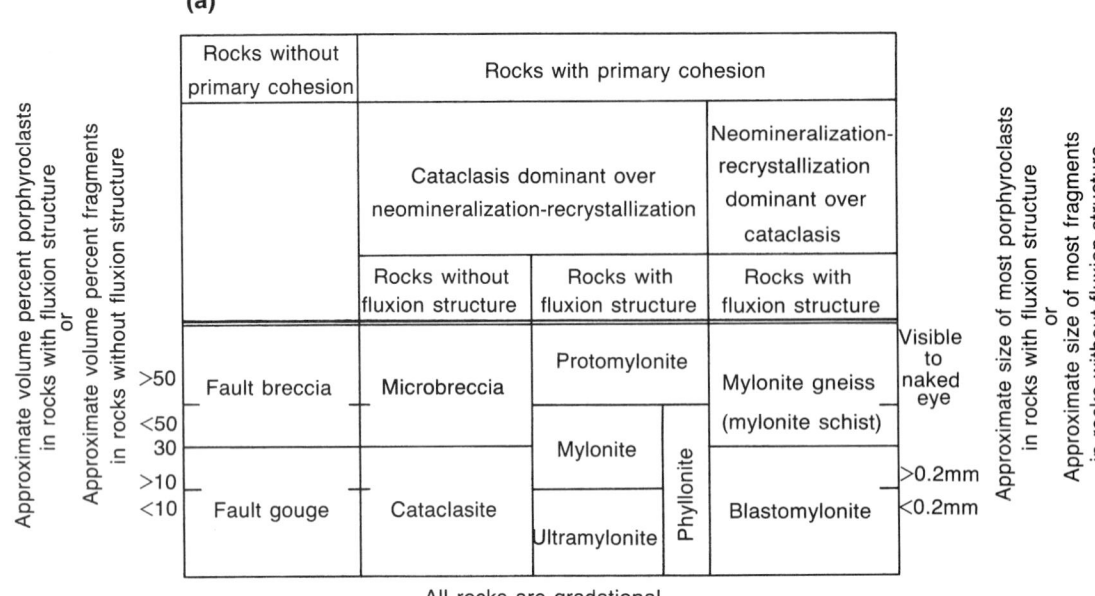

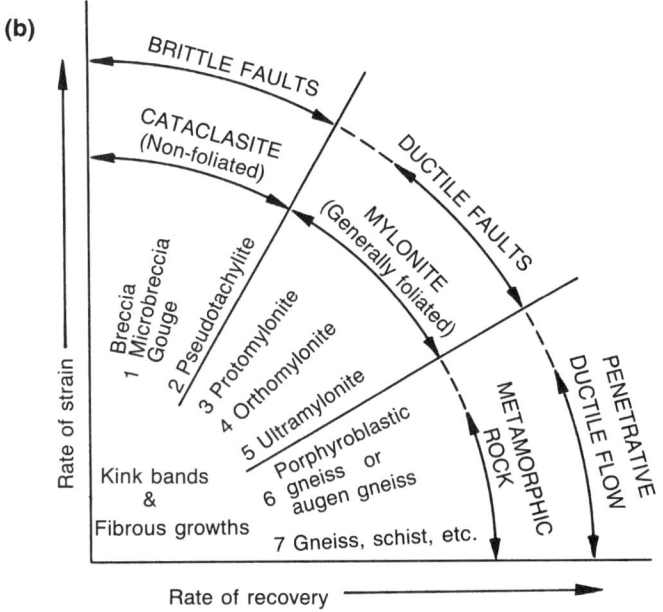

Figure 30.6 Three classifications of dynamoblastic rocks. (a) *Source: M. W. Higgins, "Cataclastic Rocks" in US Geological Survey pp. 687, 1971.* (b) Classification of Wise et al. (simplified from Wise et al., 1985). (c) Raymond's classification (1993).

by Raymond (1993) (figure 30.6c) is based on observable criteria alone. Nonfoliated dynamoblastic rocks, i.e., rocks with cataclastic textures, are subdivided into those with glass in the matrix, called **pseudotachylites**, those dominated by grains <1/16 mm in diameter called **gouges**, and those dominated by grains > 1/16 mm in diameter, **cataclastic breccias**(see figure 30.1a). Mylonitic (foliated) rocks are subdivided on the basis of the percentage of porphyroclasts plus microlithons they contain. **Protomylonites** contain > 50% porphyroclasts and microlithons (see figure 30.1b), **orthomylonites** contain between 10 and 50% (see figure 30.1c), and **ultramylonites** contain < 10% (see figure 30.1d). As is common in nature, there is a continuum between rocks of the two major textural and rock types. **Quasimylonites**, characterized by incipient foliations, some crystalline matrix materials, and microlithons or broken grains, mark the transition between cataclasites and mylonites.[19]

MINERALS AND FACIES OF DYNAMOBLASTIC ROCKS

Mineralogically, there is nothing unique about most dynamoblastic rocks. The rocks contain minerals typical of the various facies to which they belong. Although it is theoretically possible for dynamoblastic rocks to form under the P-T conditions of any facies, they most typically represent the Zeolite, Greenschist, or Amphibolite facies. Zeolite and Greenschist facies minerals are common in the cataclasites; Zeolite, Prehnite-Pumpellyite, Greenschist, and Blueschist facies assemblages occur in quasimylonites; and Greenschist and Amphibolite Facies phase assemblages usually characterize the mylonites (Cloos, 1983; Kamineni, Thivierge, and Stone, 1988; G. A. Davis, 1988; Hyndman and Myers, 1988).

The most unique phase assemblages in the dynamoblastic rocks are those containing glass and those containing high-pressure, shock-metamorphosed minerals, such as coesite and stishovite. Glass and the high-pressure phases occur in impact metamorphic rocks (Chao, Shoemaker, and Madsen, 1960; Grieve et al., 1987). Assemblages with glass also characterize the pseudotachylites (Sibson, 1975; R. H. Maddock, 1984). A representative assemblage in a pseudotachylite formed in an Amphibolite Facies quartz-feldspar gneiss is

> quartz–plagioclase–hornblende–biotite–magnetite–
> zircon–sphene–glass

(R. H. Maddock, 1984).

PROCESSES DURING DYNAMOBLASTIC ROCK FORMATION

Several process operate during the petrogenesis of dynamoblastic rocks. These include cataclasis, recrystallization, neocrystallization, pressure solution, and metasomatism. During the formation of various rock types, one or another of these process dominates.

Cataclasis

Cataclastic rocks form by brittle failure (breaking) of rock and mineral materials. Such failures occur under conditions in which the rocks (and mineral grains) are strained beyond the limits of their strength, typically under conditions of high deviatoric stress and high strain rate but relatively low temperature.[20] When the rocks or mineral grains can withstand no more strain, they break. Breaking begins as a series of fractures that extend across each grain from one point of contact with an adjoining grain to another point of contact (Blenkinsop and Rutter, 1986) (figure 30.5, upper right circle). Continued application of stress produces angular fragments. Additional grinding of the weakened and broken rock fragments between more rigid adjoining rock masses results in the milling of the broken clasts, eventually yielding a gouge (figure 30.5, upper left circle). Quasimylonites develop a foliation *either* where pervasive fracturing is accompanied by some ductile deformation or neocrystallization *or* where deformation induces alignment of inequant mineral grains or microlithons (J. C. Moore et al., 1986; Evans, 1988; Agar, Prior, and Behrmann, 1989; Babaie et al., 1991).

Different minerals fail under different conditions. Phyllosilicates tend to be weak and subject to flow at relatively low values of temperature and strain rate. In contrast, quartz and feldspars have greater strength. Feldspars may experience brittle failure even under temperatures of 800° C, strain rates of 2×10^{-5}, and confining pressures of 1.0 Gpa (10kb) (Marshall and McLaren, 1977). In dry rocks, quartz is quite strong, but water promotes weakening and plastic behavior (Blacic and Christie, 1984).

In general, neither fluids nor heat play a major role in cataclasis. Fluids may, and commonly do, transport materials that are precipitated to cement breccia fragments, but this is commonly a post-tectonic process. In some cases, elevated temperatures result in local melting and the formation of pseudotachylites. The processes are similar to those involved in the formation of breccias, but in addition to brittle failure, frictionally induced heating elevates the temperature of the deforming rocks enough to produce local melting (Sibson, 1975; 1977; R. H. Maddock, 1983; 1984). Sibson (1975) estimated the temperature of pseudotachylite formation in a Scottish fault zone to be about 1100° C. Pseudotachylites form under dry conditions.

Mylonitization

The formation of mylonites is more complex than the formation of cataclastic rocks and involves successive stages of deformation, recovery, and recrystallization. During deformation, pressure solution may contribute to fabric development (Vauchez, Maillet, and Songy, 1987), but deformation processes are basically mechanical in nature. Mylonitization also involves the chemical processes of metasomatism and neocrystallization. In these, as well as in the deformation processes, fluids are important. Other variables that control the nature of the mylonitization include the mineral composition of the protolith, the confining pressure, the temperature, and the homogeniety and continuity of the rock mass.

The deformation processes involved in mylonitization include microfracturing, twinning, dislocation glide, and grain-boundary sliding.[21] Microfracturing is a process in which microscopic fractures develop within and between grains, in response to applied stress; i.e. fractures are intragranular or intergranular, respectively (Nicolas and Poirier, 1976, pp. 43–44). In minerals with cleavage, the intragranular fractures may follow the cleavage. Feldspars, in particular, and zircon tend to fracture during mylonitization, even at high temperatures,[22] and in some cases, quartz, calcite, olivine, pyroxene, and biotite do so as well.[23]

Twinning is another mechanism by which crystals may reflect strain. The term twinning has its usual mineralogical meaning; it refers to the process in which one or more parts of a crystal lattice assume an orientation different from other parts. *Dislocation glide* refers to a shift in the position of a defect (i.e., a dislocation) within a crystal lattice.[24] The defect may change size or may simply change positions. Such dislocations are revealed in crystals by features such as deformation bands (see figures 30.1c and 23.15b).[25] *Grain-boundary sliding* is a process in which grains, in response to applied stress, shift positions relative to adjoining grains, with the shift occurring along the grain boundary.[26] All of these processes are granular adjustments made within rocks to accommodate an applied stress. The adjustments result in a foliated rock, shortened perpendicular to the foliation and lengthened parallel to it.

In addition to the mechanical processes of deformation involved in mylonite formation, three physical-chemical processes—recovery, recrystallization, and neocrystallization—and the chemical process of metasomatism are important in the development of the character of these rocks. **Recovery** is a process in which deformed grains containing a relatively high proportion of dislocations (and therefore a high-energy state) reduce the amount of intracrystalline deformation (and therefore the strain energy within the crystal) by internal reorganization or reduction of dislocations.[27] Recovery commonly results in the elimination of deformation bands, twins, and other deformation features.

Recrystallization is the process in which strain energy is reduced by the nucleation and growth of new crystals within and at the margins of host crystals.[28] The new crystals have orientations that differ from those of the deformed crystals and are typically small, irregular, and polygonal in shape. The process by which these new crystals form, called polygonization, is very common and is evidenced by mosaics of small polygonal crystals rimming and/or relacing porphyroclasts (figure 30.7). Recrystallization that occurs during deformation is referred to as *syntectonic recrystallization* or *dynamic recrystallization*. *Static recrystallization* occurs following deformation. Quartz, olivine, calcite, and micas commonly undergo syntectonic recrystallization during mylonite formation.

Neocrystallization, the process of new mineral formation, requires diffusion of new chemical species into the area of grain growth.[29] Where those chemical elements migrate in from regions outside of the local domain (or migrate out of the domain), metasomatism is considered to have occurred. The new minerals that form may simply be equilibrating with new P-T conditions or they may be responding to chemical potential gradients created by changes in the fluid composition.

Fluid flow in fault zones and ductile deformation zones is significant in promoting mechanical deformation, recrystallization, and neocrystallization.[30] Shearing associated with deformation may be accompanied by fluid flow that (1) facilitates deformation processes, such as dislocation glide, (2) transports heat, reducing the metamorphic temperature in the mylonite zone, and (3) changes the composition of the local fluid phase, promoting neocrystallization. As deformation opens channels for fluid flow, the fluids promote deformation, recrystallization, and neocrystallization, which, in turn, aid in the growth of the deformation zone. Thus, a kind of feedback loop exists. Another common consequence of fluid migration, the transportation of heat out of mylonite zones, results in the formation of retrograde effects along the

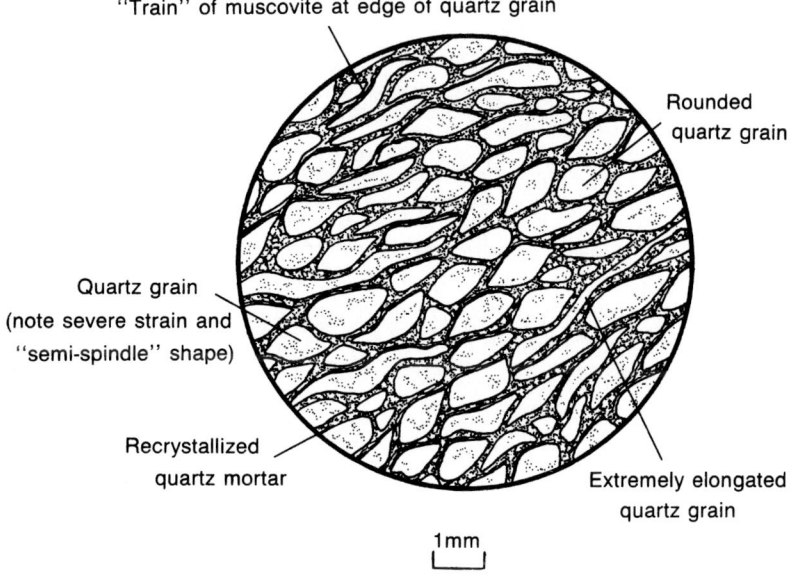

"Train" of muscovite at edge of quartz grain

Rounded quartz grain

Quartz grain (note severe strain and "semi-spindle" shape)

Recrystallized quartz mortar

Extremely elongated quartz grain

1mm

Figure 30.7 Sketch of micaceous metaquartz arenite protomylonite from the Weaverton Formation of the Blue Ridge Belt, Virginia.

(Source: M. W. Higgins, "Cataclastic Rocks" in US Geological Survey pp. 687, 1971.)

zones. For example, Amphibolite Facies mylonite zones may develop in Granulite Facies terranes, or Greenschist Facies mylonite zones may form in Amphibolite Facies terranes (Drury, 1974; Beach, 1980; Erslev and Sutter, 1990).[31] Major metasomatic effects are also produced by fluids.[32] For example, K. O'Hara (1988, 1990b) and Glazner and Bartley (1991) suggest that fluids have removed more than 60% of the volume of material in some mylonite zones, particularly by removing Si and alkali elements. Pressure solution promotes some of this volume loss. Reactions such as

5 plagioclase (andesine) + 2 microcline + 24 H_2O — clinozoisite + 2 muscovite + 10 H_4SiO_4 + 3 Na^+ + 3 $(OH)^-$

may explain the conversion of feldspar to white mica, the loss of soluble silica, and a loss of the alkali element sodium.[33] Tobisch et al. (1991), however, demonstrate that mylonitization of some granitoid rocks is accompanied by *gains* in Si, while alkalis are lost.

Together, combinations of the processes described above yield mylonitic rocks. The particular combination of processes that produces the specific fabric elements and mineral composition of any given mylonite is a function of the rock and fluid composition and the P-T and strain rate histories. Consequently, each mylonite requires individual study.

EXAMPLE: THE BREVARD ZONE, SOUTHERN APPALACHIAN OROGEN

The Brevard Zone is a major zone of deformation in the Southern Appalachian Orogen.[34] The zone extends from North Carolina to Alabama and generally separates the Blue Ridge Belt from the Inner Piedmont Belt (figure 30.8).

Definitive evidence indicates that the Brevard Zone is polymetamorphic and structurally complex. An early Amphibolite Facies metamorphic event was followed by a later Greenschist Facies event. These recrystallization events were accompanied by the formation of dynamoblastic rocks. More than thirty structural interpretations of the nature of the Brevard Zone have been proposed (Bobyarchick, Edelman, and Horton, 1988), including interpretations of the zone as a fold, a thrust fault, a normal fault, a strike-slip fault, and combinations of these various structures. Not only has there been disagreement about the nature of the zone, but there is no general agreement about the ages of the deformation events that produced the various features of the zone. What *is* clear is that the Brevard zone is characterized by mylonites.

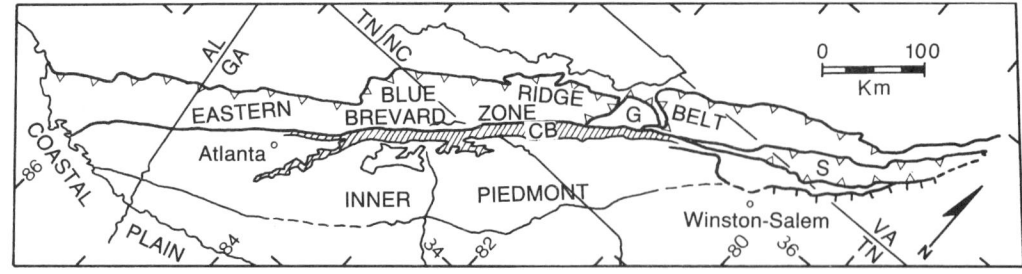

Figure 30.8   Map showing the location of the Brevard Zone in the southern Appalachian Orogen. CB = Chauga Belt, G = Grandfather Mountain Window, S = Smith River Allochthon.
(Modified from Bobyarchick, Edelman, and Horton, 1988)

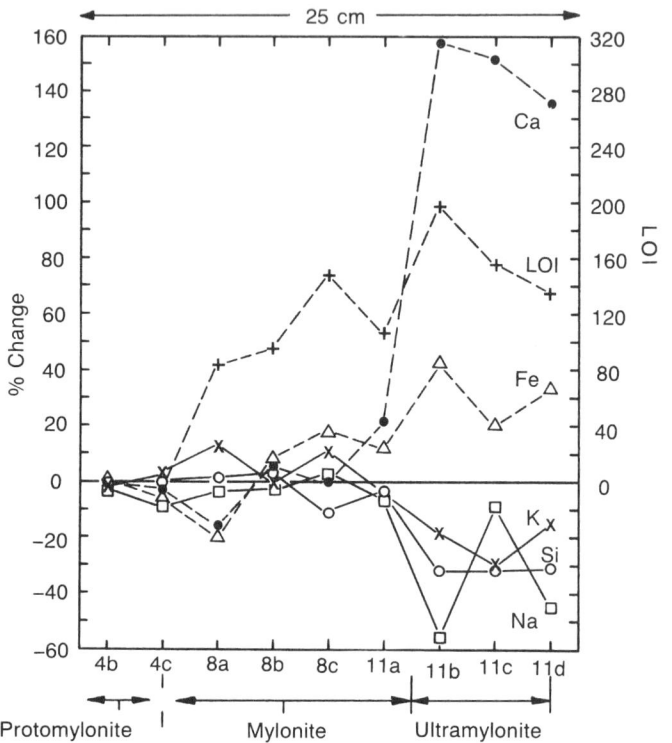

Figure 30.9
Chemical change as percentage change in oxides across a Brevard Zone mylonite sample. Values normalized to constant Al_2O_3, as measured in a subvolume of the mylonite.
(Modified from Sinha, Hewitt, and Rimstidt, 1986)

Traverses across the zone, as well as drilling, reveal that rock types and deformation styles vary within the zone (Hatcher, 1971; Bobyarchick, Edelman, and Horton, 1988; Christensen and Szymanski, 1988). Rock types vary from protomylonites to white mica ultramylonites (figure 30.1d). In the less deformed rocks, Greenschist Facies assemblages such as

quartz–white mica–biotite–plagioclase–epidote–chlorite–magnetite and

quartz–white mica–chlorite–pyrite–graphite

are typical (Bryant and Reed, 1970a; Bobyarchick, Edelman, and Horton, 1988). Such assemblages overprint older kyanite- and garnet-bearing assemblages.

Sinha, Hewitt, and Rimstidt (1986) obtained a 25 cm × 25 cm block of rock from the Brevard Zone near Rosman, North Carolina, that showed a complete gradation from protomylonite to ultramylonite, a range of lithologic change typical of the 15-km-wide Brevard Zone as a whole. Chemical analyses of small parts of this block (4b, 8c, etc., in figure 30.9) reveal that significant chemical changes occur across the block and are especially notable in ultramylonite (11b to 11d), where textural reconstitution was greatest. Fluid flow, concentrated in the ultramylonite, promoted losses of SiO_2, Na_2O, and K_2O, and gains of CaO, FeO, and H_2O. The amount of silica lost requires a fluid/rock weight ratio of 250. These data suggest that the ultramylonite zones serve as major channels for fluid migration during mylonitization and that there may be a feedback loop between fluid flow and recrystallization as the mylonite zone develops.

SUMMARY

Local to regional dynamic metamorphism is an important process in many orogenic belts and deformation zones throughout the world. Dynamic metamorphism, caused primarily by deviatoric stresses yielding moderate to high strain rates, results in the formation of dynamoblastic rocks of three major types—the nonfoliated cataclasites, the foliated quasimylonites, and the foliated mylonites. Cataclasites form at relatively shallow depths through physical breaking of grains. Coarser-grained cataclastic breccia and finer grained gouge result from this process. Where temperatures during milling are high enough, local melting occurs and pseudotachylite is formed. Fluids may facilitate postcataclasis cementation.

At increased temperatures or lower strain rates, quasimylonites form. Breaking of stronger grains (e.g., quartz and feldspar), recrystallization or neocrystallization of phyllosilicates, and shear dislocation to yield slickensides contribute to the formation of these rocks. Fluids may facilitate shearing and neocrystallization.

Temperature, confining pressure, fluids, and strain rate are important variables affecting mylonitization. Within developing mylonites, grains deform via various mechanisms, including microfracturing, twinning, dislocation glide, pressure solution, and grain-boundary sliding. The textures of the resulting rocks vary and depend on the impact of various recovery, recrystallization, and neocrystallization processes that also operate during mylonite formation. Most of these processes depend on diffusion and may be facilitated by fluid flow. In some cases, mylonite zones serve as fluid channel-ways. The fluids in these channel-ways function to transport chemical species into and out of the mylonite zone, yielding metasomatic effects. These fluids may also transport heat, reducing the temperature of the zone, with the result that retrograde metamorphism occurs. This retrograde metamorphism typically yields Greenschist and Amphibolite Facies mylonite zones within higher-grade terranes.

EXPLANATORY NOTES

1. Ductile behavior does not necessarily imply that crystal-plastic deformation has occurred (Rutter, 1986; Chester, 1988). "Cataclastic flow," or semibrittle flow, involves movement along mesoscopic to microscopic zones (rather than discrete planes) as a result of some combination of grain breaking, plastic deformation of grains, and recrystallization and neocrystallization.

2. This term is derived from the roots *dynamo,* meaning power (or in this case, stress), and *blasto,* meaning formative. Hence, *dynamoblastic* means formed by stress.

3. Soft sediments deform locally by cataclasis and recrystallization. Where this is the case, they fit the definition of metamorphic rocks. For more information on soft sediment deformation, see J. C. Moore (1986), for a collection of papers dealing with this and related subjects, and also see the references cited therein.

4. Refer to additional papers in Crittenden et al. (1980), and to Kerrich and Hyndman (1986), La Tour and Barnett (1987), J. L. Anderson (1988), Lister and Davis (1989), Guerin, Brun, and Vanden Driessche (1990), Hacker, Yin, and Christie (1990), Hodges and Walker (1990), and Palais and Peacock (1990).

5. See Raymond (1984c) for a series of papers dealing with various aspects of melanges and Raymond and Terranova (1984) for a review of melange distribution.

6. Metamorphic rocks associated with impact structures are unique and rare rocks, not treated in this text. For more information consult works such as Shoemaker (1960), Chao, Shoemaker, and Madsen (1960), Bunch and Cohen (1964), Englehardt and Stoffler (1966), N. M. Short (1966), Currie (1967), French and Short (1968), Floran et al. (1978), Simonds et al. (1978), and Grieve et al. (1987).

7. See various discussions in E. M. Moores and Vine (1971), Ave Lallemant (1976), Nicolas and Poirier (1976, ch.11), R. G. Coleman (1977), Juteau et al. (1977), Salisbury and Christensen (1978), Nicolas and Violette (1982), Boudier and Nicolas (1982a, 1982b, 1985), and Girardeau and Mercier (1988).

8. See Hebert, Bideau, and Hekinian (1983), and Honnorez, Mevel, and Montigny (1984) for examples from the ocean basins.

9. For additional mention and discussion of fault rocks of the San Andreas Fault System in California, see Waters and Campbell (1935), L. F. Noble (1954), K. J. Hsu (1955), C. R. Allen (1957), Proctor (1962), R. V. Sharp (1967), Higgins (1971), Sieh (1984), and Chester, Friedman, and Logan (1985).

10. Raymond (1974, 1977a,b), Horne (1979), E. H. Brown (1986), Horton et al. (1986), Lacazette (1986), Higgins et al. (1989), Lacazette and Rast (1989), and Raymond, Yurkovich, and McKinney (1989).

11. *Franciscan Melanges*—K. J. Hsu (1968, 1969); Berkland et al. (1972), Suppe (1973), Christensen (1973), Raymond (1973a), Cowan (1974, 1978, 1982a), J. C. Maxwell (1974), Crawford (1975), Cowan and Page (1975), Gucwa (1975), Page (1978), Aalto (1982), and Cloos (1982). *Dunnage Melange*—Horne (1969), Kay (1976), and Jacobi (1984). *Gwna Melange*—Greenly (1919), C. P. Wood (1974), and D. Wood and Schuster (1978). *Italian melanges*—Beneo (1956), Maxwell (1959), Boccaletti, Bortolotti, and Sagri (1966), Elter and Trevisan (1973), Naylor (1981, 1982), Agar, Prior, and Behrmann (1989).

12. Higgins (1971) provides an annotated bibliography of older literature and reviews of several occurrences of mylonites and other dynamoblastic rocks. In addition, see the following for particular references to regions mentioned in this text: *Rocky Mountain region* (G. A. Davis et al., 1980; Hyndman, 1980; Bykerk-Kauffman, 1986), *Peninsular Ranges region* (Southern California Batholith) of California (Alf, 1948; R. V. Sharp, 1967; C. Simpson, 1985; D. K. O'Brien et al., 1987), *San Gabriel Mountains of California* (K. J. Hsu, 1955); *Brevard Zone of the Southern Appalachian Orogen* (Higgins, 1971; Sinha, Hewitt, and Rimstidt, 1986), *Grenville Front Tectonic Zone in Ontario* (Themistocleous and Schwerdtner, 1977), *the Moine Thrust of the Scottish Highlands* (Higgins, 1971; Barr, Holdsworth, and Roberts, 1986; Blenkinsop and Rutter, 1986), *southern Sweden* (Zeck and Malling, 1974), *southern Norway* (Roy, 1977; Starmer, 1980), *the Pyrenees Mountains* (Carreras, Julivert, and Santanach, 1980; Lamouroux et al., 1980; McCaig, 1984), *the Armorican region of France* (Nicolas et al., 1977; Berthe and Brun, 1980; M. J. Watts and Williams, 1983; Vauchez, Maillet, and Songy, 1987), *massifs of the Alps* (Behr, 1980; Kerrich et al., 1980; C. Simpson, 1983), *the African craton* (Wakefield, 1977), *Woodroffe Thrust of central Australia* (T. H. Bell, 1979).

13. For example, see Lucas and Moore (1986), Kamineni, Thivierge, and Stone (1988), and Evans (1988).

14. Also see J. C. Moore et al. (1986), Evans (1988), Kano and Sato (1988), and Babaei et al. (1991).

15. Also see discussions or illustrations in Vauchez, Maillet, and Songy (1987), Hoogerduijn Strating (1988), T. H. Bell and Cuff (1989), K. A. Carlson, van der Pluijm, and Hanmer (1990), de Roo and Williams (1990), Ji and Mainprice (1990), K. O'Hara (1990a, 1990b), H. R. Wenk and Pannetier (1990), and van der Pluijm (1991).

16. The quasimylonites are referred to as *quasiplastic mylonites* by some workers.

17. K. J. Hsu (1955), Higgins (1971), Zeck (1974), D. U. Wise et al. (1984).

18. For discussions of the terminology of D. U. Wise et al. (1984), see Mawer (1985) and Raymond (1985).

19. See J. L. Anderson, Osborne, and Palmer (1983) and J. C. Moore et al. (1986) for descriptions of these kinds of rocks. One additional rock type, not shown on the classification, is the *blastomylonite*, a mylonitic rock that shows significant post-deformational crystal growth, yielding porphyroblasts or porphyroblastic overgrowths on porphyroclasts.

20. Consult standard texts on structural geology (G. H. Davis, 1984; Dennis, 1987; Suppe, 1985; Hatcher, 1990; Twiss and Moores, 1992), as well as Turcotte and Schubert (1982) and Sibson (1977), for more details and references on rock deformation studies relevant to the structural behavior of materials in fault zones. Also see Evans (1988) for an example.

21. For example, see T. H. Bell and Etheridge (1973), Boullier and Gueguen (1975), Nicolas and Poirier (1976), Roy (1977b), T. H. Bell (1979), Etheridge and Wilkie (1979), Beach (1980), Poirier (1980), C. J. L. Wilson (1980), M. J. Watts and Williams (1983), C. Simpson (1985), D. K. O'Brien et al. (1987), Vauchez (1987b), Dell'Angelo and Tullis (1989), de Roo and Williams (1990), and Ji and Mainprice (1990).

22. Alf (1948, pl. 3, fig. 2), Marshall and McLaren (1977), Boullier (1980), C. Simpson (1985), Vauchez (1987b), Vauchez, Maillet, and Songy (1987), Evans (1988).

23. Boullier and Gueguen (1975), Kerrich et al. (1980); Wojtal and Mitra (1986). Additional discussions of crack formation and annealing may be found in several papers in Chemical Effects of Water on the Deformation and Strengths of Rocks [special issue], *Journal of Geophysical Research*, v. 89, no. B6 (1984).

24. A dislocation is the *boundary* between areas of a crystal lattice in which a shift (or slip) of a set of atoms or atomic bonds has occurred and an area in which no such slip has occurred (Nicolas and Poirier, 1976, pp. 72–74). The slips may be due to the presence of point, linear, or planar defects that form initially because of the occurrence of extra atoms, the absence of atoms, or a stress-induced shift of part of the lattice. For more information, see the discussions of dislocations, slip, and twinning by Nicolas and Poirier (1976), Barber (1985), and Van Houtte and Wagner (1985).

25. See examples and discussions of experimentally and naturally deformed rocks in Carter, Christie, and Griggs (1964), Raleigh (1967), H. W. Green (1967), Avé Lallemant and Carter (1970), H. W. Green and Radcliffe (1972), J. Tullis, Christie, and Griggs (1973), C. J. L. Wilson (1973, 1980), Nicolas and Poirier (1976), Gueguen (1979a,b), Zeuch and Green (1984a,b), Mercier (1985), G. P. Price (1985), Toriumi and Karato (1985), Wenk (1985), Dell'Angelo and Tullis (1986, 1989), and Ji and Mainprice (1990).

26. For additional discussions, see Nicolas and Poirier (1976), Etheridge and Wilkie (1979), and van der Pluijm (1991).

27. Nicolas and Poirier (1976, p. 129ff.), Gottstein and Mecking (1985, p. 183).

28. Recrystallization is described in more detail by Nicolas and Poirier (1976), Bell (1978b), Gottstein and Mecking (1985), Dell'Angelo and Tullis (1989), R. W. Carlson, van der Pluijm, and Hanmer (1990), and van der Pluijm (1991).

29. For more detailed discussions of neocrystallization and metasomatic effects during mylonitization, see Drury (1974), Etheridge and Hobbs (1974), Wakefield (1977), Kerrich et al. (1977, 1980), Beach (1980), Spray and Roddick (1981), Watts and Williams (1983), McCaig (1984), Winchester and Max (1984), Sinha, Hewitt, and Rimstidt (1986), Hammond (1987), and K. O'Hara (1988).

30. The roles of fluids in promoting deformation and metamorphism are discussed in more detail by Drury (1974), Fyfe, Price, and Thompson (1978), Etheridge et al. (1984), Kerrich, La Tour, and Willmore (1984), and other authors whose work is published in the *Journal of Geophysical Research*, v. 89, no. B6 (1984), and by Ferry (1986), Wood and Walther (1986), Ridley and Thompson (1986), K. O'Hara (1988), and Glazner and Bartley (1991).

31. Also see Sinha and Glover (1978), Behr (1980), and Carreras, Julivert, and Santanach (1980).

32. Sinha, Hewitt, and Rimstidt (1986); K. O'Hara (1988, 1990a,b), T. H. Bell and Cuff (1989), K. O'Hara and Blackburn (1989), de Roo and Williams (1990), Erslev and Sutter (1990), Glazner and Bartley (1991), Tobish et al. (1991).

33. Compare this reaction to that of Bryant (1966) and see K. O'Hara (1988) for an additional reaction for the breakdown of alkali feldspar.

34. This review is based primarily on the works of Sinha, Hewitt, and Rimstidt (1986); and Bobyarchick, Edelman, and Horton (1988), and to a lesser extent on the works of J. C. Reed and Bryant (1964), Bryant and Reed (1970a), Hatcher (1971), Roper and Justus (1973), Sinha and Glover (1978), F. A. Cook et al. (1979), Horton and Butler (1986), Edelman, Lin, and Hatcher (1987), and Christensen and Syzmanski (1988). For additional references to the voluminous literature on the Brevard Zone, see Bobyarchick, Edelman, and Horton (1988).

PROBLEMS

30.1. Mylonitization and uplift of metamorphic core complexes in western North America along late detachment faults that separate a more brittle cover sequence from the underlying core may be rapid (G. A. Davis, 1988). (a) If mylonitization began at a pressure of about 0.515 Gpa (5.15 kb) at 25 m.y.b.p. and fission track, isotopic, and geobarometric analyses show that it was overprinted by a brittle fabric (mylonitization was completed) at about 23 m.y.b.p., at a pressure of about 3.30 kb, what was the rate of uplift in mm/year during this time? (b) If conglomerates reveal that the mylonites were exposed at the surface at 21 m.y.b.p., what was the rate of uplift in mm/year during the time interval 23 m.y. to 21 m.y.? (c) Considering that the mineral assemblage in the mylonite zone is

quartz–K-feldspar–oligoclase–hornblende–biotite–garnet

and the overprint assemblage is

quartz–K-feldspar–albite–white mica–biotite–chlorite,

what metamorphic facies does each assemblage represent? (d) Assuming the rock originally formed as an intrusive rock at 0.6 Gpa (6 kb) and 770° C at 26 m.y.b.p., the temperature of mylonitization was 575° C, and the overprint temperature was 350° C, sketch the P-T-t path for the core rocks of the complex on a copy of figure 24.6.

31

Alpine Ultramafic Rocks and the Mantle

INTRODUCTION

Alpine ultramafic rock bodies are irregular to elliptical bodies of ultramafic rock that occur in mountain belts (Benson, 1926; H. H. Hess, 1955; Moores and MacGregor, 1972). The rocks that comprise these bodies may form initially as (1) magmatic crystal cumulates (differentiates), (2) crystallized or recrystallized products of mantle diapirs, or (3) mantle tectonites. Thus, the bodies either form in the crust as crystallization products of mafic magmatic intrusions,[1] or they are emplaced into the crust by faulting, as mantle slabs, and by solid intrusion, as mantle diapirs. Small to large fragments of mantle or crustal rocks of any of these types may be incorporated into melanges (Moores, 1973). Evidence of these early formative and emplacement events are commonly obscured in alpine-type ultramafic rocks by subsequent metamorphism, making the histories of the rocks difficult to decipher.

Because alpine ultramafic rocks are derived from the mantle or form early in the developmental history of an orogen, they provide knowledge of generally obscure events. Mantle ultramafic rocks record histories of mantle deformation and recrystallization, mantle metasomatism, and mantle depletion resulting from partial melting, Alpine ultramafic rocks of crustal heritage may reveal the early histories of

intrusion and crystallization, metamorphism, and deformation in the mountain belt as well as a synorogenic history, which is also portrayed by the textures, structures, and minerals of the more common rocks of the belt. These attributes provide ample incentive for the study of alpine ultramafic rocks.

OCCURRENCES OF ALPINE ULTRAMAFIC ROCKS

Alpine ultramafic rocks are found worldwide in Phanerozoic mountain belts.[2] In some orogens (e.g. in the Cordilleran Orogen of western North America), they define two or more parallel, linear zones. They also occur in Precambrian terranes.[3] Young alpine ultramafic rocks form and occur in Cenozoic orogenic belts, in volcanic arcs of both continental and oceanic type, and along spreading centers and transform faults, where they constitute layered segments of the oceanic crust and mantle (Dietz, 1963; Moores and Jackson, 1974).[4] In short, alpine-type ultramafic rocks appear at present and former plate boundaries of all types.

Occurrences of alpine ultramafic rocks are not evenly distributed among the three petrogenetic types. Mantle diapirs rarely enter the crust, and thus the associated types of alpine ultramafic rocks are rare. Mantle slabs, without an associated crustal carapace, are not abundant. Most commonly, mantle rocks, with their overlying crustal rocks, are emplaced by obduction into or onto the crust (Dewey and Bird, 1970, 1971; R. G. Coleman, 1971a, 1977a). Where that crust and mantle is oceanic, the alpine ultramafic rock bodies are recognized as ophiolites. Recall that the ophiolites contain both a basal mantle tectonite and an overlying, differentiated sequence of ultramafic to silicic magmatic rocks. Thus, inasmuch as they contain both mantle rocks and crustal differentiates, they exhibit a range of characteristics. Ultramafic crustal intrusions that occur in mountain belts, especially those that are subsequently deformed, may also be considered to be alpine ultramafic bodies.

Specific and notable occurrences of alpine ultramafic rocks are found at the Bay of Islands, Newfoundland; at Thetford Mines, Quebec; in the Roxbury District and at Grafton in central and southern Vermont; in the North Carolina Blue Ridge Belt at Daybook, Webster-Addie, and Buck Creek; in the California Coast Ranges, notably at Burro Mountain, Del Puerto Canyon, and Point Sal; in the eastern and western parts of the Klamath Mountains of northern California and Oregon; in the Cascade Range of Washington, in the Twin Sisters, Darrington, and Sultan areas; at the head of the Blue River in British Columbia; at Ronda, southern Spain; at Lizard, Cornwall, England; in the Swiss Alps; at Almklovdalen, Norway; and at Dun Mountain, New Zealand.[5] These specific occurrences include mantle slabs and mantle fragments in melanges, diapiric ultramafic bodies, ophiolites, and ultramafic bodies of unknown origin. In spite of the variations in their modes of occurrence, the alpine ultramafic rocks share certain characteristics that distinguish them from the various types of igneous ultramafic bodies.

DISTINGUISHING FEATURES OF ALPINE ULTRAMAFIC ROCKS AND ROCK BODIES

The primary characteristic of alpine ultramafic rock bodies is that they occur in orogenic belts. Typically, they are deformed. Because almost any kind of ultramafic rock may be emplaced into a mountain belt, the characteristics of alpine ultramafic bodies are diverse and may overlap those of igneous ultramafic-mafic complexes. In general, however, the rock bodies are characterized by (1) olivines and orthopyroxenes with moderate to high Mg numbers, (2) rocks with tectonite fabrics, (3) lenticular to lensoid shapes, and, with some exceptions, (4) a lack of chilled margins and contact metamorphism (Thayer, 1960; E. D. Jackson and Thayer, 1972; Moores, 1973). Perhaps more than the rocks of other ultramafic bodies, alpine ultramafic rocks have undergone serpentinization. Each of the individual petrogenetic types of alpine ultramafic body—mantle slab, mantle diapir, and magmatic differentiate—is somewhat different chemically, mineralogically, texturally, and/or structurally (tables 31.1 and 31.2).

Minerals and Textures

Mineralogically, the alpine ultramafic rocks contain the same minerals as do the ultramafic igneous rocks,[6] but in addition, they may contain or consist entirely of distinctly metamorphic minerals. Typical minerals inherited from igneous protoliths or present in corresponding igneous rocks include olivine, orthopyroxenes, clinopyroxenes, Ca-rich plagioclase, and chromite and other spinels. These minerals are also stable in the mantle. Additional minerals, which occur in alpine ultramafic rocks and are stable in the mantle or are produced by crustal metamorphism, include Ca- and Mg-amphiboles (e.g., tremolite, anthophyllite, or hornblende), the phyllosilicates (talc, chlorite, the serpentines, and phlogopite), garnets, carbonate minerals (e.g., magnesite, dolomite, calcite, and aragonite), magnetite, and a host of less abundant minerals such as brucite, quartz, pyrrhotite, and garnierite.

The textures of alpine-type ultramafic rocks range from relict cumulate textures to mylonitic textures (Wicks, 1984a, 1984b; Nicolas and Poirier, 1976; Beslier, Girardeau, and Boillot, 1990).[7] Typically, alpine ultramafic tectonites dominated by olivines and pyroxenes have allotrioblastic-granular to semischistose textures, including the following:

1. *Protogranular texture*, a typically coarse, nonfoliated texture characterized by sinuous grain boundaries and intergrowths (figure 31.1a);
2. *Equigranular-mosaic texture*, a nonfoliated to very weakly foliated metamorphic texture characterized by equant polygonal grains with slightly curved to straight (equilibrium) grain boundaries, which meet at about 120° angles (figures 31.1b and 31.2a);
3. *Equigranular-tabular texture*, a semischistose texture characterized by elongate, deformed polygonal grains typified by deformation bands and related features (figure 31.2b); and
4. *Porphyroclastic texture*, a nonfoliated to semischistose texture characterized by a bimodal distribution of grain sizes, with larger, locally deformed porphyroclasts surrounded by a matrix of fine-grained minerals of the same type(s) as the porphyroclasts (figures 31.1c and 31.2c)(Mercier and Nicolas, 1975).[8]

Rocks dominated by phyllosilicates (e.g., talc and antigorite) range from diablastic to schistose in texture.[9]

Magmatic Differentiates

Magmatic differentiates that become alpine-type ultramafic rock bodies have the characteristics of the type of igneous protolith from which they were derived. These rocks are distinguished by (1) dominant or relict igneous textures, including

Table 31.1   Characteristics of Alpine-type Ultramafic Rocks

Characteristic	Magmatic Differentiates	Mantle Slabs	Mantle Diapirs
Typical Rock Types	Harzburgite, lherzolite, dunite, pyroxenites, talc-amphibole schist, serpentinite, chlorite schist	Lherzolite, harzburgite, dunite, serpentinite, talc-Amphibole schists, chlorite schist	Lherzolite, harzburgite, spinel and garnet pyroxenites, serpentinite, talc-amphibole schist, chlorite schist
Peridotite Chemistry	$Al_2O_3 = 0.1\text{–}3.5$ $CaO = 0.1\text{–}17.4$ Mg no. $= 0.68\text{–}0.85$	$Al_2O_3 = 0.1\text{–}4.4$ $CaO = 0.0\text{–}7.7$ Mg no. $= 0.80\text{–}0.91$	$Al_2O_3 = 0.6\text{–}6.4$ $CaO = 0.4\text{–}5.7$ Mg no. $= 0.58\text{–}0.92$
Typical Minerals	Orthopyroxene, clinopyroxene, olivine, talc, plagioclase, chlorite, hornblende, actinolite, tremolite, anthophyllite, serpentines, chromite, magnetite, phlogopite	Orthopyroxene, clinopyroxene, olivine, talc, hornblende, chlorite, actinolite, tremolite, anthophyllite, chromite, magnetite, serpentines	Orthopyroxene, clinopyroxene, olivine, plagioclase, garnet, chromite, spinel, magnetite, chlorite, serpentines, talc, tremolite, anthophyllite
Mineral Chemistry	Opx $(Al_2O_3) = 0.0\text{–}1.4$ Cpx $(Al_2O_3) = 0.7\text{–}4.5$ Olivine (Fo) $= 53\text{–}95$ Opx (MgO) $= 43\text{–}92$ Cr Spinel (Cr/Cr+Al) $= 0.00\text{–}0.85$	Opx $(Al_2O_3) = 0.0\text{–}6.0$ Cpx $(Al_2O_3) = 0.3\text{–}4.5$ Olivine (Fo) $= 84\text{–}95$ Opx (MgO) $= 85\text{–}94$ Cr Spinel (Cr/Cr+Al) $= 0.05\text{–}0.90$	Opx $(Al_2O_3) = 0.5\text{–}6.7$ Cpx $(Al_2O_3) = 0.5\text{–}8.5$ Olivine (Fo) $= 75\text{–}92$ Opx (MgO) $= 88\text{–}91$ Cr Spinel (Cr/Cr+Al) $= 0.05\text{–}0.50$
Typical Textures	Hypidiomorphic-granular, ophitic, diabasic, cumulate, diablastic, schistose	Protogranular, equigranular-mosaic, equigranular-tabular, porphyroclastic, mylonitic, diablastic, schistose	Porphyroclastic, equigranular-mosaic, protogranular, schistose, diablastic
Typical Structures	Primary layering, foliation, isoclinal folds, dikes (in some bodies), shear zones	Isoclinal folds, flow layering, foliation, lineation, podiform chromite bodies, dikes, ductile shear zones	Isoclinal folds, flow layering, foliation, lineation, dikes, ductile shear zones

Sources: See note 6.

cumulate, ophitic, diabasic, or hypidiomorphic-granular textures, (2) generally low amounts of alumina in the peridotites and their included orthopyroxenes, and, (3) primary layering in some bodies (table 31.1). Mafic to felsic differentiates are usually associated with these kinds of ultramafic rocks and many bodies may contain numerous dikes. In addition, some bodies consist of rocks with a wide range of olivine, orthopyroxene, and clinopyroxene compositions.[10]

The most common alpine-type, ultramafic, differentiated magmatic body is the basal cumulate section of ophiolites. Within such sections, dunite, lherzolite, harzburgite, wehrlite, and/or pyroxenites form interlayered sequences.[11]

Table 31.2 Chemical Analyses of Alpine Ultramafic Rocks

	1	2	3	4	5	6	7
SiO_2	27.2[a]	39.97	40.67	41.62	43.61	44.36	50.69
TiO_2	0.95	0.02	0.01	nd	0.06	0.12	0.06
Al_2O_3	19.6	0.37	0.75	1.08[b]	2.93	3.49	1.56
Fe_2O_3	35.3[c]	12.41[c]	1.15	9.28[c]	1.09	—	5.58[c]
FeO	—	—	6.56	—	6.74	8.29[c]	—
MnO	0.22	0.17	0.12	nd	0.11	0.14	0.11
MgO	10.4	46.75	48.77	46.40	39.86	39.38	23.99
CaO	0.60	0.26	0.00	0.20	2.66	2.96	16.21
Na_2O	0.05	0.09	0.00	nd	0.26	0.28	0.09
K_2O	0.04	0.00	0.03	nd	0.01	0.01	0.00
P_2O_5	0.03	0.00	0.02	nd	0.00	0.01	0.02
H_2O+	5.30	—	1.38	1.95	7.48	2.52	—
CO_2	0.10	—	0.12	nd	0.49	0.43	—
Other	tr	11.00[d]	0.71	nd	0.83	tr	4.17[d]
Total	99.8	100.70	98.91	100.53	100.05	101.99	99.01

Sources:

1. Magnetite-garnet-chlorite schist, Greer Hollow Ultramafic Body, Blue Ridge Belt, North Carolina. Analyst: XRAL (L. A. Raymond, previously unpublished data).

2. Dunite (cumulate), North Arm Mountain, Bay of Islands Ophiolite, Newfoundland (Komor et al., 1985).

3. Dunite, Day Book Dunite, North Carolina, Analyst: R. Stokes (Kulp and Brobst, 1954).

4. Harzburgite ("saxonite") sample A-1, Addie, North Carolina. (C. E. Hunter, 1941).

5. Plagioclase lherzolite, sample R224, Rhonda peridotite, Spain. Analyst: E. Jurosewich (Dickey, 1970).

6. Garnet lherzolite, sample R255, Rhonda peridotite, Spain, (F. A. Frey, Suen, and Stockman, 1985). $SiO_2 \rightarrow P_2O_5$ recalculated and listed here on a volatile free basis.

7. Clinopyroxene-rich wehrlite (cumulate), North Arm Mountain, Bay of Islands Ophiolite, Newfoundland (Komor et al., 1985).

[a]Values in weight percent

[b]Includes minor Cr_2O_3 and TiO_2.

[c]Total iron as this value.

[d]Loss on ignition.

nd - not determined.

tr - trace.

Mantle Slabs

Subcrustal mantle, faulted into the crust to form alpine-type ultramafic rock bodies, is characterized by high Mg numbers in both the peridotites (0.80–0.88) and in the constituent olivines and pyroxenes (table 31.1). Lherzolite and harzburgite are the most common rock types, along with their serpentinized equivalents. Dunites occur primarily as pods or dikes (Moores, 1969; Quick, 1981b; Nicolas, 1989). Similarly, minor associated mafic rocks occur only as dikes.

Many mantle slabs represent the subcrustal parts of ophiolites that were separated by faulting from the overlying oceanic crustal sections, when the rocks were obducted onto a continental margin. In some cases, fragments of the crustal rocks remain attached to the ultramafic rocks, revealing the ophiolitic character of the ultramafic body. Other slabs may represent mantle fragments thrust up between colliding continental masses (Nicolas and Poirier, 1976, ch. 11). Mantle slabs of either type are locally incorporated into melanges,

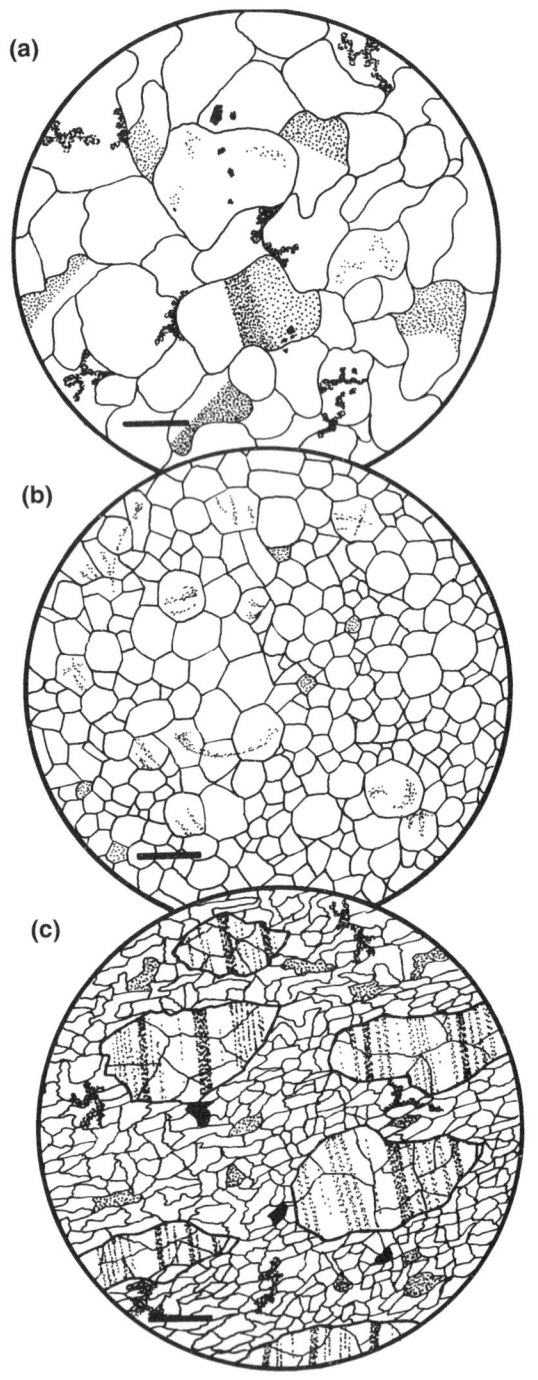

either during thrusting or during later submarine landsliding (Gansser, 1974; B. M. Page, 1981; Sarwar and DeJong, 1984).

The rocks of the mantle slabs may show a full range of tectonite fabrics, including those dominated by protogranular textures, equigranular-mosaic textures, equigranular-tabular textures, porphyroclastic textures, mylonitic textures, diablastic textures, and schistose textures (Den Tex, 1969; Nicolas and Violette, 1982). These textures are found in rocks with flow layering, lineation, lattice preferred orientations, foliations, isoclinal folds, faults, or combinations of these features (Nicolas and Boudier, 1975; Christensen and Lundquist, 1982; Nicolas, 1986b; 1989).

Mantle Diapirs

Garnet pyroxenites, garnet peridotites, and spinel peridotites are distinctive rock types in the mantle diapirs. Lherzolites of various types are also characteristic. The diapiric peridotites tend to be more aluminous than other alpine-type peridotites, and hence, so are the minerals. Pyroxenes are relatively aluminous and the chromium spinels are high in aluminum. In addition, the olivines and orthopyroxenes have a high Mg content. Typically, the diapiric rocks have porphyroclastic textures.

The mantle diapirs are distinct from most other alpine ultramafic rock bodies in that they are emplaced at high temperatures and produce a contact metamorphic aureole (D. H. Green, 1964; Loomis, 1972a). Their diapiric nature may be revealed by their structural relationships with surrounding rocks, by their internal structures, and by the mineral and rock chemistry. Internal structures may include flow layering, foliation, lineation, isoclinal folds, and dikes.

THE NATURE OF THE UPPER MANTLE: A BRIEF SURVEY

Several features indicate that many alpine ultramafic rock bodies represent fragments of the upper mantle.[12] Mantle samples are also provided by the ultramafic xenoliths that occur in some mafic magmas (Mercier and Nicolas, 1975; Basu, 1977b). Together with various geochemical and geophysical analyses, these rocks yield an image of the character of the upper parts of the mantle, which is structurally, texturally, and mineralogically complex.[13]

The uppermost mantle is a heterogeneous mass of rock dominated by peridotites.[14] Lherzolite and harzburgite are the main rock types, but dunite, "pyrolite," and eclogite may

---

Figure 31.1 Line drawings of some microtextures found in alpine ultramafic rocks, (a) Protogranular texture (XN), (b) Equigranular-mosaic texture (PL), (c) Porphyroclastic texture (XN). Scale bars are 2 mm.

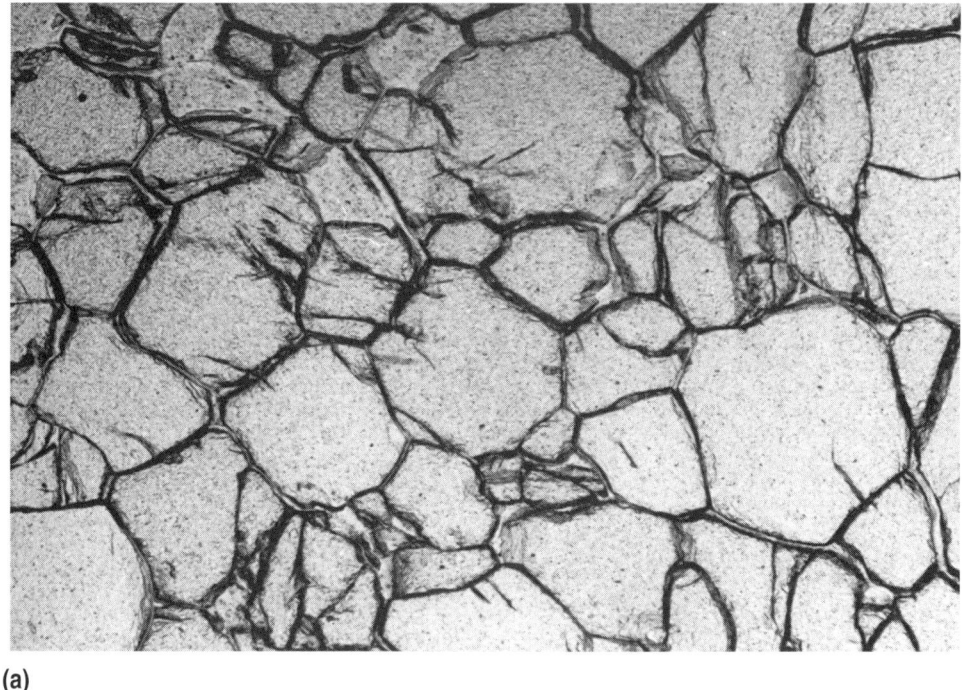

(a)

(b)

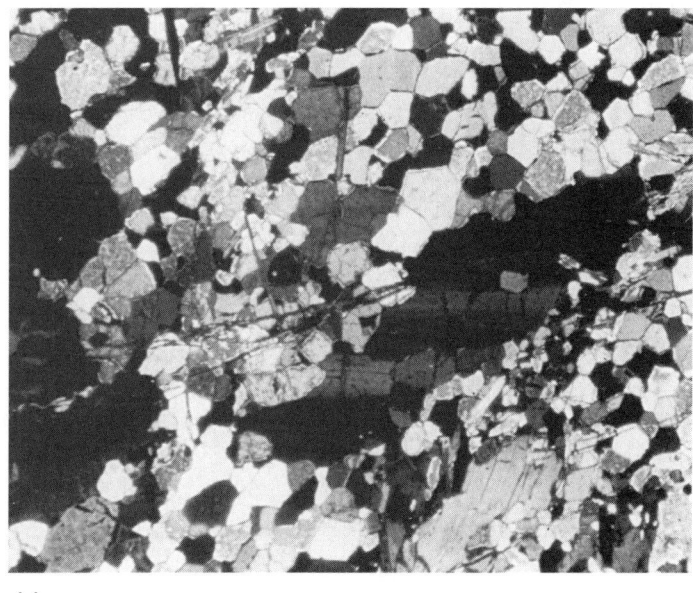

(c)

Figure 31.2 Photomicrographs of some textures found in alpine ultramafic rocks. (a) Equigranular-mosaic texture in dunite, Day Book Dunites, North Carolina. (PL). (b) Equigranular-tabular texture in dunite, Corundum Hill Ultramafic Body, North Carolina.(XN). (c) Porphyroclastic texture in dunite, Day Book Dunites, North Carolina, (XN). Long dimension of photo (a) is 3.25 mm, (b) is 2.75 mm, and (c) is 2.75 mm.

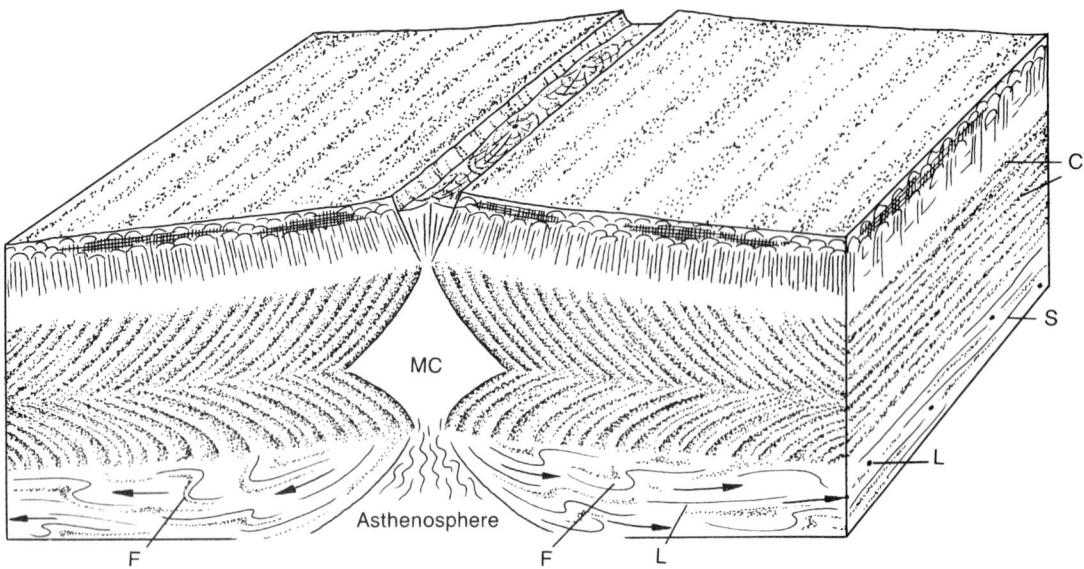

Figure 31.3
Schematic diagram showing orientations of structures in mantle rocks at and near a spreading center. C = cumulate layering, F = folds, L = lineation, MC = magma chamber, S = foliation.
(Based on Juteau et al., 1977; Girardeau and Nicolas, 1981; Christensen and Lundquist, 1982; Girardeau and Mercier, 1988; Casey et al., 1983; and Nicolas, 1986)

constitute parts of the upper mantle. Two main peridotite "subtypes" are recognized, the harzburgite subtype and the lherzolite subtype, (E. D. Jackson and Thayer, 1972; Boudier and Nicolas, 1985; Nicolas, 1986b; 1989). The harzburgite subtype is associated with ophiolites and represents depleted mantle rock that underlies fast-spreading ridges. The lherzolite subtype, less common in ophiolites, characterizes mantle diapirs and slow-spreading ridges, Dunite and chromite pods, dikes, and layers are largely confined to the upper (near-crustal) parts of the harzburgite subtype.

Textures of the mantle rocks range from protogranular to porphyroclastic and mylonitic. Foliations and lattice preferred orientations are commonly subhorizonatal to moderately inclined (Juteau et al., 1977; Christensen and Lundquist, 1982; Nicolas, 1986b; 1989), but in some areas are steeply inclined (Nicolas and Violette, 1982). Both layering and foliations may be folded (Nicolas and Boudier, 1975). Isoclinal folds are predominant. All of these features apparently form below spreading ridges. Lineations that form in mantle rocks away from the axial regions of spreading ridges may parallel and indicate the spreading direction (figure 31.3).

SERPENTINIZATION

Ultramafic rocks are prone to low-temperature metamorphism and alteration. This is the case, because their constituent high P-T phases, at near-surface and low-temperature conditions, are far from their fields of stability. In addition, mantle rocks are rather dry, and the upper levels of the crust are generally permeated with water, which readily combines with the anhydrous phases of the mantle rocks to produce new hydrous phases.

General discussions of regional and contact metamorphism of ultramafic rocks were presented in preceding chapters. In those kinds of metamorphism, hydrous phases such as tremolite, talc, anthophyllite, and serpentines are produced by metamorphism of ultramafic rock types. Although serpentine minerals were mentioned among these minerals previously—because they are the dominant phase, comprising more than 90% of the modes of most ultramafic rocks equilibrated at temperatures below about 500° C, and because serpentine is a nearly ubiquitous constituent of crustal ultramafic rocks—the processes of serpentinization warrant further comment.[15]

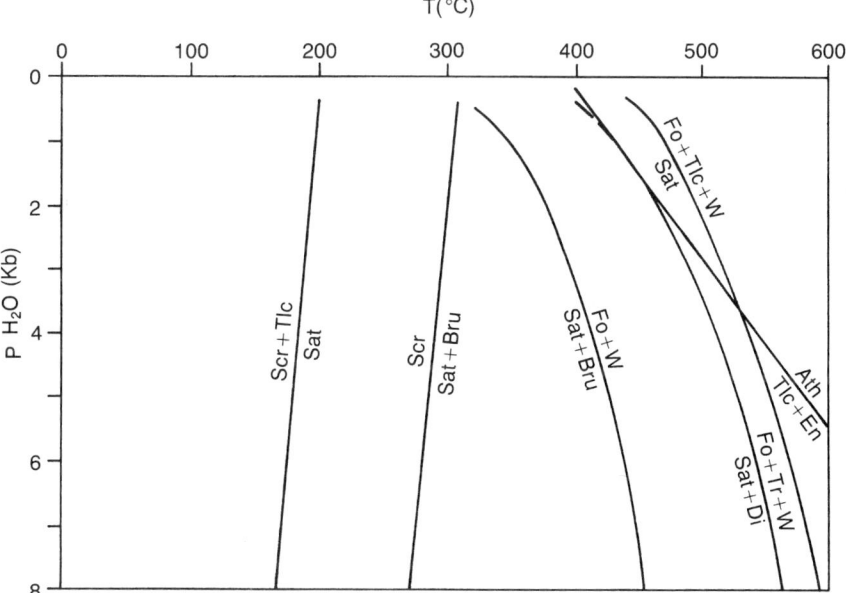

T(°C)

Figure 31.4 Petrogenetic grid for low- temperature, CO_2-free assemblages in ultramafic rocks.
Ath - anthophyllite, Bru - brucite, Di - diopside, En - enstatite, Sat - serpentine (antigorite), Scr - serpentine (chrysotile), Tlc - talc, Tr - tremolite, W - water.

(Modified from Mellini, Trommsdorf, and Compagnoni, 1987; based in part on Evans and Trommsdorf, 1970; Evans et al., 1976; and Day, Chernosky, and Kumin, 1985)

Serpentinization is the process or processes by which ultramafic or mafic rocks are transformed into serpentinites. These processes produce the three main varieties of serpentine that comprise the serpentinites—lizardite, chrysotile, and antigorite.[16] Lizardite, a planar-structured serpentine, and chrysotile, a cylindrically structured serpentine, are polymorphs with the composition $Mg_3Si_2O_5(OH)_4$. Antigorite is compositionally (as well as structurally) distinct, with Mg/Si values somewhat less than the 1.5 of lizardite and chrysotile.[17] Antigorite is generally stable at higher temperatures than the latter two phases, as discussed in preceding chapters and revealed by the petrogenetic grid in figure 31.4.

Inasmuch as alpine ultramafic rocks are composed primarily of one or more of the Mg-rich varieties of olivine, orthopyroxene, and clinopyroxene, serpentinization may result essentially from the addition of water. If that is the case, the serpentinization may be an isochemical, or *constant-chemical*, process. In serpentinization in which there is "constant chemical composition," the structural changes in the minerals require an increase in the volume of the ultramafic body. For example, in the equation

olivine + silica in solution → serpentine

$$3Mg_2SiO_4 + H_4SiO_4 + 2H_2O \rightarrow 2Mg_3Si_2O_5(OH)_4$$

(131 cc)   introduced   (220 cc)

there is a significant increase in volume.[18] For any given body, evidence suggestive of isochemical serpentinization may include:

1. Internal fracturing of the body (indicating deformation associated with a volume increase);
2. Deformation of the surrounding country rock (indicating deformation associated with a volume increase);
3. Identical chemistry of ultramafic protoliths and serpentinized equivalents (except for water content); and
4. The absence of evidence of metasomatism in the surrounding country rocks.[19]

Based on chemical evidence from the Burro Mountain, California, ultramafic body, from which only CaO was lost, R. G. Coleman and Keith (1971) argued that serpentinization there was nearly isochemical.

In contrast, serpentinization may occur as an allochemical, *constant-volume* process, in which Mg, Ca, Fe, Si, and other elements are transported by the serpentinizing fluids. Such a process is represented by the equation

olivine + water → serpentine + Mg ion + hydroxyl ion + silica in solution

$$5Mg_2SiO_4 + 10H_2O \rightarrow 2Mg_3Si_2O_5(OH_4) + 4Mg^{++} +8(OH) + H_4SiO_4$$

(219 cc) introduced   (220 cc)   removed in solution

in which the aqueous fluid that facilitates the serpentinization removes excess ions (Turner and Verhoogen, 1960, pp. 318–319). Spring waters currently emanating from serpentinites do, in fact, carry a number of ionic species. I. Barnes and O'Neil 1969 I. Barnes, Rapp, and O'Neil 1972 . This observation is not only consistent with the hypothesis that some serpentinization is a constant-volume process, it suggests that serpentinization is an ongoing process, even at ambient temperatures at the Earth's surface. Evidence cited in support of constant-volume serpentinization includes:

1. Pseudomorphic replacement of olivine and pyroxene by serpentine;[20]
2. Density decrease with concomitant porosity increase in the serpentinized ultramafic body, relative to the unserpentinized rock;
3. Bulk rock chemical differences between serpentinites and their unserpentinized protoliths;
4. Precipitation of transported ions in metasomatically altered rock at a distance from the point of dissolution (i.e., the point of serpentinization); and
5. The presence of undisturbed primary layering.[21]

Condie and Madison (1969) demonstrated that serpentinization of the ultramafic body at Webster-Addie, North Carolina, involved loss of FeO and MgO and gains in SiO_2.

Serpentinizing fluids have a complex history that results in a variety of changes in the invaded rocks. Both major element and isotopic studies reveal that these solutions are of variable compositions and sources and may begin as ocean water, connate water, metamorphic water, or meteoric water.[22] During serpentinization the fluids may be highly reducing, as evidenced by the presence of native nickel-iron ($FeNi_3$), the mineral awaruite, which occurs in some serpentinites.[23] The pH values of fluids emanating from serpentinites are high, ranging from 8 to 12.

Metasomatism and vein formation in rocks adjacent to serpentinites is well known.[24] Notably, **rodingites,** calcium-metasomatized gabbros containing diopside and grossularite garnet, are distinctive metasomatically altered rocks formed in contact with serpentinite. Sandstones and granitoid rocks may experience similar metasomatism, and, in fact, serpentinites may even be remetamorphosed by fluids that have changed character during serpentinization.

Serpentinites commonly are texturally and structurally complex (O'Hanley, 1987, 1988).[25] They may have lepidoblastic, nematoblastic, and porphyroblastic schistose textures, diablastic textures, and, in some cases, blastoporphyritic textures. Mesh texture, in which olivine grains are replaced from grain margins and fractures inward, is common (figure 31.5). Cross-cutting veins of serpentine typify many serpentinites. Comb structure-like veins, with chrysotile grown perpendicular to the vein walls, are widespread. These complex textural and structural relations make interpretation of serpentinite metamorphic histories difficult.

EXAMPLES OF ALPINE-TYPE ULTRAMAFIC ROCKS

The Bay of Islands Ophiolite

The Cambrian-Ordovician Bay of Islands Ophiolite Complex[26] is an ophiolite that is considered to have formed at an oceanic spreading center.[27] The complex consists of four massifs—the Table Mountain, North Arm Mountain, Blow-Me-Down Mountain, and Lewis Hills massifs—exposed in the west-central coastal region of Newfoundland (figure 31.6). The North Arm Mountain and Blow-Me-Down Mountain massifs have a relatively complete ophiolite stratigraphy. Of the two incomplete sections, the Table Mountain section is the least complete, but it contains a thick section of ultramafic rock. The Bay of Islands Complex rocks are only moderately deformed, but were thrust into their present site in the orogen. To the west, along the coast itself, is the Coastal Complex, a highly deformed ophiolite complex thought to have been deformed in a transform fault environment adjacent to the Bay of Islands spreading center (Karson and Dewey, 1978; J. F. Casey et al., 1983).

Like all complete ophiolites, the Bay of Islands Ophiolite Complex contains both basal ultramafic tectonite and ultramafic cumulate (plutonic) sections (figure 31.7). The ultramafic tectonite is dominantly harzburgite, but includes lherzolite near the base, abundant dunite near the top, and local orthopyroxenite, websterite, and chromitite within the section (Christensen and Salisbury, 1979; Komor, Elthon, and Casey, 1985).[28] Serpentinization is widespread. The cumulate sections contain interlayered dunite, wehrlite, lherzolite, websterite, clinopyroxenite, minor harzburgite, and chromitite, with progressively more gabbro at stratigraphically higher levels. Some ultramafic rocks intrude the cumulates (Bedard, 1991). Both serpentinization and penetrative deformation have affected rocks at and near the base of the cumulate complex.

The basal harzburgite tectonite and associated dunite and lherzolite are interpreted to be residual mantle rocks. Mesoscopic tight to isoclinal folds, characteristic of other mantle peridotites, are present within this mantle section.[29] The rocks are lineated and foliated, with porphyroclastic, equigranular-mosaic, equigranular-tabular, and, especially near the base, mylonitic textures (Girardeau and Nicolas, 1981). Olivines show a limited range of composition (Fo_{89-92}) and orthopyroxenes are similarly Mg-rich (En_{88-92}) (J. Malpas, 1978; Suen, Frey, and Malpas, 1979).

Figure 31.5 Photomicrograph of mesh texture (above) and a vein of serpentine (v) in a serpentinite from the Webster-Addie Ultramafic Body, North Carolina. (XN). Long dimension of photo is 1.27 mm.

In the plutonic complex, cumulate textures are replaced near the base by equigranular-mosaic (equivalent to xenoblastic-granular) and porphyroclastic textures. Thus, the rocks have become tectonites. In local areas, mylonitic textures also occur. The olivines and pyroxenes that define these various textures have compositions that span a compositional range typical of differentiated igneous rocks, a range slightly greater than that of the residual tectonites.[30] Forsterite components of olivine range from Fo_{85} to Fo_{92}. Orthopyroxene Mg numbers range from 0.79 to 0.92 and

clinopyroxenes show a similar range (0.85–0.94). The peridotites have correlative magnesium numbers (0.77–0.91) (Elthon, Casey, and Komor, 1982). The Cr numbers of spinels range from 17 to 70.

The structures and textures of the Bay of Islands Complex rocks suggest that the complex was formed at an ocean spreading center. Major and trace element chemistry of the mafic and ultramafic rocks corroborate that interpretation.[31] Depleted mantle harzburgites and related rocks formed a foundation upon which the igneous rocks of the ophiolite formed.

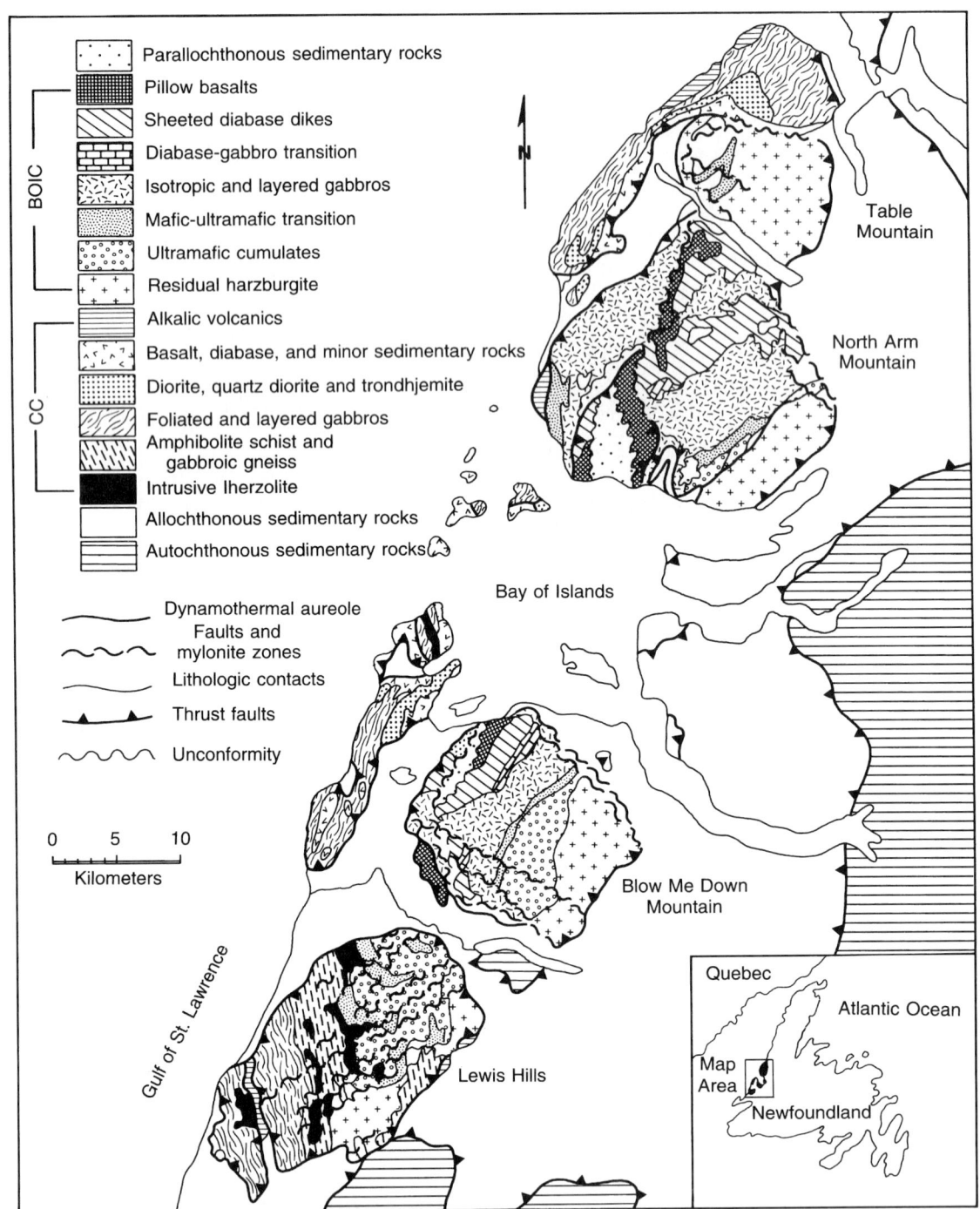

Figure 31.6
Map of the Bay of Islands Ophiolite Complex and surrounding areas, Newfoundland.
(From Casey et al., 1983)

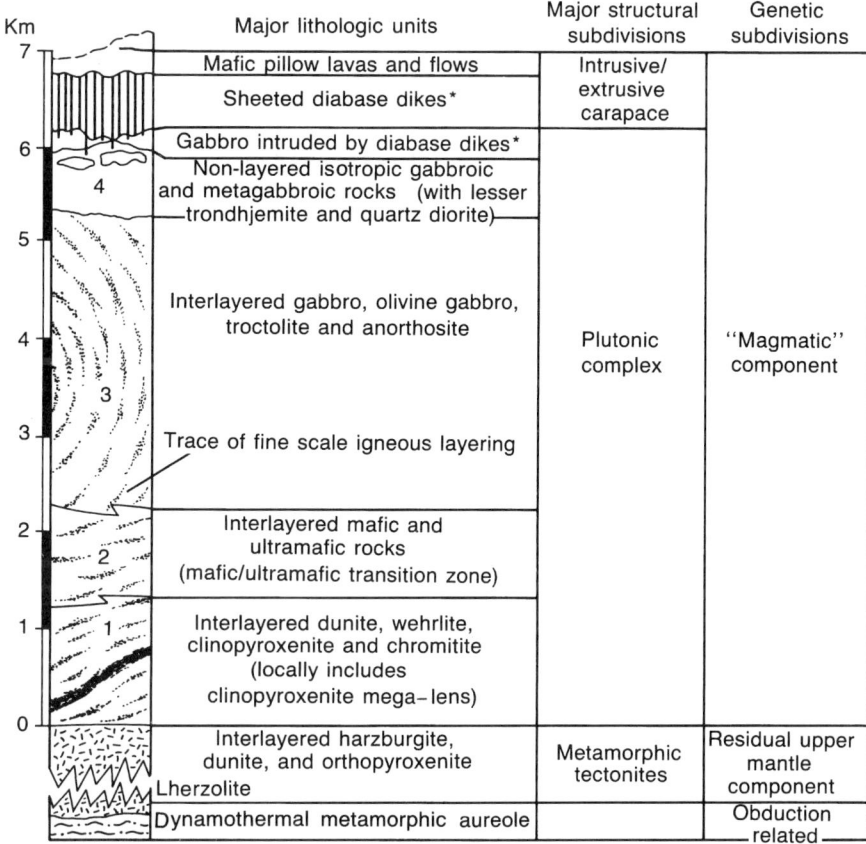

Figure 31.7 Generalized columnar section through the Bay of Islands Ophiolite Complex. *(Redrawn from Casey and Karson, 1981)*

After it was formed, the Bay of Islands Ophiolite Complex was thrust faulted (obducted) into the Newfoundland segment of the Appalachian Orogen (Dewey and Bird, 1970; W. R. Church and Stevens, 1971; Harold Williams, 1975), making it an alpine ultramafic-mafic complex. Clearly, some alpine ultramafic complexes represent oceanic crust and mantle. To the extent that ophiolites also represent both back-arc crust and mantle *and* island arc crust and mantle, rocks derived from these settings may also become alpine ultramafic rock bodies.

Alpine Ultramafic Mantle Slabs in the North Carolina Blue Ridge Belt

Alpine ultramafic rocks occur as scattered, small to large slabs and lenses (<100 m to >10 km) in two broad bands within the southern Appalachian Orogen (H. H. Hess, 1955; Larrabee, 1966; Misra and Keller, 1978). The western band, of interest here (figure 31.8), occurs primarily in the Eastern Blue Ridge Belt, but ultramafic rocks are also present in the Western Blue Ridge and Inner Piedmont Belts.[32] A number of studies have shown that the ultramafic rock bodies are polydeformed dunites, peridotites (primarily harzburgites), pyroxenites, serpentinites, and various talc, anthophyllite, tremolite, and chlorite schists.[33]

None of the North Carolina Blue Ridge bodies shows evidence of magmatic emplacement. There are no documented apophyses or dikes extending from the bodies into the country rocks, the bodies do not have chilled margins, and contact metamorphism has not been demonstrated. Instead, the rocks are generally concordant to slightly discordant masses with internal fabrics consistent with polydeformation and multiple periods of metamorphism, primarily of a retrograde nature (tables 31.3 and 31.4).[34] Isoclinal folds revealed locally by chromite and magnetite layers and an almost complete range of mantle slab textures have been recognized (Abbott and Raymond, 1984). Although some bodies are intimately associated with mafic rocks, including metatroctolites and amphibolites, few can be demonstrated to be ophiolites.[35] Virtually all olivine- and pyroxene-rich bodies have a reaction wallrock ("blackwall") of chlorite-talc-amphibole schist, like those described by R. F. Sanford (1982). Especially in the northern

667

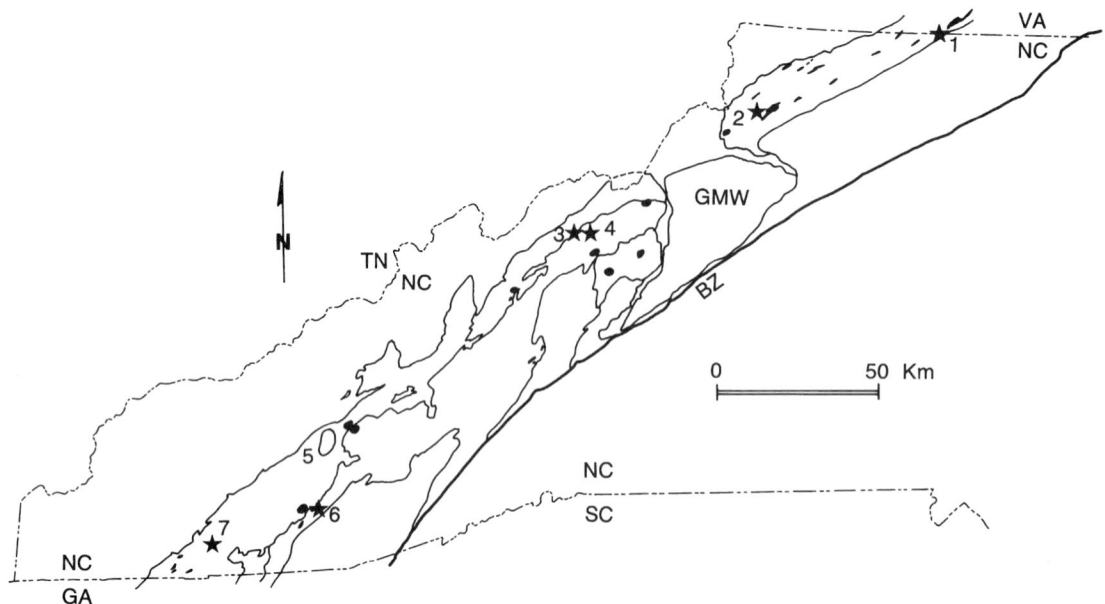

Figure 31.8 Map showing the locations of selected ultramafic bodies (stars, black dots, ellipses, and lines) scattered across the North Carolina segment of the eastern Blue Ridge Belt of the southern Appalachian Orogen. BZ = Brevard Fault Zone, GMW = Grandfather Mountain Window. Size of bodies is generally exaggerated. Many more ultramafic bodies exist than are shown. Numbered bodies (stars) correspond to those for which data are presented in table 31.4.

(Modified from Larrabee, 1966; Z. A. Brown et al., 1985)

part of North Carolina and in Virginia, olivines and pyroxenes in the bodies are extensively replaced by chlorite, serpentine, talc, tremolite, and anthophyllite. Representative CMS phase diagrams for the earlier and later phase assemblages are shown in figure 31.9a.

Chemically, the Blue Ridge peridotites are low in Al_2O_3 and CaO and relatively high in MgO. The latter results in high Mg numbers, typical of olivine-rich rocks and mantle rocks. The range of Mg numbers is small. Alkalis are nearly absent. These criteria are consistent with the rock masses being mantle slabs.

The high magnesium numbers of the peridotites, dunites, and pyroxenites are reflected in the compositions of the major minerals. Olivines have a limited range and high values of forsterite component (Fo = 84–95).[36] Orthopyroxenes are similarly magnesium-rich. Chromites, which are iron-rich and aluminum-poor, plot in a wide field that extends from the field of alpine-type chromites along a metamorphic chromite trend, indicating that metamorphic re-equilibration of the chromites and their compositions has occurred (figure 31.9b) (Lipin, 1984). Using the Evans and Frost (1975) olivine-spinel geothermometer, Lipin (1984) estimates that this metamorphic re-equilibration occurred at a temperature of about 700° C.

All of the features observed are consistent with the Blue Ridge ultramafic bodies being alpine ultramafic bodies, probably predominantly of the mantle slab type.[37] Although lherzolites and pyroxenites exist, where olivine and pyroxene mineralogies are preserved, it is evident that the dominant rock types are harzburgite and dunite. Metamorphism, however, has obscured and complicated the history of these rocks.

The bodies at four localities reflect the variations in petrology and structure within this suite of alpine ultramafic rocks (figure 31.10, table 31.4). In the north, the Greer Hollow body (locality 2 in figure 31.8) is a small, folded, concordant body composed predominantly of anthophyllite and chlorite schists. The Day Book bodies (locality 3) are dominantly dunite. The Webster-Addie body (locality 5), for which websterite is a namesake, is a folded, concordant, composite sheet containing abundant dunite and subordinate pyroxenites, serpentinites, and other rocks. In the south, the Buck Creek body (locality 7) is a part of the Chunky Gal Mountain Mafic-Ultramafic Complex.

The *Greer Hollow Ultramafic Body*, described by Raymond et al. (1988), consists of crudely teardrop-shaped mass of rock (figure 31.10a). Two mappable units exist within the body: (1) a chlorite schist unit consisting of interlayered magnetite + chlorite schist and anthophyllite ± tremolite + chlorite schist, and

Table 31.3 Mineral Associations in Southern Appalachian Blue Ridge Ultramafic Rocks

Association A-1 (Oli + Chr ± Opx ± Cpx ± Hbl ± Mtt)

1a	Olivine ± Chromite	
1b	Olivine + Orthopyroxene ± Chromite	
1c	Olivine ± Orthopyroxene ± Clinopyroxene ± Chromite	
1d	Olivine + Clinopyroxene + Hornblende + Chromite + Magnetite (Mtt)	

Association A-2 (Oli ± Opx ± Cpx ± Tre ± Chl ± Chr)

2a	Olivine + Chromite + Chlorite	
2b	Olivine + Tremolite + Chlorite	
2c	Olivine + Orthopyroxene + Tremolite	
2d	Clinopyroxene + Tremolite ± Chlorite + Chromite	

Association A-3 (± Ant ± Tre ± Tlc ± Chl ± Phl ± Mtt ± Mgt ± Grt)

3a	Anthophyllite ± Magnetite	
3b	Talc ± Magnetite	
3c	Anthophyllite + Talc ± Magnetite	
3d	Anthophyllite ± Talc ± Tremolite ± Chlorite ± Magnetite ± Magnesite (Mgt)	
3e	Talc + Chlorite ± Tremolite ± Magnetite ± Magnesite	
3f	Tremolite + Chlorite ± Magnetite ± Magnesite	
3g	Talc ± Tremolite + Magnesite + Phlogopite	
3h	Chlorite ± Magnetite ± Garnet (Grt)	

Association A-4 (± Sat ± Mtt ± Chl ± Bru ± Tlc ± Tr)

4a	Serpentine (antigorite = Sat) ± Magnetite ± Chlorite	
4b	Serpentine (antigorite) + Brucite ± Magnetite[1]	
4c	± Serpentine (antigorite) ± Talc ± Tremolite ± Chlorite ± Magnetite	

Association A-5 (± Slz ± Scr ± Mtt ± Tlc ± Chl ± Tr ± Silica ± Mgt ± others)

5a	± Serpentine (lizardite, Slz) ± Serpentine (chrysotile, Scr) ± Magnetite	
5b	± Serpentine (lizardite, Slz) ± Serpentine (chrysotile, Scr) ± Talc ± Tremolite ± Chlorite ± Magnetite	
5c	± Serpentine (lizardite, Slz) ± Serpentine (chrysotile, Scr) ± Silica minerals ± Magnetite	
5d	Silica minerals ± Magnesite ± Chlorite	
5e	Miscellaneous vein minerals (e.g., aragonite, garnierite)	

Sources: See note 33.

[1]This association was reported by Swanson et al. (1985) from the Day Book body, as an early association occurring between A-2 and A-3, but the author has been unable to confirm its existence.

Table 31.4   Observed Metamorphic Associations in Blue Ridge Ultramafic Rock Bodies

Localities[1]	Assemblages					Sources
	A-1[2]	A-2	A-3	A-4	A-5	
1. Edmonds	1a		3b, 3d, 3h	4a? 4c?		Scotford and Williams (1983), Raymond, unpublished data
2. Greer Hollow	1a, 1b, 1c,	2b	3a, 3d, 3h	4a? 4c?	5a	Raymond et al. (1988) Raymond (unpublished data)
3. Day Book	1a, 1b	2a, 2b	3b, 3e, 3g	4a, 4b	5a, 5d, 5e	S. E. Swanson (1981), Swanson et al. (1985), Raymond (unpublished data)
4. Woody	1a, 1c		3a, 3d, 3e	4a	5a? 5d	Kingsbury and Heimlich (1978), Raymond (unpublished data)
5. Webster-Addie	1a, 1c,	2b	3a, 3b, 3c, 3f	4a	5a, 5d	R. Miller (1951a), Raymond (unpublished data)
6. Corundum-Hill	1a, 1b		3d	4a?	5a	Yurkovich (1977), Raymond, (unpublished data)
7. Buck Creek	1a, 1d		3d	4a?	5a	McElhaney and McSween (1983), Raymond, unpublished data

[1] Localities correspond to those shown in figure 31.8.
[2] Assemblages are those listed in table 31.3.

(2) a porphyroblastic garnet ± magnetite + chlorite schist unit (analysis 1, table 31.2). Isolated pods of metamorphosed dunite and peridotite, boudins of hornblendite and epidote-hornblende schist, and veins of serpentinite also occur within the body. At least four, and possibly as many as six, stages of metamorphic recrystallization are reflected by various phase assemblages and textures. The olivines and pyroxenes in early-formed porphyroclastic dunites, harzburgites (?), and lherzolites (?) were replaced by tremolite, then by anthophyllite + chlorite and garnet + chlorite, and finally by magnetite + serpentine assemblages (tables 31.3 and 31.4). Available chemical analyses indicate Mg numbers for the rocks between 50 and 90.[38]

The Greer Hollow body is concordant and is infolded with hornblende schists and gneisses. At least three fold generations are evident in the body. An early isoclinal fold set has experienced postfolding extension and boudinage associated with the development of the dominant foliation in both the body and the country rock. That foliation has been refolded by two open to tight fold sets. Near its eastern margin, hornblende schist and gneiss, metagabbro tectonites, and a small mass of kyanite-bearing pelitic schist abut the body. Here and elsewhere, a thin talc + actinolite (?) selvage separates the ultramafic rocks from the country rock.

The P-T conditions of metamorphism can be estimated, assuming that the anthophyllite did not crystallize out of its field of stability and that the dominant foliation of the kyanite-bearing pelitic schist is coeval with the dominant foliation in the ultramafic rock. Overlap of the fields of stability of these phases suggests P-T conditions of about 0.7 Gpa (7 kb) and 700–800° C.

The *Day Book dunites* (there are two adjacent bodies) occur in a terrane composed of hornblende schists and gneisses, kyanite-bearing pelitic schists and gneisses, and quartz-feldspar semischists (figure 31.10b) (Brobst, 1962; S. E. Swanson, 1981; Raymond, unpublished data).[39] The two bodies are composed primarily of olivine, with minor amounts of orthopyroxene and chromite, plus a wide variety of other minerals formed during later metamorphic events. Thus, dunite is the dominant rock type, but minor harzburgite, chromitite, serpentinite, talc-anthophyllite schists, and phlogopite-talc-magnesite-olivine granoblastites also occur. Along the northwest margin of the larger of the two bodies, a granodiorite pegmatite has intruded along the contact between the country rocks and the dunite. Small dikes branch from the larger dike and cross-cut both dunite bodies. Along these dikes, as well as along the margins of the bodies, metasomatic reactions

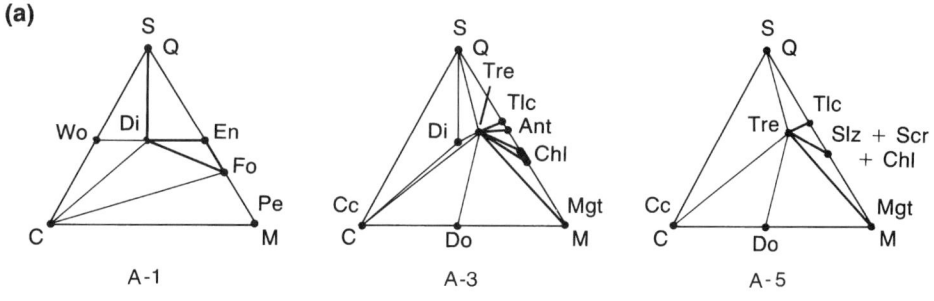

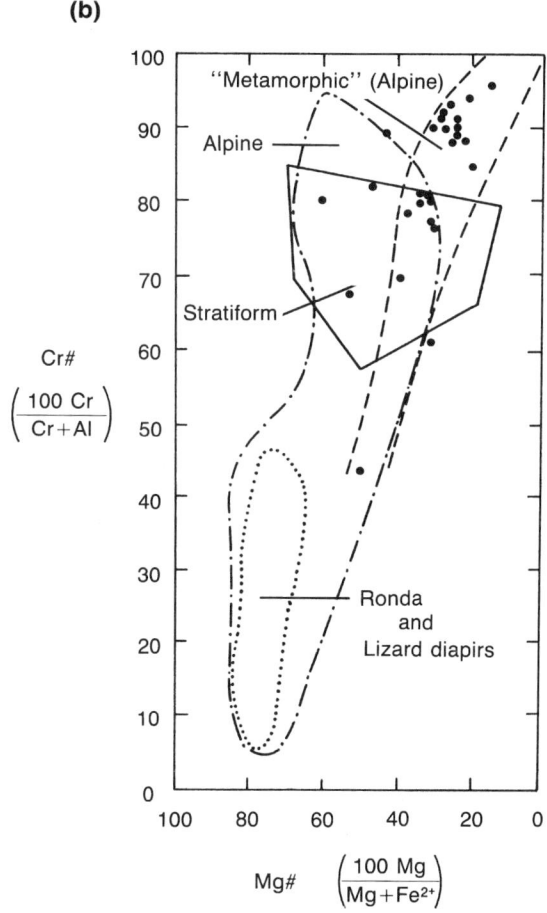

Figure 31.9 Selected phase diagrams and compositions of chromites from ultramafic rock bodies. (a) CMS phase diagrams for associations A-1, A-3, and A-5 of the Blue Ridge Belt, southern Appalachian Orogen. (b) Compositions of chromium spinels from diapiric, stratiform, alpine, and metamorphosed alpine bodies. Dots represent compositions of selected chrome spinels from Blue Ridge alpine-type ultramafic bodies.

[(b) Modified from Dick and Bullen, 1984; Lipin, 1984; and Agata, 1988]

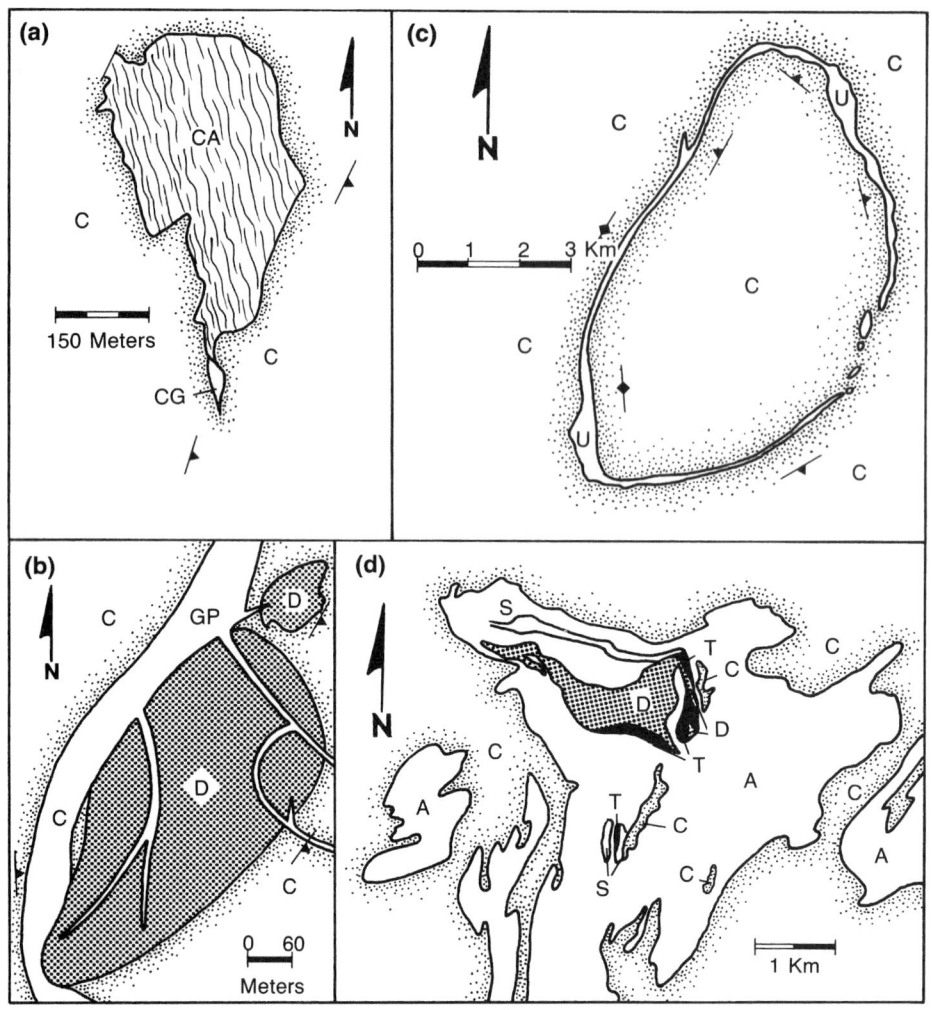

Figure 31.10   Sketch maps of Blue Ridge ultramafic bodies. (a) Map of Greer Hollow Ultramafic Body. CA = chlorite-anthophyllite unit, CG = chlorite-garnet unit, C = country rocks. (b) Map of the Day Book dunites (modified from S. E. Swanson, 1981). D = dunite, GP = granitoid pegmatite. C = country rocks. (c) Map of the Webster-Addie Ultramafic Body (modified from R. Miller, 1951a). U = ultramafic rocks, C = country rocks. (d) Map of the northern part of the Chunky Gal Mountain mafic-ultramafic complex, which includes the Buck Creek Dunite. (Modified from McElhaney and McSween, 1983). A = amphibolite, D = dunite, S = serpentinite, T = metatroctolite and related rocks, C = other country rocks.

have occurred, yielding a variety of rocks composed of such minerals as talc, anthophyllite, phlogopite, vermiculite, and chlorite (S. E. Swanson, 1981). Metasomatically altered parts of the interior of the larger dunite body contain patches of rock containing or composed of talc, tremolite, anthophyllite, chlorite, phlogopite, and magnesite. Shear zones and veins, formed later, are filled by serpentine with magnetite or less common minerals, including aragonite. Together, these minerals define a series of metamorphic assemblages formed during several successive metamorphic events (tables 31.3 and 31.4).

Structurally, the Day Book dunites are thick, concordant lenses. The earliest structures observed in the body are tabular to podiform chromite bodies. The tabular layers of chromite are locally folded into isoclinal folds. The generally massive-appearing interior of the body is characterized by a coarse porphyroclastic texture, consisting of large, cleavable, deformed, elongate olivine crystals up to 26 cm. in length (S. E. Swanson, pers. comm., 1979) in an equigranular-mosaic-textured matrix of olivine crystals that average about 0.2 mm in diameter. Disseminated subhedral to euhedral chromite and scattered orthopyroxene occur within the equigranular mosaic. Dikes, metasomatically altered zones, and shear zones are characterized by rocks of allotrioblastic-granular, diablastic, or schistose texture.

Chemical analyses of both rocks and minerals are available for the Day Book rocks. One harzburgite analysis gives an Mg number of about 0.86 (C. E. Hunter, 1941). As would be expected of rocks composed primarily of Mg-rich olivine, the dunites have high Mg numbers clustered around 0.88. The forsterite contents of olivines fall in the range 92 to 95 (J. R. Carpenter and Phyfer, 1975; S. E. Swanson, 1981). The Cr/(Cr + Al) values of the chromites fall in the range 0.7 to 0.9 at Mg numbers of 0.40 to 0.70, as is typical of alpine-type chromites (Irvine, 1967a; S. E. Swanson, 1981).[40]

The *Webster-Addie body* forms a distinctive, ring-shaped outcrop pattern (in map view) that is nearly 10 km in diameter (figure 31.10c) (R. Miller, 1951a,b).[41] The ring represents the erosional pattern of a thin sheet of ultramafic rock folded into a dome, from which the crest was removed by erosion. The body is essentially concordant with the surrounding biotite gneisses. Internally, however, the body contains compositional bands, some of which are isoclinally folded (Greenberg, 1976). As is the case at Day Book, the Webster-Addie body contains shear zones that cut across other features.

The Webster-Addie ultramafic body consists primarily of dunite, with lesser amounts of websterite and minor amounts of orthopyroxenite, harzburgite, serpentinite, and amphibole gabbro. The serpentinite and the gabbro form veins that cut the body. The dunite and serpentinized dunite exhibit a variety of textures, including equigranular-mosaic texture, equigranular-tabular texture, porphyroclastic texture, mesh texture, diablastic texture, and schistose texture. The serpentinite is dominantly schistose. Websterite has allotrioblastic-granular texture, in part similar to the equigranular mosaic of the dunite. As at Day Book and Greer Hollow, porphyroclastic and equigranular textures in rocks dominated by olivine and pyroxene are replaced successively by porphyroclastic and semischistose textures in which amphiboles are important, and schistose to diablastic textures in which amphiboles, chlorite, or serpentine dominate (tables 31.3 and 31.4).

The whole rock and mineral chemistries of the Webster-Addie ultramafic rocks are similar to those of the Day Book bodies. For example, relatively high Mg numbers (0.83–0.85) characterize the serpentinized dunites (Condie and Madison, 1969). The olivines are forsterite-rich (Fo = 84–95) and the orthopyroxenes are similarly Mg-rich (En = 87–95) (R. Miller, 1951b; Condie and Madison, 1969; Lipin, 1984). In the chromites, the Cr/(Cr + Al) values are about 0.8, but the Mg numbers are relatively low (0.3–0.6) (Lipin, 1984).

The *Buck Creek ultramafic bodies,* like the Greer Hollow ultramafite, are almost entirely enclosed within amphibole schist and gneiss (figure 31.10d) (McElhaney and McSween, 1983).[42] At the Buck Creek locality, one large dunite body is accompanied by at least four small serpentinite bodies. Minor lherzolite is present in the larger body (Kuntz and Hedge, 1981). Metatroctolites are associated with the ultramafic rocks, especially the serpentinites. The margins of the dunite and some veins within it consist of serpentinite and tremolite-talc-chlorite schist.

The Buck Creek dunite is a conformable mass within the amphibole schist and gneiss. At the contact, some shearing and minor folding have occurred, but there is no evidence of a major fault separating the dunite and the enclosing rocks.[43] The contacts and foliations in the contact zone generally parallel the foliation in the country rock. Lattice preferred orientations in the dunite (Sailor and Kuntz, 1973) apparently predate the dominant regional foliation and folding, which affected both the dunite and the country rocks during the Ordovician Taconic Orogeny (McElhaney and McSween, 1983).

Porphyroclastic, equigranular-mosaic, equigranular-tabular, mesh, and schistose textures characterize the rocks of the Buck Creek dunite. The oldest textures are porphyroclastic and equigranular textures. Mineralogically, the rocks consist of olivine (Fo = 88–89) and accessory hornblende, chromite, and/or Ca-pyroxene. Additional accessory minerals of probable retrograde origin include magnetite, pyrite, pyrrhotite, chlorite, tremolite, anthophyllite, talc, and serpentine minerals (tables 31.3 and 31.4).

Geothermometry performed on the amphibole schists and gneisses reveals that the early Taconic metamorphic event occurred at a temperature of about 725° C. Talc + chlorite + tremolite ± anthophyllite rocks may have developed at this time or during a subsequent Devonian (Acadian) metamorphic event, or both, but folding associated with the Acadian event reportedly affected the chlorite-talc schists (McElhaney and McSween, 1983). A Greenschist Facies event, which produced the serpentinites, followed the Acadian metamorphism.

In summary, all of the ultramafic bodies of the central Blue Ridge Belt apparently exhibit minerals and textures that reflect the same *sequence* of events (table 31.4). Pre- or early Taconic deformation and metamorphism in the mantle or lower crust produced porphyroclastic and equigranular textures in dunites, peridotites, and pyroxenites. Taconic and Acadian metamorphic events at Granulite and Amphibolite Facies grades produced tremolite, anthophyllite-,talc-,and chlorite-dominated assemblages that partially or completely replaced the earlier fabrics and minerals. The first of these events apparently occurred under high crustal pressures (6–9

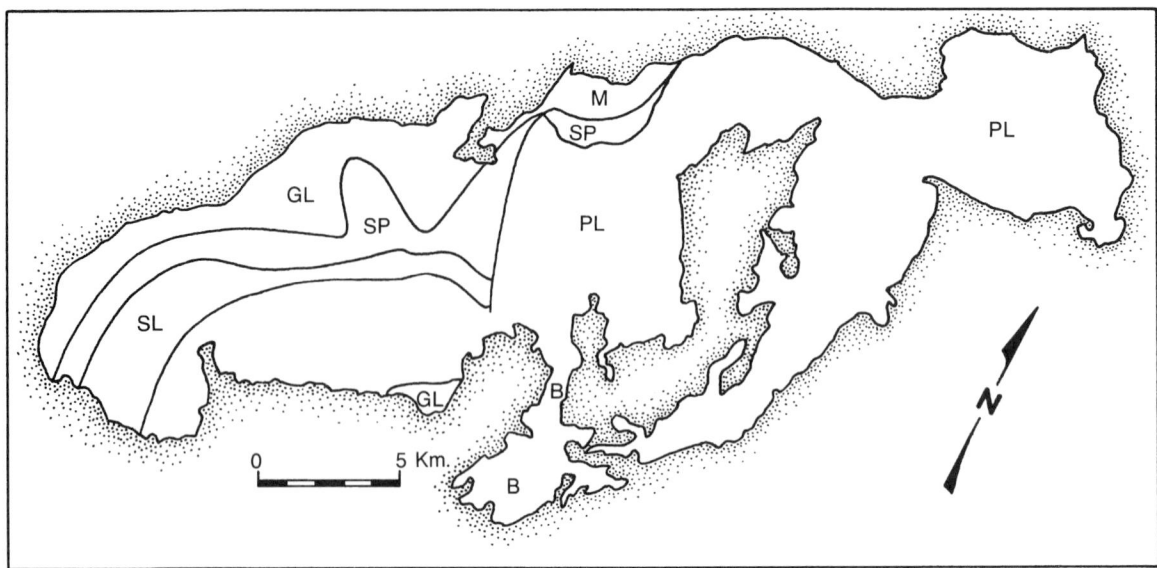

Figure 31.11 Sketch map of the Ronda Ultramafic Complex, Spain. PL = plagioclase lherzolite zone, SP = spinel pyroxenite zone, SL = spinel lherzolite zone, GL = garnet lherzolite zone, M = melange zone, B = breccia zone. *(Modified from Suen and Frey, 1987; after Lundeen, Obata, and Dickey, 1979; Obata, 1980; Frey, Suen, and Stockman, 1985)*

kb) and high temperatures (600–800° C). The second event occurred at somewhat lower P-T conditions. Later (late Paleozoic, Alleghenian?) Greenschist Facies metamorphism resulted in the replacement of earlier assemblages by diablastic to schistose serpentine + magnetite assemblages.

The Ronda Ultramafic Complex: Product of a Mantle Diapir

The Ronda ultramafic complex is an alpine-type peridotite exposed in the Betic Cordillera of southern Spain.[44] The mass is crudely elliptical in plan, is more than 35 km long, and is surrounded by metamorphic rocks, which, at least in part, are of the pyroxene hornfels facies (Loomis, 1972a; Lundeen, 1978). Internally, the Ronda Peridotite contains four major petrographic zones (figure 31.11) (Obata, 1980). These include a garnet lherzolite zone, a spinel pyroxenite zone (the "ariegite subfacies"), a spinel lherzolite zone (the "seiland subfacies"), and a plagioclase lherzolite zone. Reference to figure 6.2 will remind the reader that garnet peridotite, spinel peridotite, and plagioclase peridotite are stable at successively shallower depths (> 50 km, about 40km, and < 25 km, respectively) in the mantle. Consequently, the key petrographic zones, which are too thin to represent a cross section through the mantle (the total thickness of the Ronda body is < 2 km), are interpreted to represent partially equilibrated zones in an ascending mantle diapir. The zones are not homogeneous, as

variations in mineralogy result in the occurrence of harzburgites, amphibole-bearing lherzolites, and sepentinized variants of the dominant lherzolite rock types. In addition, there are melange and breccia zones within one or more of these units.

The mineral chemistry of the ultramafic and the mafic rocks reflect the complex history of the body. Although olivines have a restricted range of chemistry (Fo = 88–92), the pyroxenes and spinels are quite variable in composition (Obata, 1980; F. A. Frey, Suen, and Stockman, 1985). Orthopyroxene *porphyroclasts* have a restricted range of Mg numbers (0.89–0.91), but contain clinopyroxene blebs and exsolution lamellae and are strongly zoned in Al_2O_3. Orthopyroxene *porphyroblasts* also have a restricted range of Mg number (0.89–0.92) and show zoning in Al_2O_3. Similarly, clinopyroxenes show significant variations in Na_2O and Al_2O_3. The Cr/(Cr + Al) values of the spinels vary from about 0.05 to about 0.50 (see figure 31.9). The more aluminous varieties are present in the spinel and garnet peridotites.

Texturally and structurally, the body is complex. Textures range from protogranular and equigranular mosaic to porphyroclastic and cataclastic (Obata, 1980). Layering, foliation, and lineation characterize the metamorphic rocks. Layering in the body includes both metamorphic layers and magmatic layers of mafic to ultramafic composition, ranging from garnet pyroxenites to olivine gabbros (Dickey, 1970).

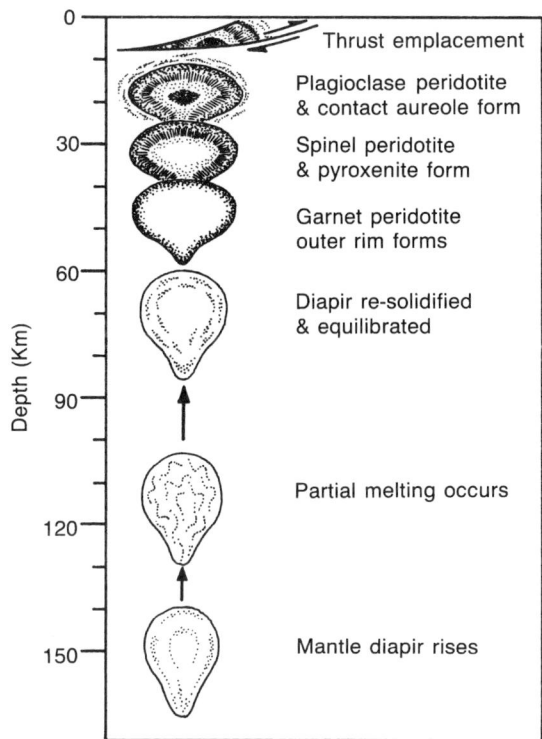

Figure 31.12 Schematic diagram depicting the history of the Ronda ultramafic complex.
(based on T. P. Loomis, 1972a; Lundeen, 1978; Obata, 1980; Frey, Suen, and Stockman, 1985; and Suen and Frey, 1987)

Overall, the Ronda ultramafic complex is now a tabular mass that constitutes part of a thrust sheet (Lundeen, 1978). Together, the data indicate a complex history for the body involving an early diapiric rise of a fertile mantle garnet lherzolite and later thrust fault emplacement into higher levels of the crust (figure 31.12). In transit from mantle depths of perhaps 200 km, the body experienced partial melting of both marginal and interior peridotites at depths of between 120 and 100 km and temperatures between 1700° C and 1500° C (F. A. Frey, Suen, and Stockman, 1985). At the margin, small degrees of partial melting left residues of garnet lherzolite, whereas in the interior, larger degrees of partial melting produced garnet-free peridotites. The melts crystallized as the diapir rose and the body equilibrated at 1100–1200° C and 2.0–2.5 Gpa at 60–75 km (Obata, 1980; Suen and Frey, 1987). Further rise of the body resulted in cooling of the body from the outside towards the interior, with consequent, incomplete textural and mineralogical re-equilibration. After emplacement in the crust and the creation of a high-temperature contact metamorphic aureole (about 22 m.y.b.p.), the body was transported and emplaced at higher levels in the crust as part of a thrust sheet (Lundeen, 1978; Reisberg, Zindler, and Jagoutz, 1989).

SUMMARY

Alpine ultramafic rock bodies are irregular to elliptical, typically deformed masses of ultramafic rock that have been incorporated into mountain belts. They are characterized by tectonite fabrics and tight to isoclinal folds. The rocks usually have high Mg numbers, as do the olivines and pyroxenes that constitute them. Very commonly, alpine-type ultramafic rocks are serpentinized, some through nearly isochemical processes and others via constant-volume processes.

Any kind of igneous mafic-ultramafic complex may be incorporated into an orogenic belt and deformed with the rocks of that mountain system. Consequently, igneous ultramafic bodies of any kind may become alpine-type masses. Nevertheless, the most common type of igneous protoliths are ophiolitic. Ophiolitic mantle and other mantle slabs may also form alpine-type ultramafic bodies. In addition, albeit rarely, mantle diapirs emplaced into the upper mantle or lower crust become alpine-type ultramafic bodies. Because these three types of occurrences—magmatic bodies, mantle slabs, and mantle diapirs—form under widely varying conditions, the rocks that constitute them, and therefore alpine-type ultramafic bodies as a group, exhibit a wide range of characteristics. Metamorphism further diversifies the mineral composition and chemistry of these rocks.

EXPLANATORY NOTES

1. Refer back to chapter 10 for a review of the various forms of magmatic ultramafic rocks and rock bodies.
2. H. H. Hess (1955) schematically depicts these orogenic occurrences of ultramafic rocks. Also see Taliaferro (1943), R. Miller (1951b), Battey (1960), Larrabee (1966), Lappin (1967), Wyllie (1967b), Menzies (1973), Gansser (1974), Livingstone (1976), Pamic and Majer (1977), Maltman (1978), Varne and Brown (1978), C. A. Hall and Bennett (1979), Ozawa (1983), R. L. Christiansen (1984), Pognante, Rosli, and Toscani (1985), O'Hanley (1987), Bodinier (1988), Bodinier, Dupuy, and Dostal (1988), Girardeau and Mercier (1988), Girardeau, G. I. Ibarguchi, and Ben Jamaa (1989), Hebert Serri (1989), Mittwede and Stoddard (1989), Seyler and Mattson (1989), Harnois, Trottier, and Morency (1990), Spell and Norrel (1990), and additional references to specific localities in the following notes (e.g., notes 5, 26, and 33).

3. For example, see Friedman (1953) and Dymek, Brothers, and Schiffries (1988). Actually, some ultramafic rocks surrounded by Precambrian rocks, such as those in the Southern Appalachian Blue Ridge Belt, occur in mountain belts formed during the Phanerozoic eon.

4. Alpine ultramafic rocks formed at present and former plate boundaries are described by Bonatti, Honnorez, and Ferrara (1970), J. F. Dewey and Bird (1970), Moores and MacGregor (1972), C. G. Engel and Fisher (1975), Harold Williams (1975), DeWit et al. (1977), Sinton (1979), Hamlyn and Bonatti (1980), Nicolas and Le Pichon (1980), J. F. Casey and Karson (1981), Quick (1981a,b), Hebert, Bideau, and Hekinian (1983), Dick, Fisher, and Bryan (1984), Harper (1984), Kimball, Spear, and Dick (1985), Nicolas (1986), Shibata and Thompson (1986), Cabanes and Briqueu (1987), R. B. Miller and Mogk (1987), Girardeau and Mercier (1988), and Mevel et al. (1991). Also see note 26. See Beslier, Girardeau, and Boillot (1988, 1990) for descriptions of a deformed spreading-center ultramafic complex.

5. *Bay of Islands* (C. M. Smith, 1958; also see note 26); *Thetford Mines, Quebec* (Dresser, 1909; Cooke, 1937; Laurent, 1975; Laurent and Hebert, 1979; Cogulu and Laurent, 1984; O'Hanley, 1987; Hebert and Laurent (1989); Harnois, Trottier, and Morency, 1990); *Vermont* (Chidester, 1962; Jahns, 1967; Sanford, 1982); *North Carolina* (C. E. Hunter, 1941; R. Miller, 1951a, b; Kulp and Brobst, 1954; Condie and Madison, 1969; Misra and Keller, 1978; S. E. Swanson, 1981; McElhaney and McSween, 1983; Abbott and Raymond, 1984); *California Coast Ranges*, including Burro Mountain (Taliaferro, 1943; Burch, 1968; N. J. Page, 1967; R. G. Coleman and Keith, 1971; E. H. Bailey and Blake (1974); Evarts, 1977; 1978; Lienert and Wasilewski, 1979; C. A. Hopson, Mattinson, and Pessagno, 1981); *Klamath Mountains* (Irwin and Lipman, 1962; Ramp, 1975; Dick, 1977a; Lindsley-Griffin, 1977; Quick, 1981a; Harper, 1984; S. M. Peacock, 1987b); *Washington Cascade Range* (Ragan, 1967; M. R. W. Johnson, Dungan, and Vance 1977; Vance and Dungan, 1977); *Blue River, British Columbia* (Pinsent and Hirst, 1977); *Ronda, Spain* (Dickey, 1970; Obata, 1980; Suen and Frey, 1987; also see note 23); *Swiss Alps* (Trommsdorff and Evans, 1974; Pfeifer, 1978, 1981; Stille and Oberhansli, 1987); *Lizard, Cornwall, England* (D. H. Green, 1964); Norway (Lappin, 1967); *Dun Mountain, New Zealand* (F. J. Turner, 1942; Battey, 1960; Lauder, 1965; N. I. Christensen, 1984).

6. Refer to chapter 10 for a review of this mineralogy. The chemistry of the minerals and other characteristics of alpine-type ultramafic bodies, listed in table 31.1, are based on the following sources: R. Miller (1951a,b); H. H. Hess (1955), E. D. Jackson (1961), D. H. Green (1964), Irvine (1967b), Wager and Brown (1968), Den Tex (1969), Moores (1969), Kornprobst (1969), Dickey (1970), O. B. James (1971), Moores and Vine (1971), Astwood, Carpenter, and Sharp (1972), Loomis (1972b), Medaris (1972), Moores and MacGregor (1972), E. D. Jackson and Thayer (1972),

Menzies (1973), J. Pike (1974), P. Henderson (1975), E. D. Jackson, Shaw, and Bargar (1975), Nicolas and Boudier (1975), Dribus et al. (1976), Loney and Himmelberg (1976), R. G. Coleman (1977b), Evarts (1977, 1978), Greenbaum (1977), Schubert (1977), Kingsbury and Heimlich (1978), Varne and Brown (1978), Dick and Sinton (1979), Ernst and Piccardo (1979), Nicolas and LePichon (1980), Obata (1980), Raymond and Swanson (1981), S. E. Swanson (1981), Garcia (1982), Hebert, Bideau, and Hekinian (1983), Ozawa (1983), Abbott and Raymond (1984), N. I. Christiansen (1984), Dick and Bullen (1984), Dick, Fisher, and Bryan (1984), Elthon, Casey, and Komor (1984), F. A. Frey, Suen, and Stockman (1985), Ishiwatari (1985), Kimball, Spear, and Dick (1985), Leroy (1985), Pognante, Rosli, and Toseani (1985), Nicolas (1986), R. B. Miller and Mogk (1987), Suen and Frey (1987), L. A. Raymond (unpublished data); Bodinier (1988), Bodinier, Dupuy, and Dostal, (1988), Hebert, Serri, and Hekinian (1989), O'Hanley (1989).

7. Also see the references in notes 8 and 9.

8. These textures are widely recognized in mantle xenoliths, as well as in ultramafic rocks derived from the mantle. For additional examples and descriptions, see Nicolas et al. (1971), Helmstaedt et al. (1972), Dribus et al. (1976), Nicolas and Poirier (1976), Loney and Himmelberg (1976), Nielson Pike and Schwarzman (1976), Basu (1977), Christensen and Salisbury (1979), Honeycutt and Heimlich (1980), Raymond and Swanson (1981), Boudier and Nicolas (1982a,b), Christensen and Lundquist (1982), Hebert, Bideau, and Hekinian (1983), Wicks (1984a), and R. B. Miller and Mogk (1987).

9. Maltman (1978), Wicks and Plant (1979), A. J. Williams (1979), Raymond and Swanson (1981), Wicks (1984a,b).

10. See chapter 10 for a discussion of these mineralogical variations.

11. The Bay of Islands Ophiolite Complex, described below, is a good example. See Ringwood (1966a, 1977), K. Ito (1974), Hervig, Smith, and Steele, (1980), Maaloe and Steel (1980), D. L. Anderson and Bass (1984), Menzies (1984), Arai (1987), Irifune and Ringwood (1987), Nicolas (1989), and recent issues of the *Journal of Geophysical Research* and *Earth and Planetary Science Letters* for more information on mantle compositions and structure.

12. An in-depth discussion of the nature of the mantle is beyond the scope of this text.

13. For example, see D. L. Anderson and Bass (1984) and Nicolas (1986, 1989). For discussions of mantle metasomatism, see Menzies and Hawkesworth (1987) and E. M. Morris and Pasteris (1987a).

14. For example, see Bonatti, Honnorez, and Ferrara (1970), Nicolas and Poirier (1976, ch. 11), Ernst and Piccardo (1979), Hamlyn and Bonatti (1980), Christensen and Lundquist (1982), Dick, Fisher, and Bryan (1984), Shibata and Thompson (1986).

15. In the early and middle parts of the twentieth century, a debate raged in the geologic community about whether serpentinite is a magmatic or a metamorphic rock. H. H. Hess (1938, 1955) championed the cause of a magmatic origin, largely on the basis of field relationships. Bowen and Tuttle (1949) demonstrated, via laboratory studies of the system $MgO\text{-}SiO_2\text{-}H_2O$, that serpentine was only stable at low temperatures and that magmas did not develop in that system, even at temperatures of about 900° C. Experimental evidence demonstrates that temperatures of 1800–2000° C are required to produce magmas from pure olivine compositions (Bowen and Anderson, 1914, Ohtani and Kumazawa, 1981). Additional references from the literature that provide a basis for our present understanding of serpentinization include Vaugnat (1963), Nauman and Dresher (1966), Thayer (1966), N. J. Page (1967, 1968), Whittaker and Wicks (1970), R. G. Coleman and Keith (1971), Wicks and Whittaker (1975), Moody (1976), DeWit et al. (1977), Vance and Dungan (1977), Caruso and Chernosky (1979), Coats and Buchan (1979), Dungan (1979a,b), Labotka and Albee (1979), Laurent and Hebert (1979), Ikin and Harmon (1983), Komor et al. (1985), Mellini, Trommsdorff, and Compagnoni (1987), and Mellini and Zanazzi (1987). Also see Wicks and O'Hanley (1988), O'Hanley, Chernosky, and Wicks (1989), and the review of Chernosky, Berman, and Bryudzig (1988).

16. Wicks and Whittaker (1975) and Moody (1976) list the various polytypes of serpentine minerals and Mellini, Trommsdorff, and Compagnoni (1987) discuss polytypism in antigorite. O'Hanley, Chernosky, and Wicks (1989), using a chemographic analysis, assess the stability of lizardite and chrysotile. Also see Middleton and Whittaker (1979), Yada (1979), Mellini and Zanazzi (1987), and Wicks and O'Hanley (1988) for additional information on the structures of serpentines. A. C. D. Newman and Brown (1987) discuss the chemistry of serpentines.

17. Whittaker and Wicks (1970), Wicks and Whittaker (1975), Mellini, Trommsdorff, and Compagnoni (1987).

18. Modified from F. J. Turner and Verhoogen (1960, pp. 318–319).

19. Hostetler, Coleman, and Evans (1966), R. G. Coleman (1971a), R. G. Coleman and Keith (1971), Moody (1976), Komor et al. (1985).

20. Serpentine pseudomorphs after pyroxenes, which typically mimic the pyroxene cleavage, are called "bastites" (see Dungan, 1979b; Wicks and Plant, 1979).

21. Thayer (1966, 1967), N. J. Page (1967), Condie and Madison (1969), Coleman (1971a), Moody (1976), Leach and Rodgers (1978).

22. I. Barnes and O'Neil (1969), I. Barnes, Rapp, and O'Neil (1972), Wenner and Taylor (1973), M. R. W. Johnson, Dungan, and Vance (1977), Janecky and Seyfried (1986), Peacock (1987b).

23. For example, E. H. Nickel (1959, 1961) and B. R. Frost (1985).

24. R. G. Coleman (1967, 1977a), Leach and Rodgers (1978), Schandl, O'Hanley, and Wicks (1989). Recall that Gresens (1969) argued that the highly reducing fluids involved in serpentinization produced Blueschist Facies rocks by metasomatic alteration of country rocks surrounding serpentinites (see chapter 28).

25. The textures of serpentinites are also discussed and/or depicted by Maltman (1978), Cressey (1979), Dungan (1979b), Morandi and Felice (1979), Wicks and Plant (1979), A. J. Williams (1979), Ikin and Harmon (1983), Wicks (1984a,b,c), Peacock (1987b), and A. E. Gates and Kambin (1990).

26. The Bay of Islands Complex has been described to varying degrees by a number of workers. This review, which focusses on the ultramafic rocks, is based on the works of C. M. Smith (1958), W. R. Church and Stevens (1971), Dewey and Bird (1970), Harold Williams (1973, 1975), J. Malpas (1978), Salisbury and Christensen (1978), Christensen and Salisbury (1979), S. B. Jacobson and Wasserburg (1979), Suen, Frey, and Malpas (1979), Casey and Karson (1981), Casey et al. (1981), Girardeau and Nicolas (1981). Christensen and Lundquist (1982), Elthon, Casey, and Komor (1982), Casey et al. (1983), Elthon, Casey, and Komor (1984), Casey et al. (1985), and Komor et al. (1985a,b).

27. For example, see W. R. Church and Stevens (1971), S. B. Jacobson and Wasserburg (1979), and Casey et al. (1985). Others have argued that the Bay of Islands Complex represents the crust of a marginal basin (Dewey and Bird, 1970).

28. Also see Girardeau and Nicolas (1981) and Casey et al. (1981).

29. Although not described in detail, such structures are mentioned by Casey et al. (1983).

30. Elthon, Casey, and Komor (1982); Elthon, Casey, and Komor (1984), Komor et al. (1985).

31. For example, see Jacobson and Wasserburg (1979) and Casey et al. (1985).

32. See Larrabee (1966), Rankin, Espenshade, and Neuman (1972), Hatcher (1978), Misra and Keller (1978), Scotford and Williams (1983), Abbott and Raymond (1984), Raymond and Abbott (1985), Butler (1989), Mittwede and Stoddard (1989), and Spell and Norrell (1990).

33. Reports on the petrology, petrography, structure, and petrofabrics of these bodies include the following: J. V. Lewis (1896); J. H. Pratt and Lewis (1905), C. E. Hunter (1941), Hadley (1949), R. Miller (1951a,b, 1953), Kulp and Brobst (1954), Stose and Stose (1957), Brobst (1962), Larrabee (1966), J. R. Carpenter and Phyfer (1969, 1975), Condie and Madison (1969), Stueber (1969), Bentzen (1970, 1975), Neuhauser and Carpenter (1971), Astwood, Carpenter, and Sharp (1972), Hartley (1973), Jones et al. (1973), Neuhauser

677

(1973), Sailor and Kuntz (1973), Dallmeyer (1974), Dribus et al. (1976), Greenberg (1976), Swanson and Raymond (1976), Swanson and Whittkop (1976), Alcorn and Carpenter (1976), Bluhm and Zimmerman (1977), Hearn et al. (1977), Palmer, Heimlich, and Kolb (1977), Tien (1977), Yurkovich (1977), Kingsbury and Heimlich (1978), Misra and Keller (1978), Sharpe and Whitney (1979), Heimlich et al. (1980), Honeycutt and Heimlich (1980), S. E. Swanson (1980, 1981), Honeycutt, Heimlich, and Palmer (1981), Kuntz and Hedge (1981), Penso, Heimlich, and Palmer (1981), Raymond and Swanson (1981), Dribus, Heimlich, and Palmer (1982), Neuhauser (1982), Schiering, Heimlich, and Palmer (1982), McElhaney and McSween (1983), Scotford and Williams (1983), Abbott and Raymond (1984), Hatcher, et al. (1984), Lipin (1984), Abbott and Raymond (1985), McSween and Hatcher (1985), Raymond and Abbott (1985), Swanson et al. (1985), Conley (1987), Meen (1988), and Raymond et al. (1988).

34. For example, see Astwood, Carpenter, and Sharp (1972), Dribus et al. (1976), Kingsbury and Heimlich (1978), Honeycutt and Heimlich (1980), Swanson (1981), McElhaney and McSween (1983), Abbott and Raymond (1984), and Raymond et al. (1988).

35. For discussions of ophiolites among these ultramafic bodies, see McSween and Hatcher (1985) and Conley (1987).

36. Miller (1951), Kulp and Brobst (1954), J. R. Carpenter and Phyfer (1975), Swanson (1981), Lipin (1984), Meen (1988).

37. The Lake Chatuge Complex on the North Carolina–Georgia border has a debated history, reportedly somewhat different from that of other Blue Ridge alpine ultramafic bodies (Hartley, 1973; Dallmeyer, 1974; Meen, 1988). Meen (1988) suggests that the body originally crystallized from a magma at a pressure of 1.0–1.3 Gpa (corresponding to a deep crustal or a mantle depth) at a temperature of $T > 900°$ C and was later emplaced via tectonic processes. The dunite-serpentinite masses in the Lake Chatuge Complex contain assemblages 1a and 4a, reflecting a history partly like those of other Blue Ridge ultramafic bodies.

38. Unpublished data. The low Mg numbers may reflect metasomatic effects. Compositions like analysis 1 in table 31.2 do not represent normal igneous bulk rock compositions, although some cumulates do have similarly odd bulk rock chemistries.

39. The Day Book dunites are widely known and visited bodies. Additional reports on their petrology and mineralogy may be found in C. E. Hunter (1941), Kulp and Brobst (1954), Phyfer and Carpenter (1969), Dribus et al. (1976), Swanson and Raymond (1976), Swanson and Whittkop (1976), Tien (1977), Abbott and Raymond (1984), Raymond and Abbott (1985), and Swanson et al. (1985). *Note:* Mining operations have dramatically changed the exposures reviewed here and described by Swanson (1981).

40. Also see Bentzen (1970), Lipin (1984), and Dick and Bullen (1984).

41. Additional references to the mineralogy and petrology of the Webster-Addie body include C. E. Hunter (1941), Condie and Madison (1969), Bentzen (1970), Dribus et al. (1976), and Greenberg (1976).

42. Petrologic and mineralogic information on the Buck Creek ultramafic rocks is also included in C. E. Hunter (1941), Sailor and Kuntz (1973), Kuntz and Hedge (1981), and McSween and Hatcher (1985).

43. Kuntz (1964, in McElhaney and McSween, 1983), McElhaney and McSween (1983).

44. This review is based on the works of Dickey (1970), Loomis (1972ab), Menzies, Blanchard, and Jacobs (1977), Schubert (1977), Lundeen (1978), Lundeen, Obata, and Dickey (1979), Obata (1980), M. A. Frey, Suen, and Stockman (1985), Suen and Frey (1987), and Reisberg, Zindler, and Jagoutz (1989). Somewhat similar spinel peridotites are described by Bodinier, Depuy, and Dostal (1988).

PROBLEMS

31.1. If an ultramafic body consists of equigranular-tabular-textured dunite and harzburgite and the harzburgite has an Al_2O_3 content of 3.1%, a CaO content of 6.01%, and a Mg number of 86, what type of alpine ultramafic body is it likely to be?

31.2. (a) Construct CMS phase diagrams for Blue Ridge alpine ultramafic rock associations A-2 and A-4, (see table 31.3). (b) If chlorite is assumed to be a saturating phase, only one reaction is needed to represent the change from A-1 to A-2. Write and balance that reaction. (c) Does this reaction represent a prograde or retrograde reaction? Explain.

Epilogue

Igneous, sedimentary, and metamorphic rocks of various types occur together at regional to local scales. In the photo at right, for example, an andesite dike and sill intrude Precambrian gneiss and both the andesite and gneiss are locally overlain by sediments deposited as a result of glacial and gravitational activity. On the regional scale, more diverse associations of rock types define the petrotectonic assemblages characteristic of plate settings, as outlined in this part of the text . These assemblages provide a foundation for understanding the history of the Earth.

PART V

Petrotectonic Assemblages

INTRODUCTION

Information on all of the major rock types within each of the classes—igneous, sedimentary, and metamorphic—has been provided in this text. Both descriptive details and petrogenetic theories were summarized. Together, the data give a view of the histories that have produced crustal rocks.

The igneous rocks are found to be crystallized from magmas, those liquids formed in the mantle and at the base of the crust by partial melting. Various primary basaltic magmas are the principal mantle melts, although andesitic, komatiitic, and other melts are generated under special conditions. The magmas rise towards the surface, losing heat along the way. The compositions of magmas may be modified as they migrate, by assimilation of country rock, by mixing with other magmas, or by various processes of differentiation. The loss of heat, if significant, results in crystallization of plutonic rocks at depth, but if minimal, allows magmas to rise to the surface where they form volcanic rocks.

The sedimentary rocks form at the surface through various combinations of sedimentation processes. The clastic rocks, both silicate and carbonate, are formed after weathering and erosion of a source terrane produces sediment, which is subsequently deposited via sediment rain, traction currents, or submarine or subaerial debris flows and landslides. The various chemical sediments form through crystallization from aqueous solutions under varying conditions of pH and Eh. These sediments and their characteristics reflect the environments of deposition. On the continents, fluvial environments are preeminent. Deltas are the sites of the greatest accumulations of sediment at the continent-ocean interface, whereas shallow marine (shore, shelf, reef) and submarine fan environments contain great quantities of marine sediment. The sediment formed in each of these environments becomes rock through a number of diagenetic processes, including compaction, cementation, and recrystallization.

Where preexisting igneous, sedimentary, or metamorphic rocks are subjected to new conditions of pressure and temperature, especially in the presence of a fluid phase, they are metamorphosed. Such changes occur (1) in ocean crust, especially where hydrothermal activity is significant; (2) in incipient mountain ranges, where stratigraphic or structural burial adds pressure at low temperatures, yielding high P/T regional metamorphic terranes; (3) in arc and collisional mountain ranges, where intrusions of igneous rock and thermal conduction from the mantle result in heating, both local and regional, yielding contact and regional metamorphism, respectively; (4) beneath volcanoes, where magmas heat the surrounding rocks; (5) along fault zones, where dynamic metamorphism yields mylonitic and cataclastic rocks; (6) within the continental crust, both at depth and along fault zones; and (7) in the mantle. Metamorphic processes include various processes of recrystallization and neocrystallization, which generally result in both new textures and new mineral assemblages in the metamorphic rocks. Factors such as the bulk rock composition, the composition of the fluid phase, and the P-T conditions control the phase assemblages that result from these processes.

ROCKS AND PLATE BOUNDARIES
• •

All of the processes that produce and transform crustal rocks occur at specific sites within or at the margins of tectonic plates. Each site is characterized by specific conditions that control the chemistry, minerals, textures, and structures of the rocks. Therefore, those features can be used to characterize the petrogenetic sites of ancient rocks and help us develop an understanding of their histories, one of the ultimate aims of petrology.

The major rock types that occur at petrogenetic sites within a plate or at a plate boundary are shown in figure 32.1 (a recasting of figure 1.4). Each petrogenetic site within a plate or at a plate boundary may yield the rocks shown for that site. At spreading centers, oceanic tholeitic basalts, the Mid-Ocean Ridge Basalts (MORBs), form from magmas developed by partial melting of mantle peridotites at shallow depths. The MORBs form a layered and pillowed cap on the rest of the rocks of the ocean crust, which together with the basalts constitute an ophiolite sequence. Below the basalts are the sheeted dike complex, nonlayered plutonic differentiates, cumulate gabbros and ultramafic rocks, and a basal, mantle ultramafic tectonite. Because the crust is faulted at the spreading center, a basin is formed into which mafic breccias, derived from adjoining upfaulted oceanic crust, and minor pelagic sediment typically accumulate. Hydrothermal alteration, resulting from the flux of seawater through the fractured crustal rocks, produces hydrothermal (contact) metamorphism near the areas above the spreading center magma chamber. The resulting rocks belong to the Zeolite, Prehnite-Pumpellyite, or Albite-Epidote Hornfels Facies. At depth, deformation and pervasive re-equilibration of magmatic rocks to crustal temperatures result in the development of metagabbro tectonites, amphibole schists, and related rocks of the Greenschist and Amphibolite facies. Mylonites and cataclasites form along faults.

Transform faults that cut the oceanic crust contain the same rocks as the spreading centers, but here the rocks are usually deformed by deviatoric stresses. Ophiolitic melanges and mylonite zones are characteristic of such fault zones (Karson, 1984; Saleeby, 1984). Where a prominent scarp develops along the transform fault, breccias eroded from the scarp may be deposited in the fault zone. Within the oceanic transform fault zones, pelagic sediments also are deposited. On the continents, a wide range of sedimentary rocks form in transform fault zones, including slide breccias and debris flows, lacustrine shales, and fluvial conglomerate, sandstone, and shale.

Within the open ocean, pelagic sedimentation is the preeminent rock-forming process. Fine-grained sedimentary rocks such as shale, chert, and biogenic limestone are formed in environments controlled by water temperature, water depth, and proximity to land or volcanoes.

Several environments are associated with subduction zones. These include the trench, forearc basins, the subduction zone (*sensu stricto*), volcanic arcs and associated orogenic belts, back-arc basins, and back-arc passive margins. The trench and the forearc basins, although varied in nature, are characterized primarily by sedimentary rocks (Dickinson and Seely, 1979). These may include olistostromes, turbidite wackes and related rocks, and pelagic shales. The turbidites and olistostromes especially characterize submarine fans that form both in trenches and in trench-slope basins within the forearc. Sedimentary rocks that have undergone soft sediment deformation and dynamically recrystallized sedimentary rocks may also develop in the forearc region. Diapiric melanges intrude some forearc regions and may give rise to listostromes (Cloos, 1984). Rarely, magmas may intrude into forearc and trench rocks, giving rise to plutons and contact metamorphism (Echeverria, 1980).

The subduction zone (*sensu stricto*) is a region of metamorphism and anatexis. Beneath the forearc region, rocks carried down on the subducting plate are subjected to progressively higher pressures, while the temperatures remain low. Depending on the temperature, the subducted sedimentary rocks and the underlying oceanic crust are metamorphosed under low to moderate temperature conditions of the Zeolite, Prehnite-Pumpellyite, Blueschist, Greenschist, Amphibolite, or Eclogite Facies. As the rocks descend beneath the overriding plate, they are heated, resulting in dehydration. The fluid phase generated migrates into surrounding rocks. In the depth range of about 100 to 200 km, minor melting of the subducted rocks and significant, but partial, melting of the rocks of the overlying mantle, result in the formation of magmas of tholeiitic, calc-alkaline, and alkaline character.

The magmas generated in the subduction zone provide the building material for the volcanic arc. These magmas rise upwards and intrude at depth to form plutons or erupt at the surface to form volcanic rocks. In transit, they may differentiate, assimilate country rocks, or melt the base of the crust to form siliceous magmas, with which they may mix to form andesites. The siliceous magmas erupt to form rhyolites and less siliceous volcanic rocks, or, if they fail to reach the surface, crystallize to form the rocks of the calc-alkaline plutons. The latter are typified by granodiorites and quartz monzonites. Crystallization of the most siliceous magmas yields granites. Differentiation of granitoid magmas, with the development of a fluid phase, may result in the crystallization of pegmatites. Surrounding the intrusions, at depth, entire regions are heated by the thermal flux, and together with the regional

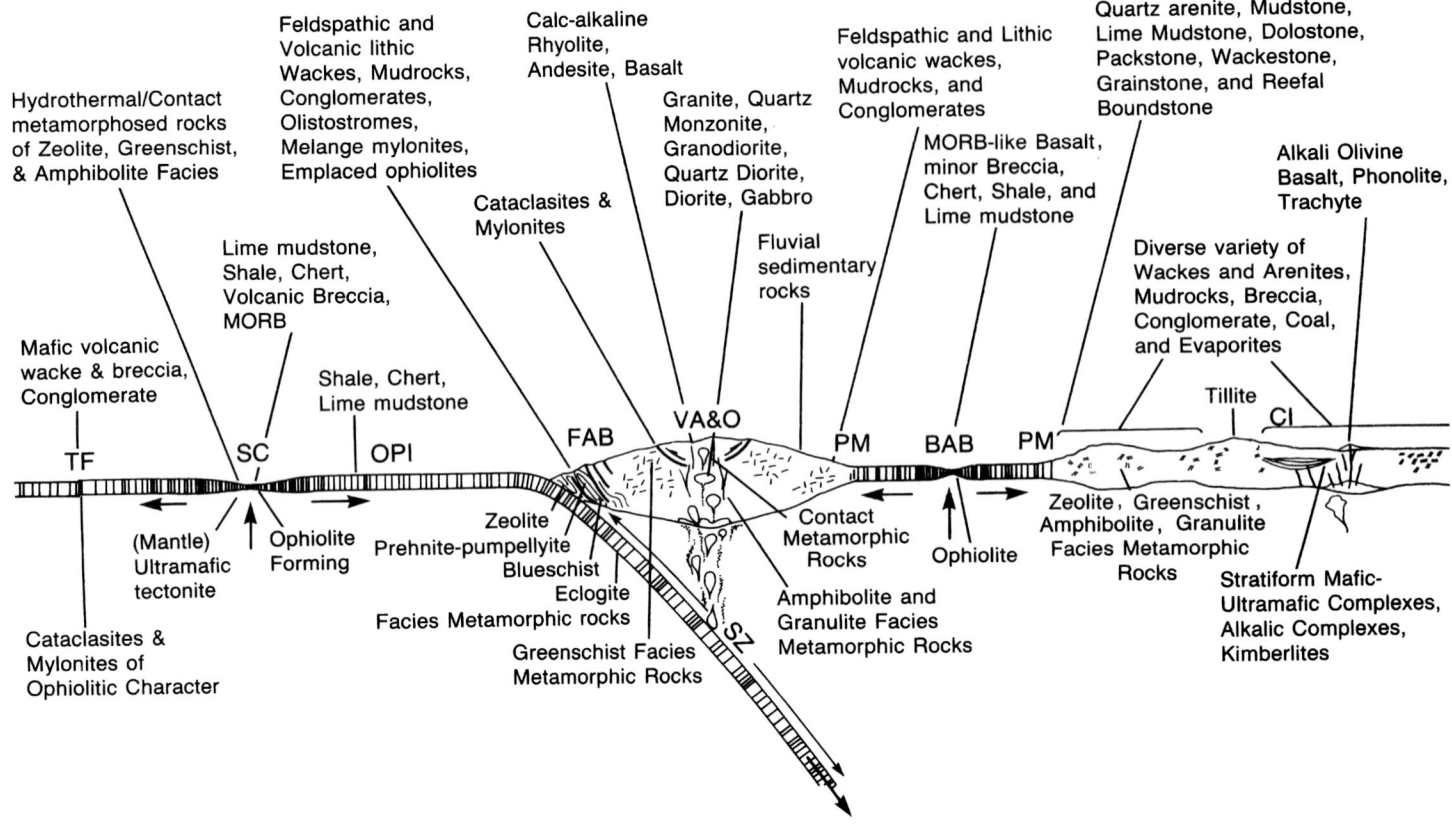

Figure 32.1 Petrotectonic assemblages of the principal plate-interior and plate-boundary sites. SC = spreading center, TF = transform fault, OPI = oceanic plate interior;

SZ = subduction zone, FAB = forearc basin, VA&O = volcanic arc and orogen, BAB = back-arc basin, PM = passive margin, CI = continental interior.

stresses caused by the plate collision, the temperatures (and pressures) yield regional terranes of dynamothermally metamorphosed rock of the Greenschist, Amphibolite, and Granulite Facies. At shallower depths, contact metamorphism associated with mesozonal and epizonal plutons produces rocks of the Zeolite and various hornfels facies. At the surface, fluvial, lacustrine, and glacial deposition will give rise to sediments, many of which are ephemeral, but some of which will be preserved and lithified to form sedimentary rocks.

The back-arc basin is formed by the rise of mantle, resulting partial melting, intrusion of magmas, and consequent spreading of the crust in the region on the side of the arc opposite the subduction zone (Karig, 1971). The principal rocks formed are similar to MORB, but differ slightly in their chemistry. Ophiolitic crust forms here, just as it does in oceanic spreading centers. Although the regional extent of back-arc spreading centers is smaller than that of oceanic spreading centers, the various rocks formed are very much like those of the mid-ocean ridge regions. Perhaps the most significant exception is that the sedimentary facies associated with the back-arc basin margins differ from the sedimentary

associations of the mid-ocean ridges. On the arc side of the basin, sedimentation (and diagenesis) produces volcanic wackes and related shales and conglomerates as the dominant lithologies (Karig and Moore, 1975). On the continent side of the basin, rocks typical of passive margins—arenites, limestones, dolostones, and shales—are typical rock types.

The passive margins of the continents, especially those with broad shelves and reefs, as well as inland seas, are the sites at which large quantities of the sedimentary rocks that occur within the present continents were formed. Here various limestones, ranging from boundstones to lime mudstones, and dolostones developed. The limestones result primarily from biochemical precipitation. The dolostones largely result from the flux of surface and near-surface fluids through limestones, altering their chemistry. Quartz arenites form from sediment deposited in vast inland seas and along strandlines. Ubiquitous shales are interlayered with these rocks. In environments transitional to the continental regions—estuaries, lagoons, and paludal deltas—precursor sediments for various mudrocks and associated sandstones, as well as coal, are deposited.

The continental interiors are diverse in rock types. Diatremes are created where carbonate-rich, ultramafic magmas blast through crustal rock to form kimberlite breccias, some of which are diamondiferous. Alkalic magmas of diverse types, differentiated from alkali olivine basalt parents, intrude the crust to form alkalic complexes that contain rocks ranging from alkali granites to jacupirangite and carbonatite. At depth within the continent, the ambient temperatures and pressures are such that Greenschist, Amphibolite, and Granulite Facies metamorphism are active. In contrast, at the surface under approximately STP conditions, sedimentary rocks develop. These include various mudrocks, wackes, arenites, evaporites, conglomerates, and breccias that are generated in the diverse environments of the continental surface.

CONCLUSION

The rock assemblages that form at various sites are distinctive *petrotectonic assemblages*. We use them, especially in conjunction with associated structures, to interpret past petrologic, tectonic, and Earth history.

Petrology, the study of rocks, is thus, a central subdiscipline within geology. Its utility within a variety of other subfields, such as structural geology, stratigraphy, paleontology, and geophysics, will be evident to the student who pursues knowledge in those areas. The rocks preserve the record of the past and are recording the history of the future.

APPENDIX

A

MINERALS IN COMMON ROCKS

Table A.1 Minerals in Igneous Rocks

	Grantd	Rhylt	Tra/Sy	Pho/NeS	An/D	Bas/Gab	UM
Quartz	X	X	m	–	o	o	–
Feldspars							
K-rich alkali	X	X	+	X	o	–	–
Sanidine	–	+	+	–	–	–	–
Anorthoclase	–	o	+	+	–	–	–
Na-plagioclase	X	+	+	+	X	o	–
Ca-plagioclase	X	–	–	–	+	X	m
Feldspathoids							
Nepheline	–	–	o	X	–	o	o
Leucite	–	–	–	o	–	o	–
Sodalite Group	–	–	–	+	–	–	–
Analcite	–	–	–	o	–	o	–
Zeolites	–	–	–	–	–	o	–
Olivines	o	–	o	o	o	+	X
Pyroxenes							
Orthopyroxene	o	–	o	–	o	+	X
Augite	o	o	o	–	+	X	+
Pigeonite	–	–	–	–	+	+	+
Diopside	–	–	–	–	+	+	+
Na-pyroxene	o	o	+	+	–	–	–

	Grantd	Rhylt	Tra/Sy	Pho/NeS	An/D	Bas/Gab	UM
Amphiboles							
Hornblende	+	o	+	−	+	+	o
Na-amphibole	o	o	o	+	−	−	o
Micas							
Muscovite	+	−	−	−	−	−	−
Biotite	+	+	+	+	+	+	−
Phlogopite	−	−	−	o	−	o	m
Zircon	m	m	m	m	m	m	−
Sphene	m	m	m	m	m	m	o
Garnets							
Pyralspites (Fe-Al)	o	o	−	−	−	−	o
Ugrandites (Ca)	−	−	−	o	−	o	−
Sil/Ky/Andalusite	o	−	−	−	−	−	−
Epidote Group	m	o	−	−	−	−	−
Cordierite	o	−	−	−	o	−	−
Tourmaline	o	−	−	−	−	−	−
Chlorites (sec.)	m	o	o	o	o	o	o
Clay minerals (sec.)	m	m	m	m	m	m	m
Carbonate Minerals							
Calcite (sec.)	o	o	o	o	o	o	o
Aragonite (sec.)	−	−	−	−	−	o	o
Magnetite	m	m	m	m	m	+	o
Ilmenite	m	−	m	−	o	m	o
Chromite	−	−	−	−	−	o	+
Spinel	−	−	−	−	−	o	+
Apatite	m	m	m	m	m	m	m
Pyrite (sec.)	o	o	o	o	o	o	o
Other sulfides	o	−	−	−	o	+	o

Grantd = granitoid rocks; Rhylt = rhyolite and other siliceous volcanic rocks; Tra/Sy = trachyte and syenite; Pho/NeS = phonolite, nepheline syenite and re-lated rocks; An/D = andesite and diorite; Bas/Gab = basalt and gabbro; UM = ultramafic rocks; sec = secondary minerals.
X = abundant, key minerals; + = common minerals; m = minor minerals; o = occasionally present or uncommon minerals; − = generally absent.

Table A.2 Minerals in Sedimentary Rocks

	Mudrx	Wacke	Arenite	Ls	Dolost	Other
Quartz	m	X	X	m	m	m→X
Feldspars						
K-rich alkali	o	+	+	o	o	o
Na-plagioclase	o	X	+	o	o	o
Ca-plagioclase	–	o	o	–	–	–
Zeolites	o	–	–	–	–	o
Olivines	–	–	–	–	–	–
Pyroxenes	–	o	o	–	–	–
Amphiboles	–	o	o	–	–	o
Clay minerals	X	+	m	–	–	o
Micas						
White micas	m	m	m	–	–	o
Glauconite	o	o	–	o	–	o
Biotite	o	m	o	–	–	–
Chlorites	X	m	o	–	–	o
Zircon	–	m	m	–	–	–
Sphene	–	o	o	–	–	–
Garnets	–	o	o	–	–	–
Tourmaline	–	o	o	–	–	–
Staurolite	–	o	o	–	–	–
Aluminosilicates	–	o	o	–	–	–
Epidote Group	–	o	o	–	–	–
Carbonate Minerals						
Calcite	o	o	m	X	o	o
Aragonite	–	–	–	o	–	o
Dolomite	o	–	–	o	X	o
Magnetite	–	m	m	–	–	o→X
Hematite	o	o	o	o	o	o→X
Ilmenite	–	m	m	–	–	–
Apatite	–	m	m	–	–	– →X
Gypsum	o	–	–	o	–	– →X
Pyrite	o	–	–	–	–	– →o
Other sulfides	o	–	–	o	o	o

Mudrx = mudrocks; Wacke = wacke and polymict lithic conglomerates; Arenite = arenites, cherts, and quartz conglomerates; Ls = limestones; Dolost = dolostones; Other = other rocks, particularly precipitates.

X = abundant, key minerals; + = common minerals; m = minor minerals; o = occasionally present or uncommon minerals; – = generally absent.

Table A.3 Minerals in Metamorphic Rocks

	S	SAC	A	C	MB	UB
Quartz	X	X	X	m	m	–
Feldspars						
K-rich alkali	o	+	+	o	o (LTP)	–
Na-plagioclase	o	+	+	–	X	–
Ca-plagioclase	–	+	+	o	X	o
Analcite	–	+ (LTP)	–	–	+ (LTP)	–
Feldspathoids	–	o	–	–	–	–
Zeolites	–	X (LTP)	+ (LTP)	–	X (LTP)	–
Olivines	–	–	–	o	m (HT)	X (HT)
Pyroxenes						
Orthopyroxenes	–	m (HT)	m (HT)	m (HT)	+ (HT)	+ (HT)
Diopside	–	–	–	+	o	o
Augite	–	–	–	–	o (HT)	–
Na-pyroxenes	m (HP)	X (HP)	–	–	+ (HP)	o (HP)
Wollastonite	–	–	–	+	–	–
Amphiboles						
Hornblende	–	m (HT)	m (HT)	o (HT)	X	o
Actinolite	–	o	–	o	X	+
Tremolite	–	–	–	+	–	+
Orthoamphiboles	–	o	–	–	o	+
Na-amphiboles	m (HP)	m (HP)	X (HP)	–	X (HP)	–
Micas						
White micas	m	X	X	o	m	–
Biotite	o	+	X	–	+	–
Phlogopite	–	–	–	m	o	m
Stilpnomelane	m	m	m	–	m	–
Clay minerals	o (LTP)	+ (LTP)	X (LTP)	m (LTP)	o (LTP)	–
Prehnite	–	m (LTP)	o (LTP)	o (LTP)	X (LTP)	–
Chlorites	o	+	X	o	X	+
Serpentines	–	–	–	o	o	X
Talc	–	–	–	o	o	X

Table A.3 Continued

	S	SAC	A	C	MB	UB
ZZircon	m	m	m	–	–	–
Sphene	–	m	o	o	m	–
Garnets	+	+	X	o	o	o
Tourmaline	–	o	o	–	–	–
Staurolite	–	–	X	–	o	–
Chloritoid	–	–	m	–	–	–
Aluminosilicates						
Andalusite	o (HT)	–	+ (HT)	–	–	–
Kyanite	+	o	+	–	–	–
Sillimanite	+ (HT)	o (HT)	+ HT	–	–	–
Epidotes	m	m	m	o	+	–
Lawsonite	–	+ (HP)	m (HP)	–	+ (HP)	–
Pumpellyite	–	+	o	–	+	–
Cordierite	+ (HT)	o (HT)	+ (HT)	–	–	–
Carbonate Minerals						
Calcite	o	o	o	X	o	o
Aragonite	o (HP)	+ (HP)	o (HP)	X (HP)	o (HP)	o
Dolomite	–	–	–	X	–	o
Magnetite	m	m	m	o	m	m
Ilmenite	o	–	m	–	–	–
Chromite	–	–	–	–	–	m
Spinel	–	–	o	o	–	o
Apatite	o	m	m	o	m	–
Sulfides	o	o	o	m	m	o

S = siliceous rocks; SAC = siliceous-alkali-calcic rocks; A = aluminous (including pelitic) rocks; C = carbonate rocks; MB = metabasites; UB = ultrabasic rocks.
X = abundant, key minerals; + = common minerals; m = minor minerals; o = occasionally present or uncommon minerals; – = generally absent.
HT = in high-temperature rocks only; HP = in high-pressure rocks only; LPT = in low-temperature and low-pressure rocks only.

APPENDIX

B

CIPW NORMS

•••

A rock norm is a list of standard minerals and their percentages calculated from the chemical analysis of a rock. The CIPW method of calculating a norm, developed by Cross, Iddings, Pirsson, and Washington (1903), has been described and modified by a number of petrologists (see Johannsen, 1931, p. 88ff.; Barth, 1962; Bickel, 1979). The procedure described here generally follows the original method, but is modified and slightly abbreviated in part, following Johannsen (1931) and Bickel (1979). The chemical analysis used in the norm calculation must be in weight percent oxides (e.g., weight percent SiO_2;

(a)

Step		SiO_2	TiO_2	Al_2O_3	Fe_2O_3	FeO	MnO	MgO	CaO	Na_2O	K_2O	P_2O_5	CO_2	Other Oxides		H_2O / %Min.
A	Wt.%															
	Mol. Wt.	60.09	79.90	101.96	159.69	71.85	70.94	40.31	56.08	61.98	94.20	141.95	44.01			
B	Moles															
C	Comb. M.						↵									
D	Mineral —															
	=															
	—															
	=															
	—															
	=															
	—															
	=															
	—															
	=															
	—															
	=															
	—															
	=															
E	—															
G	=															
	—															
	=															
	—															
	=															
L							Total Wt. % Oxides =				≅Total Wt. % Minerals + Wt. % H_2O =					

••••••••••••

Figure B.1

Charts for calculating CIPW norms. (a) Main chart for calculations (modified from and courtesy of Charles E. Bickel). (b) Chart for calculating diopside, hypersthene, and olivine. (c) Chart for calculating normative mineral percentages. See text for explanation of steps A–L.

(b)

Step	
F	$PRF = \dfrac{\text{Moles FeO}}{\text{Moles (FeO + MgO)}} =$ \qquad $PRM = \dfrac{\text{Moles MgO}}{\text{Moles (FeO + MgO)}} =$
I	Wt. % Wo in Diopside = (Mol. CaO in Diop.)(Mol. Wt. Wo) = ( )(116.17) = Wo = _____ Wt. % En in Diopside = [Mol. (Fe, Mg)O in Diop.] (PRM) (Mol. Wt. En) = [ ] ( ) (100.40) = En = _____ Wt. % Fs in Diopside = [Mol. (Fe, Mg)O in Diop.] (PRF) (Mol. Wt. Fs) = [ ] ( ) (131.94) = Fs = _____ Total Diopside = Wo + En + Fs =
J	Wt. % En in Hypersthene = [(Fe, Mg)O in Hyp] (PRM) (Mol. Wt. En) = [ ] ( ) (100.40) = En = _____ Wt. % Fs in Hypersthene = [(Fe, Mg)O in Hyp] (PRF) (Mol. Wt. Fs) = [ ] ( ) (131.94) = Fs = _____ Total Hypersthene = En + Fs =
K	Wt. % Fo in Olivine = [(Fe, Mg)O in Ol] (PRM) (Mol. Wt. Fo/2) = [ ] ( ) (70.355) = Fo = _____ Wt. % Fa in Olivine = [(Fe, Mg)O in Ol] (PRF) (Mol. Wt. Fa/2) = [ ] ( ) (101.895) = Fa = _____ Total Olivine = Fo + Fa =

············

Figure B.1
Charts for calculating CIPW norms. (a) Main chart for calculations (modified from and courtesy of Charles E. Bickel). (b) Chart for calculating diopside, hypersthene, and olivine. (c) Chart for calculating normative mineral percentages. See text for explanation of steps A–L.

weight percent CaO, except for Cl, F, and S). Figure B.1 provides blank forms that may be copied for use in calculating a norm and figure B.2 shows a simple example. Standard minerals, their symbols, and their chemistries are shown in figure B.1c, a chart for calculating normative mineral percentages. The norm is calculated following the list of steps (A–L) listed here.

STEP A. List the values of the oxides from the chemical analysis in the column under the oxide heading. For minor element oxides, use the blanks at the end of the row.

STEP B. Calculate the number of moles of each oxide by dividing the weight percent of that oxide by the molecular weight of the oxide. Enter the value in the row marked B ("Moles").

STEP C. Combine certain molecular values of oxides that typically substitute for major oxides with those major oxide molecular values. Moles of MnO and NiO are added to the value of FeO, and are considered as FeO throughout the calculation. Moles of BaO and SrO are added to CaO. Moles of

Cr_2O_3 are added to moles of Fe_2O_3. In cases where these elements are abundant and may represent unusual minerals in a rock, they may be kept separate and used to calculate special mineral molecules (e.g., Cr_2O_3 may be used to calculate chromite in ultramafic rocks).

STEP D. Calculate the amounts of moles used for each standard mineral. Throughout this part of the calculation, the word "amount" refers to the number of moles. In general, two or three oxides are combined for the formation of a mineral. Each time a mineral is formed, one of the oxides will be completely used up. The amounts of the other oxides used are subtracted from the amounts available. Figure B.1a contains blanks for both the amount used (preceded by a negative sign) and the amount remaining (preceded by an equals sign). The amount remaining will determine, in many cases, which of two calculations will be completed for the formation of the next mineral. These alternative calculations are labelled a and b. *Never* do both calculations; always do one or the other. Complete the numbered steps in order. If an oxide is absent or has been used up, skip to the next successive instruction.

(c)

				Step H - Calculate Wt. % normative mineral		
Mineral	Symbol	Formula	Reference oxide	$\left(\dfrac{\text{Formula}}{\text{weight}}\right)$ X	$\left(\dfrac{\text{Moles ref.}}{\text{oxide}}\right)$ =	Wt. % mineral
Quartz	Q	SiO_2	SiO_2	60.9		
Orthoclase	or	$K_2O \bullet Al_2O_3 \bullet 6SiO_2$	K_2O	556.70		
Albite	ab	$Na_2O \bullet Al_2O_3 \bullet 6\ SiO_2$	Na_2O	524.48		
Anorthite	an	$CaO \bullet Al_2O_3 \bullet 2\ SiO_2$	CaO	278.22		
Leucite	lc	$K_2O \bullet Al_2O_3 \bullet 4\ SiO_2$	K_2O	436.52		
Nepheline	ne	$Na_2O \bullet Al_2O_3 \bullet 2\ SiO_2$	Na_2O	284.12		
Kaliophilite	kp	$K_2O \bullet Al_2O_3 \bullet 2\ SiO_2$	K_2O	316.34		
Na-Metasilicate	ns	$Na_2O \bullet SiO_2$	Na_2O	122.07		
K-Metasilicate	ks	$K_2O \bullet SiO_2$	K_2O	154.29		
Acmite	ac	$Na_2O \bullet Fe_2O_3 \bullet 4\ SiO_2$	Na_2O	462.03		
Diopside	di	$\longrightarrow$ Obtain from Fig. B.1 (b) [Step I] =				
Hypersthene	hy	$\longrightarrow$ Obtain from Fig. B.1 (b) [Step J] =				
Olivine	ol	$\longrightarrow$ Obtain from Fig. B.1 (b) [Step K] =				
Wollastonite	wo	$CaO \bullet SiO_2$	CaO	116.17		
Larnite	la	$2\ CaO \bullet SiO_2$	SiO_2	172.25		
Sphene	sp	$CaO \bullet TiO_2 \bullet SiO_2$	CaO	196.07		
Perofskite	pf	$CaO \bullet TiO_2$	CaO	135.98		
Rutile	ru	TiO_2	TiO_2	79.90		
Ilmenite	il	$FeO \bullet TiO_2$	FeO	151.75		
Magnetite	mt	$FeO \bullet Fe_2O_3$	FeO	231.54		
Hematite	hm	Fe_2O_3	Fe_2O_3	159.69		
Chromite	cm	$FeO \bullet Cr_2O_3$	FeO	223.84		
Pyrite	pr	FeS_2	FeO	120.01		
Fluorite	fr	CaF_2	CaO	78.08		
Calcite	cc	$CaO \bullet CO_2$	CaO	100.09		
Halite	hl	$NaCl$	Na_2O	58.44		
Na-Carbonite	nc	$Na_2O \bullet CO_2$	Na_2O	105.99		
Thenardite	th	$Na_2O \bullet SO_3$	Na_2O	142.04		
Apatite	ap	$3\ (3\ CaO \bullet P_2O_5) \bullet CaF_2$	P_2O_5	336.22		
Corundum	c	Al_2O_3	Al_2O_3	101.96		
Zircon	z	$ZrO_2 \bullet SiO_2$	ZrO_2	183.31		

Figure B.1
Charts for calculating CIPW norms. (a) Main chart for calculations (modified from and courtesy of Charles E. Bickel). (b) Chart for calculating diopside, hypersthene, and olivine. (c) Chart for calculating normative mineral percentages. See text for explanation of steps A–L.

Do not skip steps unless the instructions expressly direct you to do so. Silica-poor rocks may result in negative values for SiO_2 early in the norm calculation. This is acceptable and will be corrected later in the calculation. *Do not* generate negative amounts for oxides other than SiO_2. Use four decimal places for your calculations, but use only the correct number of significant figures in reporting the final mineral percentages.

1. If Cr_2O_3 is present in more than trace amounts, allot all of the Cr_2O_3 and an equal amount of FeO to the formation of chromite. If Cr_2O_3 is absent or present only in trace amounts, skip to step 2 (next).

2. If TiO_2 is not present, skip to step 3. If TiO_2 exceeds the amount of FeO, do calculation 2a. If the amount of FeO exceeds that of TiO_2, do calculation 2b.
   2a. Allot all of the FeO and an equal amount of TiO_2 to the formation of ilmenite.
   2b. Allot all of the TiO_2 and an equal amount of FeO to the formation of ilmenite.

3. Allot all of the P_2O_5 and 3.333 times this amount of CaO to the formation of apatite. If F or Cl are present, use 3.000 times the amount of P_2O_5 and 0.333 times the amount of F plus Cl. If P_2O_5 is absent, skip to step 4.

4. If F is present, use the amount of F and one-half this amount of CaO to form fluorite. If F is absent or has been used up, skip to step 5.

5. If Cl is present or remains after making apatite, combine the amount remaining with an equal amount of Na_2O to form halite. If Cl is absent or used up, skip to step 6.

6. If SO_2 is present, allot all of it and an equal amount of Na_2O to thenardite. If SO_2 is absent, skip to step 7.

7. If S is present, allot all the S and one-half this amount of FeO to make pyrite. If S is absent, skip to step 8.

8. If CO_2 is present, allot all the CO_2 and an equal amount of CaO to calcite, or, if the rock is silica-undersaturated, allot the CO_2 and an equal amount of Na_2O to sodium carbonate. If CO_2 is absent, skip to step 9.

691

(a)

Step		SiO₂	TiO₂	Al₂O₃	Fe₂O₃	FeO	MnO	MgO	CaO	Na₂O	K₂O	P₂O₅	CO₂	Other Oxides			H₂O %Min.
A	Wt.%	72.06	0.28	12.38	1.00	1.44	0.04	0.35	0.96	2.88	5.74	0.04	0.01				2.68
	Mol. Wt.	60.09	79.90	101.96	159.69	71.85	70.94	40.31	56.08	61.98	94.20	141.95	44.01				—
B	Moles	1.1992	.0035	.1214	.0063	.0200	.0006	.0087	.0171	.0465	.0609	.0003	.0002				—
C	Comb. M.	—		—		.0206	↵	—	—	—		—	—				—
D	Mineral il =		.0035 / -0-			.0035 / .0171											0.53
	ap =								.0010 / .0161			.0003 / -0-					0.1
	cc =								.0002 / .0159				.0002 / -0-				0.02
	or =	.3654 / .8338		.0609 / .0605							.0609 / -0-						33.90
	ab =	.2790 / .5548		.0465 / .0140						.0465 / -0-							24.39
	an =	.0280 / .5268		.0140 / -0-					.0140 / .0019								3.90
	mt =				.0063 / -0-	.0063 / .0108											1.5
E	✕ =					(Fe,Mg)O .0195											
G	di =	.0038 / .5230				.0019 / .0176		.0019 / -0-									0.44
	hy =	.0176 / .5054				.0176 / -0-											2.08
	Q =	.5054 / -0-															30.37
L									Total Wt. % Oxides = 99.86		≅Total Wt. % Minerals + Wt. % H₂O = 99.9						

Figure B.2

Example of a simple norm calculation. (a), (b), and (c) as in figure B.1, but with illustrative numbers filled in appropriate blanks.

9. If ZrO₂ is present, allot all of it and an equal amount of SiO₂ to zircon. If ZrO₂ is absent, skip to step 10.

10. If the amount of K₂O exceeds the amount of Al₂O₃, do calculation 10a. If the amount of Al₂O₃ exceeds the amount of K₂O, do calculation 10b.

   10a. Allot all of the Al₂O₃, an equal amount of K₂O, and six times this amount of SiO₂ to provisional orthoclase. (The orthoclase is provisional until it is determined that the rock is not silica-deficient. In the latter case, leucite rather than orthoclase may need to be created.)

   10b. Allot all of the K₂O, an equal amount of Al₂O₃, and six times this amount of SiO₂ to the formation of provisional orthoclase. (The orthoclase is provisional until it is determined that the rock is not silica-deficient. In the latter case, leucite rather than orthoclase may need to be created.) If K₂O is absent, skip to step 12.

11. Use any remaining K₂O and an equal amount of SiO₂ to form potassium metasilicate. If potash is absent or if no potash remains, skip to step 12.

12. If the amount of Al₂O₃ remaining exceeds the amount of Na₂O, do calculation 12a. If the amount of Na₂O exceeds the amount of Al₂O₃, do calculation 12b.

   12a. Allot all the Na₂O, an equal amount of Al₂O₃, and six times this amount of SiO₂ to the formation of provisional albite.

   12b. Allot all of the Al₂O₃, an equal amount of Na₂O, and six times this amount of SiO₂ to the formation of provisional albite. Skip to step 15.

13. If the amount of Al₂O₃ remaining exceeds the amount of CaO remaining, do calculation 13a. If the amount of CaO remaining exceeds the amount of Al₂O₃ remaining, do calculation 13b.

   13a. Allot all of the CaO, an equal amount of Al₂O₃, and twice this amount of SiO₂ to the formation of anorthite.

(b)

Step	
F	$$PRF = \frac{\overset{.0108}{\text{Moles FeO}}}{\underset{.0195}{\text{Moles (FeO + MgO)}}} = 0.55 \quad PRM = \frac{\overset{.0087}{\text{Moles MgO}}}{\underset{.0195}{\text{Moles (FeO + MgO)}}} = 0.45$$
I	Wt. % Wo in Diopside = (Mol. CaO in Diop.)(Mol. Wt. Wo) $\qquad = (.0019)(116.17) = $ Wo = __0.22__ Wt. % En in Diopside = [Mol. (Fe, Mg)O in Diop.] (PRM) (Mol. Wt. En) $\qquad = [.0019] (0.45)(100.40) = $ En = __0.08__ Wt. % Fs in Diopside = [Mol. (Fe, Mg)O in Diop.] (PRF) (Mol. Wt. Fs) $\qquad = [.0019] (0.55)(131.94) = $ Fs = __0.14__ $\qquad\qquad$ Total Diopside = Wo + En + Fs = __0.44__
J	Wt. % En in Hypersthene = [(Fe, Mg)O in Hyp] (PRM) (Mol. Wt. En) $\qquad = [.0176] (0.45)(100.40) = $ En = __0.80__ Wt. % Fs in Hypersthene = [(Fe, Mg)O in Hyp] (PRF) (Mol. Wt. Fs) $\qquad = [.0176] (0.55)(131.94) = $ Fs = __1.28__ $\qquad\qquad$ Total Hypersthene = En + Fs = __2.08__
K	Wt. % Fo in Olivine = [(Fe, Mg)O in Ol] (PRM) (Mol. Wt. Fo/2) $\qquad = [\quad] (\quad)(70.355) = $ Fo = _____ Wt. % Fa in Olivine = [(Fe, Mg)O in Ol] (PRF) (Mol. Wt. Fa/2) $\qquad = [\quad] (\quad)(101.895) = $ Fa = _____ $\qquad\qquad$ Total Olivine = Fo + Fa = _____

Figure B.2
Example of a simple norm calculation. (a), (b), and (c) as in figure B.1, but with illustrative numbers filled in appropriate blanks.

13b. Allot all of the Al_2O_3, an equal amount of CaO, and twice this amount of SiO_2 to the formation of anorthite. Skip to step 15.

14. If Al_2O_3 remains, the Al_2O_3 is assigned to corundum. If none remains, skip to step 15.

15. If TiO_2 is absent or has been used up, skip to step 17. If TiO_2 remains and the amount exceeds that of CaO, do calculation 15a. If TiO_2 remains, but the amount of remaining CaO exceeds the amount of TiO_2, do calculation 15b.

   15a. Allot all of the CaO, an equal amount of TiO_2, and an equal amount of SiO_2 to the formation of provisional sphene.

   15b. Allot all of the TiO_2, an equal amount of CaO, and an equal amount of SiO_2 to the formation of sphene.

16. If TiO_2 remains, it is assigned to rutile. If no TiO_2 remains, skip to step 17.

17. If all of the Na_2O is used up, skip to step 18. If Na_2O remains and the amount exceeds the amount of Fe_2O_3, do calculation 17a. If the amount of Na_2O is less than the amount of Fe_2O_3, do calculation 17b.

17a. Allot all of the Fe_2O_3, an equal amount of Na_2O, and four times this amount of SiO_2 to the formation of acmite.

17b. Allot all of the Na_2O, an equal amount of Fe_2O_3, and four times this amount of SiO_2 to the formation of acmite. Skip to step 19.

18. If Na_2O remains, assign the remaining amount and an equal amount of SiO_2 to the formation of sodium metasilicate. If no soda remains, skip to step 19.

19. Compare the amounts of remaining FeO and Fe_2O_3. If the amount of FeO exceeds the amount of Fe_2O_3, do calculation 19a. If the amount of FeO is less than the amount of Fe_2O_3, do calculation 19b. If no Fe_2O_3 remains, skip to Step E.

   19a. Use all of the remaining Fe_2O_3 and an equal amount of FeO to make magnetite.

   19b. Use all of the remaining FeO and an equal amount of Fe_2O_3 to make magnetite.

20. Assign any remaining Fe_2O_3 to hematite. If no Fe_2O_3 remains, skip to Step E.

(c)

Step H - Calculate Wt. % normative mineral						
Mineral	Symbol	Formula	Reference oxide	(Formula weight) X	(Moles ref. oxide) =	Wt. % mineral
Quartz	Q	SiO_2	SiO_2	60.9	.5054	30.37
Orthoclase	or	$K_2O \bullet Al_2O_3 \bullet 6SiO_2$	K_2O	556.70	.0609	33.90
Albite	ab	$Na_2O \bullet Al_2O_3 \bullet 6\ SiO_2$	Na_2O	524.48	.0465	24.39
Anorthite	an	$CaO \bullet Al_2O_3 \bullet 2\ SiO_2$	CaO	278.22	.0140	3.90
Leucite	lc	$K_2O \bullet Al_2O_3 \bullet 4\ SiO_2$	K_2O	436.52		
Nepheline	ne	$Na_2O \bullet Al_2O_3 \bullet 2\ SiO_2$	Na_2O	284.12		
Kaliophilite	kp	$K_2O \bullet Al_2O_3 \bullet 2\ SiO_2$	K_2O	316.34		
Na-Metasilicate	ns	$Na_2O \bullet SiO_2$	Na_2O	122.07		
K-Metasilicate	ks	$K_2O \bullet SiO_2$	K_2O	154.29		
Acmite	ac	$Na_2O \bullet Fe_2O_3 \bullet 4\ SiO_2$	Na_2O	462.03		
Diopside	di	⟶ Obtain from Fig. B.1 (b) [Step I] =				0.44
Hypersthene	hy	⟶ Obtain from Fig. B.1 (b) [Step J] =				2.08
Olivine	ol	⟶ Obtain from Fig. B.1 (b) [Step K] =				
Wollastonite	wo	$CaO \bullet SiO_2$	CaO	116.17		
Larnite	la	$2\ CaO \bullet SiO_2$	SiO_2	172.25		
Sphene	sp	$CaO \bullet TiO_2 \bullet SiO_2$	CaO	196.07		
Perofskite	pf	$CaO \bullet TiO_2$	CaO	135.98		
Rutile	ru	TiO_2	TiO_2	79.90		
Ilmenite	il	$FeO \bullet TiO_2$	FeO	151.75	.0035	0.53
Magnetite	mt	$FeO \bullet Fe_2O_3$	FeO	231.54	.0063	1.5
Hematite	hm	Fe_2O_3	Fe_2O_3	159.69		
Chromite	cm	$FeO \bullet Cr_2O_3$	FeO	223.84		
Pyrite	pr	FeS_2	FeO	120.01		
Fluorite	fr	CaF_2	CaO	78.08		
Calcite	cc	$CaO \bullet CO_2$	CaO	100.09	.0002	0.02
Halite	hl	$NaCl$	Na_2O	58.44		
Na-Carbonite	nc	$Na_2O \bullet CO_2$	Na_2O	105.99		
Thenardite	th	$Na_2O \bullet SO_3$	Na_2O	142.04		
Apatite	ap	$3\ (3\ CaO \bullet P_2O_5) \bullet CaF_2$	P_2O_5	336.22	.0003	0.1
Corundum	c	Al_2O_3	Al_2O_3	101.96		
Zircon	z	$ZrO_2 \bullet SiO_2$	ZrO_2	183.31		

Figure B.2
Example of a simple norm calculation. (a), (b), and (c) as in figure B.1, but with illustrative numbers filled in appropriate blanks.

STEP E. Combine FeO and MgO. At this point, all of the remaining amounts of FeO and MgO are added together and are called (Fe,Mg) O.

STEP F. Determine the porportions of each of the two oxides in (Fe,Mg)O. The proportion of each of the oxides in the combined oxide (Fe,Mg)O is calculated by dividing the amount of each oxide by the sum of the two oxides (figure B.1. The proportion of FeO is called PRF and that of MgO is called PRM. PRF + PRM = 1.0000.

STEP G. Continue calculating mineral percentages.

21. Compare the amounts of (Fe,Mg)O and CaO. If the amount of (Fe,Mg)O exceeds the amount of CaO, do calculation 21a. If the amount of (Fe,Mg)O is less than the amount of CaO, do calculation 21b. If no CaO remains, skip to step 22.

   21a. Allot all of the CaO, an equal amount of (Fe,Mg)O, and twice this amount of SiO_2 to the formation of diopside.

   21b. Allot all of the (Fe,Mg)O, an equal amount of CaO, and twice this amount of SiO_2 to the formation of diopside. Skip to step 23.

22. Assign the remaining (Fe,Mg)O and an equal amount of SiO_2 to the formation of provisional hypersthene. If no (Fe,Mg)O remains, skip to step 23.

23. Allot any remaining CaO and an equal amount of SiO_2 to provisional wollastonite.

24. If silica remains at this point (i.e., there is a positive value for moles of SiO_2), it is assigned to quartz. If SiO_2 has a negative value, skip to step 26.

25. If quartz is created in step 24, skip to Step H.

26. If SiO_2 has a negative value, it will be necessary to break down one or more of the provisional minerals to make minerals that contain less silica (the negative value must be eliminated). Provisional hypersthene will be the first mineral to be broken down. If no hypersthene was created in step 22, skip to step 28. If hypersthene was created in step 22, move on to step 27.

27. Break down the hypersthene, adding the SiO_2 back into the SiO_2 column and the $(Fe,Mg)O$ back into the $(Fe,Mg)O$ column. If the amount of SiO_2 available now is a positive amount and that amount exceeds or equals one-half the amount of $(Fe,Mg)O$, do calculation 27a. If the amount of silica is still negative or is less than half of the value of $(Fe,Mg)O$, do calculation 27b.

  27a. Olivine and hypersthene will be created using the available SiO_2 and $(Fe,Mg)O$. Use the following equations in which x is the amount of $(Fe,Mg)O$ allotted to hypersthene, y is the amount of $(Fe,Mg)O$ allotted to olivine, M is the amount of available $(Fe,Mg)O$, and S is the amount of available SiO_2.

$$x = 2S - M \qquad\qquad y = M - x$$

  The amount of SiO_2 allotted to hypersthene is equal to x. The amount of SiO_2 allotted to olivine is $1/2(y)$. Go to Step H.

  27b. Create olivine using all of the available $(Fe,Mg)O$ and one-half this amount of SiO_2.

28. If the amount of SiO_2 is still negative, break down the provisional sphene. Add the amount of SiO_2 back into the silica column. The remaining CaO and TiO_2 becomes (is assigned to) perofskite. If no sphene was created in step 15, skip to step 29.

29. Break down the provisional albite. Add the amounts of SiO_2, Na_2O and Al_2O_3 back into their respective columns. If the amount of available silica is positive and exceeds or equals twice the amount of Na_2O, do calculation 29a. If the amount of SiO_2 is still negative or is less than twice the amount of Na_2O, do calculation 29b.

  29a. Create nepheline and albite using the equations below, in which x is the amount of Na_2O allotted to albite, y is the amount of Na_2O allotted to nepheline, N is the amount of available Na_2O, and S is the amount of available SiO_2.

$$x = (S - 2N)/4 \qquad\qquad y = N - x$$

  Albite is composed of x, an equal amount of Al_2O_3, and $6x$ SiO_2. Nepheline is composed of y, an equal amount of Al_2O_3, and $2y$ SiO_2. Go to Step H.

  29b. Create nepheline by combining all of the remaining Na_2O, an equal amount of Al_2O_3, and twice the amount of SiO_2.

30. If there is still a negative amount of SiO_2, break down the provisional orthoclase. Add the amounts of SiO_2 back into the SiO_2 column and the Al_2O_3 and K_2O into their respective columns. If the amount of available SiO_2 is now positive and exceeds or equals four times the amount of K_2O, do calculation 30a. If the amount of SiO_2 is still negative or is less than four times the amount of K_2O, do calculation 30b.

  30a. Create leucite and orthoclase using the equation below, in which x is the amount of K_2O used to form orthoclase, y is the amount of K_2O used to form leucite, K is the amount of available K_2O, and S is the amount of available SiO_2.

$$x = (S - 4K)/2 \qquad\qquad y = K - x$$

  Orthoclase consists of x, an equal amount of Al_2O_3, and $6x$ SiO_2. Leucite consists of y, an equal amount of Al_2O_3, and $4y$ SiO_2. Go to Step H.

  30b. Form leucite from all of the K_2O, an equal amount of Al_2O_3, and four times that amount of SiO_2.

31. If there is still a negative amount of SiO_2, additional minerals must be broken down. If wollastonite was produced in step 23, break down the wollastonite, adding the silica thus obtained to the silica column and the lime to the CaO column. If silica is now positive and equal to one-half or more of the CaO, do calculation 31a. If the amount of SiO_2 is still negative or is less than one-half the amount of CaO, do calculation 31b. If wollastonite was not created in step 23, skip to step 32.

  31a. Create larnite and wollastonite using the equations below, in which x is the amount of CaO allotted to wollastonite, y is the amount of CaO allotted to larnite, C is the amount of available CaO, and S is the amount of available SiO_2.

$$x = 2S - C \qquad\qquad y = C - x$$

  Wollastonite consists of x and an equal amount of SiO_2. Larnite consists of y and one-half this amount of SiO_2. Go to Step H.

  31b. Larnite is formed from all the CaO and one-half that amount of SiO_2.

32. If the amount of SiO_2 is still negative, break down all the provisional diopside. The silica is added to the silica column and the CaO and $(Fe,Mg)O$ are added to their respective columns. If the amount of available silica is positive and equal to more than the available amount of CaO, do calculation 32a. If the amount of silica is still negative or is less than the amount of CaO, do calculation 32b.

  32a. Form olivine, diopside, and larnite using the equations below, in which x is the amount of CaO allotted to diopside, y is the amount of CaO allotted to larnite, C is the amount of available CaO, and S is the amount of available SiO_2.

$$x = S - C \qquad\qquad y = C - x$$

Diopside consists of x, an equal amount of $(Fe,Mg)O$, and $2x$ SiO_2. Larnite consists of y and one-half this amount of SiO_2. Olivine consists of an amount of $(Fe,Mg)O$ equal to y and 1/2 that amount of SiO_2. Add the larnite to any larnite produced in step 21. Add the olivine to any olivine produced in step 27. Go to Step H.

32b. Form larnite and olivine. Allot to larnite all of the CaO and one-half that amount of SiO_2. Allot to olivine all the $(Fe,Mg)O$ and one-half that amount of SiO_2. Add the larnite and olivine to any larnite and olivine formed previously.

33. If the amount of SiO_2 is still negative, break down the leucite. Add the SiO_2, K_2O, and Al_2O_3 to their respective columns. Form kaliophilite and leucite using the equations below, in which x is the amount of K_2O allotted to leucite, y is the amount of K_2O allotted to kaliophilite, K is the amount of available K_2O, and S is the amount of available SiO_2.

$$x = (S - 2K)/2 \qquad y = (4K - S)/2$$

Leucite consists of x, an equal amount of Al_2O_3, and $4x$ SiO_2. Kaliophilite consists of y, an equal amount of Al_2O_3, and $2x$ SiO_2.

STEP H. Calculate the normative weight percentages of each of the minerals in the norm. Multiply the formula weight of each mineral times the number of moles of the reference oxide for that mineral (figure B.1c). The weight percentages of diopside, hypersthene, and olivine cannot be calculated in this way because these minerals contain both FeO and MgO. Compute the weight percentages of these minerals following Steps I, J, and K, in which the wollastonite (wo), enstatite (en), ferrosilite

(fs), forsterite (fo), and fayalite (fa) components of the respective minerals are calculated using PRF and PRM determined in Step F.

STEP I. Calculate the weight percent of diopside using the formulae in figure B.1b and summing wo + en + fs.
Wt. % wo in di = [amount CaO in di](mol. wt. wo)
Wt. % en in di = [amount $(Fe,Mg)O$ in di](PRM)(mol. wt. en)
Wt. % fs in di = [amount $(Fe,Mg)O$ in di](PRF)(mol. wt. fs)

STEP J. Calculate the weight percent of hypersthene using the formulae in figure B.1b and summing the en + fs.
Wt. % en in hy = [amount $(Fe,Mg)O$ in hy](PRM)(mol. wt. en)
Wt. % fs in hy = [amount $(Fe,Mg)O$ in hy](PRF)(mol. wt. fs)

STEP K. Calculate the weight percent olivine using the formulae in figure B.1b and summing fa + fo.
Wt. % fo in ol = [amount of $(Fe,Mg)O$ in ol](PRM)(mol. wt. fo/2)
Wt. % fa in ol = [amount of $(Fe,Mg)O$ in ol](PRF)(mol. wt. fa/2)

STEP L. List the mineral weight percentages opposite the mineral symbol in the column marked L **and sum the minerals plus the water**. Use values with correct significant figures for the mineral weight percents. The total should be equal to the total of the original oxides within a few hundredths of a percent (rounding error). If the two totals do not match, you have made an error in the calculation.

APPENDIX

C

PETROGENETIC GRIDS FOR METAMORPHIC ROCKS

Petrogenetic grids are pressure-temperature graphs that show reaction curves along which one mineral or mineral assemblage changes to another. Curves are determined experimentally or by calculation. Reactions for all curves are listed below (tables C.2, C.3, C.4, and C.5) and the mineral abbreviations are listed in table C.1. The equations are not balanced. These grids allow us to estimate approximately the metamorphic conditions for particular mineral assemblages found in metamorphic rocks.

Table C.1 Mineral Abbreviations Used in Petrogenetic Grids and Equations

A	=	Antigorite	Ez	=	Zoisite	P	=	Plagioclase
Ab	=	Albite	F	=	Fluid (H_2O)	Pe	=	Periclase
Ac	=	Acmite	Fo	=	Forsterite	Ph	=	Phlogopite
Af	=	Alkali feldspar	G	=	Garnet	Pr	=	Prehnite
Ag	=	Augite	Gl	=	Glaucophane	Pu	=	Pumpellyite
Ak	=	Akermanite	H	=	Hornblende	Py	=	Pyrophyllite
Am	=	Amphibole	Hm	=	Hematite	Q	=	Quartz
At	=	Actinolite	Hu	=	Heulandite	R	=	Rankinite
An	=	Anorthite	Hy	=	Hypersthene	S	=	Sphene
And	=	Andalusite	I	=	Ilmenite	Sa	=	Serpentine (antigorite)
Anl	=	Analcite	Id	=	Idocrase	Sb	=	Stilbite
Ant	=	Anthophyllite	Jd	=	Jadeite	Sc	=	Serpentine (chrysotile)
Ar	=	Aragonite	Jdpx	=	Jadeitic pyroxene	Sd	=	Sanidine
B	=	Brucite	K	=	Kaolinite	Sil	=	Sillimanite
Bio	=	Biotite	Ky	=	Kyanite	Sl	=	Serpentine (lizardite)
C	=	Chrysotile	L	=	Liquid = melt	Sp	=	Spinel
Ca	=	Carpholite	La	=	Larnite	Spp	=	Sapphirine
Cc	=	Calcite	Lm	=	Laumontite	St	=	Staurolite
Cd	=	Cordierite	Lw	=	Lawsonite	Stp	=	Stilpnomelane
Ch	=	Chlorite	M	=	"Mica"	Su	=	Spurrite
Cp	=	Clinopyroxene	Me	=	Merwinite	T	=	Talc
Ct	=	Chloritoid	Mg	=	Magnesite	Ti	=	Tillyite
CV	=	Chlorite-Vermiculite	Mm	=	Montmorillonite	Tr	=	Tremolite
Di	=	Diopside	Mo	=	Monticellite	W	=	Wairakite
Do	=	Dolomite	Mt	=	Magnetite	Wm	=	White mica
E	=	Epidote	Ol	=	Olivine	Wo	=	Wollastonite
Ec	=	Clinozoisite	Om	=	Omphacite	X	=	Extra components
En	=	Enstatite	Op	=	Orthopyroxene			

Table C.2 Equations and References for Petrogenetic Grid for SAC and Aluminous Rocks

1. $Q + Anl = Ab + F$	A. S. Campbell and Fyfe (1965)
2. $Hu = Lm + Q + F$	Cho, Maruyama, and Liou (1987)
3. $Pr + Lm + Ch = Pu + Q + F$	Liou et al. (1987)
4. $Hu = Lw + Q + F$	Liou et al. (1987)
5. $Cc = Ar$	Composite based on Jamieson (1953), Clark (1957), Crawford and Hoersch (1972), Irving and Wyllie (1975), Salje and Viswanathan (1976), and Brar and Schloessin (1979)
6. $Ab + x = Jdpx + Q$	Newton and Smith (1967)
7. $Ab = Jd + Q$	Newton and Smith (1967)
8. $Lm + Pr = Ec + Q + F$	Liou et al. (1987)
9. $Lm = W + F$	Liou (1971b)
10. $Lm = Lw + Q + F$	Nitsch (1968, in Liou, 1971)
11. Stability field of Gl	Maresch (1977)
$Gl = M + T$	
$Gl = M + T + Ab$	
$Gl = T + Ch$	
12. $Pr = Ez + G + Q + F$	Liou, Maruyama, and Cho (1985, 1986)
13. $W = An + Q + F$	Liou (1970, 1971b)
14. $Pr + Pu + Q + Ch = At + Pu + Q + Ch$	Nitsch (1971)
15. $Pr + E + Ch = Pu + Q$	Liou et al. (1987)
16. $Py = And$	Spear and Cheney (1989)
17. $W = Lw + Q$	Liou (1971b)
18. $Py = Ky + Q + F$	Haas and Holdaway (1973)
19. $Lw = An + F$	Newton and Kennedy (1963)
20. $K + Q = Py + F$	Chatterjee, Johannes, and Leistner (1984)
21. $Lw + Ab = Ez + Wm + Q + F$	Heinrich and Althaus (1988)
22. $Py = Ky + F$	Spear and Cheney (1989)
23. $Lw = Ez + Ky + Q + F$	Newton and Kennedy (1963)
Between 0.5 and 1.0 Gpa, the same curve approximates the position of the reaction	
$Ca = Ky + Ch$	Guiraud, Holland, and Powell (1990)
24. $Lw + Jd = Wm + Ez + Q + F$	Heinrich and Althaus (1988)
25. Ab out ($2P = P + x$)	Maruyama, Liou, and Suzuki (1982)
26. $And = Ky$	Holdaway (1971)
27. $Wm + G + Ch = St + Bi + Q + F$	Spear and Cheney (1989)
28. $Ct + Bi + Q + F = G + Ch + Wm$	Spear and Cheney (1989)
29. $Ez + Ky + Q = An + F$	Newton and Kennedy (1963)
30. $Ct + Ch + Wm = St + Bi + F$	Modified from Spear and Cheney (1989)
31. $And + Bi + Q = G + Cd + Af + F$	Spear and Cheney (1989)
32. $Wm + Q = Af + Sil + F$	Chatterjee and Johannes (1974)
33. $Ph + Q = En + Sd + F$	Helgeson et al. (1978)
34. $And = Sil$	Hemingway et al. (1991)
35. $St + Q = G + Sil + F$	Dutrow and Holdaway (1989); 0.5–0.8 Gpa extension projected by the author
36. $Cd + G = Hy + Q + Ky$	Henson and Green (1973)
37. $St + Q = Ky + G + F$	Pigage and Greenwood (1982)
38. $Gl + Wm + G = Ab + Ch$	Gl stability from Guiraud, Holland, and Powell (1990)
$Gl + Ch = T + G + Wm$	
39. $Cd = Sil + Q + Gt + F$	Richardson (1968)
40. $Cd + G = Bi + Sil$	Spear and Cheney (1989)
41. $Ez + Q = An + G + F$	Newton (1966)
42. $Ky = Sil$	Composite curve based on Holdaway (1971), Bohlen, Montana, and Kerrick (1991), and Hemingway et al. (1991)

43. Cd = Sil + Q + Sp   Richardson (1968)
44. Cd + G = Ol + Q + Sp   Henson and Green (1973)
45. Af + Cd = Sil + Bi + Q   A. B. Thompson (1976a)
46. Cd + G = Hy + Q + Sil   Cd stability from Henson and Green (1973)
    Cd + G = Hy + Q + Spp
47. Cd + Ol = Hy + Q + Sp   Henson and Green (1973)
48. Cd + G = Hy + Q + Sp   Henson and Green (1973)
49. Minimum melting curve of muscovite granite   W.-L. Huang and Wyllie (1981)

Table C.3 Equations and References for Petrogenetic Grid for Metabasic Rocks

1. Hu = Lm + Q   Cho, Maruyama, and Liou (1987)
2. Pr + Lm + Ch = Pu + Q + F   Liou, Maruyama, and Cho (1985, 1987)
3. Sb = Lm + Q + F   Liou (1971c)
4. Cc = Ar   Composite based on Jamieson (1953), Clark (1957), Crawford and
                                                  Hoersch (1972), Irving and Wyllie (1975), Salje and Viswanathan
                                                  (1976), and Brar and Schloessin (1979)
5. Q + Anl = Ab + F   Liou (1971a)
6. Ab + x = $Jd_{50}Di_{50}$ + Q   Maruyama and Liou (1988)
7. Ab = Jd + Q   Newton and Smith (1967)
8. Lm + Pr = Ec + Q + F   Liou, Maruyama, and Cho (1987)
9. Pu + Hm + Ch + Q = Ec + Am + Ch   Moody, Meyer, and Jenkins (1983)
10. Lm = Lw + Q + F   Liou (1971b)
11. Lm + Pu = Ec + Ch + Q + F   Liou, Maruyama, and Cho (1985, 1987)
    Pu + Q = Ec + Pr + Ch + F   Liou, Maruyama, and Cho (1985, 1987)
12. Pr + Ch + Q = Pu + Tr + F   Liou, Maruyama, and Cho (1985, 1987)
13. Pu + Ch + Q = Ec + Tr + F   Evans (1990)
14. Pu + Ch + Ab = Ec + Gl + F   Liou, Maruyama, and Cho (1987)
15. Lw + Jd/Di = Ec + Gl + Q + F   Evans (1990)
16. W = An + Q + F   Liou (1970)
17. W = Lw + Q + F   Liou (1971b)
18. Pu + Ch + Q = Ec + Tr + F   Liou, Maruyama, and Cho (1987)
19. Ab = Om + Q   Newton (1986b)
20. Pr + Ch + Q = Ec + Tr + F   Liou, Maruyama, and Cho (1987)
21. Ab + E + Ch + S + Hm = H + P + E + Ch + I   Moody, Meyer, and Jenkins (1983)
22. Lw = An + F   Newton and Kennedy (1963)
23. Lw = Ez + Ky + Q + F   Newton and Kennedy (1963)
24. Tr + Ch + Ab = Ec + Gl + Q + F   Evans (1990)
25. Tr + Ch + Ab = Ec + Gl + Q + F   Maruyama, Cho, and Liou (1986)
26. Ez + Q = An + G + F   Newton (1966), Boetcher (1970), Helgeson et al. (1978)
27. Ec + Ch + Q = G + Tr + F   Evans (1990)
28. Ab out (2P = P + x)   Maruyama, Liou, and Suzuki (1982)
29. Ec + G = An + Tr + F   Evans (1990)
30. Ch + Q = T + Ky + F   Massone (1989)
31. Gl + Ec + Q = G + Tr + Ab + F   Evans (1990)
32. G + Ky + Q = An   Hariya and Kennedy (1968)
33. Gl + Ec = G + Q + Jd/Di + F   Evans (1990)
34. Cp + An + Ch + F = Fo + An + F   Yoder (1967)
35. Tr = En + Di + Q + F   Jenkins, Holland, and Clare (1991)
36. Ez + Ky + Q = An + F   Newton and Kennedy (1963)
37. H + P + I = Cp + H + P + I [+ F]   Spear (1981b)
38. H + Cp + P + I = Op + H + Cp + P + I   Spear (1981b)

699

39. Am + Fo = Cp + Op +An + F — Obata and Thompson (1981)
40. Ch + Cp + An + F = Ch + Cp + An + Am + F — Yoder (1966)
41. Ch + Cp + An + Am + F = Sp + Cp + An + Am + F — Yoder (1966)
42. Ch = Op + Fo + Sp + F — Jenkins (1982), Jenkins and Chernosky (1986)
43. H + Cp + Op + Ol + P + I =
    Cp + Op + Ol + P + I [+ F] — Spear (1981)
44. An + Fo = An + Sp + Cp + Op — Yoder (1966)
45. An + Sp + Cp + Am + F = An + Sp + Cp + Op + F — Yoder (1966)
46. Solidus (initial melting curve) for "wet" gabbro — Lambert and Wyllie (1982)

Table C.4 Equations and References for Petrogenetic Grid for Metaultramafic Rocks

1. C + T = A — B. W. Evans et al. (1976)
2. C = A + B — O'Hanley, Chernosky, and Wicks (1989)
3. A + B = Fo + F — O'Hanley, Chernosky, and Wicks (1989)
4. C = Fo + T + F — Chernosky (1982)
5. B + L = Fo + Ch — O'Hanley, Chernosky, and Wicks (1989)
6. T + En = Ant — Greenwood (1971)
7. A = Fo + T + F — O'Hanley, Chernosky, and Wicks (1989)
8. T + En = Ant — Day, Moores, and Tuminas (1985)
9. A = Fo + T + F — B. W. Evans et al. (1976)
10. A = Fo + T — Johannes (1975)
11. L = Fo + T + Ch + F — O'Hanley, Chernosky, and Wicks (1989)
12. T + En = Ant — Hemley et al. (1977)
13. T + Fo = Ant + F — Chernosky, Day, and Caruso (1985)
14. T + Mg = Fo + F — Johannes (1969)
15. Ant + Fo = En + F — Chernosky, Day, and Caruso (1985)
16. Ant + Fo = Op (En) + F — Jenkins (1981)
17. Ant + Fo = En + F — Hemley et al. (1977)
18. Ant = En + Q + F — Chernosky, Day, and Caruso (1985)
19. Ch = Op + Fo + Sp + F — Jenkins and Chernosky (1986)
20. Tr + Fo = Op + Cp + F — Jenkins (1983)
21. T + En = Ant — Chernosky, Day, and Caruso (1985)
22. Mg = Pe + F — Trommsdorf and Connolly (1990)
23. Tr + Fo = Op + Cp + Ch + F — Jenkins (1983)
24. T = En + Q + F — Chernosky, Day, and Caruso (1985)
25. Ch = Op + Fo + Sp + F — Jenkins and Chernosky (1986)
26. Tr = Op + Cp + Q + F — Jenkins, Holland, and Clare (1991)
27. Tr + Fo = Op + Cp + Sp +F — Jenkins (1983)
28. B = Pe + F — Day, Chernosky, and Kumin (1985)

Table C.5 Equations and References for Petrogenetic Grid for Metamorphosed Carbonate Rocks

1. $Cc = Ar$	Composite based on Jamieson (1953), Clark (1957), Crawford and Hoersch (1972), Irving and Wyllie (1975), Salje and Viswanathan (1976), and Brar and Schloessin (1979)
2. $Mg + Fo + F = T + F$ ($X_{CO_2} = 0.5$)	Johannes (1969)
3. $Do + Q + F = T + Cc + F$ ($X_{CO_2} = 0.5$)	Eggert and Kerrick (1981)
4. $Do + Q + F = T + F$ ($X_{CO_2} = 0.5$)	Metz and Puhan (1970)
5. $T + Cc + Q = Di + F$ ($X_{CO_2} = 0.5$)	Skippen (1974)
6. $Tr + Cc = Do + Di + F$	Slaughter, Kerrick, and Wall (1975)
7. $Tr + Cc + Q = Di + F$ ($X_{CO_2} = 0.2$)	Dachs and Metz (1988)
8. $Do + Q = Di + F$	Slaughter, Kerrick, and Wall (1975), Trommsdorf and Connolly (1990), Tracy and Frost (1991)
9. $Tr + Do = Fo + Cc + F$ ($X_{CO_2} = 0.3$)	Metz (1976)
10. $Di + Do = Fo + Cc + F$	Kase and Metz (1980)
11. $Di + Do = Fo + Cc + F$	Trommsdorf and Connolly (1990), Tracy and Frost (1991)
12. $Cc + Q = Wo + F$	Trommsdorf and Connolly (1990)
13. $An + Wo + Cc = G + F$	Zharikov, Shmulovich, and Bulatov (1977)
14. $Mg = Pe + F$	Trommsdorf and Connolly (1990)
15. $Do = Pe + Cc + F$	R. I. Harker and Tuttle (1955), Tracy and Frost (1991)
16. $Cc + Wo = Ti + F$	Joesten (1974)
17. $R + Su = La + F$	Joesten (1974)
$Ti = Su + F$	Joesten (1974)
18. $Cc + Di = Ak + F$	Tracy and Frost (1991)
19. $Ak + Cc = Me + F$	Zharikov, Shmulovich, and Bulatow (1977)
20. $Di + Fo + Cc = Mo + F$	Zharikov, Shmulovich, and Bulatow (1977)
21. $Di + Cc = Ak + F$	Zharikov, Shmulovich, and Bulatow (1977)
22. $Su = Cc + La$	Joesten (1974)

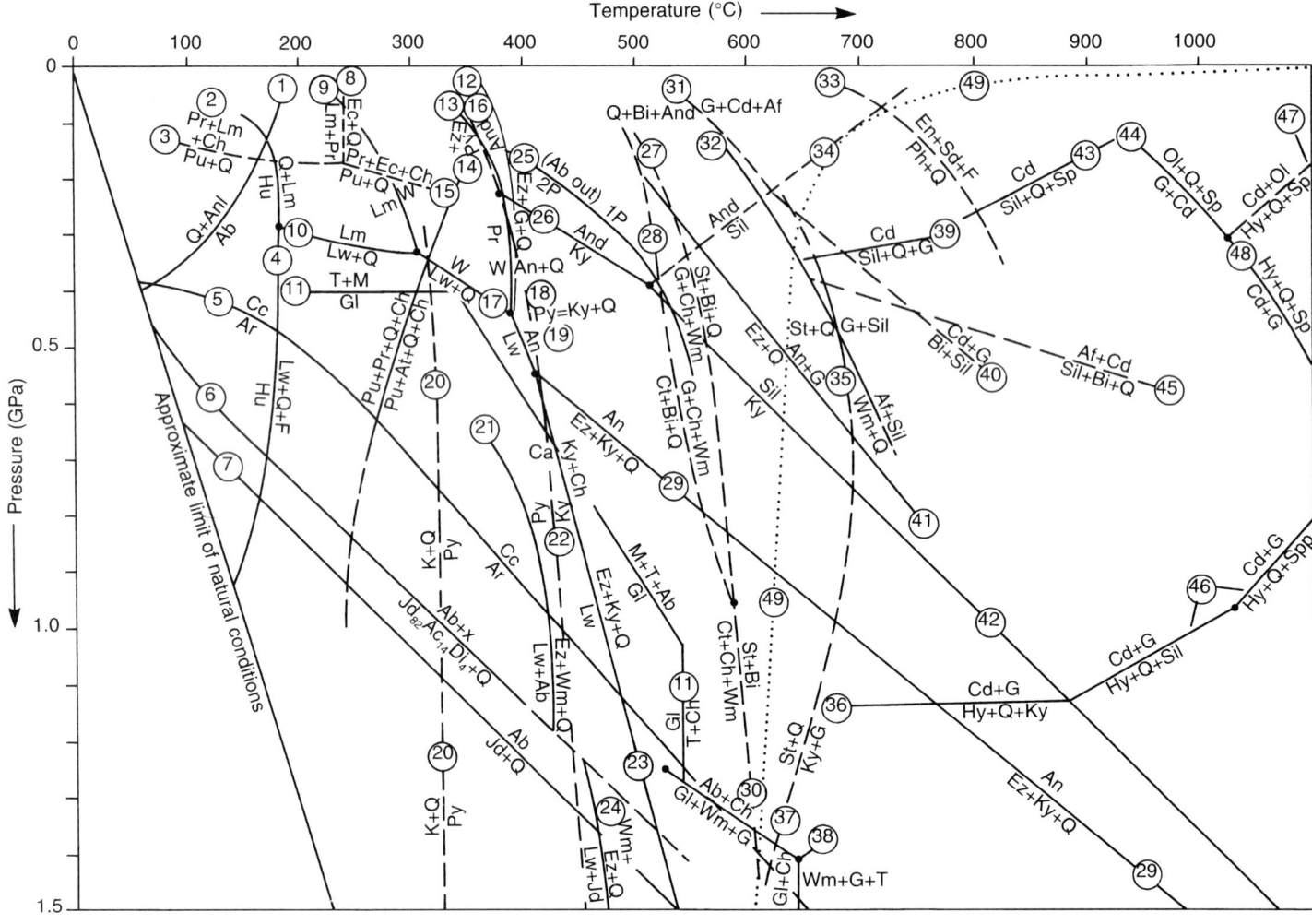

Petrogenetic grid for SAC and aluminous rocks

Figure C.1
Petrogenetic grid for SAC and aluminous rocks. See text for numbered reactions and symbols.

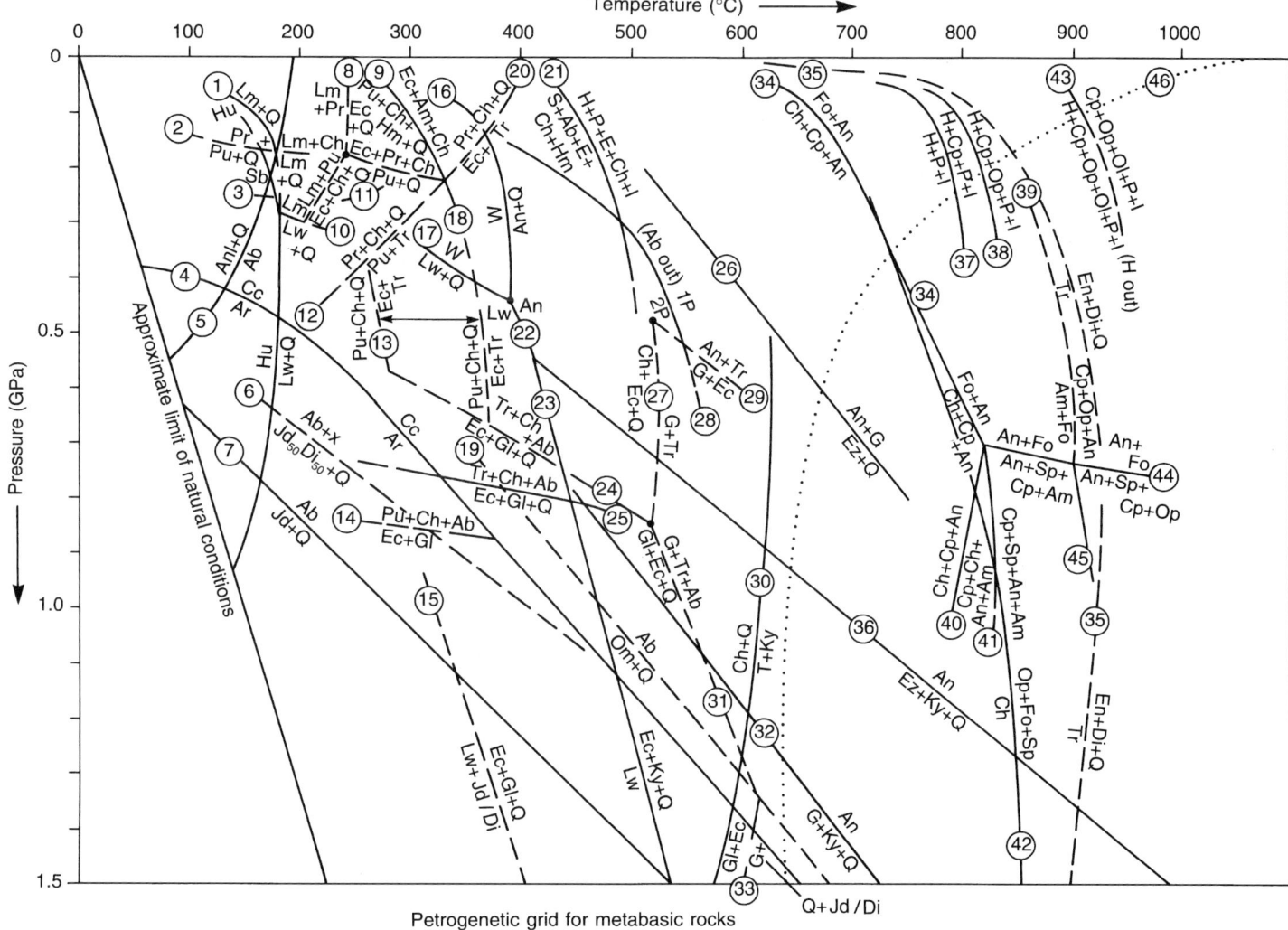

Figure C.2
Petrogenetic grid for metabasic rocks. See text for numbered reactions and symbols.

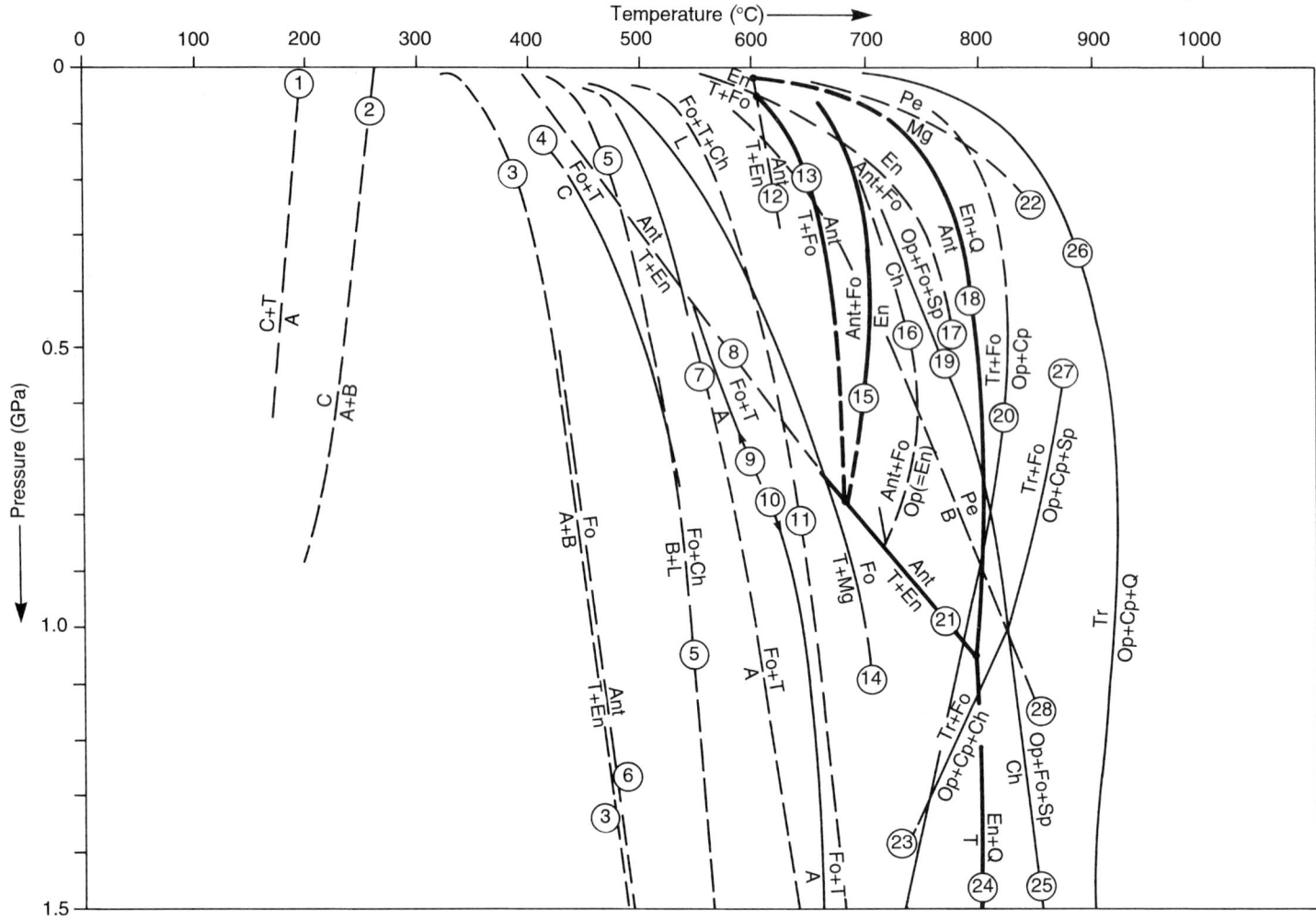

Figure C.3
Petrogenetic grid for metaultramafic rocks. See text for numbered reactions and symbols.

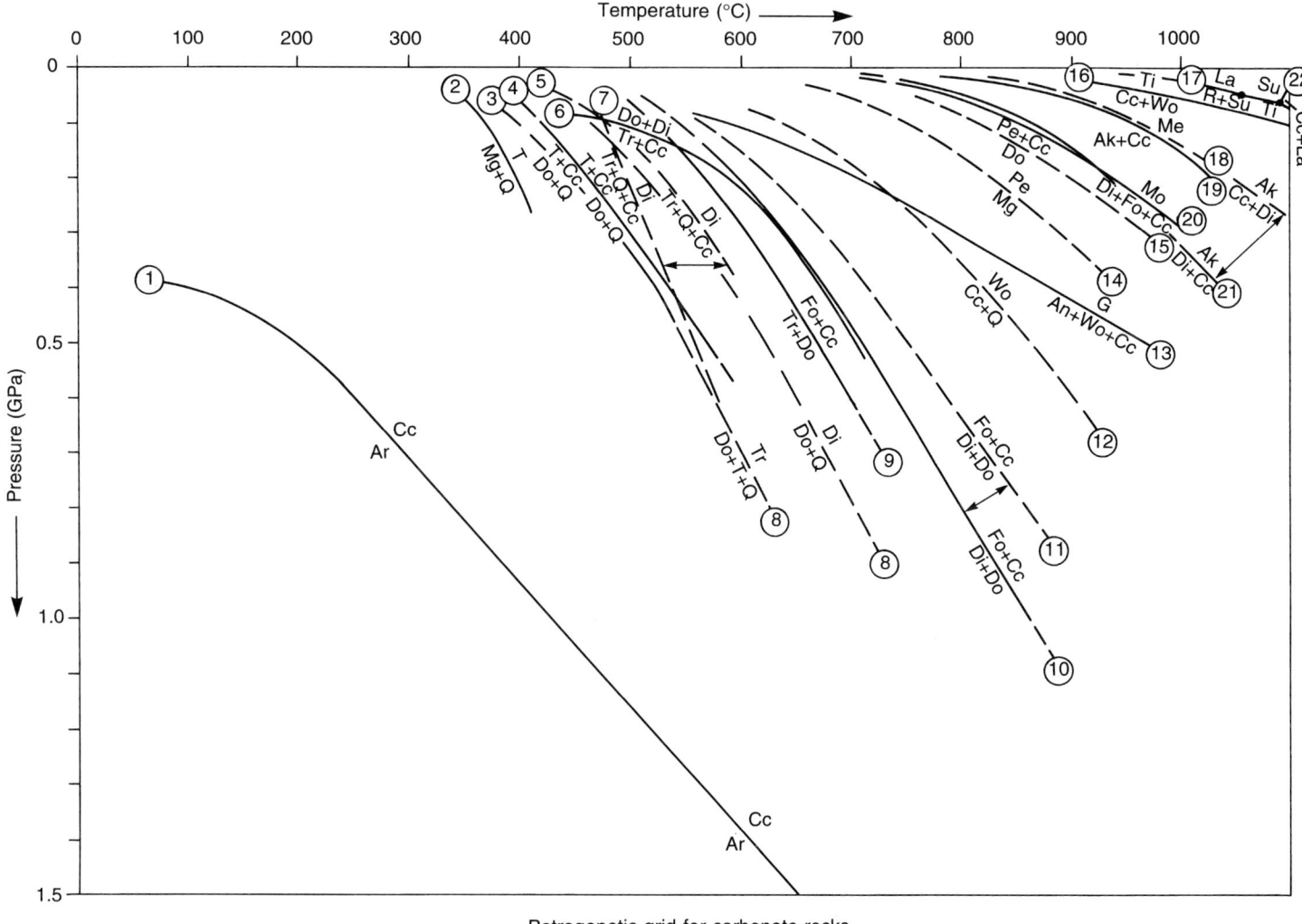

Figure C.4
Petrogenetic grid for metamorphosed carbonate rocks. See text for numbered reactions and symbols.

GLOSSARY

This glossary is based on definitions and references in the text. Refer there for references to original sources. For additional information, definitions of additional terms, and sources of terms, the reader is encouraged to consult Bates and Jackson (1987) and Mitchell (1985).

A

aa lava   A type of volcanic flow, typically basaltic in composition, characterized by sharp, angular fragments. Aa flows have flow tops that are rough and irregular. The rock of aa flows is referred to as aa lava.

abyssal   Of or referring to depth. Abyssal igneous rocks form at depth in the crust. In the ocean, abyssal depths are depths of greater than about 914 m (500 fathoms).

abyssal plain   The flatter, deep regions of the ocean floor extending between steeper slopes of the continental rises and other features such as mid-ocean ridges.

accessory mineral   A mineral in an igneous or metamorphic rock that is not used in deriving the rock name from a rock classification chart.

ACF diagram   A type of triangular pseudo-phase diagram used in metamorphic petrology, in which the A corner represents aluminous phases or compositions, the C corner represents CaO-bearing phases or compositions, and the F corner represents iron- and magnesium-rich phases or compositions.

adcumulate   An igneous rock with an adcumulate texture.

adcumulate texture   An igneous texture formed by fractional crystallization and consisting of a framework of touching crystals that continued to grow from the magma to the extent that less than 5% of the rock consists of *other* minerals crystallized from the intercrystalline liquid.

adcumulus   Of or relating to adcumulate texture.

adiabatic   A term used to describe a thermodynamic process in which rocks undergo neither a gain nor a loss of heat while undergoing changes in pressure.

aerobic layer   The upper, oxygenated layer of water in a water column.

AFM diagram   In igneous petrology, a triangular chemical plot of alkalis, iron oxide, and magnesia. In metamorphic petrology, a triangular phase diagram in which the respective corners represent aluminous, iron-rich, and magnesium-rich phases and compositions.

agglomerate   A fragmental volcanic rock dominated by rounded clasts larger than 64 mm in diameter or length.

agmatite   A type of migmatite that consists of angular fragments in a matrix; also called migmatite breccia.

A'KF diagram   A pseudo-phase diagram used in metamorphic petrology, in which aluminous, potash-rich, and iron- and magnesium-rich phases and compositions are plotted.

Alaska-type mafic-ultramafic rock body A crudely ellipsoidal, zoned igneous complex containing a variety of silica-poor rocks including two-pyroxene gabbros, peridotites, hornblende pyroxenites, and dunites.

alaskite   A name applied by some petrologists to granitoid igneous rocks that contain less than 5% mafic phases.

Albite-epidote Hornfels Facies   The set of all of the metamorphic rocks containing phase assemblages indicative of conditions of crystallization in the general range of $P = 0$–0.2 Gpa (0–2 kb) and $T = 300$–$400°C$. The name is also applied to the P-T space represented by those conditions.

alkalic   An adjective used to describe both the rock suite and the rocks belonging to a rock suite with an alkali-lime index of less than 51. The term designates the rock suite consisting of basalt $\rightarrow$ hawaiite $\rightarrow$ mugearite $\rightarrow$ trachyte $\rightarrow$ phonolite.

alkali-lime index   For any suite of igneous rocks, the number determined by the intersection of a curve representing the sums of the alkalis ($K_2O + Na_2O$) for the rocks of the suite with a curve representing the Ca0 values for the rocks of the suite. The indices are used to divide rocks into one of four groups—alkalic, alkali-calcic, calc-alkali, and calcic. Also called the Peacock Index.

alkali olivine basalt   A basalt type characterized by abundant olivine and Ca-pyroxene (augite). Alkali olivine basalts are typically olivine- or nepheline-normative.

allo-   A combining form meaning other or extraneous (e.g., from an*other* place).

allochem   In a sediment or sedimentary rock, a fragment of a chemical or biochemical precipitate formed earlier in another place. These grains have been transported away from their sites of origin to the site of deposition.

allochemical metamorphism   The type of metamorphism in which, during the metamorphic process, there is a change in the bulk chemistry of the rock domain under consideration. (Literally, other-chemical metamorphism).

allolistostrome   A type of sedimentary melange, formed by submarine sliding, that is, a mappable body of rock lacking continuity of internal stata or contacts and containing both native and exotic fragments in a finer-grained matrix.

allomorphic   A process of diagenetic replacement in which an original sedimentary phase is replaced by a new phase of another crystal form. (Literally, other forms.) *Note:* There are other definitions and uses of this term (see Bates and Jackson, 1987).

allotrioblastic-granular texture A metamorphic texture characterized by approximately equant, anhedral grains.

allotrioblastic texture   A metamorphic texture characterized by anhedral grains.

allotriomorphic-granular texture   An igneous texture characterized by approximately equant anhedral grains.

alluvial   Referring to sediment deposited by running water or features associated with such sediment.

alluvial fan   A typically semicircular (in plan), wedge-shaped (in longitudinal section) accumulation of sediment formed where a stream exits a valley or canyon and enters a broad valley or plain.

alluvium   Sediment deposited by running water.

alnoite   A type of mafic, porphyritic volcanic rock (a lamprophyre) with phenocrysts of euhedral biotite or melilite and commonly containing matrix phases such as nepheline, olivine, and clinopyroxene.

alpine-type ultramafic bodies   Irregular to elliptical bodies of rock that occur in orogenic belts and contain silica-poor rocks (with <45% SiO_2) composed of dark minerals. Typically they are deformed.

alpine-type ultramafic rock   Dark-shaded rock types containing less than 45% silica that occur in mountain belts.

alteration   A process of chemical change, typically associated with ore zones, in which the bulk chemistry and the mineralogy of rock masses are changed by passing, hydrothermal fluids.

amphibolite   A name applied by some petrologists to igneous or metamorphic rocks dominated mineralogically by hornblende.

Amphibolite Facies   The set of all metamorphic rocks containing phase assemblages indicative of conditions of crystallization in the general range of $P = 0.4–1.1$ Gpa (4–11 kb) and $T = 550–750°C$. The name is also applied to the P-T space defined by those conditions.

amygdaloidal   A textural term describing volcanic rocks containing mineral-filled, elliptically shaped vesicles.

amygdule   In a volcanic rock, a vesicle that has become filled with minerals such as calcite, quartz, or zeolites.

anaerobic   Oxygen-deficient.

anaerobic layer   An oxygen-deficient, bottom layer in a water column.

anastamosing cleavage   A type of metamorphic rock structure consisting of a curvi-planar fabric resulting from braided microzones of phyllosilicates or other minerals that give the rock a tendency to break into lens-shaped pieces. It is a type of spaced cleavage.

anatexis   The process of melting preexisting rock materials.

anchimetamorphism   A term used by some petrologists to designate very low-grade metamorphism under conditions ranging from those of diagenesis to those of the Zeolite Facies.

andesite   An intermediate-silica volcanic rock, characterized by a silica content between 52 and 65%. Plagioclase feldspar is the dominant feldspar, quartz is generally absent, and typical phenocrysts consist of plagioclase, hornblende, or augite. Plagioclase is typically light and has an An content of less than 50.

anhydrite evaporite   Aphanitic to phaneritic, soft, commonly layered, anhydrite-rich rock formed by evaporation of or crystallization from a brine.

anorthosite   A plutonic igneous rock composed almost entirely of plagioclase feldspar (a hyperleucocratic gabbro).

antidune   Elongate piles of sediment (a type of sand wave) of less than 1 m to tens of meters high, which may develop in currents of wind or water and may be preserved under appropriate conditions. In streams, antidunes form in phase with surface-water waves.

aphanitic   A descriptive term meaning that the grains in a rock are too small to see or identify with either the unaided eye or a low-power lens.

aphyric   A term used to describe volcanic rocks that lack phenocrysts.

aplite   A medium- to fine-grained, allotriomorphic-granular, hyperleucocratic, quartz-alkali feldspar (plutonic) rock.

apophysis   A short, irregular dike that extends from a pluton margin into the country rock.

appinite   An igneous rock of quartz diorite composition characterized by euhedral hornblende crystals.

Appinite-type mafic-ultramafic rock body   A type of rock complex consisting of small elliptical to lenticular plutons composed of rocks such as olivine hornblendite, norite, and hornblende quartz diorite. Appinite-type complexes are typically associated with granitoid rocks.

arc   A linear or curvilinear chain of volcanoes on land or in the sea.

arenite   A type of sandstone with matrix materials comprising 5% or less of the rock.

arkose   A root name used in some sandstone classifications to designate rocks relatively rich in feldspar.

ash   Fine-grained, unconsolidated, clastic volcanic material composed of grains less than 2 mm in diameter.

ash fall   A layer of volcanic ash formed by sediment rain from the atmosphere during or following a volcanic eruption. Also the process of sedimentation of volcanic ash.

ash flow   The rock mass, composed of ash and other pyroclastic materials, formed by a nuée ardente (a hot, gaseous, flowing volcanic cloud). Also called ignimbrite.

asthenosphere   A plastic zone in the mantle of the Earth separating the lithosphere and the mesosphere.

augen   An eye-shaped crystal or porphyroblast in a gneiss or schist.

augen texture   A metamorphic texture characterized by eye-shaped, larger grains or grain clusters in a finer-grained matrix.

aureole   A zone of contact metamorphic rock surrounding and associated with a pluton.

authigenesis   The process in which new mineral phases are crystallized in a sediment or rock during diagenesis. See also *neocrystallization*.

authigenic   Adjective describing minerals formed in place (i.e., precipitated rather than transported).

autolith   A small body or inclusion of rock found within and related to an igneous rock mass. (Literally, self-stone *auto* = self, *lith* = stone), that is, a piece of the enclosing body of rock.

automorphic-granular   An igneous texture that is dominated by euhedral crystals.

B

baked zone   A zone of thermal effects, typically a brick-red zone of oxidized soil, between older and younger lava flows.

ball-and-pillow structure   A primary sedimentary structure consisting of spherical and hemispherical to elliptical masses of deformed sediment.

bank benches   A flat feature formed along the stream banks or margins over time as the stream channel migrates.

bar   A ridgelike, linear accumulation of sediment in a stream or offshore marine environment.

barrier beach complex   A linear beach (and dune) landform isolated from the land by a lagoon. Also called barrier island.

Barrovian Facies Series   A medium-pressure, high-temperature sequence of metamorphic facies, with kyanite-sillimanite-bearing pelitic rocks. It is named for rocks of the Scottish Highlands described by Barrow (1893).

basalt   A volcanic rock type that contains essential plagioclase feldspar and is characterized by a silica content between 45 and 52%. Plagioclase is characteristically dark and has an An content of more than 50. Phenocrysts are typically one or more of the phases plagioclase, augite, and olivine.

basaltic plain   A major extrusive volcanic structure that is tabular in overall form and is composed primarily of lava flows of silica-poor (basaltic) volcanic rock erupted from overlapping centers of eruption that develop shield cones.

batch melting   A process of fractional melting in which masses of magma repeatedly develop and then separate from a parent rock.

batholith   A body of plutonic rock having aerial extent of 100 km^2 or more. Alternatively, a nonlayered body of intrusive rock, predominantly phaneritic in texture, with a minimum volume of 100 km^3.

beach   An accumulation of sediment formed along the margin of a water body.

beach backshore   A supratidal zone, behind a beach, composed of a berm–a sandy, somewhat flat area above the normal high tide level–and a dune.

beach foreshore   The zone between high and low tide, including the swash zone where breaking waves run up the beach face.

beachrock   A foreshore grainstone or conglomerate cemented on the beach by calcite, soon after deposition.

bed   A layer of sediment or sedimentary rock, distinguished from adjoining layers on the basis of differences in color, texture, and composition. The most characteristic structure of sedimentary rocks.

bed form   A sedimentary structure formed at the sediment-water interface, such as ripples, sand waves, dunes, and antidunes.

bed-load transport   Movement of particles too large for extended suspension above the surface by rolling, bouncing (saltation), and temporary suspension in a current.

benmoreite   A type of intermediate volcanic rock with a saturated to undersaturated silica content, a potash: soda value of less than 1:2, and modal pyroxene, alkali feldspar, and either andesine or oligoclase.

benthic   Referring to the bottom of a water body (as in *benthic fauna*, the bottom-dwelling animals in the sea).

bentonite   A soft, plastic mudrock composed primarily of smectite-group clays and derived from weathering of volcanic ash.

binary solid solution system   A chemical system composed of two chemicals that may be mixed in any ratio to yield a single phase.

Bingham plastic material   A material that has an initial strength, or yield strength, but in which, during deformation or flow, the viscosity remains constant after the yield strength is exceeded.

biochemical precipitate   Material crystallized from a solution as a result of the activity of organisms.

biofacies   A unit of sedimentary rock that is distinguished by its biota and is representative of specific environmental conditions.

biogenic element   Any textural component of a rock formed by biological activity.

biolithic element   A part of a biolithite or boundstone formed by organisms.

biolithite   A fossiliferous reef rock formed *in situ* by biochemical precipitation and the intergrowth of organisms.

bioturbation   The postdepositional process of mixing and disruption of lamination and bedding in sediment caused by the activities of organisms.

bioturbation structure   A structure in sedimentary rock, formed where burrowing by organisms was extensive, characterized by highly swirled laminations or completely destroyed lamination.

bivariant system   A system in which there are two degrees of freedom; that is, two variables may be changed independently without changing the number or nature of phases in the system.

blastomylonite   A foliated metamorphic rock that has experienced ductile deformation and later recrystallization, so that it contains recrystallized microlithons, porphyroclasts that exhibit overgrowths, or new porphyroblasts.

block   A large rock fragment that is typically angular. In volcanic rocks, blocks are more than 64 mm in diameter.

block-in-matrix structure   A structure found in coarse, clastic sedimentary rocks and some igneous and metamorphic rocks consisting of large masses of rock of one or more types enclosed by generally finer-grained material of a different rock type.

Blueschist Facies   The set of all metamorphic rocks containing phase assemblages indicative of conditions of crystallization in the general range of $P = 0.35–0.9$ Gpa (3.5–9 kb) and $T = 50–400°C$. The name is also applied to the P-T space defined by those conditions.

bomb   A mass of volcanic rock larger than 64 mm in diameter that has a rounded form resulting from streamlining or deformation during flight, after ejection from a volcano.

boninite   A type of andesite with a high magnesium content (Mg number > 0.7).

boudin   A structure consisting of cylindrical masses of rock derived from a single bed or layer that has been stretched or pulled apart to resemble a row of sausages. (French for sausage.)

bouldery   A term applied to sedimentary rocks in which clasts larger than 256 mm in diameter comprise less than 25% of the rock.

Bouma sequence   A sedimentary unit consisting of a five-part series of beds and laminations that from bottom to top include a graded or structureless bed, a parallel-laminated layer, a convolute or cross-laminated layer, an upper parallel-laminated layer, and a mudrock layer. Bouma sequences occur in turbidites.

boundstone   A reef rock consisting of intergrown and cemented organic structures.

breccia (sed.)   A sedimentary rock composed of angular clasts larger than 2 mm in diameter in a sandy or gravelly matrix.

brittle deformation   Deformation in which grains (and rocks) are broken before the material has attained 5% strain.

Buchan Facies Series A low-pressure sequence of facies, with pelitic rocks containing andalusite to sillimanite, that was named after the Buchan area of northeastern Scotland.

burrow An irregular to cylindrical, tubular structure found in sedimentary rocks, representing a filled hole that was dug by an organism.

bysmalith An igneous intrusive structure consisting of a nearly vertical, cylindrical mass of rock bounded by faults.

C

caldera An igneous structure consisting of a large circular depression produced by the eruption-induced collapse of a volcano.

caliche A sedimentary rock that is chalky, light-colored, micritic or microsparitic and is formed by evaporation and precipitation of calcite in soil, sediment, or preexisting rock.

carbonate buildup A sedimentary mass that is depositionally thickened, fossiliferous, and composed of carbonate rock. A general term that encompasses reefs, Walsortian mounds, and other similar structures.

carbonate rocks Those rocks with 50% or more carbonate minerals (e.g., calcite, aragonite, and dolomite). The term is most commonly applied to sedimentary rocks.

carbonatite An igneous rock composed primarily of one or more carbonate minerals, such as calcite, dolomite, and ankerite. Both volcanic and plutonic carbonatites are known.

cast A sedimentary structure consisting of the filling in a depression or void.

cataclasis The process of crushing and breaking the grains in a rock.

cataclasite A nonfoliated metamorphic rock composed of angular fragments of rock in a matrix of finer-grained, but similarly composed materials.

cataclastic breccia A nonfoliated dynamoblastic metamorphic rock dominated by broken, angular fragments greater than 1/16 mm in diameter in a matrix of more finely crushed material. Cataclastic breccias form along fault zones.

cataclastic texture A generally nonfoliated, metamorphic texture characterized by broken rock and mineral grains. *Note:* Foliated-cataclastic texture is a transitional texture between (nonfoliated) cataclastic textures *sensu stricto* and (foliated) mylonitic textures.

catazonal A descriptive term applied to plutons that form deep (>9 km) in the crust. Also spelled katazonal.

cement Chemically precipitated material that fills the spaces between framework grains in sedimentary rocks.

cementation The process in which chemical precipitates, in the form of new crystals, form in the pores of a sediment or rock, binding the grains together.

chalk A soft, earthy, friable, porous, light-colored limestone.

channel The concave-up, linear depression through which a stream flows.

channel bar A cross-bedded mound of sediment within a stream channel.

characterizing accessory minerals Minerals occurring in abundances of 5% or more that are not implied by the rock name but are used to modify the name of an igneous rock. Also called varietal minerals.

chelate A compound in which a metal cation is bonded with and connects organic ring structures.

chelation A chemical process, important in weathering, that results in the formation of a chelate, a chemical complex containing metal ions, via extraction of metal cations from a mineral.

chemical precipitates In sedimentary petrology, a category of sedimentary rock that includes rocks that are crystalline-textured, generally fine-grained to aphanitic, and formed by inorganic crystallization of minerals from solution. In general, any materials formed by inorganic crystallization of minerals from solution.

chert A hard, multi- or variously colored, waxy to grainy sedimentary rock composed predominantly of silica minerals.

chilled margin A finer-grained phase of a pluton that forms within the pluton, along the pluton margin, where the magma was cooled by contact with cooler country rocks during emplacement of the pluton.

clast-supported A descriptive term for sedimentary rocks composed of clasts abundant enough to be touching several neighboring clasts.

cleavage A structure in rocks consisting of relatively closely spaced, parallel to subparallel surfaces of fracture or mineral alignment. Also, the tendency of rocks to break along parallel to subparallel surfaces. (Cleavage also refers to the tendency of minerals to break along parallel planes.)

closed system A system that cannot exchange material but can exchange energy with the surroundings.

coarse-tail grading A feature in medium- to coarse-grained clastic sedimentary rock beds in which there is a change in grain size from the bottom to the top of the bed (grading), but in which the change in grain size is defined only by a change in the sizes of the framework or coarsest clasts.

cobbly A term applied to sedimentary rocks in which clasts between 64 mm and 256 mm in diameter comprise less than 25% of the rock.

colloid Finely divided material in suspension.

color index In an igneous rock, the total percentage of minerals in which iron, magnesium, or both are essential constituents of the mineral chemistry.

columnar joint Pencil-like, polygonal mass of rock, several centimeters across and centimeters to meters in length, formed where fractures in volcanic rocks intersect.

comb texture A texture in rocks characterized by a row of crystals that have grown perpendicular to a surface, like the teeth of a comb.

compaction The process by which the volume of a rock mass is reduced, usually as a result of sediment or rock being compressed under a load of overlying materials.

complex composite pegmatite A type of very coarse-grained, zoned or unzoned plutonic rock body with veined and/or replacement structures. Complex composite pegmatites show two distinct stages of development with the later stage consisting of replacements of early-formed minerals, fracture fillings that cross-cut early-formed minerals, or both.

complex pegmatite A type of very coarse-grained plutonic rock body with a zoned, veined, or replacement structure.

component A chemical species (e.g., SiO_2, OH^-). In the Phase Rule, the number of components is defined as the smallest number of chemical species needed to define the compositions of all phases in the system.

composite cone Steep-sided volcanic structure consisting of layers of lava and pyroclastic material that radiate and thin away from a central vent area. Also called stratovolcanoes.

concordant A term referring to intrusive structures (i.e., plutons) with contacts that parallel the layers in the intruded rock.

concretion   A sedimentary structure consisting of a round to irregular mass of more resistant rock formed as a result of cement precipitating around a core material, commonly a fossil or grain of a different composition.

cone sheet   An intrusive igneous structure (a dike) that is a discordant curvi-planar layer which diverges from a point at some depth in the Earth.

congelation crystallization   A process of magmatic crystallization occurring on the walls and floor of the magma chamber which concentrates liquids enriched in residual elements in the top or center of the chamber.

conglomerate   A sedimentary rock composed of rounded clasts greater than 2 mm in diameter and comprising more than 25% of the rock. The clasts are typically enclosed in a matrix of sand (sandstone) or gravel (conglomerate).

congruent melting   A process of melting in which the melt has the same composition as the solid that is melted.

consanguineous   An adjective used to describe igneous rocks derived from the same parent magma.

contact   The boundary between two bodies of rock.

contact diablastite   A fine- to medium-grained, nonfoliated rock composed of a combination of granular, acicular, and platy grains. The term is used as a substitute for hornfels where there is a need to imply genesis with the rock name.

Contact Facies Series   A very low-pressure sequence of facies, represented by the rocks of contact metamorphic zones.

contact metamorphism   Metamorphism caused primarily by heat in country rocks at and near their contact with an igneous rock mass.

continental environment   A region of sedimentation on the Earth totally above sea level at high tide and beyond the direct influence of marine processes.

continental rise   That gently sloping part of the seafloor that occurs between the abyssal plain and the more steeply sloping continental slope.

continental sabkha   A salt flat along the margin of an ephemeral lake in a flat, vegetation-free desert basin.

continuous cleavage   In rocks, a planar fabric element with a spacing of less than 0.01 mm or one that results from a penetrative preferred orientation of phyllosilicates or other minerals.

continuous melting   A process in which melting and separation of magma from the parent rock are coeval and uninterrupted.

continuous reaction   A reaction in which the compositions of one or more of the coexisting phases changes gradually over the course of the reaction.

continuous reaction series   A group of reactions in which the compositions of phases formed earlier react with remaining liquids to change composition gradually, without abrupt changes in the phases present.

contourite   A category of sandstone deposited in the sea by currents that flow parallel to the slope (contour currents).

convective fractionation   A crystallization process in which convective flow, driven by heat within a magma chamber, feeds crystallization on the floor, walls, and roof of a magma chamber.

convolute lamination   A sedimentary structure consisting of highly contorted, folded, and disrupted layers less than 1 cm thick. *Convolute beds* have layers greater than 1 cm thick.

corona   A ring or shell of minerals, in some cases having a radial growth pattern, surrounding a core mineral grain. Some are reaction rims.

coronitic texture   A texture in which larger grains have rims of generally finer-grained minerals.

cortlandite   An ultrabasic rock consisting essentially of hornblende with poikilitically enclosed olivine grains.

cotectic line   The line on a phase diagram that separates two separate phase fields and represents the line of liquid composition, or the temperature-composition curve, along which two solid phases crystallize simultaneously.

country rock   The rock surrounding a rock body of interest, particularly, the rock surrounding an igneous intrusion.

crater   A concave depression usually less than 1 km in diameter, and generally at the crest of a volcano, directly caused by eruptive activity.

critical mineral   A mineral for which the reaction curves and stability field are known and which may be used to distinguish between one facies or zone and another.

critical mineral assemblage   A group of minerals that occur together for which the reaction curves and stability field are known and which may be used to distinguish between one facies or zone and another.

cross-beds   Layers of sedimentary rock inclined at an angle to the main plane of the stratification.

cryolacustrine environment   A sedimentary setting consisting of a lake directly associated with a glacier and characterized by the unique physical, chemical, and biological features of such a setting.

cryptic layering   Generally invisible tabular zones, within and typically parallel to the sides and bottom of plutons, that are marked by variations in the chemistries of the included minerals.

crystalline   An adjective used to describe any material consisting of an internally ordered, long range array of atoms. The term is used to distinguish certain rocks composed entirely of crystalline phases (minerals) from those partly or entirely composed of glass or other noncrystalline materials.

crystalline texture   A texture characterized by crystalline materials arranged in an interlocking array.

crystal-liquid fractionation   A process of magmatic crystallization in which crystals form and separate from their parental liquid magma.

cumulate   A rock with a cumulate texture. Also an adjective used to describe rocks and textures in which there is a framework of touching mineral grains that formed primarily through fractional crystallization.

cumulus crystals   Crystals formed by fractional crystallization from a magma. They are concentrated in layers along the floor, walls, or top of a magma chamber.

cupola   A large, convex-up bulge of plutonic rock on the roof of a magma chamber or pluton.

current deposition   Bed-load or suspended load deposition by currents as they lose their ability to transport the sediment.

D

decomposition   The general category of processes in which chemicals break down rock materials during weathering.

decussate texture   A metamorphic texture consisting of platy or acicular crystals arranged in such a way as to produce no preferred orientation. Also called diablastic texture.

degree of freedom   A condition or variable that may be changed in a system (also called variance). In the Phase Rule, the *number* of degrees of freedom is of interest and is defined as the minimum number of variables needed to define a specific state of the system; in other words, the smallest number of conditions or variables that may be changed independently without altering the number or nature of phases in the system.

delta   An accumulation of sediment—locally shaped, in map view, like the Greek letter delta—that forms in a body of water where a stream enters the body and deposits its load.

derivative magma   A magma formed from another magma by a process of separation.

dessication crack   A crack formed in sediment as a result of drying.

deviatoric stress   A stress (force/unit area) that acts in a particular direction and exceeds the hydrostatic or mean stress.

diabasic   An igneous texture consisting of rectangular (lath-shaped) plagioclase feldspar crystals with smaller, intergranular crystals of pyroxene and other minerals. Diabasic texture is one member of the ophitic-subophitic-diabasic series and is a coarse-grained equivalent of the intergranular textures found in volcanic rocks.

diablastic texture   A metamorphic texture in which platy and acicular minerals are intergrown in a nonfoliated, interlocking, and locally radiating manner.

diablastite   A metamorphic rock with a diablastic texture.

diagenesis   A general term designating all of the surface-to-subsurface physical, chemical, and biological processes that collectively result in transformation of sediment into sedimentary rock and modification of the texture and mineralogy of a rock.

diamictite   A sedimentary rock composed of 25% or more clasts larger than 2 mm in diameter (or length) enclosed in a finer-grained matrix. Diamictites have rounded clasts, angular clasts, or both in a matrix dominated by mud.

diapir   A spherical, elliptical, or tail-down drop-shaped mass that rises towards the surface as a result of its low density compared to surrounding rocks. Diapirs may be solid or liquid.

diapiric   An adjective referring to a process in which a mobile core of material rises up (generally due to its low density), forcing the overlying materials into a domal or antiformal structure. The adjective is also used to describe features formed by diapirism.

diatomaceous chert   Diatomite that is well cemented.

diatomite   An aphanitic, light-colored, soft, friable, siliceous rock composed of diatoms (the fragments of certain plant cell walls). Where well cemented, diatomite is called diatomaceous chert.

diatreme   A cylindrical to dike-like volcanic structure generally composed of ultramafic to carbonate breccias.

dictytaxitic texture   An extrusive rock texture consisting of vesicular rock with vesicles that have crystals projecting into the cavity from the vesicle walls.

differentiation   A general term referring to all of those processes in which more than one rock type is formed from a magma.

diffuse   The action of diffusion.

diffusion   A process in which chemical species migrate between two sites, in a solvent phase, under the influence of a chemical potential gradient.

dike   A generally tabular, intrusive structure that is discordant to surrounding layers or that cross-cuts massive rocks.

dimensional preferred orientation   Also called shape preferred orientation (SPO). A texture in which inequant grains display some degree of parallel alignment.

discontinuous reaction   A reaction in which there is an abrupt change in the stable phase or phase assemblage present. In metamorphic petrology, both terminal and nonterminal reactions that involve univariant curves may be discontinuous.

discontinuous reaction series   A group of reactions in which reaction of earlier-formed crystals with a liquid phase results in abrupt changes in the phases present.

discordant   An adjective referring to igneous structures that cut across layering.

disintegration   The general category of processes of physical breakdown of rock materials during weathering.

dish structure   A sedimentary structure, typically found in sandstone, consisting of thin, oval- to circular-shaped, concave-up concentrations of finer-grained material.

dismicrite   A fine-grained limestone containing bird's-eye structures composed of sparry calcite and having less than 1% allochems.

disorganized bed   A type of layer composed of gravel or conglomerate that lacks internal grading or stratification.

distal   Located at a considerable distance from the source.

distal environment   A depositional setting on the surface of the Earth that is located at a considerable distance from the source of sediment.

distributary   A branch of a stream that occurs on a delta or alluvial fan where a major stream splits to form a series of smaller streams that do not reenter the main stream.

distribution grading   A change in grain size within a bed involving the gradual change of all of the grain sizes; contrasts with coarse-tail grading.

dolograinstone   A dolomite-rich rock composed of larger grains in a grain-supported array and a matrix of less than 5% fine-grained materials.

dolomitization   The process by which rocks are converted to dolostone. The process is primarily one of replacement.

dolomudstone   A rock composed of dolomite in grains of less than 1/16 mm.

dolopackstone   A dolomite-rich rock composed of larger grains in a grain-supported array with a fine-grained matrix.

dolosparite   A rock composed of dolomite in grains greater than 1/16 mm.

dolostone   A general term that refers to all aphanitic to phaneritic rocks composed dominantly of dolomite.

dolowackestone   A "mud"-supported, dolomite-rich rock consisting of scattered larger grains in a finer-grained matrix.

domain   A volume of rock being considered for a particular purpose or study.

double-diffusive convection   A process of material transport that occurs in magma chambers and involves the combined effects of physical movement (convection) and chemical transport (diffusion).

ductile   A descriptive term for deformation in which plastic behavior and flow occur via structural changes within and between grains.

ductile deformation   Deformation that is predominantly plastic in nature. Ductile materials experience strains of 5% or more before failing.

dune   A convex-up accumulation of sediment, deposited by wind or water currents, that may be domal, linear, curvilinear, stellate, elliptical, or irregular in shape.

dynamic metamorphism   A process of change in rocks caused primarily by deviatoric stress.

dynamoblastic rock   A metamorphic rock with micolithons or porphyroclasts surrounded by a deformed matrix in a texture resulting from the operation of deviatoric stress (i.e., from dynamic metamorphism).

dynamothermal metamorphism   A process of change in rocks caused primarily by elevated temperatures and pressures.

dysaerobic layer   An intermediate layer in a water column–between the surficial aerobic layer and the bottom anaerobic layer–that has variable oxygen content, salinity, and density.

E

eclogite   A plagioclase-free, silica-poor rock composed primarily of Mg,Fe garnet and the sodic clinopyroxene omphacite.

Eclogite Facies   The set of all metamorphic rocks containing phase assemblages indicative of conditions of crystallization in the general range of P = 0.9–2.0 Gpa (9–20 kb) and T = 150–900°C. The name is also applied to the P-T space defined by those conditions.

Eh   A measure of the ability of a solution to produce oxidation or reduction. Also called redox potential.

enclave   An inclusion in a plutonic rock, specifically, a xenolith, xenocryst, or autolith.

en echelon   A descriptive term referring to a group of features that form a diagonal array with each feature offset slightly from the next.

englacial environment   A depositional setting within a glacier (e.g., within tunnels and channels within the ice).

entrained   Carried by; as in, rock was *entrained* by a glacier.

eogenesis   The process of early diagenesis that occurs at or near the surface between deposition and burial.

epeiric sea   A large shallow sea that occurs within or along the margin of a continent.

epiclastic texture   A general term for all sedimentary textures formed at the surface (*epi-*) and consisting of accumulated rounded to angular grains (clasts) packed together. Grains are derived by normal processes of surficial weathering, erosion, and abrasion, and are bound together through recrystallization of grain boundaries, cementation, or matrix-grain amalgamation.

epilimnion   A layer of water in a lake consisting of less dense fresh water that overlies more dense saline water.

epimatrix   A type of intergrain material in sandstones and conglomerates consisting of a polymineralic mix of minerals produced by neocrystallization and alteration of framework grains.

epitaxial   An adjective describing a relationship between minerals in which one mineral overgrows another in such a way that some atomic (crystallographic) structural elements of the overgrowth parallel similar or identical elements in the core grain.

epizonal   Of or relating to the epizone.

epizone   A depth category, generally less than 6.5 km deep, in which generally composite, discordant plutons, with chilled margins and contact metamorphism, are characteristic.

equigranular   A textural term referring to textures in which the grains are all about the same size.

equigranular-mosaic texture   A metamorphic texture in which the grains are equant to subequant, polygonal, and fine-grained.

equigranular-tabular texture   A metamorphic texture characterized by aligned, elongate to tabular, subequant grains.

erg   A large, sand-covered desert terrain.

escape structure   A small, generally cylindrical structure, usually found in sandstones, consisting of a tube, now filled with sand or sandstone, that served as the path by which water or an organism escaped from a lower layer after being buried.

essential minerals   Minerals used in determining the root name of an igneous rock.

estuary   The wide mouth of a river that has been drowned by the sea, typically characterized by mixed fresh and marine waters.

eutectic   An invariant point representing the melt composition and the lowest temperature at which a mixture of two or more solid phases will melt. Initial melting does not change the composition of the solid phases. The eutectic point is the lowest point on the liquidus curve.

euxinic basin   A depression in the crust, filled at depth with anerobic (oxygen-deficient) water of limited circulation (i.e., stagnant water).

evaporite   A sedimentary rock formed by the crystallization of salts from concentrated solutions produced by evaporation of a watery solvent.

exotic block   A mass of rock that occurs in a lithologic association foreign to that in which the mass formed. Commonly, exotic blocks are also tectonic blocks.

extensive variable   A property of a system capable of change, the value of which is dependent on the mass of the system.

externally buffered   A descriptive term for reactions that have components that are provided by outside sources (e.g., a fluid phase) in enough abundance that the reaction is not controlled or limited by the availability of those components.

extrusive   A descriptive term for igneous bodies that erupt at the surface of the Earth.

extrusive structures   Structures of igneous rocks that are formed where magmas are forced out onto the surface of the Earth.

F

fabric   The sum of the structural and textural features of a rock that, taken together, define its geometrical character.

facies, metamorphic   See *metamorphic facies*.

facies, sedimentary   A body of sediments or sedimentary rock characterized by a set of physical, chemical, and biological features that together reflect a particular environment of formation.

fault   A fracture in rock or other material across which there is a total loss of cohesion and along which there has been significant movement parallel to the surface of failure.

feldspathic rocks   A collective term referring to igneous rocks characterized by the presence of feldspars in amounts greater than 10% and in which feldspars plus quartz or feldspathoids are the essential minerals.

felsic   Light-colored. Also refers to feldspar, feldspathoid, and silica minerals in rocks.

felty   A volcanic texture characterized by unoriented, microscopic crystals of plagioclase in a matrix of other minerals.

fenestrae   Holes, typically of bird's-eye shape, that occur in carbonate sediment. In ancient rocks, fenestrae are usually filled with calcite.

ferromagnesian rocks   Igneous rocks containing less than 10% feldspars and an abundance of minerals rich in iron and magnesium. Olivines, pyroxenes, amphiboles, and some micas are the essential minerals.

fibroblastic texture   A metamorphic texture consisting of nonaligned, radiating to randomly oriented acicular minerals of about the same size (i.e., a diablastic texture consisting of equally sized acicular minerals).

filter pressing   A process that consists of the separation of crystals from a magma via tectonic squeezing of the magma chamber, which drives off the magma, leaving the crystals behind.

fissility   The tendency of some rocks to split into thin pieces.

flame structure   A soft-sediment deformation feature consisting of a curved set of fine laminae that project into an overlying (sand) layer and have the general form of a small flame blowing in a breeze.

floatstone   A matrix-supported carbonate sedimentary rock in which large bioclasts comprise more than 10% of the rock.

flood basalt   Basalt formed by fissure eruptions in units that cover large regions. Also called plateau basalt.

floodplain   The flat geomorphic feature that occurs adjacent to the stream channel in a stream valley and on which floods occur.

flow-banding   Colored bands in volcanic rocks formed by flow of the magma and resulting from concentrations of crystals, vesicles, or inclusions.

flow differentiation   A process of separation of crystals and liquid in a magma in which crystals are concentrated in the center of a horizontal or inclined dike (or sill) by grain-dispersive pressure generated during flow of the magma.

fluid escape structure   A small, generally cylindrical structure, usually found in sandstones, consisting of a tube, now filled with sand or sandstone, that served as the path by which water escaped from a lower layer after being buried.

flute   A primary sedimentary structure consisting of an elongate or lobate depression or groove. Flutes occur on the top surface of beds and are probably produced by erosion during turbulent flow across the surface. Flutes are preserved in rocks where the flute is filled (usually by sand) to form a flute cast, a resistant mass on the sole of the overlying bed.

fluvial   A descriptive term referring to features or processes associated with streams and rivers.

flysch   A term used most commonly in Europe to refer to thin-bedded, calcareous sandstones and shales that form thick sequences. Turbidites are present in flysch, especially where the term has been applied to like sequences of noncalcareous rocks.

foid   A feldspathoid mineral.

foidite   An igneous rock in which feldspathoids make up 60–100% of the essential minerals of the rock. The term is used only for volcanic rocks by some petrologists.

foidolite   In the IUGS classification of igneous rocks, a plutonic rock in which feldspathoids constitute more than 60% of the essential minerals.

fold   A bend in a composition band or other planar structure in a rock.

foliated   A term used to describe rocks with an alignment of mineral grains that imparts to the rock a planar character or fabric.

foliated cataclastic texture   A texture consisting of broken grains of rock or mineral materials in elongate segregations (i.e., meso- to microlithons) arranged in preferred orientations and separated by anastamosing to subparallel shear surfaces or microfoliated laminae.

foliated texture   A texture characterized by an alignment of mineral grains in such a way as to give the rock the appearance of or the tendency for splitting into layers or flat pieces.

foliation   A planar feature of rocks that results from the parallel to subparallel alignment of inequant mineral grains or fractures.

forereef   The seaward, generally steep side of a reef.

foreset   A type of sedimentary bed formed in deltas on delta fronts and consisting of inclined layers between horizontally bedded bottomset and topset beds.

formation   A mappable body of rock of distinctive lithology or lithologies and unique stratigraphic position.

fossil   Any prehistoric evidence of past life.

fossiliferous   A descriptive term for rocks that contain fossils. The term is usually applied to rocks in which the fossils are abundant.

fractional crystallization   A process of magmatic differentiation in which a part of a magma is crystallized and the minerals thus formed are separated or isolated from further reaction with the remaining magma.

fractional fusion   A melting process in which a rock is melted in stages and the melt produced by heating is separated from the remaining rocks, preventing further reaction between the melt and the residual rock (the restite).

fractional melting   See *fractional fusion.*

framework (grains)   The larger grains in a clastic sediment or sedimentary rock (as distinguished from the matrix, the finer material).

Franciscan-type Facies Series   A high-pressure, very low temperature sequence of facies, represented by the rocks of the Franciscan Complex of California, in which jadeitic pyroxene and lawsonite are the key or critical minerals.

free energy   A thermodynamic characteristic of a system representing a measure of the internal energy of or the amount of work that can be done by the system.

friable   A descriptive term, usually applied to sedimentary rocks, meaning easily crumbled or broken.

funnel   A solid, cone-shaped, layered pluton in which the apex is down in the Earth.

fusion   Melting.

G

gabbro   A phaneritic igneous rock lacking essential quartz and containing plagioclase feldspar in excess of 90% of the feldspar. The An component of the plagioclase is greater than 50.

gabbroid   A descriptive term referring to those rocks of gabbroic or gabbro-like composition. Used as a field term in the IUGS classification.

gabbronorite   A rock name in the IUGS classification used to designate rocks containing 0–5% quartz, plagioclase in excess of 90% of the feldspar in the rock, plagioclase more calcic than An_{50}, and significant amounts of pyroxene, olivine, and/or hornblende (see the IUGS classification charts for numerical limits).

gas fluxing   A magmatic process in which upward-streaming gases in a magma chamber promote melting of wall rocks.

gas streaming   A magmatic differentiation process in which moving gases facilitate separation of magma and crystals.

geobarometry   A quantitative study used to determine the pressures existing during metamorphic or igneous processes.

geothermal gradient   The ratio of temperature to depth, typically plotted as a curve in P-T space.

geothermometry   A quantitative study used to determine the temperatures existing during metamorphic or igneous processes.

glassy   A textural term used to designate igneous rocks in which there are no crystals, only supercooled magma.

glomeroporphyritic   A textural term applied to porphyritic textures in which the phenocrysts are grouped together in clusters that are scattered through the rock.

gneiss   A name for phaneritic, foliated metamorphic rocks characterized by alternating bands of contrasting mineralogical composition, at least some of which are characterized by the preferred orientation of the included minerals. *Note:* Alternative definitions are in use (see note 34 in chapter 23).

gneissic structure   A layered or banded structure in metamorphic rocks in which layers are between 1 mm and 1 m thick.

gneissose texture   A metamorphic texture characterized by alternating bands of contrasting mineralogical composition, at least some of which are characterized by the preferred orientation of the included minerals.

Goldschmidt's Mineralogical Phase Rule The Phase Rule that indicates that under certain conditions, the number of phases in a rock will equal the number of components in the rock system ($P = C$).

gouge   A soft, nonfoliated dynamoblastic rock dominated by grains less than 1/16 mm in diameter and characterized by cataclastic texture. Gouges are formed by extreme grinding and crushing of grains along a fault zone.

graded bedding   A sedimentary structure in which individual beds show a gradual decrease in grain size from the bottom to the top of the bed.

graded stratified bed   A sedimentary bed in which the lower part shows a gradual change in grain size from bottom to top and in which the top (finer-grained) part exhibits a series of layers.

grain flow   A transport process (and the name given to the deposit formed by that process) in which a high-density, incohesive mix of sand and water moves downslope under the influence of gravity.

grain shape   The dimensional descriptor for sedimentary grains in which the form may be classified as equant (subequal length, width, and height), tabular (two larger dimensions and one substantially smaller dimension), or rod-shaped (two short dimensions and one large one). Alternative descriptions of grain shape exist.

grainstone   A type of limestone composed of clastic carbonate grains with little or no matrix.

granite   A phaneritic igneous rock composed of essential quartz and feldspar. Alkali feldspar comprises two-thirds or more of the feldspar.

granitization   A process of rock formation in which phaneritic rocks of hypidiomorphic-granular texture containing essential quartz and alkali feldspar are produced by metamorphic processes (without melting) via diffusion of ions.

granitoid   A general term applied to all phaneritic plutonic rocks containing essential quartz and feldspar.

granoblastic texture   A metamorphic texture consisting of equant to subequant grains in an aggregate.

granoblastic-polygonal texture   A metamorphic texture characterized by polygonally shaped crystals of equal or nearly equal size with straight to slightly curved grain boundaries.

granoblastic-polysutured texture   A metamorphic texture characterized by crystals of equal or nearly equal size with lobate or serrate grain boundaries.

granoblastite   A metamorphic rock with granoblastic texture.

granophyre   An igneous rock with granophyric texture.

granophyric   An adjective used to describe rocks with irregular, microscopic intergrowths of quartz and alkali feldspar.

granular   In sedimentary rocks a term applied, especially to sandstones, in which grains between 2 and 4 millimeters comprise less than 25% of the rock.

Granulite Facies   A name used to designate the set of all metamorphic rocks containing phase assemblages indicative of conditions of crystallization in the general range of $P = 0.3$–1.2 Gpa (3–12 kb) and $T = 700$–900°C. The name is also applied to the P-T space defined by those conditions.

grapestone   A sediment or rock composed of rounded clusters of carbonate grains cemented together in the form of clusters like bunches of grapes by micrite.

graphic   A term describing phaneritic, granitoid igneous rocks with regular, poikilitic intergrowths of triangular or linear-angular quartz grains in larger alkali feldspar grains.

gravitational separation   A process of magmatic differentiation in which crystals are separated from the liquid by sinking (or flotation) due to the relatively higher (or lower) density of the crystals.

graywacke   A somewhat obsolete term used to designate sandstones in which matrix material composed of micas, chlorite, or clays typically constitutes more than 10 or 15% of the rock.

Greenschist Facies   A name used to designate the set of all metamorphic rocks containing phase assemblages indicative of conditions of crystallization in the general range of $P = 0.3$–1.0 Gpa (3–10 kb) and $T = 300$–550°C. The name is also applied to the P-T space defined by those conditions.

greenstone   A term used by some geologists to designate low-temperature metabasites, typically consisting of chlorite and other minerals.

groundmass   The fine-grained material between larger grains.

gypsum evaporite   Aphanitic to phaneritic, soft, commonly layered, gypsum-rich rock formed via evaporation of a saline solution.

H

halite evaporite   Aphanitic to phaneritic, halite-rich rock formed via evaporation of a saline solution.

hardground   A hard, lithified layer of sediment a few centimeters thick that forms on the seafloor during periods of nondeposition.

hawaiite   A type of volcanic rock with a potash: soda value of less than 1:2, a moderate to high color index, and a modal composition that includes essential andesine and accessory olivine.

helicitic texture   A metamorphic texture consisting of bands of small inclusions arranged in spiral patterns or folds within porphyroblasts or other poikilitic grains.

hematitite   A rock in which hematite either constitutes more than 50% of the mineralogy of the rock or one in which hematite is the principle component and comprises more than 33% of the rock.

hemipelagic mud   Fine-grained to aphanitic marine sediment containing more than 25% material of greater than 5-micron size derived from silicate continental sources, volcanoes, or the shallow levels of the sea.

heterogeneous nucleation   The process in which a new crystal growth-center appears on a preexisting surface.

heterogranoblastic texture   A metamorphic texture characterized by equant or subequant minerals of various sizes.

heterogranular texture   A general metamorphic textural term for textures in which the grains are of notably different sizes.

HFSE   High field strength elements, a group of elements with a high charge to radius ratio, including elements such as Ti, Ni, Hf, Ta, and Y.

holocrystalline   A textural term applied to igneous rocks in which all of the physical components of the rock are crystalline. Contrast with *holohyaline* and *hypocrystalline*.

holohyaline   A textural term applied to igneous rocks composed entirely of glassy (noncrystalline) constituents.

homogeneous nucleation   The process in which new crystal growth-centers form within a melt spontaneously and independent of any preexisting crystal or surface.

homogranular texture   A general category of metamorphic texture in which all the grains in a rock are approximately the same size.

Hornblende Hornfels Facies   A name used to designate the set of all metamorphic rocks containing phase assemblages indicative of conditions of crystallization in the general range of $P = 0.0–0.3$ Gpa (0–3 kb) and $T = 400–575°C$. The name is also applied to the P-T space defined by those conditions.

hornfels   Fine-grained, usually dark-shaded rock with diablastic or granoblastic texture; produced by contact metamorphism.

HP-contact metamorphism   Metamorphism in which temperature is the dominant agent of metamorphism, but in which intrusions that cause the metamorphism are emplaced into high-pressure (generally low-temperature) zones of the crust.

HREE   Heavy rare earth elements, a group of metallic elements of high atomic number (65–71) and low abundance in the Earth's crust, including Dy, Yb, and Lu.

hybridization   An igneous process of magma modification and petrogenesis in which assimilation of country rocks alters magma compositions and consequent rock chemistries. Some petrologists use the term for processes involving magma mixing as well as assimilation.

hydration   A process of chemical weathering in which water combines with other components to yield a new phase.

hydrolysis   A process of chemical weathering in which excess H^+ or OH^- are produced in the associated solution.

hypabyssal   An archaic term referring to igneous rocks intruded at shallow depths. (The rocks are considered to be texturally intermediate between extrusive and intrusive rocks and characteristically have porphyritic textures.)

hyperleucocratic   A descriptive term for igneous rocks with a color index of 10 or less. (Literally, dominantly very light-colored; *hyper* = very, *leuco* = light-colored, *cratic* = ruled by.)

hypermelanocratic   A descriptive term for igneous rocks with a color index of 90 or more. (Literally, very dark-colored.)

hypidioblastic texture   A combining textural term used for metamorphic rocks to indicate that the grains are dominantly subhedral.

hypidiomorphic-granular texture   An igneous rock texture for phaneritic rocks in which equant to subequant, unaligned grains of subhedral character dominate the rock.

hypocrystalline   An igneous rock texture applied to volcanic rocks that consist of a mixture of crystals and glass.

hypolimnion   The oxygen-deficient, saline, bottom-water layer of a lake.

I

idioblastic texture   A combining textural term used for metamorphic rocks to indicate that the grains are dominantly euhedral.

igneous rock   A rock formed by crystallization or solidification of a magma.

ignimbrite   A volcanic rock formed by the lithification of ash flow deposits.

ijolite   A phaneritic igneous rock composed of abundant feldspathoids and 30–70% ferromagnesian minerals.

imbrication   A sedimentary structure consisting of disk-shaped, ovoid, or other inequant clasts stacked in inclined piles at an angle to the bedding.

incongruent melting   A process of melting in which heating of a solid phase produces a new solid phase plus a melt of a composition different from the composition of the original solid phase.

intensive variable   A property of a system that is capable of being changed and the value of which is independent of the mass (size) of the system (e.g., temperature).

inter-   A combining form meaning between.

intercumulus crystals   Crystals formed late in a fractional crystallization sequence from liquid trapped between early-formed cumulus crystals.

intercumulus liquid   The magmatic liquid trapped between crystals and with which postcumulus crystals interact when they crystallized or recrystallized to form cumulate textures.

intergranular   A holocrystalline, volcanic texture consisting of granular pyroxenes and other equant to subequant crystals filling the spaces between tabular plagioclase feldspar crystals. It is the volcanic equivalent of diabasic texture.

internally buffered   A description for systems or reactions in which the mineral phases control the chemistry of the fluid phase.

intersertal   A volcanic rock texture consisting of crystals, especially phenocrysts, in a matrix of glass.

intraclast   A fragment of sand to gravel size of a penecontemporaneous sediment or sedimentary rock included within (deposited to form a part of) a limestone.

intrusive   A descriptive term denoting magmatic bodies that invade other rocks below the surface of the Earth.

intrusive structures   Structures of igneous rocks that are formed below the surface of the Earth.

715

invariant A descriptive term denoting conditions in which there are no degrees of freedom.

invariant system A system in which there are no degrees of freedom; that is, no variable may be changed without altering the number or nature of phases in the system.

inversely graded bed A layer of sedimentary (or volcanic) rock in which particle size varies from fine at the bottom to coarse at the top.

inversely graded disorganized bed A layer of sedimentary rock in which the lower part shows a gradual change from fine to coarser material and the upper part lacks any grading or layering.

inversely-normally graded bed A layer of sedimentary rock that exhibits a gradual change from finer sediment at the base to coarser sediment in the middle and back to finer sediment at the top.

iron formation A cherty, aphanitic to phaneritic, thin-bedded, typically red to black, iron-rich rock containing 20% or more total iron oxides ($FeO + Fe_2O_3$).

ironstone An aphanitic to phaneritic, massive to bedded, commonly oolitic, yellow to maroon, silver, or black, iron-rich, noncherty sedimentary rock containing 20% or more total iron oxides ($FeO + Fe_2O_3$).

island arc A linear or curvilinear row of volcanic mountains that form islands in the sea.

isochemical metamorphism Metamorphism (a pressure-, temperature-, or fluid-induced change in mineralogy, chemistry, or structure of a rock) that occurs without change in the chemistry of the rock domain being considered. (Iso = equal; therefore, of equal or the same chemistry.)

isograd A line on a map representing the first appearance or the disappearance of any particular mineral. Originally, isograds were thought to represent all points connecting the exact same grade (*iso-* = same, *grad* = grade) of metamorphism, a supposition now generally known to be false.

isolated system A system that exchanges neither energy nor matter with its surroundings.

isotope An element that varies from others of the same atomic number in having more or less neutrons, and therefore a higher or lower mass number. (*Iso* = same, *tope* = place; hence, having the same place on the periodic chart.)

J

jacupirangite A phaneritic ultramafic rock composed of clinopyroxene and magnetite, with some nepheline.

joint A fracture in rock along which there has been a total loss of cohesion, across which there is some separation, and parallel to which there has been no significant movement.

K

kelphytic A descriptive term for a special corona texture involving garnet or olivine or the rock texture in which such mineral textures occur.

kelphytic rim A corona texture in which garnet or olivine are rimmed by amphibole, pyroxene, or both of these minerals.

keratophyre A sodic, metavolcanic rock, chemically equivalent to andesite and similar rocks, and containing phases such as albite, calcite, quartz, epidote, and chlorite.

kerogen A fine, brown to black, insoluble material composed of hydrogen, carbon, oxygen, and nitrogen, with or without sulfur, found in sedimentary rocks.

kimberlite A porphyritic ferromagnesian (ultramafic) rock composed of olivine and phlogopite phenocrysts in a groundmass of serpentine, chlorite, calcite, olivine, phlogopite, and other minerals.

kink band A small-scale, tight, mesoscopic fold developed in a rock such as a phyllite or schist that already has a fabric.

komatiite (1) A term used to designate a suite of MgO-rich igneous rocks containing ultramafic volcanic rocks. (2) An ultramafic volcanic rock characterized by spinifex texture; also called uv-komatiite.

L

laccolith A moderate-sized, concordant plutonic structure with a convex-up roof and a diameter/thickness ratio of less than 10.

lacustrine A descriptive term referring to lakes and associated environments.

lagoon A shallow water body along a coast that is separated from the sea by a small barrier, such as a bar.

lamination A layer in a sedimentary rock that is less than 1 cm thick.

lamprophyre A group of dark, aphanitic-porphyritic volcanic rocks containing an abundance of euhedral, ferromagnesian mineral phenocrysts (e.g., biotite, hornblende) in matix of similar minerals, plus feldspars and/or feldspathoids.

landslide A general category of subaerial mass wasting or slope failure that includes a wide variety of phenomena including earthflows, slumps, debris flows, rockslides, and other related processes and features. Also, the deposits produced by these processes.

lapilli Pyroclastic fragments 2–64 mm in diameter. (Singular *lapillus*.)

lapilli tuff A volcaniclastic (pyroclastic) rock composed of lapilli in a matrix of ash.

latite A volcanic rock with essential feldspar, little or no quartz or feldspathoid minerals, and an alkali feldspar to total feldspar ratio of 0.67–0.33.

lattice preferred orientation (LPO) A descriptive term for a metamorphic texture in which there is a special arrangement of mineral grains reflected by alignment of optical and crystallographic axes of those grains.

lava (1) Magma that has lost dissolved gases when it erupted at the surface. (2) The rock produced by solidification of magma that has lost its dissolved gases.

lava flow Tabular to lobate masses of erupted and solidified magma that has lost dissolved gases prior to or during eruption.

lava plateau A major, tabular, constructional volcanic structure consisting of a layered accumulation of extensive, tabular, generally basaltic lava flows dominantly erupted through fissures.

layering A group of tabular features, characterized by distinct mineralogical, textural, structural, or color attributes, that occur in igneous, sedimentary, or metamorphic rocks.

lepidoblastic texture A strongly foliated metamorphic texture characterized by aligned platy or sheet-structured minerals.

leucocratic (1) A descriptive term for light-shaded igneous rocks. (2) The color index designation for rocks with color indices between 11 and 50. (Literally, dominantly light-colored.)

leucosome   A light-shaded, granitoid rock that occurs in the high-grade metamorphic rocks called migmatites. It is produced by partial melting and/or the local injection and crystallization of a lower-temperature (eutectic) melt between layers or foliation planes in a metamorphic rock.

liesegang rings   Variously colored, highly regular rings that occur in weathered rocks and form via oxidation or reduction. Typically, they are dominated by yellow to brown iron oxides.

lime mudstone   A calcite-dominated sedimentary rock (i.e., a limestone) in which the grains are less than 1/16 mm in diameter.

limestone   A sedimentary rock dominated mineralogically by calcite.

lineation   A structural feature of rocks, particularly metamorphic rocks, resulting from the alignment of acicular minerals, the intersection of planar features, the alignment of minor fold axes or kink-band axes, or parallel arrangement of other elongate structural elements.

lipids   Long-chain carboxylic acids, particularly plant and animal fats, found in sediments.

liquidus   (1) A line (or surface, in three dimensions) representing all points of equilibrium between particular solid phases and liquids of specific compositions and specific temperatures or pressures. (2) The curve or surface in a phase diagram separating a field containing only liquid (or melt) from one containing some solid phases.

liquidus curve   See *liquidus*.

lithofacies   A body of sediments or sedimentary rocks with distinct physical and chemical characteristics—rock types, textures, and structures—that represent a particular sedimentary environment.

lithosphere   The outer, more brittle layer of the Earth, consisting of the crust and uppermost mantle.

littoral   A descriptive term for the marine environment or zone between high and low tide, or features of that zone.

load cast   A sole mark on a sedimentary bed consisting of a bulge or elongate knob formed during compaction.

local metamorphism   Metamorphism that affects relatively small volumes of rock (less than 100 km^3).

loess   A porous, friable, commonly calcareous siltstone that forms blanket deposits. Loess is generally formed by deposition of wind-blown material of glacial origin.

lopolith   A layered pluton with a dish-shaped or concave-up shape.

LP-contact metamorphism   Metamorphism in which temperature is the dominant agent of metamorphism and in which intrusions that cause the metamorphism are emplaced into low-pressure, low-temperature zones of the crust.

LPO   See *lattice preferred orientation*.

LREE   Light rare earth elements, notably those Lanthanides with atomic numbers between 57 and 64.

L-tectonite   A rock dominated by linear features with a fabric that reflects the history of deformation during flow (i.e., ductile deformation).

M

mafic   Dark-colored; a descriptive term used for igneous and metamorphic rocks and magnesium- and iron-rich minerals, which are typically dark in color and low in silica.

magma   Melted rock material, with included gases, crystals, and rock fragments.

magnetitite   A rock with more a than 50% magnetite or one with magnetite as the dominant phase and constituting more than 33% of the rock.

macrosparite   A crystalline carbonate rock with grains larger than 0.06 mm in diameter.

major elements   The most abundant elements that function as cations in the crust and mantle of the Earth, including Si, Al, Ti, Fe, Mn, Mg, Ca, Na, K, P, and H.

manganese nodule   A globular mass of aphanitic, black to brown manganese oxide minerals typically formed on the seafloor by precipitation of manganese oxides and associated elements.

manganolite   An aphanitic, black to brownish-black rock composed of manganese oxide minerals.

marine environment   The environment in the sea below low tide level.

marl   An old name used to refer to sediments or rocks composed of subequal amounts of clay minerals and calcite, that is, friable argillaceous limestones and calcareous mudrocks, which are intermediate between limestone and shale in composition.

marlstone   A name used by some workers to refer to rocks composed of subequal amounts of clay minerals and calcite (also see marl), especially if the rock is well lithified.

mass deposition   A process of deposition in which materials are deposited by various types of subaerial and submarine landslides (*sensu lato*) and grain flows.

matrix   A term used to designate the finer-grained material that occurs between the coarser grains in a rock. In sedimentary rocks, matrix is *clastic* material. Matrix commonly consists of clay with silt-size particles of quartz and other minerals.

matrix-supported   A descriptive term used to characterize sediments and sedimentary rocks containing clasts in a matrix, but in which the clasts are separated by matrix and do not generally contact one another.

meandering stream   A stream in which water flows in one main channel that moves back and forth across the floodplain in a looping, snakelike fashion.

mechanical transportation   Transportation by physical movement of solid materials via suspension, sliding, bouncing, rolling, dragging, or other related processes.

melange   A body of rock mappable at a scale of 1:24,000 or smaller, characterized by a lack of internal continuity of contacts or strata and by the inclusion of fragments and blocks of all sizes, both exotic and native, embedded in a fragmental matrix of finer-grained material.

melanocratic   (1) A descriptive term used to refer to igneous rocks that are dark in shade or color. (2) A term referring to a color index category with color indices of 50 to 89. (Literally, dominated by dark-shaded.)

melanosome   The dark-shaded, ferromagnesian mineral-rich part of the high-grade metamorphic rock migmatite.

member   A subdivision of a formation characterized by distinctive lithologic character and stratigraphic position.

mesocumulate   A rock composed of crystals formed via fractional crystallization plus small amounts of minerals formed by crystallization of melt in the spaces between these crystals.

mesocumulate texture   An igneous texture consisting of cumulate crystals formed via fractional crystallization with small amounts of crystals formed by crystallization of melt between these cumulate crystals. Mesocumulate textures are intermediate between orthocumulate textures and adcumulate textures.

mesogenesis   The middle-stage process of conversion of sediment to rock (diagenesis) that occurs relatively soon after burial.

mesoscopic structures   Handspecimen- to outcrop-scale, nonpenetrative features of rocks.

mesosome   A metamorphic rock of intermediate shade associated with migmatites.

mesosphere   The more rigid part of the mantle that occurs below the plastic asthenosphere.

mesozonal   A descriptive term referring to a category of plutons and other features formed at intermediate depths (about 6 and 16 km).

metaluminous   A descriptive term applied to igneous rocks in which the mole percent alumina is less than the mole sum of alkalis and lime but more than the mole sum of alkalis.

metamorphic differentiation   The metamorphic process or processes that produce banded or lenticular segregations of minerals from an initially homogeneous rock.

metamorphic facies   A set of rocks representing the full range of possible rock chemistries, with each rock characterized by an equilibrium assemblage of minerals that reflects a specific, but limited, range of metamorphic conditions.

metamorphic facies series   The progression of facies across a metamorphic belt or across a petrogenetic grid.

metamorphic rock   A rock that formed originally as igneous or sedimentary rock but which has been changed mineralogically, texturally, or both—without undergoing melting—in response to heat, pressure, directed stress, or chemically active fluids or gases. Metamorphic rocks have textures or minerals, or both, that reflect cataclasis, recrystallization, or neocrystallization in response to conditions that differ from those under which the rock formed and that lie between those of diagenesis and anatexis.

metamorphism   A process or set of processes that affect rocks in such a way as to produce textural changes, mineralogical changes, or both under conditions in the Earth between those of diagenesis and weathering (at the lower limit) and melting (at the upper limit).

metasomatism   Metamorphism dominated by chemical changes induced primarily by chemically active fluids.

metastable system   A system that has no tendency to change if influenced by small perturbations, but which is not at the lowest possible energy state and is, therefore, subject to change if an energy barrier is overcome.

miarolitic cavity   A cavity in a plutonic rock into which crystals project.

micrite   (1) Sedimentary calcite in grains smaller than 0.004 mm in diameter; microcrystalline ooze. (2) Lime mudstone, a sedimentary rock composed of grains smaller than a specified threshhold value.

microdolostone   A rock composed predominantly of dolomite of crytocrystalline to aphanitic grain size.

micrographic   An igneous texture, visible only through a microscope, that consists of regular, angular crystals of quartz poikilitically enclosed in feldspar.

microlite   A crystal in a volcanic rock that is microscopic in size.

microlithon   A small mesoscopic to microscopic lens or piece of rock, typically granular, relatively undeformed, and quartz and feldspar-rich, that occurs between folia in metamorphic rocks that have cleavage, especially in quasimylonites and mylonites.

microlitic   A descriptive term for an igneous matrix texture consisting of microlites (very small fibrous crystals) with glassy or cryptocrystalline intercrystalline material.

microporphyritic   An igneous texture characterized by microscopic phenocrysts set in a finer-grained matrix.

microsparite   (1) Crystalline calcite grains between 0.004 and 0.06 mm in diameter. (2) A rock composed of calcite grains between 0.004 and 0.06 mm in diameter.

migmatite   A complex, mixed, high-grade metamorphic rock consisting of various proportions of dark, ferromagnesian mineral-rich rock and light, quartz- and feldspar-rich rock.

minette   A type of volcanic rock belonging to the lamprophyre group and composed of euhedral biotite phenocrysts in a groundmass of biotite and alkali feldspar.

minor accessory mineral   An igneous mineral that constitutes less than 5% of the rock and the presence of which is not implied by the name of the rock.

modal analysis   An analysis to determine the kinds and amounts of minerals in a rock by counting each type of mineral grain to determine the percentage of each.

mode   The list of minerals observed in a rock, with their volume percentages.

Moho   Short for Mohorovicic Discontinuity.

Mohorovicic Discontinuity   The seismic boundary between the mantle and the crust.

mold   A depression on a sedimentary surface, especially one that retains the detail of the surface of the structure or object that left the depression.

MORB   Abbreviation for Mid-Ocean Ridge Basalt, a type of tholeiitic basalt characteristic of mid-ocean ridges.

mortar texture   A nonfoliated metamorphic texture consisting of larger broken grains in a matrix of smaller broken grains.

mosaic texture   A metamorphic texture characterized by polygonally shaped crystals of equal or nearly equal size with straight to slightly curved grain boundaries.

mound   A small hill-like sedimentary structure built on the seafloor by organisms.

mudcrack   A crack formed in muddy sediment as a result of drying. Mudcracks typically form polygonal patterns on sedimentary surfaces. If filled by sediment different from the cracked sediment, the mudcracks may be preserved in sedimentary rocks and can indicate facing (the top of a bed).

mudrock   A sedimentary rock dominated by materials less than 0.06 mm in diameter.

mudstone   A sedimentary rock dominated by particles of less than 0.004 mm in diameter and lacking laminations.

mugaerite   A type of volcanic rock composed of essential alkali feldspar and oligioclase, plus the mafic accessory phases olivine and clinopyroxene.

mullion   A columnar metamorphic structure in which the cylindrical columns are 2 cm to 2 m in diameter. The columns are composed of the same material as the surrounding rocks and may exhibit internal folding.

mylonite   A metamorphic rock characterized by textures that typically show both plastic deformation features and evidence of syntectonic recrystallization and recovery.

mylonitic texture   A foliated to porphyroclastic-foliated texture characterized by porphyroclasts in a fine-grained to very fine-grained matrix of materials that show syntectonic recrystallization and/or recovery textures.

myrmekitic texture   An igneous texture characterized by an intergrowth of sodic plagioclase feldspar and wormlike grains of quartz.

N

nematoblastic texture   A foliated metamorphic texture consisting of layers of needle-like crystals.

neocrystallization   The process in which new minerals (that did not previously exist in a rock) are formed. Neocrystallization occurs during diagenesis and metamorphism (neo = new; thus new crystal formation).

neomorphic   An adjective referring to a process of diagenesis in which minerals either recrystallize or change to a polymorph.

neosome   A complex rock, including both a leucosome and a melanosome, formed through migmatization (a process involving partial melting and deformation). (Literally, new body.)

nepheloid layer   In submarine environments, a turbid bottom-water layer in which mud is suspended.

Newtonian fluid   A fluid that has no strength and does not change in viscosity as the shear rate increases.

nodularblastic texture   A metamorphic texture consisting of nodular clusters of small grains of one or two minerals in a matrix of a different composition.

nodule, igneous   A fragment of phaneritic igneous rock within an igneous matrix.

nodule, sedimentary   A small rounded to irregular concretion characterized by a bumpy surface.

non-Newtonian fluid   A fluid that lacks strength but will change in viscosity as the shear rate increases or decreases.

nonphyllosilicate cement   Precipitated intergranular material that is not composed of sheet-structured minerals. Nonphyllosilicate cements consist of minerals such as calcite, quartz, dolomite, gypsum, hematite, phosphate minerals, manganese oxides, and zeolites.

nonterminal reaction   A metamorphic reaction in which one mineral pair becomes stable as another pair becomes unstable, resulting in a tie-line flip on an appropriate phase diagram.

norm   A list of hypothetical minerals and their percentages calculated from the chemical analysis of a rock.

normally graded bed   A sedimentary bed in which there is a gradual change in grain size from coarse at the base to fine at the top of the bed.

normative analysis   The procedure for calculating a list of hypothetical minerals and their percentages from a chemical analysis of a rock.

nucleation   The formation of nucleii, a process in which a few atoms assume the same relationship to one another as they would have in a solid.

nuée ardente   A hot cloud of volcanic ash and gas that erupts from and flows rapidly down the flanks of a volcano.

O

obsidian   Volcanic glass; supercooled, solidified magma. High viscosity and rapid cooling combine during eruptions of high-silica magmas to produce its glassy texture.

olistostromal flow   A high-density, high-viscosity, rapid, submarine, downslope movement of a cohesive mass of mud and rock fragments.

olistostrome   (1) A submarine landslide or debris flow; an olistostromal flow. (2) The diamictite deposit of a submarine landslide or debris flow.

oncolite   A small (generally <10 cm), concentrically laminated, spherical to irregular body formed during deposition by biochemical precipitation and trapping of carbonate mud by algae.

ooid   A spherical to oval, concentrically layered particle between 0.25 and 2 mm in diameter that is composed of carbonate minerals or replacement phases.

oolite   (1) An ooid. (2) A rock composed predominantly of ooids.

oolith   Ooid.

oolitic   A descriptive term applied to rocks containing ooids or ooidlike grains.

ooze   A fine-grained sediment, like a clay in having less than 25% siliciclastic material of greater than 5-micron size, but different in containing more than 50% biogenic material.

opal-A   Amorphous opaline silica that is a biochemical precipitate.

opal-CT   A metastably crystallized, interlayered cristobalite-tridymite form of silica.

open system   A system that exchanges both matter and energy with its surroundings.

ophiolite   A type of igneous mafic-ultramafic body of specific structure consisting of an ultramafic tectonite at the base, a plutonic ultramafic-gabbro sequence that includes bodies of more siliceous differentiates, a sheeted dike complex composed of diabase, and an overlying basaltic lava flow sequence.

ophitic   An igneous texture that occurs in gabbros and consists of large poikilitic pyroxenes enclosing smaller tabular plagioclase crystals.

orthocumulate   A rock composed of cumulate crystals formed via fractional crystallization plus significant amounts of minerals formed by crystallization of melt between these cumulate crystals.

orthocumulate texture   A texture in which crystals formed via fractional crystallization are interspersed with significant amounts of minerals formed by crystallization of melt that existed between these crystals.

orthomatrix   Intergranular material in sandstones consisting of recrystallized, detrital, clay-rich mud.

orthomylonite   A foliated metamorphic rock in which microlithons, porphyroclasts, or both, in a finer-grained, phyllosilicate-bearing matrix, comprise 10 to 50% of the rock.

orthomylonitic texture   A metamorphic texture that is foliated to porphyroclastic-foliated and includes 10% to 50% porphyroclasts in a matrix of materials that are syntectonically crystallized or that show recovery textures.

oversaturated   In igneous petrology, a term used to designate rocks that have free silica (silica in excess of that needed to compose feldspars) and hence contain quartz or another silica mineral.

oxidation   A process of chemical weathering in which the valence of the cation increases.

P

packstone   A limestone consisting of touching allochems (or other framework grains) in a matrix of lime mud.

pahoehoe lava   Solidified volcanic flow rock with a relatively smooth but vesicular and ropy surface.

paired metamorphic belts   In mountain systems, two roughly parallel, linear zones of metamorphic rock, one of higher P and lower T and the other of higher T and lower P. The inner belt (continent side) typically exhibits a Buchan- or Barrovian-type Facies Series, whereas the outer belt (ocean side) exhibits a Franciscan- or Sanbagawa-type Facies Series.

parental magma   A magma that has given rise to other magmas via some process of differentiation.

partial fusion   The process of partially melting a rock. See also *fractional fusion*.

partial melting   See *partial fusion*.

pebbly   A descriptive term applied to sedimentary rocks that contain less than 25% pebbles.

pebbly mudstone   A rock name used by some geologists for rocks with rounded, pebble-size clasts in a muddy matrix.

pegmatite   (1) A textural term used by some petrologists to designate igneous rocks dominated by large crystals (generally >3 cm in length). (2) A term used by some geologists to designate a granitic rock characterized by a very coarse-grained texture.

pegmatitic   A descriptive term for igneous rocks dominated by crystals greater than 3 cm in length.

pelagic clay   A type of muddy marine sediment with less than 25% silicate detritus of more than 5 microns and less than 50% biogenic material.

pelagic environment   A tectonically passive environment that lies below the deep sea.

pelagic mud   A type of muddy marine sediment with less than 50% biogenic material and more than 25% material of greater than 5-micron size derived from silicate continental sources, volcanoes, or the shallow levels of the seas.

pelitic   A descriptive term used to designate mudrocks or rocks derived from mudrocks. Pelitic metamorphic rocks are typically aluminum-rich rocks characterized by abundant micas and containing aluminum silicates such as kyanite or andalusite.

pelitic schist   A phaneritic, foliated, nonbanded metamorphic rock derived from a mudrock.

pellet   A small (< 0.2 mm) rounded mass of very fine calcareous material (micrite). Pellets are peloids that were feces excreted by mud-eating worms, shrimp, and other organisms.

pelloid   A small, rounded, carbonate allochem generally less than 1/4 mm in diameter.

penecontemporaneous fault   A fracture, along which there has been movement, that occurs in and was formed during the depositional history of a rock sequence.

peralkaline   A descriptive term used to designate igneous rocks in which the combined molecular percent of alkalis exceeds the molecular percent of alumina.

peraluminous   A descriptive term used to designate igneous rocks in which the molecular percent of alumina exceeds the total combined molecular percentages of alkalis and lime.

peri-platform sediments   Carbonate slope and basin margin sediments.

peritectic   An inflection point on a liquidus curve representing the melt composition and temperature at which a discontinuous reaction (i.e., incongruent melting or crystallization) occurs.

peritectic system   System exhibiting incongruent melting, or the melting of one solid phase to produce a liquid plus another solid phase.

permeability   A measure of how well fluid will flow through a rock.

petrofacies   Stratigraphic subdivisions based on similarity of sandstone petrology.

petrogenesis   The study of the histories and origins of rocks.

petrogenetic grid   A graph of subdivided P-T space with regions characterized by particular phases or phase assemblages.

petrography   The study of the description and classification of rocks (also called lithology).

petrology   The overall study of rocks, including petrography and petrogenesis.

petrotectonic assemblage   A distinctive suite of rocks characterizing a type of plate boundary or a plate setting.

pH   The negative logarithm of the hydrogen ion concentration of a solution.

phacolith   A concordant, sill-like plutonic structure of generally modest size, that is lenticular and is emplaced into a site along a fold axis.

phaneritic   A descriptive term applied to crystalline materials in which grains can be discerned without the aid of a microscope.

phase   A homogeneous material that, because of its physical properties, can be separated by mechanical means from other phases with which it may occur.

Phase Rule   An equation relating the number of components, the phases, and the variance of a system: $F = C - P + 2$, where F is the number of degrees of freedom, C is the number of components, and P is the number of phases.

phenocryst   A larger grain surrounded by a population of grains of significantly smaller size; found in porphyritic (volcanic) textures.

phosphatic rock   A rock that is a conglomerate, sandstone, or mudrock; is commonly dark and variously colored; and contains an apatite cement.

phosphorite   (1) An aphanitic to phaneritic, typically brown to black rock that is oolitic, laminated, nodular, or fossiliferous, with more than 50% apatite. (2) A sedimentary rock containing more than 19.5% P_2O_5.

phyllitic texture   An aphanitic to fine-grained metamorphic texture characterized by the presence of a crenulation cleavage, kink bands, or microfolds.

phyllosilicate cement   A precipitated or recrystallized cement composed of one or more of the following minerals: smectites, chlorites, mixed-layer phyllosilicate minerals, chlorite-vermiculite, kaolinites, celadonite, illite, and muscovite.

phyric   An adjective used to describe volcanic rocks with phenocrysts.

picritic basalt   Olivine-rich basalt.

pillow lava   Lava, extruded under water, that has tubular to elliptical structures.

pilotaxitic A volcanic texture consisting of randomly arranged, microscopic laths of plagioclase or other minerals.

pipe A crudely cylindrical channel, typically filled with breccia and lava, through which magmas have risen beneath the central vent of a volcano. *Note:* Other definitions are listed in Bates and Jackson, (1987).

pisolite A spherical, concentrically layered structure (similar to an oolite) that is larger than 2 mm in diameter. Pisolites occur in sedimentary and volcaniclastic rocks.

plate A major fragment of the lithosphere, thousands of kilometers in areal extent and about 100 km thick.

playa A flat, vegetation-free desert basin that occasionally contains an ephemeral lake.

playa lake An ephemeral, generally very shallow lake that forms in a playa.

pluton A general term denoting a rock body composed of plutonic rock.

plutonic rock Igneous rock that crystallized well below the surface of the Earth.

poikilitic A descriptive term designating an igneous texture in which a large crystal (an oikocryst) encloses irregularly scattered, smaller crystals of another mineral.

poikiloblastic texture A metamorphic texture in which a larger grain encloses several smaller grains.

poikilotopic A sedimentary texture in which framework grains are enclosed by larger crystals of cement materials (e.g., calcite).

porosity A measure of the amount of empty space existing between grains in a rock.

porphyritic An igneous texture in which there is a bimodal grain distribution; that is, a number of larger grains, called phenocrysts, are surrounded by a population of grains of significantly smaller size, which constitute the groundmass.

porphyroblastic-foliated texture A metamorphic texture in which there is a bimodal distribution of grains, consisting of some larger grains (the porphyroblasts) surrounded by a larger number of substantially smaller grains, including acicular or platey grains that are aligned to give a layered character to the rock.

porphyroblastic texture A metamorphic texture in which there is a bimodal distribution of grains, consisting of some larger grains (the porphyroblasts) surrounded by a larger number of substantially smaller grains.

porphyroclastic texture A weakly to strongly foliated metamorphic texture characterized by a bimodal distribution of grains, with larger, typically deformed protolith grains (the porphyroclasts) enclosed in a matrix of finer-grained, typically deformed grains, derived from protolith minerals. In some cases the matrix has recrystallized (recovered) and is strain free.

postcumulus crystal A crystal that crystallized from, or recrystallized through interaction with, an intercumulus liquid. Postcumulus crystals surround the cumulus crystals in cumulate-textured rocks.

preferred orientation The textural alignment of mineral grains. Two types of preferred orientations exist: (1) LPO, or lattice preferred orientation, a typically invisible preferred orientation resulting from the like orientation of lattices within the constituent crystals of a rock, and (2) SPO, or shape preferred orientation, the visible alignment of elongate or tabular mineral grains.

Prehnite-Pumpellyite Facies The set of all metamorphic rocks containing phase assemblages indicative of conditions of crystallization in the general range of $P = 0.1-0.3$ Gpa (1–3 kb) and $T = 200-350°C$. The name is also applied to the P-T space defined by those conditions.

pressure shadow A zone of lower stress and equant grain growth adjacent to a porphyroblast and generally parallel to the foliation in a metamorphic rock.

pressure solution A process in which pressure is concentrated at the point of contact between two grains, causing the contact area to dissolve and allowing for subsequent migration (diffusion) of ions or molecules away from the point of contact.

primary magma A chemically unchanged, anatectic melt derived from any kind of preexisting rock.

primary mineral A mineral that forms when a rock first forms.

primitive magma An unmodified magma, which forms through anatexis of mantle rocks that have not been melted or otherwise changed in composition since they formed.

proglacial aeolian environment A site of deposition downwind of a glacier, where winds deposit fine sediment.

proglacial fluvial environment A site of deposition where streams of meltwater flowing out from a glacial terminus deposit sediments originally derived from the glacier.

prograde metamorphism Metamorphism that follows a P-T path generally away from surface pressure and temperature conditions.

protolith The parent rock from which a metamorphic rock was derived. (Literally, first stone.)

protomatrix In a sedimentary rock, a type of matrix, or detrital, intergranular material, consisting of clay-rich mud.

protomylonite A foliated metamorphic rock composed of more than 50% porphyroclasts and microlithons enclosed in a matrix of finer-grained, typically phyllosilicate-bearing, deformed to syntectonically recrystallized material.

protomylonitic texture A porphyroclastic, foliated metamorphic texture consisting of more than 50% porphyroclasts and microlithons in a finer-grained matrix of deformed to syntectonically recrystallized minerals.

provenance The source area from which sediment is derived.

proximal A descriptive term referring to a site near to either the source of sediment or another reference point.

proximal environment A site of deposition near a source of sediment.

pseudomatrix Fine-grained, intergranular material in sandstones that looks like primary clastic material but is derived from deformed and recrystallized lithic fragments; false matrix.

pseudomorphic A descriptive term used to denote mineral replacements in which the new phase mimics the external crystal form of the replaced phase.

pseudo-phase diagram A triangular or other-shaped diagram that appears to be a phase diagram, but which does not meet the requirements of the Phase Rule, usually because of the way components are combined in defining the corners of the diagram.

pseudoplastic material A material in which viscosity varies during flow.

pseudotachylite A nonfoliated, cataclastic metamorphic rock that contains glass in the matrix. Such rocks form along faults.

P-T-t path Pressure-temperature-time path used in describing metamorphic rock histories.

pumiceous A term used to describe a texture or structure in volcanic rocks characterized by elongate, fine tubular vesicles.

pycnocline   A zone in a water body (or column) characterized by variable oxygen content, salinity, and density. Also called dysaerobic zone.

pyroclast   A clastic volcanic fragment produced by the eruption of material into the air (or water).

pyroclastic   A term used to describe rocks or textures that are characterized by pyroclasts.

pyroclastic breccia   A clastic volcanic rock dominated by angular clasts larger than 64 mm in diameter.

pyroclastic cone   A small, steep-sided structure composed largely of pyroclasts of various sizes with little, if any, lava. Also called cinder cone.

pyroclastic rock   Rock composed of fragments (clasts) of volcanic rock formed by explosive eruption.

pyroclastic sheet   A tabular accumulation, generally of silica-rich, fragmental volcanic rock, that forms where particles settle out of the atmosphere during and after explosive volcanic eruptions.

pyrolite   A hypothetical ultramafic mantle rock composed of one part basalt and three or four parts peridotite or dunite.

Pyroxene Hornfels Facies   The set of all metamorphic rocks containing phase assemblages indicative of conditions of crystallization in the general range of $P = 0.0$–0.4 Gpa (0–4 kb) and $T = 550$–$750°C$. The name is also applied to the P-T space defined by those conditions.

Q

quartzine   An optically length-slow form of chalcedony.

quartz keratophyre   A sodic, metavolcanic rock chemically equivalent to rhyolite or other siliceous volcanic rocks and characterized by phases such as quartz, albite, calcite, and epidote.

quasimylonite   A weakly to moderately foliated metamorphic rock characterized by incipient foliations, some crystalline matrix materials, and microlithons or broken grains. Aligned phyllosilicates or other mineral grains engulf the microlithons or porphyroclasts and provide the foliation. Evidence of cataclastic (grain-breaking) and crystal-plastic deformation is present. (*quasi* = to a degree; hence, mylonitic to a degree.)

R

radiolarian chert   An aphanitic to fine-grained sedimentary rock composed predominantly of silica minerals and containing significant numbers of radiolaria.

rain print   The preserved impact mark developed where a raindrop strikes a muddy surface.

rapikivi   A texture found in plutonic igneous and metaigneous rocks consisting of relatively large alkali feldspar grains, with cores of K-rich feldspar and rims of Na-rich (plagioclase) feldspar, surrounded by a matrix of finer-grained minerals.

reaction space   A hypothetical volume delimited by specific metamorphic reactions and containing one specific reaction path that was followed by a rock during metamorphism. The method of analysis used to determine the reaction space consists of an examination of the chemical data, minerals, and textures of a rock.

recovery   A process in which grains containing large amounts of strain (e.g., with deformation bands) reduce the amount of strain and intracrystalline deformation by internal reorganization via recrystallization.

recrystallization   A process of ion migration and lattice reorganization in which physical or chemical conditions induce a reorientation of intergrain relationships and the crystal lattices of mineral grains without accompanying breaking of grains.

redox potential   See *Eh*.

reduction   The weathering process in which the valence of cations in a mineral is decreased.

reefs   Domal to elongate, massive to bedded sedimentary structures built, during carbonate deposition, by organisms that biochemically precipitate carbonate minerals.

regional metamorphism   Metamorphism that affects thousands of cubic kilometers of rock.

relict structures   Primary structures, inherited from parent rocks, that remain in metamorphic rocks after metamorphism. Also called palimpsest features.

relict textures   Primary epiclastic, pyroclastic, or crystalline textures originally formed in a protolith and preserved in the derivative metamorphic rock.

replacement   An *in situ* diagenetic or alteration process in which a new mineral takes the place of an originally crystallized or sedimented phase.

replacement chert   A chert formed by a diagenetic process in which silica minerals take the place of preexisting minerals of various types in a rock.

restite   A residual (depleted) rock remaining after partial melting and melt extraction have affected a mass of rock. The term is applied to the rock remaining after leucosome has been removed from a protolith during the process of migmatization.

resurgent caldera   A large, generally circular, volcanic collapse structure within which later uplift of a central domal region occurred.

retrograde metamorphism   Metamorphism that follows a P-T path generally towards the conditions existing at the surface of the Earth.

reverse grading   A structure consisting of a layer exhibiting an increase in grain size from bottom to top.

rhyolite   Generally, siliceous volcanic rock that contains more than 69 weight percent silica and, where porphyritic, phenocrysts of quartz ± sanidine or another alkali feldspar. Various definitions exist.

rhythmic layering   A structure of plutonic igneous rocks that consists of repeated layer couplets, each layer of which is composed of a mineral assemblage with a different ratio of plagioclase to ferromagnesian minerals. Rhythmic layering is most distinct where the layers form dark and light bands.

ring dike   A special type of arcuate, sheetlike, discordant igneous intrusion, often large, that is vertical and cylindrical in orientation and form.

ripple marks   A series of regularly undulating shapes on bedding surfaces created either by oscillating water or by wind or water currents.

rip-up   A chip of rock that has been removed by erosion from an underlying bed and is preserved as a distinct fragment in the overlying bed, usually near its base.

rise, continental   See *continental rise*.

rock   (1) A solid aggregate of mineral grains. (2) A solid, naturally occurring mass of matter composed of mineral grains, glass, altered organic matter, or combinations of these components.

rock salt   An aphanitic to phaneritic, usually light-colored, salty-tasting, soft rock dominated by the mineral halite.

rod   A metamorphic structure that is long and cylindrical and is composed of segregated or introduced material (i.e., dike or vein material, such as quartz).

rodingite A calcium-rich metamorphic rock, typically containing diopside and grossular garnet, formed by metasomatism of gabbro.

roof pendant A mass of rock that formed part of the roof over a pluton and hung down into the magma at the time of intrusion but was later isolated from other connected roof rocks by erosion.

rubble Angular sediment dominated by clasts of coarse size (> 2mm).

S

sabkha An arid to semiarid, coastal plain sedimentary environment that lies above the normal high tide level but is subject to periodic tidal flooding and subsequent evaporite formation.

salinastone A rock composed of saline minerals of unspecified origin.

salinite An evaporite-like rock that forms via processes other than evaporation of saline solutions, e.g., by chemical precipitation from supersaturated solutions. Compare to *evaporite*.

salt marsh A flat, vegetated area, periodically flooded by salt water, that adjoins an estuary or lagoon.

Sanbagawa Facies Series A high-pressure, moderate-temperature sequence of metamorphic facies, characterized by rocks containing both blue sodic amphiboles and epidote at middle to high grades. The series is named for rocks of the Sanbagawa Metamorphic Belt of Japan.

sandstone dike A tabular, discordant structure composed of sandstone formed by the filling of a joint by fluidized sand and the lithification of the sand.

Sanidinite Facies The set of all metamorphic rocks containing phase assemblages indicative of conditions of crystallization in the general range of P = 0.0–0.3 Gpa (0–3 kb) and T = 750–1100°C. The name is also applied to the P-T space defined by those conditions.

saturated A descriptive term for igneous rocks with neither quartz nor an undersaturated mineral (e.g., a feldspathoid), but typically rich in feldspar.

schist A phaneritic, foliated metamorphic rock with abundant phyllosilicate or acicular minerals.

schistose texture A fine-grained to very coarse-grained metamorphic texture characterized by a subparallel arrangement of tabular, acicular, or flaky minerals.

schlieren Tabular, disklike concentrations of minerals with diffuse boundaries that occur within an igneous rock mass. Also called flow layers.

schlieren arch In a mesozonal pluton, an archlike pattern of aligned minerals and xenoliths.

schlieren dome A dome-shaped pattern of aligned minerals and xenoliths within a mesozonal pluton.

secondary mineral A mineral that forms after a rock has first formed, via alteration or weathering.

secondary porosity Porosity developed via postsedimentation processes.

sediment Material transported in water, air, or ice that accumulates and lithifies to form sedimentary rock.

sediment rain The falling of sediment, including siliciclastic, allochemical, and precipitated sediment, from the water column onto the floor of a depositional basin or other depositional site.

sedimentary environment A surface region of the lithosphere where sedimentation occurs, which is either above or below sea level and is distinguished by a particular set of chemical, physical, and biological characteristics.

sedimentary rock A rock that forms under surface conditions and consists of accumulations of (1) chemical precipitates, (2) biochemical precipitates, (3) fragments or grains of rocks, minerals, and fossils, or (4) combinations of these kinds of materials.

semischist A phaneritic metamorphic rock that is weakly foliated, generally because acicular to flaky minerals are poorly aligned or because such minerals, even though aligned, are a minor constituent of a quartz and/or feldspar dominated rock.

semislate An aphanitic, weakly foliated metamorphic rock with a poorly developed cleavage.

seriate A textural term used to describe igneous rocks consisting of grains of a wide range of sizes that grade from one size to another.

serpentinite A rock composed primarily of serpentine minerals produced by metamorphism of the magnesium-rich minerals of ultramafic or mafic rocks.

serpentinization The process or processes by which ultramafic or mafic rocks are transformed into serpentinites.

shale A laminated or fissile mudrock.

Shape preferred orientation (SPO) The textural alignment of elongate or tabular mineral grains.

sheaf texture A metamorphic texture resulting from an array of clusters of diverging acicular or platy grains.

sheeted dike complex An assemblage of dikes that are distinctly parallel in strike and may have only one chilled margin, indicating that successive dikes intruded earlier-formed dikes parallel to a regional fracture pattern.

shelf The part of the seafloor that occurs above the shelf-slope break at depths of about 124 m but below the normal wave base.

shield cone Flat cone-shaped accumulation of lava containing minor amounts of interlayered pyroclastic materials.

shoshonite An intermediate, porphyritic volcanic rock with phenocrysts of augite and olivine in a groundmass of leucite, labradorite with rims of alkali feldspar, olivine, augite, and glass.

sideritestone A sedimentary rock, the major component of which is siderite.

silexite A rock that is composed primarily of quartz and has a very high silica content.

silicate fragment A detrital or terrigenous grain or clast, including gravel-, sand-, silt-, and clay-size fragments of preexisting silicate minerals or rocks.

siliceous sinter An aphanitic to fine-grained, typically layered, variously colored, hard, typically porous, siliceous rock deposited by groundwater or surface water at or near a hot spring or geyser.

siliciclastic rock A general term for rock composed of silicate mineral and rock fragments and related clasts.

sill A concordant intrusive igneous structure of generally modest size and a tabular to elliptical shape, with a width to thickness ratio greater than 10.

siltstone A sedimentary rock composed of substantial, but variable, amounts of silt-size grains. The name is applied to rocks of varying structure, including laminated, fissile, nonlaminated, and nonfissile rocks.

simple pegmatite An igneous rock body which typically consists of (1) very coarse-grained areas within finer-grained granitoid plutons or (2) very coarse-grained lenses in high-grade (high T-P) metamorphic rocks.

skeletal   A descriptive term used to designate rounded to angular fragments of biochemically precipitated shell or hard parts of invertebrate organisms. Most skeletal fragments have a carbonate composition.

skeleton   The biochemically precipitated external shell (exoskeleton) or internal support (endoskeleton) of an organism.

skree   Angular sediment typically formed at the base of a cliff or steep slope and dominated by clasts of coarse size (>2mm).

slaty texture   An aphanitic, foliated metamorphic texture conditioned by the alignment of flaky (phyllosilicate) or acicular grains, which gives the rock a tendency to break into very flat, relatively smooth pieces.

soft-sediment fault   A break in sedimentary layers that forms where sediments undergo deformation before they are fully lithified.

soft-sediment fold   A bend in sedimentary layers formed before the sediments were fully lithified.

sole mark   A cast of a filled groove or flute, which is exposed on the base (sole) of an overlying bed when the weak underlying beds are eroded or removed.

solidus   (1) A line (or surface, in three dimensions) defined as the locus of all points for which a correspondence exists between specific temperatures (and/or pressures) and a particular liquid that is in equilibrium with one or more solid phases. (2) On a phase diagram, the line below which all materials are solid.

solute transportation   Stream transportation of dissolved materials that are moved in solution by the moving water.

solution   The process in which compounds are broken down and release ions.

solvus   A curve connecting the points representing compositions of pairs of coexisting minerals (notably the alkali feldspars) in a solid solution system.

sorting   A size parameter used to describe the textures of clastic rocks by relating the variation in grain sizes within the rock. In geology, well-sorted sediment has grains that are nearly all of identical size, whereas poorly sorted sediment consists of grains of varied sizes. In engineering, these definitions are reversed.

spaced cleavage   Rock cleavage (the tendency to break into tabular to lenticular pieces) with domain spacings of > 0.01 mm and which lacks a penetrative fabric.

spar   Crystalline calcite grains larger than 0.004 mm. In handspecimen study, spar may be defined as calcite grains larger than 0.16 mm.

sparite   A carbonate sedimentary rock composed of spar (crystalline calcite).

spherulitic   A descriptive term for a texture in which fibers that radiate from a point are clustered in spherical groups.

spheruloblastic texture   A metamorphic texture in which fibers that radiate from a point are clustered in spherical groups.

spilite   A metamorphosed basalt rich in sodium and containing such phases as albite, calcite, epidote, chlorite, sphene, and quartz. Spilites form at low pressures and temperatures of metamorphism.

spinifex   An igneous texture, characteristic of the komatiites, that consists of needle-like olivine grains forming a cross-hatched pattern enclosing intergranular pyroxenes.

spreading center   In plate tectonic theory, a kind of plate boundary where plates pull apart and new crust is formed by solidification of intruded and extruded magmas. Also called a divergent plate boundary.

stable system   A system that, under the specified conditions, is in its lowest energy state; i.e., the system has no tendency to change.

static metamorphism   Metamorphism that occurs primarily in response to pressure.

stock   A plutonic rock body with an aerial exposure of less than 100 km^2.

strand-plain   A wide beach.

stromatactis   A millimeter- to centimeter-scale, layered, plano-convex-up, lens-shaped structure consisting of lighter sparry calcite layers between darker carbonate layers. The structures are formed in carbonate rocks after deposition.

stromatolite   A sedimentary structure consisting of algal-generated, millimeter-scale micritic carbonate layers in flat, domal, columnar, conical, or nearly spherical forms larger than 10 cm in diameter.

structure   In rocks, a visible feature larger than a grain (but not penetrative at the handspecimen or larger scale) that results from the physical arrangement of grains, holes, fractures, or other entities in the rock mass.

stylolite   An irregular surface of dissolution that commonly appears as a dark, jagged line on exposed surfaces of carbonate rock (and rarely on other sedimentary rock types).

subduction zone   In plate tectonic theory, a zone of plate convergence and collision where one plate descends beneath another.

subglacial environment   A site of deposition beneath a glacier where sediments are formed as a glacier advances and retreats, depositing materials at its base.

submarine slope   The steep part of the seafloor that extends from the shelf break to the rise.

subophitic   An igneous texture in gabbros in which the augite and plagioclase assume similar sizes, with augite encompassing the ends of some plagioclase laths.

supraglacial environment   An environment of sedimentation along the sides and at the end of a glacier.

symplectic texture   A general category of texture in which wormy (vermicular) or irregular intergrowths of one mineral occur within another.

syntectonic recrystallization   A process of intracrystalline structural reorganization that occurs during (and as a result of) deformation of a rock and involves dissolution of detrital phyllosilicates and other minerals and recrystallization or neocrystallization of phyllosilicate and other grains parallel to the cleavage direction. Also called dynamic recrystallization.

system   Any part of the universe selected for study.

T

talus   Sediment dominated by angular clasts of coarse size (> 2mm) that accumulates at the base of cliffs and very steep, rocky slopes.

tectonic overpressure   Pressure in excess of P_{load} in metamorphic rocks. Tectonic overpressures typically develop beneath thrust faults that have shale or other impermeable rocks in the hanging wall.

tectonite   A rock with a fabric that reflects the history of its deformation.

tectonite fabric   A texture or structure, or both, that reflects a history of deformation.

telogenesis   Late-stage diagenesis that occurs after reexposure of formerly buried rocks.

tepee A small (centimeter- to meter-scale) angular anticline produced by compression of a layer towards a fracture in such a way as to produce two opposing, concave-up layers that come to a point like a Native American tepee.

terminal reaction (1) A metamorphic reaction in which a new phase is added to a phase diagram or a phase already present disappears from the phase diagram. (2) A metamorphic reaction in which one or more phases ceases to be possible on one side of the reaction.

texture The microscopic to small-scale mesoscopic character of a rock imparted by the size, shape, orientation, and distribution of grains and the intergrain relationships.

tholeiite A main type of basalt that is hypersthene-normative and contains little or no modal olivine but does contain modal Ca-poor pyroxene (hypersthene or pigeonite).

tidal flat A low-lying, subhorizontal geomorphic feature and sedimentary environment, which includes marshes and sabkhas, occurring along the edge of the continent.

tie-line In a phase diagram, a line drawn within a phase field connecting the phases coexisting (in equilibrium) at a fixed T (or P).

till Glacially deposited, very poorly sorted sediment (a diamict) composed of coarse, angular to rounded, locally faceted and striated clasts in a finer-grained matrix that typically consists of clay- to sand-size rock and mineral fragments.

tillite The rock that is equivalent to till; a glacial diamictite with a muddy or rock flour matrix enclosing larger clasts.

tool mark A groove on the top of a bed at the sediment-water interface made by currents carrying pebbles, sticks, shells, or other large objects. Tool marks may be filled with sediment and preserved as sole marks.

trace elements All elements other than the major elements.

trachyte A volcanic rock of intermediate silica content that is characterized modally by an abundance of alkali feldspar and a general lack of quartz and feldspathoid minerals.

trachytic A textural term for a volcanic rock texture in which the feldspars are aligned in subparallel arrangements.

trachytoidal A textural term for a plutonic rock texture in which the feldspars are aligned in subparallel arrangements.

tracks and trails Fossils that result from preservation of marks left by various organisms as they walk, crawl, or otherwise move across a sediment surface.

transform fault In plate tectonic theory, a shear boundary, where plates slide past one another; a major type of plate boundary.

transitional environment A sedimentary environment that occurs between the levels of the marine and continental environments and is influenced by both marine and continental agents (e.g., fresh and salt water, wind and wave action).

transposition A process in which primary layering is transected at a significant angle by a cleavage and through a history of movement and recrystallization is converted to compositional layering that parallels the cleavage direction.

travertine An aphanitic to phaneritic, layered carbonate rock, usually light-colored and concretionary, deposited by ground and surface waters emanating from springs.

triple junction In plate tectonic theory, a site where three plate boundaries intersect.

tufa Cellular travertine; a layered carbonate rock with holes, formed by springwater.

tuff Pyroclastic rock dominated by grains less than 2 mm in average diameter.

turbidite A rock unit deposited from moving, heavy sediment-water mixtures.

turbidity current A current of water (or air) that moves downslope by virtue of the gravitational influence on its mass, which is greater (denser) than that of the surrounding water (or air). In geology, significant turbidity currents develop in submarine environments, where sediment-rich waters, which have greater density than the surrounding waters, flow downslope.

U

ultrabasic rock A rock with an SiO_2 content of less than 45%.

ultramafic rock A dark rock with a color index of 90–100. Also called ferromagnesian rock.

ultramylonite A foliated metamorphic rock bearing significant evidence of crystal-plastic deformation and characterized by less than 10% microlithons and porphyroclasts in a fine-grained, typically phyllosilicate-bearing matrix.

ultramylonitic texture An aphanitic to phaneritic, foliated to porphyroclastic-foliated metamorphic texture with <10% microlithons and porphyroclasts, and evidence of crystal-plastic deformation or syntectonic recrystallization.

undercooling The difference between the temperature of the liquidus and the actual temperature of crystal growth for a given phase, expressed as $\Delta T = T_1 - T_c$.

undersaturated The term used to describe the chemistry of low-silica rocks containing minerals such as nepheline that are incompatible with quartz or other silica minerals.

uniform layering In igneous rocks, layering lacking conspicuous differences in mineralogy.

univariant A term designating a system with one degree of freedom or a reaction in which only one variable may be changed without changing the number or nature of phases.

univariant system A system with one degree of freedom.

unstable An adjective used to describe systems that are not in equilibrium; that is, they have a tendency to change spontaneously.

upwelling A process of vertical circulation in which deep waters, moving as currents, flow up to the surface in equatorial regions and along coastlines.

uv-komatiite High-Mg, low-Ti, olivine-rich, ultramafic volcanic rocks characterized by a distinctive herringbone-like texture called a spinifex texture.

V

varves Rhythmically repeated sediment couplets, each representing a year, consisting of silty, light-shaded summer layers and dark, clay-rich, organic winter layers.

varvite A laminated rock exhibiting varves, typically containing dropstones; diagnostic of a cryolacustrine environment of deposition.

vein A crudely tabular body that forms fracture fillings composed of one or more minerals.

vent   The site on a volcano from which magma erupts.

vesicle   A hole left when gas escapes from a lava flow.

vesicular   The adjective used to describe a volcanic rock in which the escape of gas has left holes.

vitriclastoblastic texture   A nonfoliated metamorphic texture consisting of broken fragments of rocks or minerals in a matrix of frictionally generated glass.

vitrophyric   A volcanic texture characterized by discrete phenocrysts in a glassy groundmass.

volcanic dome   A small, steep-sided structure shaped like an inverted cup or cone and composed of volcanic rocks.

volcanic neck   The eroded and exposed dike system that served as the feeder channel in a volcanic cone. Also called a pipe or vent.

volcanic rock   A rock produced by crystallization of magma, where magmas approach and break through to the surface in volcanic eruptions.

volcaniclastic rock
A sedimentary rock composed of fragments of volcanic material.

volcaniclastic texture   A texture consisting of angular grains of volcanic rock fragments, feldspar, quartz, and/or other minerals, generated by volcanic processes, that are cemented or otherwise stuck together.

vug   Cavities lined with crystals.

W

wacke   (1) A type of sandstone characterized by a matrix of 5% or more. (2) A type of sandstone characterized by a matrix of 10% or more.

wackestone   A type of limestone that contains more than 10% grains but is mud-supported (i.e., the larger grains do not generally touch one another).

weakly foliated   A texture in which some alignment of grains (usually phyllosilicate grains) yields a faint layered quality to the rock.

weathering   The general term assigned to the group of processes that result in the transformation of rock into soil.

welded tuff   A type of volcanic rock composed predominantly of fragments of volcanic material smaller than 2 mm in diameter, annealed soon after deposition by the heat still contained in the grains.

working   A term used to describe the condition of sediments that have experienced extensive transportation and abrasion.

X

xenocryst   A foreign crystal (*xeno* = foreign) formed in one rock but later incorporated into another by magmatic erosion or assimilation of the original rock.

xenolith   A piece of foreign rock (*xeno* = foreign, *lith* = stone) occurring as an inclusion within a plutonic or volcanic rock mass.

xenomorphic-granular   An igneous texture that is dominated by anhedral crystals.

xenotopic   A crystalline sedimentary texture dominated by anhedral crystals.

Y

yttrium   A trace element; the transition element of atomic number 39.

Z

Zeolite Facies   The set of all metamorphic rocks containing phase assemblages indicative of conditions of crystallization in the general range of P = 0.05–0.4, Gpa (0.5–4 kb) and T = 75° – 300°C. The name is also applied to the P-T space defined by those conditions.

zone, metamorphic   A specific mineral assemblage within a metamorphic facies and the region of geography and P-T space over which that assemblage is stable.

zoned   An adjective describing a texture in a mineral grain that exhibits a layered character resulting from changes in composition from the core to the rim of the grain.

zoned pegmatite   A very coarse-grained part of a plutonic body containing layers, lenses, shells, or irregular masses of rock of distinctive composition.

zoned texture   A texture characterized by zoned minerals.

CREDITS

Photo Credits

All photos not credited on-page are courtesy of the author, Loren A. Raymond.

Line Art Credits

Chapter 1

Figure 1.1 From: *Inside the Earth* by Bruce A. Bolt. Copyright © 1982 by W. H. Freeman and Company. Reprinted with permission;
Figure 1.3 From B. L. Isacks, et al., in *Journal of Geophysical Research*, 73:1968. Copyright © American Geophysical Union.

Chapter 2

Figure 2.1 From L. A. Raymond, *Petrology Laboratory Manual*, Vol. 1, Handspecimen Petrography. Copyright © 1984 GEOSI. Reprinted by permission; **Figure 2.3** From R. Greeley, "The Snake River Plain, Idaho: Representation of a New Category of Volcanism" in *Journal of Geophysical Research*, 87:2705–12, 1982. Copyright © 1982 American Geophysical Union; **Figure 2.11 A** From Arthur G. Sylvester, et al., "Papoose Flat Pluton: A granite blister in the Inyo Mountains, California" in *Geological Society of America Bulletin*, 89:1205–1219, 1978. Copyright © 1978 Geological Society of America. Reprinted by permission of the author; **Figure 2.11 B** From T. A. Richards and K. C. McTaggart, "Granatic rocks of Southern Coast Plutonic Complex and Northern Cascades of British Columbia" in *Geological Society of America Bulletin*, 87:935–53, 1976. Copyright © 1976 Geological Society of America. Reprinted by permission of the author; **Figure 2.11 C** From C. W. Williams and M. P. Billings, "Petrology and structure of the Franconia Quadrangle: New Hampshire" in *Geological Society of America Bulletin*, 49:1011–44, 1938. Copyright © 1938 Geological Society of America. Reprinted by permission of the author; **Figure 2.11 E** From W. A. Bothner, "Gravity study of Exeter Pluton: Southeastern New Hampshire" in *Geological Society of America Bulletin*, 85:51–6, 1974. Copyright © 1974 Geological Society of America. Reprinted by permission of the author; **Figure 2.11 F** From P. Kearey, "An interpretation of the gravity field of the Morin Anorthosite Complex, Southwest Quebec" in *Geological Society of America Bulletin*, 89:467–475, 1978. Copyright © 1978 Geological Society of America. Reprinted by permission of the author; **Figure 2.11 G** From J. F. Sweeney, "Subsurface distribution of Granitic rocks, South-central Maine" in *Geological Society of America Bulletin*, 87:241–249, 1976. Copyright © 1976 Geological Society of America. Reprinted by permission of the author; **Figure 2.12 A** From Arthur G. Sylvester, et al., "Papoose Flat Pluton: A granite blister in the Inyo Mountains, California" in *Geological Society of America Bulletin*, 89:1205–1219, 1978. Copyright © 1978 Geological Society of America. Reprinted by permission of the author; **Figure 2.12 B&C** From A. F. Buffington, "Granite Emplacement with Special Reference to North America" in *Geological Society of America Bulletin*, 70:641–47, 1959. Copyright © 1959 Geological Society of America; **Figure 2.19** From L. A. Raymond, *Petrology Laboratory Manual*, Vol. 1, Handspecimen Petrography. Copyright © 1984 GEOSI. Reprinted by permission; **Figure 2.25 B** Reprinted by permission from Samuel E. Swanson, "Relation of nucleation and crystal-growth rate to the development of granitic textures" in *American Mineralogist*, 62:(9/10):966–78, 1977. Copyright by the Mineralogical Society of America.

Chapter 3

Figure 3.2 B From T. H. Dixon and R. J. Stern, "Petrology, chemistry, and isotopic composition of submarine volcanoes in the Southern Mariana Arc" in *Geological Society of America Bulletin*, 94:1159–72, 1983. Copyright © 1983 Geological Society of America. Reprinted by permission of the author; **Figure 3.4 B** From A. Miyashiro, "Volcanic Rock Series and Tectonic Setting" in *Annual Review of Earth and Planetary Sciences*, 3:251–69, 1975. Copyright © 1975 Annual Reviews, Inc., Palo Alto CA. Reprinted by permission.

Chapter 4

Figure 4.1 A From L. A. Raymond, *Petrology Laboratory Manual*, Vol. 1, Handspecimen Petrography. Copyright © 1984 GEOSI. Reprinted by permission; **Figure 4.1 B** From D. W. Peterson, "Descriptive model classification of Igneous Rocks." Reprinted with permission from *Geotimes*, table 1, p. 32; March 1961; **Figure 4.1 C** From A. Streckeisen, "To Each Plutonic Rock its Proper Name" in *Earth Science Reviews*, 12:1–33, 1976. Copyright © 1976 Elsevier Science Publishers BV, Amsterdam, Netherlands. Reprinted by permission; **Figure 4.3** From A. Streckeisen, et al., "Plutonic rocks: Classification and Nomenclature Recommended by the IUGS Subcommission on the systematics of Igneous Rocks." Reprinted with permission from *Geotimes*, pp. 26–30, p. 28; October, 1973; **Figure 4.4** From A. Streckeisen, et al., "Plutonic rocks: Classification and nomenclature recommended by the IUGS Subcommission on the systematics of Igneous Rocks." Reprinted with permission from *Geotimes*, pp. 26–30, fig. 3, p. 27; October, 1973; **Figure 4.5** From L. A. Raymond, *Petrology Laboratory Manual*, Vol. 1, Handspecimen Petrography. Copyright © 1984 GEOSI. Reprinted by permission; **Figure 4.6** From A. Streckeisen, et al., "Plutonic rocks: Classification and nomenclature recommended by the IUGS Subcommission on the systematics of Igneous Rocks." Reprinted with permission from *Geotimes*, pp. 26–30, fig. 2, p. 26; October, 1973; **Figure 4.7 C** From L. A. Raymond, *Petrology Laboratory Manual*, Vol. 1, Handspecimen Petrography. Copyright © 1984 GEOSI. Reprinted by permission; **Figure 4.8** From M. A. Peacock, "Classification of Igneous Rock Series" in *Journal of Geology*, 39:54–67, 1931. Copyright © 1931 University of Chicago Press. Reprinted by permission; **Figure 4.10** From A. Miyashiro, "Volcanic Rock Series and Tectonic Setting" in *Annual Review of Earth and Planetary Sciences*, 3:251–69, 1975. Copyright © 1975 Annual Reviews, Inc., Palo Alto CA.

Chapter 16

Figure 16.2 From T. R. Nardin, et al., "Review of mass movement processes, sediment and acoustic characteristics, and contrasts in slope and base-of-slope systems versus canyon-fan-basin floor systems" in L. H. Doyle and O. H. Pilkey, Eds., *Geology of Continental Slopes*. *Journal of Sedimentary Petrology Special Publication* 27:61–73:1979. Copyright © 1979 Society of Economic Paleontologists and Mineralogists, Tulsa, Oklahoma. Reprinted by permission; **Figure 16.3** From H. Blatt, et al., *Origin of Sedimentary Rocks*, 2e, © 1980, p. 187. Reprinted by permission of Prentice-Hall, Englewood Clifs, New Jersey; **Figure 16.5** From W. C. Krumbein and R. M. Garrels, "Origin and classification of chemical sediments in terms of pH and oxidation-reduction potentials" in *Journal of Geology*, 60:1–33, 1952. Copyright © 1952 University of Chicago Press. Reprinted by permission; **Figure 16.8** From B. B. Henshaw, et al., "A geochemical hypothesis for dolomitization by ground water" in *Economic Geology*, 66: 710–24, 1971. Copyright © 1971 Economic Geology Publishing Company. Reprinted by permission; **Figure 16.9** From K. Badiozamani, "The Doprag dolomitization model-application to the Middle Ordovician of Wisconsin" in *Journal of Sedimentary Petrology*, 43:965–84, 1973. Copyright © 1973 Society of Economic Paleontologists and Mineralogists, Tulsa, Oklahoma. Reprinted by permission; **Figure 16.13** From K. P. Helmold and P. C. van de Kamp, "Diagentic mineralogy and controls on albitization . . ." in *AAPG Memoir 37*, 1984. Copyright © 1984 American Association of Petroleum Geologists, Tulsa OK. Reprinted by permission.

Chapter 17

Figure 17.7 A From G. E. Reinson, "Barrier island and associated strand-plain systems" in R. G. Walkder, Ed., *Facies Models*, 2d ed. Copyright © 1984 Geological Association of Canada, Sudbury, Ontario, Canada. Reprinted by permission; **Figure 17.7 B–D** From S. D. Heron, et al., "Holocene sedimentation of a wave-dominated barrier-island shoreline: Cape Lookout, North Carolina" in *Marine Geology*, 60:413–34, 1984. Copyright © 1984 Elsevier Science Publishers B V, Amsterdam. Reprinted by permission; **Figure 17.9 A** From L. J. Doyle and T. N. Sparks, "Sediments of the Mississippi, Alabama, and Florida (MAFLA) Continental Shelf" in *Journal of Sedimentary Petrology*, 50:905–16, 1980. Copyright © 1980 Society of Economic Paleontologists and Mineralogists, Tulsa, Oklahoma. Reprinted by permission; **Figure 17.9 B** From B. D. Bornhold and C. J. Yorath, "Surficial geology of the continental shelf, Northwestern Vancouver Island" in *Marine Geology*, 57:89–112, 1984. Copyright © 1984 Elsevier Science Publishers, B V, Amsterdam, Netherlands. Reprinted by permission; **Figure 17.11 A** From "Paleoenvironment of late Mississippian fenestrate bryozoans, Eastern United States" by F K. McKinney and H. W. Gault from *Lethaia*, 13:127–46, 1980, by permission of Scandinavian University Press; **Figure 17.11 B** From W. E. Galloway and D. K. Hobday, *Terrigenous Clastic Depositional Systems: Applications to petroleum, coal, and uranium exploration*. Copyright © 1983 Springer-Verlag, New York. Reprinted by permission; **Figure 17.13 A** From R. V. Ingersoll, "Submarine fan facies of the Upper Cretaceous Great Valley Sequence, Northern and Central California" in *Sedimentary Geology*, 21:205–50, 1978. Copyright © 1978 Elsevier Science Publishers, BV, Amsterdam. Reprinted by permission; **Figure 17.13 B** From J. A. May, et al., "Role of submarine canyons on shelf break critical interference on continental margins" in *Journal of Sedimentary Petrology Special Publication* 33:315–82, 1983. Copyright © 1983 Society of Economic Paleontologists and Mineralogists, Tulsa, Oklahoma. Reprinted by permission; **Figure 17.13 D** From *AGI Reprint Series* 3, by E. Mutti and F. Ricci Lucchi, translated by Tor H. Milson; copyright © 1978, p. 141, figure 14. Courtesy of the American Geological Institute.

Chapter 18

Figure 18.3 From P. E. Potter, et al. *Sedimentology of Shale*. Copyright © 1980 Springer-Verlag, New York. Reprinted by permission; **Figure 18.5** From T. A. Davies and D. S. Gorsline, "Oceanic sediments and sedimentary procenes" in R. Riley, Ed., *Chemical Oceanography*, 1976. Copyright © 1976 Academic Press Ltd., London. Reprinted by permission; **Figure 18.7** From M. Ewing, et al., "Sediments and topography of the Gulf of Mexico" in L. G. Weeks, ed., *Habitat of Oil*. Copyright © 1958 American Association of Petroleum Geologists, Tulsa OK. Reprinted by permission; **Figure 18.8** From C. W. Holmes, "Geochemical indices of fine sediment transport, Northwest Gulf of Mexico" in *Journal of Sedimentary Petrology*, 52:307–321, 1982. Copyright © 1982 Society of Economic Paleontologists and Mineralogists, Tulsa, Oklahoma. Reprinted by permission; **Figure 18.9** From J. A. Lineback, et al. "Glacial and postglacial sediments in Lakes Superior and Michigan" in *Geological Society of America Bulletin*, 90:781–91, 1979. Copyright © 1979 Geological Society of America. Reprinted by permission of the author; **Figure 18.12** From Cook, T. D., et al., *Stratigraphic Atlas of North and Central America*. Copyright © 1975 Princeton University Press, Princeton NJ. Reprinted by permission; **Figure 18.14** From F. Ettonsohn and T. D. Elam, "Defining the nature and location of a Late Devonian-Early Mississippian pycnocline in Eastern Kentucky" in *Geological Society of America Bulletin*, 96:1313–21, 1985. Copyright © 1985 Geological Society of America. Reprinted by permission of the author.

Chapter 19

Figure 19.3 A From W. R. Dickinson and C. A. Suczek, "Plate Tectonics and Sandstone Compositions" in *AAPG Bulletin*, 63:2164, 1979. Copyright © 1979 American Association of Petroleum Geologists, Tulsa OK. Reprinted by permission; **Figure 19.3 B** From W. R. Dickinson, et al., "Provenance of North American phanerozoic sandstone in relation to tectonic setting" in *Geological Society of America Bulletin*, 94:222–35, 1983. Copyright © 1983 Geological Society of America. Reprinted by permission of the author; **Figure 19.6** Copyright © L. A. Raymond. Reprinted by permission; **Figure 19.13** Copyright © L. A. Raymond. Reprinted by permission; **Figure 19.14** From W. R. Dickenson, et al., "Provenance of Franciscan graywackes in Coastal California" in *Geological Society of America Bulletin*, 93:95–107, 1982. Copyright © 1982 Geological Society of America. Reprinted by permission of the author.

Chapter 20

Figure 20.2 From Loren A. Raymond, "Classification of melanges" in L. A. Raymond, Ed., "Melanges: Their nature, origin and significance" in *Geological Society of America Special Paper* 198, 1984. Copyright © 1984 Geological Society of America. Reprinted by permission of the author; **Figure 20.4 A–C** Unpublished data, courtesy of L. A. Raymond; **Figure 20.4 D–F** From John S. Schlee, "Upland gravels of Southern Maryland" in *Geological Society of America Bulletin*, 68:1371–1410, 1957. Copyright © 1957 Geological Society of America. Reprinted by permission of the author; **Figure 20.5 A** Figure 6.5, page 159 from *Sedimentary Rocks*, 2nd ed. by Francis J. Pettijohn. Copyright © 1949, 1957 by Harper & Brothers, renewed 1985 by Francis J. Pettijohn. Reprinted by permission of HarperCollins Publishers, Inc.; **Figure 20.5 B** From W. Nemec and R. J. Steel, "Alluvial and coastal conglomerates: Their significant features and some comments on gravelly mass-flow deposits" in E. H. Koster and R. J. Steel, Eds., *Sedimentology of Gravels and Conglomerates, Memoir 10*. Copyright © 1984 Canadian Society of Petroleum Geologists, Calgary, Alberta, Canada. Reprinted by permission; **Figure 20.11** Copyright © L. A. Raymond. Reprinted by permission; **Figure 20.15** From F. J. Hein and R. G. Walker, "The Cambro-Ordovician Cap Enrage Formation,

Bulletin, 70:879–920, 1959. Copyright © 1959 Geological Society of America. Reprinted by permission of the author; **Figure 26.15** From C. W. Burnham, "Contact metamorphism of magnesian limestones at Crestmore, California" in *Geological Society of America Bulletin*, 70:879–920, 1959. Copyright © 1959 Geological Society of America. Reprinted by permission of the author.

Chapter 27
Figure 27.8 From J. Hoffman and J. Hower, "Clay mineral assemblages as low grade metamorphic geothermometers: Application to the thrust faulted disturbed belt of Montana, U.S.A." in A. Scholle and P. R. Schluger, Eds., *Aspects of Diagenesis. Journal of Sedimentary Petrology Special Publication*, 26:1979. Copyright © 1979 Society of Economic Paleontologists and Mineralogists, Tulsa, Oklahoma. Reprinted by permission; **Figure 27.13 A** From R. D. Hatcher, Jr., et al., "Guidebook for Southern Appalachian Field Trip in the Carolinas, Tennessee, and Northeastern Georgia," Project 27 Caledonide Oregon. International Geol Correl Program (IUGS), 1979. Copyright © 1979 North Carolina Geological Survey, Raleigh NC. Reprinted by permission; **Figure 27.13 B** From Richard N. Abbott, Jr., and Loren A. Raymond, "The Ashe Metamorphic Suite, Northwest North Carolina: metamorphism and observations on geologic history" in *American Journal of Science*, 284:1986. Copyright © 1986 American Journal of Science, New Haven, Connecticut. Reprinted by permission; **Figure 27.16** From K. R. Mehnert, *Migmatites and the Origin of Granitic Rocks*. Copyright © 1968 Elsevier Science Publishers, BV, Amsterdam, Netherlands. Reprinted by permission; **Figure 27.17** From S. N. Olsen, "A Quantative Approach to Load Mass Balance in Migmitites" in M. P. Atherton and C. D. Gribble, Eds., *Migmitites, Meeting and Metamorphism*. Copyright © 1983 Shiva Publishing Ltd. Reprinted by permission of the author.

Chapter 28
Figure 28.1 Sources: R. G. Coleman, "Blueschist Metamorphism and Plate Tectonics," 24th International Geological Congress, Rept Sec. 2, 1972 and R. W. Wood, "A Re-evaluation of the Blueschist Facies" in *Geological Magazine*, 116:24–25, 1979; **Figure 28.5** From E. H. Brown, "Phase equilibria among pumpellyite, lawsonite, epidote and associated minerals in low grade metamorphic rocks" in *Contributions to Mineralogy and Petrology*, 64:123–36, 1977. Copyright © 1977 Springer-Verlag GmbH & Co. KG, Berlin. Reprinted by permission; **Figure 28.8** From W. G. Ernst, "Blueschist metamorphism and P-T regimes in active subduction zones" in *Tectonophysics*, 17:255–72, 1973. Copyright © 1973 Elsevier Science Publishers B V, Amsterdam, Netherlands. Reprinted by permission.

Chapter 29
Figure 29.2 From K. Yokoyama, "Regional ecologite facies in the high-pressure metamorphic belt of New Caledonia" in B. W. Evans and E. H. Brown, Eds., *Blueschists and Eclogites*, in *Geological Society of America Memoir*, 164, 1986. Copyright © 1986 Geological Society of America. Reprinted by permission of the author; **Figure 29.6** From

R. A. Jamieson, "Metamorphism of an Early Paleozoic continental margin: Western Baie Veseta Peninsula, Newfoundland" in *Journal of Metamorphic Geology*, 8:269–288, 1990. Copyright © 1990 Blackwell Scientific Publications, Inc., Cambridge MA. Reprinted by permission.

Chapter 30
Figure 30.6 B From D. U. Wise, et al., "Fault-related rocks: Suggestions for terminology," in *Geology*, 12:391–94, 1984. Copyright © 1984 Geological Society of America. Reprinted by permission of the author; **Figure 30.6 C** From L. A. Raymond, *Petrology Laboratory Manual*, Vol. 1, Handspecimen Petrography, 2d ed. Copyright © 1993 GEOSI. Reprinted by permission; **Figure 30.8** From A. R. Bobyarchick, et al., "The role of dextral strike-slip in the displacement history of the Brevard Zone" in D. T. Secor, Jr., Ed., *Southeastern Geological Excursions*, 1988. Copyright © 1988 South Carolina Survey, Columbia, South Carolina. Reprinted by permission; **Figure 30.9** From A. K. Sinha, et al., "Fluid interaction and element mobility in the development of ultramylonites" in *Geology*, 14:883–86, 1986. Copyright © 1986 Geological Society of America. Reprinted by permission of the author.

Chapter 31
Figure 31.4 Based, in part, on Evans and Trommsdorf, 1970; Evans, et al., 1976; and Day, et al., 1985; **Figure 31.6** From John M. Casey, "Reconstruction of the geometry of accretion during formation of the Bay of Islands Ophiolite Complex" in *Tectonics*, 2:509–28, 1983. Copyright © 1983 American Geophysical Union; **Figure 31.7** From J. F. Casey and J. A. Karson, "Magma chamber profiles from the Bay of Islands ophiolite complex" in *Nature*, 292:295–301, 1981. Copyright © 1981 Macmillan Magazines Limited. Reprinted by permission; **Figure 31.10 B** From S. E. Swanson, "Mineralogy and petrology of the Day Book dunite and associated rocks, Western North Carolina" in *Southeastern Geology*, 22:53–77, 1981. Copyright © 1981 Southeastern Geology. Reprinted by permission; **Figure 31.10 C** Reprinted by permission from Rosewell Miller, III, "The Webster-Addie ultramafic ring, Jackson County, North Carolina, and secondary alteration of its chromite" in *American Mineralogist*, 38(1/12):1134–47, 1951. Copyright by the Mineralogical Society of America; **Figure 31.10 D** From M. S. McElhaney and H. Y. McSween, Jr., "Petrology of the Chunky Gal Mountain mafic-ultramafic complex, North Carolina" in *Geological Society of America Bulletin*, 94:855–74, 1983. Copyright © 1983 Geological Society of America. Reprinted by permission of the author; **Figure 31.11** From C. J. Suen and F. A. Frey, "Origins of the mafic and ultramafic rocks in the Rhonda peridotite" in *Earth and Planetary Science Letters*, 85:183–202, 1987. Copyright © 1987 Elsevier Science Publishers, Amsterdam. Reprinted by permission.

Appendix
Appendix B.1 A–C From Charles E. Bickel, unpublished manuscript. Reprinted by permission of the author.

INDEX